2nd Fully Revised and Enlarged Edition

Agricultural Engineering Explorer All In One

NIPA GENX ELECTRONIC RESOURCES & SOLUTIONS P. LTD.

New Delhi-110 034

2nd Fully Revised and Enlarged Edition

Agricultural Engineering Explorer All In One

Er. Amandeep Godara
Assistant Engineer (WD&SC), Rajasthan
Bachelor of Engineering (M.P.U.A.T., Udaipur, Rajasthan)
GATE 2014 Qualified (34th A.I.R.)
ICAR-PG Exam 2014 Qualified (182nd A.I.R.)
DST-Rajasthan 2014 Scholarship Holder

NIPA GENX ELECTRONIC RESOURCES & SOLUTIONS P. LTD.
New Delhi-110 034

NIPA GENX ELECTRONIC RESOURCES & SOLUTIONS P. LTD.

101,103, Vikas Surya Plaza, CU Block
L.S.C.Market, Pitam Pura, New Delhi-110 034
Ph : +91 11 27341616, 27341717, 27341718
E-mail:newindiapublishingagency@gmail.com
www: www.nipabooks.com

For customer assistance, please contact
Phone: + 91-11-27 34 17 17 Fax: + 91-11- 27 34 16 16
E-Mail: feedbacks@nipabooks.com

ISBN: 978-93-94490-16-1

This is the revised edition of our earlier published book: "Question Bank Agricultural Engineering" Er Amandeep Godara ISBN 9789385516122, published in 2016.

Composed and Designed by NIPA.

|| Om Hrih Mahamayange Mahasarasvatyai Namah ||
Saraswati Namastubhyam Varde Kamroopani
Vidya Aarambham Karishyami Siddhir-bhavatu May Sada ||

सरस्वती नमस्तुभ्यं वरदे कामरूपिणी ।
विध्यारम्भ करिष्यामी सिद्धीर्भवतु मे सदा ।।

This Book is Dedicated To My
Param Pujaniya
Guru Uttam Nath Ji Maharaj
&
Beloved Wife Priyanka
&
Loving Parents
Adv. Prahlad Godara-Mrs. Kalawati Devi
&
Er. Sachin Godara-Mrs. Nitu (Brother-Sister in law).

Foreword

Nowadays competition has increased upto zenith level in agricultural engineering stream. To crack these tough competitive examinations, aspirants require a book which develops fundamental and thoughtful approaches in simplified manner without remembering lots of complex formulas.

The "Agricultural Engineering Explorer: All In One by Er. Amandeep Godara" is the only first book of its kind that has covered almost all the questions asked in various university and government competitive examinations solved and explained in such an easily understandable way for an agricultural engineering graduate and postgraduate.

The book is divided into 7 parts such as (i) Engineering Mathematics, (ii) Farm Power, (iii) Farm Machinery, (iv) Soil-Water Conservation and Irrigation Engineering, (v) Agricultural Process Engineering, (vi) General English, Aptitude and Reasoning (vii) Miscellaneous. This book may be helpful for almost all competitive and government recruitment examinations like UPSC IFoS, GATE, ARS/NET, SRF, JRF, LBPS Specialist Officer and various State PSCs examinations.

I appreciate the efforts made by Er. Amandeep Godara for bringing out this book which I hope will be a precious study resource and useful for students, teachers related to the agricultural engineering field.

7/6/2022

Er. Amar Singh Yadav
Superintending Engineer
Watershed Develop. & Soil Conservation
Government of Rajasthan

Preface

The book "Agricultural Engineering Explorer – All In One" is an attempt to provide detailed solutions of question papers of UPSC IFoS, GATE and Various State PSC Examinations in Agricultural Engineering in a concise and simplified manner to facilitate the aspirants. The book is intended to be a workbook that will help the students to practice solving numerical problems in agricultural engineering. The students whoever refer this book will be able to get a good concept and problem solving approaches.

The book is endowed with a whole lot of unique short cuts and thought processes. This feature makes the book a must have as part of your preparation material to crack the crucial examinations like UPSC IFoS, GATE and Various State PSC Examinations. This book will also helpful for the aspirants preparing for various Government Examinations such as UPSC IFoS, ARS/NET, SRF, JRF, IBPS Specialist Officer. Salient features of the book are given as:

- It Includes detailed solutions of 15 years GATE (A.E.) 2007 to 2021 question papers arranged topic-wise.
- It Includes solved papers from Union Public Services Commission(UPSC) and various State (Andhra Pradesh, Madhya Pradesh, Uttrakhand, Punjab, Rajasthan, Orisa, Chatisgarh, Kerla) PSC Examinations in Agricultural Engineering.
- It Includes ready reference official syllabus of UPSC IFoS 2022 and GATE 2022.
- It have short notes for important/ specific points.

I am extremely thankful to my elder brother Er. Sachin Godara (SSE, PSTCL, Punjab) and my family for their support and encouragement. My special thanks to Mr. Neeraj Kumar Kamra for his valuable guidance throughout the period of writing the book. My sincere thanks also go to Er. Jitendra Singh Rajpurohit (Branch Manager, Andhra Bank) for his aid in material collection & proof reading of the book.

I also wish to express thanks Mr. Atmaram Swami (Computer Operator) for his invaluable help during the preparation of manuscript and finalizing the text.

Unnecessary and repeated questions from various examinations are removed to prevent the book being bulky. Despite of our best efforts, some errors may have inadvertently crept into the book. Constructive comments and suggestions from readers for further improvement will be highly appreciated and gratefully acknowledged.

I feel that this book will be sufficient and highly useful for all the competitive examinations conducted for agricultural engineering graduates. I also would like to thank all the readers who loved and appreciated my other books.

With best wishes to all aspirants of agricultural engineering examination.

Er. Amandeep Godara

Contents

I. Engineering Mathematics

II. Farm Power

III. Farm Machinery

IV. Soil-Water Conservation and Irrigation Engineering

V. Agricultural Process Engineering

VI. General English, Aptitude and Reasoning

VII. Miscellaneous

GATE 2022 Agricultural Engineering Syllabus

Engineering Mathematics

Linear Algebra: Matrices and determinants, systems of linear equations, Eigen values and eigen vectors. Calculus: Limit, continuity and differentiability; partial derivatives; maxima and minima; sequences and series; tests for convergence; Fourier series, Taylor series.

Vector Calculus: Gradient; divergence and curl; line; surface and volume integrals; Stokes, Gauss and Green's theorems.

Differential Equations: Linear and non-linear first order Ordinary Differential Equations (ODE); Higher order linear ODEs with constant coefficients; Cauchy's and Euler's equations; Laplace transforms; Partial Differential Equations - Laplace, heat and wave equations.

Probability and Statistics: Mean, median, mode and standard deviation; random variables; Poisson, normal and binomial distributions; correlation and regression analysis; tests of significance, analysis of variance (ANOVA).

Numerical Methods: Solutions of linear and non-linear algebraic equations; numerical integration - trapezoidal and Simpson's rule; numerical solutions of ODE.

Farm Machinery

Machine Design: Design and selection of machine elements – gears, pulleys, chains and sprockets and belts; overload safety devices used in farm machinery; measurement of force, torque, speed, displacement and acceleration on machine elements.

Farm Machinery: Soil tillage; forces acting on a tillage tool; hitch systems and hitching of tillage implements; functional requirements, principles of working, construction and operation of manual, animal and power operated equipment for tillage, sowing, planting, fertilizer application, inter-cultivation, spraying, mowing, chaff cutting, harvesting, threshing and transport; testing of agricultural machinery and equipment; calculation of performance parameters

- field capacity, efficiency, application rate and losses; cost analysis of implements and tractors.

Farm Power

Sources of Power: Sources of power on the farm - human, animal, mechanical, electrical, wind, solar and biomass; bio-fuels.

Farm Power: Thermodynamic principles of I.C. engines; I.C. engine cycles; engine components; fuels and combustion; lubricants and their properties; I.C. engine systems – fuel, cooling, lubrication, ignition, electrical, intake and exhaust; selection, operation, maintenance and repair of I.C. engines; power efficiencies and measurement; calculation of power, torque, fuel consumption, heat load and power losses.

Tractors and Power-tillers: Type, selection, maintenance and repair of tractors and powertillers; tractor clutches and brakes; power transmission systems – gear trains, differential, final drives and power take-off; mechanics of tractor chassis; traction theory; three point hitches- free link and restrained link operations; mechanical steering and hydraulic control systems used in tractors; tractor tests and performance. Human engineering and safety in design of tractor and agricultural implements.

Soil and Water Conservation Engineering

Fluid Mechanics: Ideal and real fluids, properties of fluids; hydrostatic pressure and its measurement; hydrostatic forces on plane and curved surface; continuity equation; Bernoulli's theorem; laminar and turbulent flow in pipes, Darcy- Weisbach and Hazen-Williams equations, Moody's diagram; flow through orifices and notches; flow in open channels.

Soil Mechanics: Engineering properties of soils; fundamental definitions and relationships; index properties of soils; permeability and seepage analysis; shear strength, Mohr's circle of stress, active and passive earth pressures; stability of slopes.

Hydrology: Hydrological cycle and components; meteorological parameters, their measurement and analysis of precipitation data; runoff estimation; hydrograph analysis, unit hydrograph theory and application; stream flow measurement; flood routing, hydrological reservoir and channel routing.

Surveying and Leveling: Measurement of distance and area; instruments for surveying and leveling; chain surveying, methods of traversing; measurement of angles and bearings, plane table surveying; types of leveling; theodolite traversing; contouring; computation of areas and volume.

Soil and Water Erosion: Mechanics of soil erosion, soil erosion types, wind and water erosion, factors affecting erosion; soil loss estimation; biological and engineering measures to control erosion; terraces and bunds; vegetative waterways; gully control structures, drop, drop inlet and chute spillways; earthen dams.

Watershed Management: Watershed characterization; land use capability classification; rainwater harvesting structures, check dams and farm ponds.

Irrigation and Drainage Engineering

Soil-Water-Plant Relationship: Water requirement of crops; consumptive use and evapo-transpiration; measurement of infiltration, soil moisture and irrigation water infiltration.

Irrigation Water Conveyance and Application Methods: Design of irrigation channels and underground pipelines; irrigation scheduling; surface, sprinkler and micro irrigation methods, design and evaluation of irrigation methods; irrigation efficiencies.

Agricultural Drainage: Drainage coefficient; planning, design and layout of surface and sub-surface drainage systems; leaching requirement and salinity control; irrigation and drainage water quality and reuse.

Groundwater Hydrology: Groundwater occurrence; Darcy's Law, steady flow in confined and unconfined aquifers, evaluation of aquifer properties; groundwater recharge.

Wells and Pumps: Types of wells, steady flow through wells; classification of pumps; pump characteristics; pump selection and installation.

Agricultural Processing Engineering

Drying: Psychrometry – properties of air-vapors mixture; concentration and drying of liquid foods – evaporators, tray, drum and spray dryers; hydrothermal treatment; drying and milling of cereals, pulses and oilseeds.

Size Reduction and Conveying: Mechanics and energy requirement in size reduction of granular solids; particle size analysis for comminuted solids; size separation by screening; fluidization of granular solids-pneumatic, bucket, screw and belt conveying; cleaning and grading; effectiveness of grain cleaners; centrifugal separation of solids, liquids and gases.

Processing and By-product Utilization: Processing of seeds, spices, fruits and vegetables; By-product utilization from processing industries.

Storage Systems: Controlled and modified atmosphere storage; perishable food storage, godowns, bins and grain silos.

Dairy and Food Engineering

Heat and Mass Transfer: Steady state heat transfer in conduction, convection and radiation; transient heat transfer in simple geometry; working principles of heat exchangers; diffusive and convective mass transfer; simultaneous heat and mass transfer in agricultural processing operations; material and energy balances in food processing systems; water activity, sorption and desorption isotherms.

Preservation of Food: Kinetics of microbial death – pasteurization and sterilization of milk and other liquid foods; preservation of food by cooling and freezing; refrigeration and cold storage basics and applications.

Reference: Official Website of GATE-2022.

UPSC IFoS Agricultural Engineering 2022 Syllabus

PAPER - I

SECTION A

1. Soil and Water Conservation: Scope of soil and water conservation. Mechanics and types of erosion, their causes. Mechanics and types of erosion, their causes. Rainfall, runoff and sedimentation relationships and their measurement. Soil erosion control measures - biological and engineering including stream bank protectionvegetative barriers, contour bunds, contour trenches, contour stone walls, contour ditches, terraces, outlets and grassed waterways. Gully control structures - temporary and permanent - design of permanent soil conservation structures such as chute, drop and drop inlet spillways. Design of farm ponds and percolation ponds. Principles of flood control-flood routing. Watershed Management - investigation, planning and implementation - selection of priority areas and water shed work plan, water harvesting and moisture conservation. Land development - leveling, estimation of 19 earth volumes and costing. Wind Erosion process - design for shelter belts and wind brakes and their management. Forest (Conservation) Act.

2. Aerial Photography and Remote Sensing: Basic characteristics of photographic images, interpretation keys, equipment for interpretation, imagery interpretation for land use, geology, soil and forestry. Remote sensing - merits and demerits of conventional and remote sensing approaches. Types of satellite images, fundamentals of satellite image interpretation, techniques of visual and digital interpretations for soil, water and land use management. Use of GIS in planning and development of watersheds, forests including forest cover, water resources etc.

SECTION B

3. Irrigation and Drainage : Sources of water for irrigation. Planning and design of minor irrigation projects. Techniques of measuring soil moisture - laboratory and in situ, Soil-water plant relationships. Water requirement of crops. Planning conjunctive use of surface and ground water. Measurement of irrigation water, measuring devices - orifices, weirs and flumes. Methods

of irrigation - surface, sprinkler and drip, fertigation. Irrigation efficiencies and their estimation. Design and construction of canals, field channels, underground pipelines, head-gates, diversion boxes and structures for road crossing. Occurrence of ground water, hydraulics of wells, types of wells (tube wells and open wells) and their construction. Well development and testing. Pumps-types, selection and installation. Rehabilitation of sick and failed wells. Drainage causes of water logging and salt problem. Methods of drainage- drainage of irrigated and unirrigated lands, design of surface, sub-surface and vertical drainage systems. Improvement and utilization of poor quality water. Reclamation of saline and alkali soils. Economics of irrigation and drainage systems. Use of waste water for irrigation - standards of waste water for sustained irrigation, feasibility and economics.

4. Agricultural Structures : Site selection, design and construction of farmstead - farm house, cattle shed, dairy bam, poultry shed, hog housing, machinery and implement shed, storage structures for food grains, feed and forage. Design and construction of fences and farm roads. Structures for plant environment – green houses, poly houses and shade houses. Common building materials used in construction - timber, brick, stone, tiles, concrete etc and their properties. Water supply, drainage and sanitation system.

PAPER-II

SECTION A

1. Farm Power and Machinery : Agricultural mechanization and its scope. Sources of farm power - animate and electro-mechanical. Thermodynamics, construction and working of internal combustion engines. Fuel, ignition, lubrication, cooling and governing system of IC engines. Different types of tractors and power tillers. Power transmission, ground drive, power take off (p.t.o.) and control systems. Operation and maintenance of farm machinery for primary and secondary tillage. Traction theory. Sowing transplanting and interculture implements and tools. Plant protection equipment – spraying and dusting. Harvesting, threshing and combining equipment. Machinery for earth moving and land development - methods and cost estimation. Ergonomics of man-machine system. Machinery for horticulture and agro-forestry, feeds and forages. Haulage of agricultural and forest produce.

2. Agro-energy : Energy requirements of agricultural operations and agro-processing. Selection, installation, safety and maintenance of electric motors for agricultural applications. Solar (thermal and photovoltoic), wind and bio-gas energy and their utilization in agriculture. Gasification of biomass for running IC engines and for electric power generation. Energy efficient cooking

stoves and alternate cooking fuels. Distribution of electricity for agricultural and agro-industrial applications.

SECTION B

3. Agricultural Process Engineering : Post harvest technology of crops and its scope. Engineering properties of agricultural produces and by-products. Unit operations - clearing grading, size reduction, densification, concentration, drying/dehydration, evaporation, filtration, freezing and packaging of agricultural produces and by- 20 products. Material handling equipment belt and screw conveyors, bucket elevators, their capacity and power requirement. Processing of milk and dairy products - homogenization, cream separation, pasteurization, sterilization, spray and roller drying, butter making, ice cream, cheese and shrikhand manufacture. Waste and by-product utilization - rice husk, rice bran, sugarcane bagasse, plant residues and coir pith.

4. Instrumentation and computer applications in Agricultural Engineering : Electronic devices and their characteristics-rectifiers, amplifiers, oscillators, multivibrators. Digital circuits - sequential and combinational system. Application of microprocessors in data acquisition and control of agricultural engineering processesmeasurement systems for level, flow, strain, force, torque, power, pressure, vacuum and temperature. Computers - introduction, input/ output devices, central processing unit, memory devices, operating systems, processors, keyboards and printers. Algorithms, flowchart specification, programme translation and problem analysis in Agricultural Engineering. Multimedia and Audio-Visual aids.

Reference: Official Website of UPSC(www.upsc.gov.in).

I. Engineering Mathematics

1

Matrices and Determinants

Graduate Aptitude Test in Engineering - 2008

Q. 1 Eigen values of the matrix are

(A) 1 and 2 (B) 1 and 3

(C) 1 and 4 (D) 2 and 3

Graduate Aptitude Test in Engineering - 2009

Q. 2 Inverse of the matrix $\begin{bmatrix} 2 & 3 \\ 2 & 1 \end{bmatrix}$ is

(A) $\begin{bmatrix} -0.5 & 0.75 \\ 0.5 & -0.25 \end{bmatrix}$ (B) $\begin{bmatrix} -0.25 & 0.5 \\ -0.5 & 0.75 \end{bmatrix}$

(C) $\begin{bmatrix} -0.25 & 0.75 \\ 0.5 & -0.5 \end{bmatrix}$ (D) $\begin{bmatrix} -0.25 & -0.5 \\ 0.75 & 0.5 \end{bmatrix}$

Graduate Aptitude Test in Engineering - 2010

Q. 3 A system of equations represented as

$$\begin{bmatrix} 1 & -1 & 2 \\ 2 & 1 & -4 \\ 1 & 3 & 1 \end{bmatrix}\begin{bmatrix} x \\ y \\ z \end{bmatrix} = \begin{bmatrix} 4 \\ 1 \\ 3 \end{bmatrix} \text{ is}$$

(A) Consistent and has unique solution

(B) Inconsistent and has no solution

(C) Consistent and has infinite solution

(D) Inconsistent and has unique solution

Graduate Aptitude Test in Engineering - 2011

Q.4 A square matrix [A] will be lower triangular if and only if (represents an element of Mth row and N^{th} column of the matrix)

(A) $a_{MN} = 0, N>M$ (B) $a_{MN} = 0, M>N$

(C) $a_{MN} \neq 0, M>N$ (D) $a_{MN} \neq 0, N>M$

Graduate Aptitude Test in Engineering - 2012

Q. 5 The matrix is $\begin{bmatrix} 0 & 2 & -3 \\ -2 & 0 & 4 \\ 3 & -4 & 0 \end{bmatrix}$ is

(A) Diagonal (B) Symmetric

(C) Skew symmetric (D) Triangular

Q. 6 Eigen values of the matrix $A = \begin{bmatrix} 6 & 1 \\ -2 & 3 \end{bmatrix}$ are

(A) 3 and 6 (B) 1 and -2

(C) 5 and 4 (D) 1 and 6

Graduate Aptitude Test in Engineering - 2013

Q. 7 If $P = A \times B$, where $A = \begin{bmatrix} 2 & 1 \\ 3 & 0 \end{bmatrix}$ and $B = \begin{bmatrix} 1 & 3 & 0 \\ 2 & 1 & 2 \end{bmatrix}$; then P is

(A) $\begin{bmatrix} 7 & 2 \\ 9 & 0 \end{bmatrix}$ (B) $\begin{bmatrix} 4 & 7 & 2 \\ 3 & 9 & 0 \end{bmatrix}$

(C) $\begin{bmatrix} 2 & 4 \\ 3 & 9 \end{bmatrix}$ (D) $\begin{bmatrix} 2 & 4 & 7 \\ 0 & 3 & 9 \end{bmatrix}$

Q. 8 Eigen values of the matrix $\begin{bmatrix} 2 & 1 \\ 3 & 2 \end{bmatrix}$ are

(A) $\pm 2i$ (B) $2i \pm \sqrt{3}$

(C) $2 \pm i\sqrt{3}$ (D) $2 \pm \sqrt{3}$

Graduate Aptitude Test in Engineering - 2014

Q. 9 Eigen values of the matrix $A = \begin{bmatrix} 2 & 2 \\ -1 & 5 \end{bmatrix}$ are

(A) 1 and 2 (B) 2 and 3

(C) 3 and 4 (D) 4 and 5

Q. 10 Consider the following set of linear equations

$x_1 + x_2 + x_3 = 6$

$2x_1 + 2x_2 + 3x_3 = 14$

$3x_1 + x_2 + 2x_3 = 14$

The solution for this set exists only when the value of is ....

Graduate Aptitude Test in Engineering – 2015

Q. 11 Inverse of the matrix $\begin{bmatrix} 3 & 2 \\ 1 & 4 \end{bmatrix}$ is

(A) $\begin{bmatrix} 0.1 & -0.4 \\ -0.3 & 0.2 \end{bmatrix}$ (B) $\begin{bmatrix} 0.3 & -0.2 \\ -0.1 & 0.4 \end{bmatrix}$

(C) $\begin{bmatrix} 0.3 & -0.1 \\ -0.2 & 0.4 \end{bmatrix}$ (D) $\begin{bmatrix} 0.4 & -0.2 \\ -0.1 & 0.3 \end{bmatrix}$

Graduate Aptitude Test in Engineering – 2016

Q. 12 Eigen values of the matrix $\begin{bmatrix} 5 & 3 \\ 1 & 4 \end{bmatrix}$ are

(A) –6.3 & –2.7 (B) –2.3 & –6.7

(C) 6, 3 & 2.7 (D) 2.3 & 6.7

Graduate Aptitude Test in Engineering – 2017

Q. 13 Matrix $\begin{bmatrix} 0 & 0.5 & 1.5 \\ -0.5 & 0 & 2.5 \\ -1.5 & -2.5 & 0 \end{bmatrix}$ is a

(A) Diagonal Matrix (B) Orthogonal Matrix

(C) Symmetric Matrix (D) Skew-Symmetric Matrix

Q. 14 Characteristic equation of matrix $\begin{bmatrix} 2 & \sqrt{2} \\ \sqrt{2} & 1 \end{bmatrix}$ with Eigen value λ is

(A) $\lambda^2 + 3\lambda + 4 = 0$ (B) $\lambda^2 + 3\lambda - 2 = 0$

(C) $\lambda^2 - 3\lambda = 0$ (D) $\lambda^2 + 3\lambda = 0$

Graduate Aptitude Test in Engineering – 2018

Q. 15 Rank of a matrix $\begin{bmatrix} 5 & 3 & -3 & -1 \\ 3 & 2 & -2 & -1 \\ 2 & -1 & 2 & 8 \end{bmatrix}$

(A) 1 (B) 2

(C) 3 (D) 4

Graduate Aptitude Test in Engineering – 2019

Q. 16 The determinant of the matrix $A = \begin{bmatrix} 2 & 1 & 1 \\ 2 & 3 & 2 \\ 1 & 2 & 1 \end{bmatrix}$ is

(A) 1 (B) 0

(C) –1 (D) 2

Graduate Aptitude Test in Engineering – 2020

Q. 17 A linear system of equations has n unknowns. The ranks of the coefficient matrix and the augmented matrix of the linear system of equations are r1 and r2, respectively. The condition for the equations to be consistent with a unique solution is

(A) $r1 \neq r2 < n$ (B) $r1 = r2 = n$

(C) $r1 = r2 < n$ (D) $r1 \neq r2 > n$

Q. 18 If one of the two Eigen values of a matrix $\begin{bmatrix} 3 & 2 \\ 2 & 1 \end{bmatrix}$ is 4.236, then the other Eigen value is

Graduate Aptitude Test in Engineering – 2021

Q. 19 Trace of the matrix $\begin{bmatrix} 3 & 2 & 1 & 4 \\ 5 & 7 & 8 & 1 \\ 2 & 4 & 6 & 7 \\ 9 & 6 & 4 & 2 \end{bmatrix}$ is ....

Q. 20 Summation of eigen values of a matrix $\begin{bmatrix} 4 & 1 \\ 3 & 6 \end{bmatrix}$ is ....

Answers Key

1	2	3	4	5	6	7	8	9	10
C	C	A	A	C	C	B	D	C	1
11	12	13	14	15	16	17	18	19	20
D	C	D	C	C	C	B	-0.236	18	10

Explanations

Q. 1

First Method

We know that

(i) Sum of eigen values of that matrix = Trace of a matrix

(ii) Multiplication of eigen values = Determinant of that matrix

According to above properties, eigen values of $\begin{bmatrix} 2 & 1 \\ 2 & 3 \end{bmatrix}$ are 1 and 4.

Second Method

To find eigen values

$$|A-I\lambda| = 0$$

$$\begin{vmatrix} 2-\lambda & 1 \\ 2 & 3-\lambda \end{vmatrix} = 0$$

$$(2-\lambda)(3-\lambda) - 2 = 0$$

$$\Rightarrow \lambda = 1, 4$$

Q. 2

Inverse of a second order square matrix $\begin{bmatrix} a & b \\ c & d \end{bmatrix}$ is

$$= \frac{1}{ad-bc}\begin{bmatrix} d & -b \\ -c & a \end{bmatrix}$$

$$= \frac{1}{(2\times1-2\times3)}\begin{bmatrix} 1 & -3 \\ -2 & 2 \end{bmatrix} = \begin{bmatrix} -0.25 & 0.75 \\ 0.5 & -0.5 \end{bmatrix}$$

Q. 3

$$\text{Given } A = \begin{bmatrix} 1 & -1 & 2 \\ 2 & 1 & -4 \\ 1 & 3 & 1 \end{bmatrix}$$

Making Echelon Matrix

$$(A;B) = \begin{bmatrix} 1 & -1 & 2 & ; 4 \\ 2 & 1 & -4 & ; 1 \\ 1 & 3 & 1 & ; 3 \end{bmatrix}$$

$$= \begin{bmatrix} 1 & -1 & 2 & ; 4 \\ 0 & 3 & -8 & ; -7 \\ 0 & 4 & -1 & ; -1 \end{bmatrix} \quad (R_2 = R_2 - 2R_1 \text{ and } R_3 = R_3 - R_1)$$

$$= \begin{bmatrix} 1 & -1 & 2 & ; 4 \\ 0 & 3 & -8 & ; -7 \\ 0 & 0 & 29 & ; 25 \end{bmatrix} (R_3 = 3R_3 - 4R_2)$$

Here Rank(A;B) = Rank(A) – 3 – No. of variables ⇒ System of equation is consistent and has unique solution.

Q. 5

It is a skew symmetric matrix because transpose of this matrix is equal to negative of the matrix.

Q. 6

We know that-

(i) Sum of eigen values of that matrix = Trace of a matrix

(ii) Multiplication of eigen values = Determinant of that matrix

According to above properties, eigen values of $\begin{bmatrix} 6 & 1 \\ -2 & 3 \end{bmatrix}$ are 5 and 4.

Q. 7

$$\begin{bmatrix} 2 & 1 \\ 3 & 0 \end{bmatrix} \times \begin{bmatrix} 1 & 3 & 0 \\ 2 & 1 & 2 \end{bmatrix} = \begin{bmatrix} 2+2 & 6+1 & 0+2 \\ 3+0 & 9+0 & 0+0 \end{bmatrix} = \begin{bmatrix} 4 & 7 & 2 \\ 3 & 9 & 0 \end{bmatrix}$$

Q. 8

We know that-

(i) Sum of eigen values of that matrix = Trace of a matrix

(ii) Multiplication of eigen values = Determinant of that matrix

According to above properties, eigen values of $\begin{bmatrix} 2 & 1 \\ 3 & 2 \end{bmatrix}$ are $2 \pm \sqrt{3}$.

Q. 9

We know that-

(i) Sum of eigen values of that matrix = Trace of a matrix

(ii) Multiplication of eigen values = Determinant of that matrix

According to above properties, eigen values of $\begin{bmatrix} 2 & 2 \\ -1 & 5 \end{bmatrix}$ are 3 and 4.

Q. 10

Applying rank method

$$[A;B] = \begin{bmatrix} 1 & 1 & 1 ; & 6 \\ 2 & 2 & 3 ; & 14 \\ 3 & 1 & 2 ; & 14 \end{bmatrix}$$

$$= \begin{bmatrix} 1 & 1 & 1 ; & 6 \\ 0 & 0 & 1 ; & 2 \\ 0 & -2 & -1 ; & -4 \end{bmatrix} (R_2 \to R_2 - 2R_1; R_3 \to R_3 - 3R_1)$$

$$= \begin{bmatrix} 1 & 1 & 1 ; & 6 \\ 0 & -2 & -1 ; & -4 \\ 0 & 0 & 1 ; & 2 \end{bmatrix} (R_2 \leftrightarrow R_3)$$

Here,

Rank (A) = Rank (A;B) = 3 ⇒ Consistent and finite solution

On solving above matrix $x_1 = 3, x_2 = 1, x_3 = 2$

Q. 11

Inverse of a second order square matrix $\begin{bmatrix} a & b \\ c & d \end{bmatrix}$ is $= \frac{1}{ad - bc}\begin{bmatrix} d & -b \\ -c & a \end{bmatrix}$

$$= \frac{1}{(3\times4-2\times1)}\begin{bmatrix}4 & -2\\-1 & 3\end{bmatrix} = \begin{bmatrix}0.4 & -0.2\\-0.1 & 0.3\end{bmatrix}$$

Q. 12

First Method

We know that

(iii) Sum of eigen values of a matrix = Trace of the matrix

(iv) Multiplication of eigen values = Determinant of the matrix

According to above properties, option (c) suits the above conditions.

Hence, eigen values of $\begin{bmatrix}5 & 3\\1 & 4\end{bmatrix}$ are 2.7 and 6.3.

Second Method

To find eigen values

$|A\text{-}I\,\lambda| = 0$

$$\begin{vmatrix}5-\lambda & 3\\1 & 4-\lambda\end{vmatrix} = 0$$

$(5\text{-}\lambda)(4\text{-}\lambda)\text{-}3 = 0$

$\lambda = 2.7, 6.3$

Q. 13

(A) Diagonal matrix - These are square matrices that can have nonzero entries only on the main diagonal. Any entry above or below the main diagonal must be zero.

(B) Orthogonal matrix – A square matrix A is said to be orthogonal if

Transpose of matrix A = Inverse of matrix A

i.e. $A^T = A^{-1}$

Note:- Determinant of an orthogonal matrix = 1 always

(C) Symmetric matrix - matrices are square matrices whose transpose equals the matrix itself.

(D) Skew-symmetric matrix - matrices are square matrices whose transpose equals minus the matrix.

Here, transpose of the matrix is equal to the minus of the matrix. Hence, it's a Skew-symmetric matrix.

Q. 14

Characteristic equation of the matrix is $|A - \lambda I_0| = 0$

Hence, $\begin{bmatrix} 2 & \sqrt{2} \\ \sqrt{2} & 1 \end{bmatrix} = 0$

$\Rightarrow$ $\lambda^2 - 3 = 0$

Q. 15

We know that the rank of a matrix is number of non-zero rows in its transformed matrix to its row echelon form.

Given

$$[A] = \begin{bmatrix} 5 & 3 & -3 & -1 \\ 3 & 2 & -2 & -1 \\ 2 & -1 & 2 & 8 \end{bmatrix}$$

Making Row Echelon Form

$$[A] = \begin{bmatrix} 5 & 3 & -3 & -1 \\ 0 & 1 & -1 & -2 \\ 0 & -7 & 10 & 28 \end{bmatrix} \quad (R_3 \to 3R_3 - 2R_2;\ R_2 \to 5R_2 - 3R_1)$$

$$= \begin{bmatrix} 5 & 3 & -3 & -1 \\ 0 & 1 & -1 & -2 \\ 0 & 0 & 3 & 14 \end{bmatrix} \quad (R_3 \leftrightarrow R_3 - 7R_2)$$

Here,

Rank (A) = 3 $\Rightarrow$ Number of non-zero rows

Q. 16

Determinant of the matrix A = 2(3-4) - 1(2-2) + 1(4-3) = - 1

Q. 17

To illustrate this problem, let's look at the examples below.

$x_1 + 2x_2 = 1$ $\quad$ $3x_1 + 2x_2 = 3$ $\quad$ $3x_1 + 2x_2 = 3$

$3x_1 + x_2 = -2$ $\quad$ $-6x_1 - 4x_2 = 0$ $\quad$ $-6x_1 - 4x_2 = -6$

The augmented matrices for these systems are, respectively,

$$\left\langle \begin{array}{cc|c} 1 & 0 & -1 \\ 0 & 1 & 1 \end{array} \right\rangle \quad \left\langle \begin{array}{cc|c} 1 & 2/3 & 0 \\ 0 & 0 & 1 \end{array} \right\rangle \quad \left\langle \begin{array}{cc|c} 1 & 2/3 & 1 \\ 0 & 0 & 0 \end{array} \right\rangle$$

Applying the row-reduction algorithm yields the row-reduced form of each of these augmented matrices. The results are, again respectively,

Hence,

System	Rank[A]	Rank[A|b]	n	# of solutions
First	2	2	2	1 (consistent)
Second	1	2	2	0 (inconsistent)
Third	1	1	2	∞

Q. 18

First Method

We know that

(i) Sum of eigen values of that matrix = Trace of a matrix

(ii) Multiplication of eigen values = Determinant of that matrix

According to above properties, eigen values of $\begin{bmatrix} 3 & 2 \\ 2 & 1 \end{bmatrix}$ are 4.236 and -0.236.

Second Method

To find eigen values $|A\text{-}I\lambda| = 0$

$$\begin{vmatrix} 3-\lambda & 2 \\ 2 & 1-\lambda \end{vmatrix} = 0$$

$$(3\text{-}\lambda)\,(1\text{-}\lambda)\text{-}\,4 = 0$$

$\Rightarrow$ $\lambda = -0.236,\ 4.236$

Q. 19

We know that-

Trace i.e. sum of its diagonal components of a matrix = 3+7+6+2 = 18

Q. 20

We know that-

Sum of eigen values = Trace i.e. sum of its diagonal components of a matrix

$$= 6 + 4 = 10$$

2

Differential Equations

Graduate Aptitude Test in Engineering - 2007

Q. 1 Taking y (0) = 0 and using Euler's method with size h = 0.1 solution of the differential equation $\frac{dy}{dx} = 2xy + 1$ gives the value of y(0.3) as

(A) 0.3101 (B) 0.3142

(C) 0.6202 (D) 4.080

Q. 2 Integrating the function $f(x) = 1 + e^{-x} \sin(4x)$ over the interval [0, 1] using Simpson's 1/3rd rule gives

(A) 1.021 (B) 0.091

(C) 1.321 (D) 2.642

Graduate Aptitude Test in Engineering - 2008

Q. 3 Solution of the ordinary differential equation $\frac{dy}{dx} = \frac{x^2 + 2}{y}$ is

(A) $y = \sqrt{\left(\frac{2}{3}\right)x^3 + 4x}$ (B) $y = \sqrt{\left(\frac{2}{3}\right)x^3 + 4x + k}$

(C) $y = \sqrt{\left(\frac{2}{3}\right)x^3 - 4x + k}$ (D) $y = x^3 + 4x + k$

Q. 4 If, $\log_e(y) = -x \log_e(x)$ then the maximum value of y is

(A) e (B) e^{x^2}

(C) $e^{e^{-1}}$ (D) e^{-1}

Q. 5 A function f(x) is evaluated as 1, 1.5, 2.2 and 3.4 at four values of x having intervals of 0.5. The area under the curve f(x) using trapezoidal rule is

(A) 1.95 (B) 2.45

(C) 2.95 (D) 3.45

Graduate Aptitude Test in Engineering - 2009

Q. 6 $I = \int_0^{\pi/2} \frac{cosx\, dx}{(1+sinx)^2}$ is

(A) − 0.5 (B) 0

(C) 0.5 (D) 1

Q. 7 A curve is having the equation, r = a(1 − cos θ). The perimeter of the curve between θ = 0 to 2π is

(A) 2a (B) 4a

(C) 6a (D) 8a

Graduate Aptitude Test in Engineering - 2010

Q. 8 The value of is $\int_0^{2}\int_0^{y} xy\, dx\, dy$

(A) 0.5 (B) 1.0

(C) 2.0 (D) 4.0

Q. 9 The derivative of $y = \sqrt{x + \sqrt{x + \sqrt{x + ...}}}$ with respect to x at y = 0 is

(A) − 1 (B) 0

(C) 1 (D) 2

Q. 10 A particular solution of is $\frac{d^2y}{dx^2} + 5\frac{dy}{dx} - 3y = 6$

(A) 2.0 (B) 0.5

(C) − 0.5 (D) − 2.0

Q. 11 The partial differential equation $\frac{\partial^2 u}{\partial x^2} - 7\frac{\partial^2 u}{\partial x \partial y} + 2\frac{\partial^2 u}{\partial y^2}$ is said to be

(A) Parabolic (B) Hyperbolic

(C) Elliptic (D) Eccentric

Graduate Aptitude Test in Engineering - 2011

Q. 12 The differential equation $2\frac{d^2z}{dx^2} + \frac{dz}{dx} + 3y = \sin x$ is considered to be ordinary, as it has

(A) One dependent variable
(B) More than one dependent variable
(C) One independent variable
(D) More than one independent variable

Q. 13 The stationary points of $f(x, y) = \frac{1}{3}x^3 - xy^2 - 2y$ are

(A) (1,-1) and (-1,1) (B) (1,1) and (-1,-1)

(C) (2,-2) and (-2,2) (D) (2,2) and (-2,-2)

Q. 14 The solution of $x\frac{dy}{dx} = x^2 + 3y$ for $x > 0$ and $y(1) = 2$ is

(A) $y = x^3 + 3x^2$ (B) $y = x^2 + 3x^3$

(C) $y = -x^3 + 3x^2$ (D) $y = -x^2 - 3x^3$

Q. 15 The mean value of a function f(x) from x = a to x = b is given by

(A) $\frac{f(a)+f(b)}{2}$ (B) $\frac{f(a)+2f\left(\frac{a+b}{2}\right)+f(b)}{4}$

(C) $\int_a^b f(x)dx$ (D) $\frac{\int_a^b f(x)dx}{b-a}$

Q. 16 The highest order of polynomial integrand for which Simpson's 1/3 rule of integration is exact is

(A) First (B) Second

(C) Third (D) Fourth

Graduate Aptitude Test in Engineering - 2012

Q. 17 The integrating factor of the differential equation $(x + 1)\frac{dy}{dx} - y = \sin x$ is

(A) x (B) (x + 1)

(C) 1/ x (D) 1/(x + 1)

Q. 18 The tangent line to y = f (x) at the point (x_0, y_0), assuming $f'(x) \neq 0$, intersects the x axis At

(A) $(x_0) - [\frac{y_0}{f'(x_0)}], 0)$ (B) $(x_0) + [\frac{y_0}{f'(x_0)}], 0)$

(C) $(x_0) - [\frac{f'(x_0)}{y_0}], 0)$ (D) $(x_0) + [\frac{f'(x_0)}{y_0}], 0)$

Q. 19 The value of $\int_0^{\pi/2} \cos x\, dx$ using trapezoidal rule with two equal intervals is

(A) 0.95 (B) 1.00

(C) 1.22 (D) 1.29

Q. 20 If $f'(x) = e^x$ and $f(0) = 5$, then from Mean Value Theorem, the value of $f(1)$ lies between

(A) 2 and $(2 + e)$ (B) 3 and $(2 + e)$

(C) 3 and $(3 + e)$ (D) 6 and $(5 + e)$

Graduate Aptitude Test in Engineering - 2013

Q. 21 The general solution of the differential equation $\frac{d^2y}{dx^2} - 3\frac{dy}{dx} - 4y = 0$ is

(A) $Ae^x + Be^{-4x}$ (B) $Ae^{-x} + Be^{4x}$

(C) $Ae^x + Be^{4x}$ (D) $Ae^{-x} + Be^{-4x}$

Q. 22 $f(x) = e^{-x^2}$ for x = 1.1, 1.2, 1.3, 1.4 and 1.5, the value of $\int_{1.1}^{1.5} f(x)dx$ by Simpson's 1/3rd rule is ....

Graduate Aptitude Test in Engineering - 2014

Q. 23 A partial differential equation containing dependent variable u is given by

$$A\frac{\partial^2 u}{\partial x^2} + 2B\frac{\partial^2 u}{\partial x \partial y} + C\frac{\partial^2 u}{\partial y^2} + D\frac{\partial u}{\partial x} + E\frac{\partial u}{\partial y} + Fu + G = 0$$

Where A, B, C, D, E, F and G are constants of functions of independent variables x or y only. Also $G \neq 0$. The nature of the equation is

(A) Linear and homogeneous

(B) Non-linear and homogeneous

(C) Linear and non-homogeneous

(D) Non-linear and non-homogeneous

Q. 24 If $f(x,y) = 3x^2 - 4xy + 2y^2$, the summation of $\frac{\partial f}{\partial x}$ and $\frac{\partial f}{\partial y}$ for x = 1 and y = 2 is ....

Q. 25 The value of $\int_0^{\pi/2} \frac{\cos^2 x\, dx}{1 + \sin x}$ is

(A) 0 (B) $\frac{\pi}{2} - 1$

(C) 2 (D) $\frac{\pi}{2} + 1$

Q. 26 The value of line integral $I = \int_c \{(x^2 y)\,dx + (x-z)\,dy + (xyz)\,dz\}$, where C is the arc of the parabola $y = x^2$ in the plane z = 2 from P(0, 0, 2) to Q(1, 1, 2) is

(A) $-\frac{43}{15}$ (B) $\frac{43}{15}$

(C) $-\frac{17}{15}$ (D) $\frac{17}{15}$

Graduate Aptitude Test in Engineering – 2015

Q. 27 The series represented by $a_n = (n^2 - 2n)/(3n^2 + n)$ is

(A) Convergent (B) Divergent

(C) Asymptotic (D) Oscillatory

Q. 28 Values of the fraction $y = 1/(1 + x^2)$ are y(0) = 1; y(1) = 0.5; y(2) = 0.2; y(3) = 0.1; y(4) = 0.0588; y(5) = 0.0385; y(6) = 0.027. Using Simpson's one-third rule, the value of the integral $\int_0^6 \frac{dx}{1+x^2}$ is ....

Q. 29 The particular solution to the first order differential equation $\frac{dy}{dx} = x - e^{-x}$ for y(0) = – 1 is

(A) $2x^2 + e^{-x} - 2$ (B) $2x^2 + e^{-x} - 1/2$

(C) $x^2/2 - e^{-x} + 1/2$ (D) $x^2/2 + e^{-x} - 2$

Q. 30 The value of $\int_0^{\pi/2} sin^3 x\,dx$ is

(A) 0 (B) 1/3

(C) 2/3 (D) 1

Graduate Aptitude Test in Engineering – 2016

Q. 31 The general solution of the differential equation

$\frac{dy}{dx} = e^{3x-2y} + x^2 e^{-2y}$ is

(A) $C = \frac{1}{2}e^{2y} - \frac{1}{3}(e^{3x} + x^3)$ (B) $C = e^{2y} - \frac{1}{3}(e^{3x} + x^2)$

(C) $C = \frac{1}{2}e^{2y} - \frac{1}{3}(e^{3x} + x^3)$ (D) $C = e^{2y} - \frac{1}{3}(e^{3x} + x^3)$

Q. 32 The function $f(x) = x^2 - x - 6$ is

(A) Minimum at x = 1/2 (B) Maximum at $x = \frac{1}{2}$

(C) Minimum at x = –1/2 (D) Maximum at x = –1/2

Q. 33 Integration by trapezoidal method of $\log_{10}(x)$ with lower limit of 1 to upper limit of 3 using seven distinct values (equally covering the whole range) is ....

Q. 34 The value of the integral, $I = \int_2^4 \frac{x^2+1}{x^2-1} dx$ is ....

Q. 35 Let I, J, and K are unit vectors along the three mutually perpendicular x, y and z axes, respectively. If F = fI + gJ + hK is a continuously differentiable vector point function, then curl F is

(A) $I\left(\frac{\partial g}{\partial z}-\frac{\partial h}{\partial y}\right)-J\left(\frac{\partial f}{\partial z}-\frac{\partial h}{\partial x}\right)+K\left(\frac{\partial f}{\partial y}-\frac{\partial g}{\partial x}\right)$

(B) $I\left(\frac{\partial h}{\partial y}-\frac{\partial g}{\partial z}\right)-J\left(\frac{\partial h}{\partial x}-\frac{\partial f}{\partial z}\right)+K\left(\frac{\partial g}{\partial x}-\frac{\partial f}{\partial y}\right)$

(C) $I\left(\frac{\partial h}{\partial y}-\frac{\partial g}{\partial z}\right)+J\left(\frac{\partial h}{\partial x}-\frac{\partial f}{\partial z}\right)+K\left(\frac{\partial g}{\partial x}-\frac{\partial f}{\partial y}\right)$

(D) $I\left(\frac{\partial g}{\partial z}-\frac{\partial h}{\partial y}\right)+J\left(\frac{\partial f}{\partial z}-\frac{\partial h}{\partial x}\right)+K\left(\frac{\partial f}{\partial y}-\frac{\partial g}{\partial x}\right)$

Graduate Aptitude Test in Engineering – 2017

Q. 36 The areas of seven horizontal cross-sections of a water reservoir at intervals of 9 m are 210, 250, 320, 350, 290, 230 and 170 m^2. The estimated volume of the reservoir in m^3 using Simpsons's rule is ....

Q. 37 Divergence value of a function $x^2y)I - (z^3 - 3x)j + (4y^2)k$ at $x = 1$, $y = 2$ and $z = 3$ is ....

Q. 38 Differentiation of $\sqrt{(1+x^2}$ gives

(A) $\frac{1}{1+x^2}$ (B) $\frac{1}{\sqrt{(1+x^2}}$

(C) $\frac{\sqrt{(1+x^2}}{x}$ (D) $\frac{x}{\sqrt{(1+x^2}}$

Q. 39 $I = \int\sqrt{(a^2-x^2)}$ dx is

(A) $0.5\left[\sqrt{(a^2-x^2)} + sin^{-1}(\frac{x}{a})\right]$ (B) $0.5\left[x\sqrt{(a^2-x^2)} + sin^{-1}(\frac{x}{a})\right]$

(C) $0.5\left[\sqrt{(a^2-x^2)} + a^2 sin^{-1}(\frac{x}{a})\right]$ (D) $0.5\left[x\sqrt{(a^2-x^2)} + a^2 sin^{-1}(\frac{x}{a})\right]$

Graduate Aptitude Test in Engineering – 2018

Q. 40 The general solution to the second order linear homogeneous differential equation

$y'' - 6y' + 25y = 0$ is

(A) $e^{3x}(a \cos 4x + b \sin 4x)$ (B) $e^{3ix}(a \cos 4x + b \sin 4x)$

(C) $e^{4x}(a \cos 3x + b \sin 3x)$ (D) $e^{4ix}(a \cos 3x + b \sin 3x)$

Q. 41 Solution of $f(x) = x^4 + 2x^3 - 4x^2 + 3x - 1 = 0$ is

(A) 0.333 (B) 0.646

(C) 0.658 (D) 1.000

Q. 42 For $0 \le x \le 2\pi$, sin*x* and cos*x* are both decreasing functions in the interval....

(A) $(0, \pi/2)$ (B) $(\pi/2, \pi)$

(C) $(\pi, 3\pi/2)$ (D) $(3\pi/2, 2\pi)$

Q. 43 Integration of $I = \int_0^{D/2} \frac{2R^2 x dx}{\left(R^2 + x^2\right)^2}$ is

(A) $\frac{R^2}{R^2 + D^2}$ (B) $\frac{4R^2}{4R^2 + D^2}$

(C) $\frac{D^2}{R^2 + D^2}$ (D) $\frac{D^2}{4R^2 + D^2}$

Q. 44 The velocity (v) of a tractor, which starts from rest, is given at fixed intervals of time (t) as follows:

T(min)	0	2	4	6	8	10	12	14	16	18	20
V (m/min)	0	0.8	1.5	2.1	2.4	2.7	1.7	0.9	0.4	0.2	0

Using Simpson's 1/3rd rule, the distance covered by the tractor in 20 minutes will be ...... m.

Graduate Aptitude Test in Engineering – 2019

Q. 45 Using trapezoidal rule, the value of $I = \int_{4.0}^{5.2} ln(x) dx$ is ....

x	4.0	4.2	4.4	4.6	4.8	5.0	5.2
y = ln(x)	1.386	1.435	1.482	1.526	1.569	1.609	1.648

Q. 46 $I = \int_0^{\infty} \frac{dx}{(x^2 + 1)^2}$ has the value

(A) 0.785 (B) 0.915

(C) 1.000 (D) 1.245

Q. 47 General solution to the differential equation y"+ 4y'+ 5y = 0 is

(A) $e^{2x}(a \cos x + b \sin x)$ (B) $e^{-2x}(a \cos x + b \sin x)$

(C) $e^{x}(a \cos 2x + b \sin 2x)$ (D) $e^{-x}(a \cos 2x + b \sin 2x)$

Graduate Aptitude Test in Engineering – 2020

Q. 48 The function $f(x) = x^4 - 4x^3 + 6x^2 - 4x + 1$ has a

(A) Maxima at x = **0** (B) Minima at x = **0**

(C) Maxima at x = **1** (D) Minima at x = **1**

Q. 49 General solution to the ordinary differential equation,

$$\frac{d^3y}{dx^3} - 6\frac{d^2y}{dx^2} + 11\frac{dy}{dx} - 6y = 0 \text{ is}$$

(A) $C_1e^{-x} + C_2e^{-2x} + C_3\, e^{-3x}$ (B) $C_1e^{-x} + C_2e^{-2x} + C_3\, e^{3x}$

(C) $C_1e^{-x} + C_2e^{2x} + C_3\, e^{3x}$ (D) $C_1e^{x} + C_2e^{2x} + C_3\, e^{3x}$

Q. 50 $lim_{x\to 0}\left(\frac{e^x - e^{-x} - 2x}{x - sinx}\right)$ is ....

Q. 51 Taking six intervals, each of π/12 and using Simpson's one-third rule, the value of the definite integral $\int_0^{\pi/2} \sqrt{cosx}\, dx$ is ....

Graduate Aptitude Test in Engineering – 2021

Q. 52 If x is an integer with x > 1, the solution of

$$\lim_{x\to\infty} \left(\frac{1}{x^2} + \frac{2}{x^2} + \frac{3}{x^2} + \ldots\ldots + \frac{x-1}{x^2} + \frac{1}{x}\right) \text{ is}$$

(A) Zero (B) 0.5

(C) 1.0 (D) ∞

Q. 53 Solution of the differential equation y" + y' + 0.25 y = 0 with the initial values y(0) = 3.0 and y'(0) = - 3.5 is

(A) $y = (3 - 2x)\, e^{0.5x}$ (B) $y = (3 - 2x)\, e^{-0.25x}$

(C) $y = (3 - 2x)\, e^{-0.5x}$ (D) $y = (2 - 3x)\, e^{-0.5x}$

Answers Key

1	2	3	4	5	6	7	8	9	10
A	A	B	C	C	C	D	C	A	D
11	12	13	14	15	16	17	18	19	20
B	C	A	C	D	C	D	A	A	D
21	22	23	24	25	26	27	28	29	30
B	0.08	C	2	B	C	A	1.37	D	C
31	32	33	34	35	36	37	38	39	40
A	A	0.5608	2.25	B	14760	4	B	D	A
41	42	43	44	45	46	47	48	49	50
C	B	D	25.86	1.83	A	B	D	D	2
51	52	53							
1.1870	B	C							

Explanations

Q. 1

Given

$y' = 2xy + 1$

With, $y(0) = 0$ and $h = 0.1$

By Euler's method:

$y_1 = y_0 + h(y')_0$

	x_0	y_0	$(y')_0$
1.	0.0	0.0	1
2.	0.1	0.1	1.02
3.	0.2	0.202	1.0808
4.	0.3	0.3101	1.1861

Q. 2

Given that $f(x) = 1 + e^{-x}\sin(4x)$

Interval is [0, 1]

X	$f(x) = 1 + e^{-x}\sin(4x)$
0	1
0.25	1.0136
0.5	1.0212
0.75	1.0247
1	1.0257

According to Simpson's 1/3rd rule

$$\int_{x_1}^{x_n} f(x)\,dx = \frac{h}{3}\ [(y_0 + y_n) + 4[(y_1 + y_3 + y_5 \ldots) + 2[(y_2 + y_4 + y_6 \ldots)]$$

$$\int_0^1 [1 + e^{-x}\sin(4x)]\,dx = \frac{0.25}{3}\ [(1+1.0257)+4[(1.0136+1.0247)+2[(1.0212)]$$

$$= 1.018$$

Q. 3

Separating the variables

$$y.dv = \int (x^2 + 2)\,dx$$

Integrating both sides

$$\int y.dv = \int (x^2 + 2)dx$$

$$\frac{y^2}{2} = \frac{x^3}{3} + 2x + C$$

$$y^2 = \frac{2x^3}{3} + 4x + 2C$$

$$y = \sqrt{\frac{2x^3}{3} + 4x + k}\ \ (k = 2C)$$

Q. 4

Given

$$\log_e(y) = -x\log_e(x)$$

$$y = x^{-x}$$

Now $\dfrac{dy}{dx} = -\dfrac{ln(x)+1}{x^x} = 0$

$$x = 1/e$$

Thus, $y = e^{1/e}$

Q. 5

According to trapezoidal rule

$$\int_{x_1}^{x_n} f(x)dx = \frac{h}{2}\ [(y_0 + y_n) + 2[(y_1 + y_2 + y_3 \ldots\ldots)]$$

$$\int_0^2 f(x)dx = \frac{0.5}{2}\ [(1+3.4) + 2[(1.5+2.2)]$$

$$= 2.95$$

Q. 6

Given

$$I = \int_0^{\pi/2} \frac{cosx\,dx}{(1+sinx)^2}$$

Put, 1 + sin x = t

cos x dx = dt

Now

$$I = \int_0^{\pi/2} \frac{dt}{(t)^2}$$

$$= \left[-\frac{1}{t}\right]_0^{\pi/2}$$

Putting, t = 1 + sin x

$$= \left[-\frac{1}{(1+sin\,x)}\right]_0^{\pi/2}$$

$$= -\frac{1}{2} + 1 = 0.5$$

Q. 7

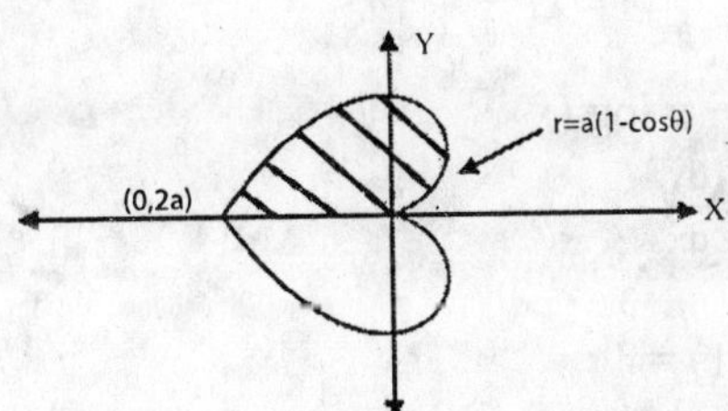

Given

$r = a(1 - \cos\theta)$

$$\frac{dr}{d\theta} = a\sin\theta$$

$$\Rightarrow \text{Length of curve} = 2\int_a^b \sqrt{r^2 + \left(\frac{dr}{d\theta}\right)^2}\, d\theta$$

$$= 2\int_0^{2\pi} \sqrt{(a - a\,cos\theta)^2 + (a\,sin\theta)^2}\, d\theta$$

$$= 2\sqrt{2}a \int_0^{2\pi} \sqrt{(1 - cos\theta)}\, d\theta$$

$$= 2\sqrt{2}a \int_0^{2\pi} \sqrt{1 - \left(1 - 2\,sin^2\frac{\theta}{2}\right)}\, d\theta$$

$$= 4a\int_0^{2\pi}\left(\sin\frac{\theta}{2}\right)d\theta$$

$$= \left[-4a\cos\frac{\theta}{2}\right]_0^{2\pi}$$

$$= 4a + 4a = 8\ a$$

Q. 8

Given

$$\int_0^2\int_0^y xy\ dx\ dy = \int_0^2 y[\frac{x^2}{2}]_0^y\ dy$$

$$= \int_0^2 \frac{y^3}{2} dy$$

$$= [\frac{y^4}{8}]_0^2 = 2$$

Q. 9

Given

$$y = \sqrt{x + \sqrt{x + \sqrt{x......}}}$$

Putting $y = \sqrt{x + \sqrt{x.....}}$

$$y = \sqrt{x+y}$$

$$y^2 = x + y$$

Now $\frac{dy^2}{dx} = 1 + \frac{dy}{dx}$

$$\frac{dy}{dx}(2y - 1) = 1$$

$$\frac{dy}{dx} = \frac{1}{(2y-1)}$$

Putting, y = 0

$$\frac{dy}{dx} = -1$$

Q. 10

Here, f(x) = Constant

Put y = k

$(D^2 + 5D - 3)k = 6$

Comparing both sides,

We get

$-3k = 6$

$\Rightarrow k = -2$

Hence, Particular Solution is − 2.0.

Q. 11

We know that

A hyperbolic partial differential equation is a type of second-order partial differential equation (PDE) of the form

$$A\frac{\partial^2 u}{\partial x^2} + 2B\frac{\partial^2 u}{\partial x \partial y} + C\frac{\partial^2 u}{\partial y^2} + D\frac{\partial u}{\partial x} + E\frac{\partial u}{\partial y} Fu + G = 0$$

that, satisfies the condition

$B^2 - AC > 0$

Here, $B = -7/2, A = 1, C = 2$

Now,

$(-7/2)^2 - 1 \times 2 > 0$

Q. 12

In mathematics, an ordinary differential equation or ODE is an equation containing a function of one independent variable and its derivatives.

Q. 13

$$f(x, y) = \frac{1}{3}x^3 - xy^2 - 2y$$

$$\frac{\partial f}{\partial x} = x^2 - y^2$$

$$\frac{\partial f}{\partial y} = -2xy - 2$$

Putting, $\frac{\partial f}{\partial x} = 0$ and $\frac{\partial f}{\partial y} = 0$

$x^2 - y^2 = 0$

and $-2xy - 2 = 0$

Solving both the equations,

$x \pm 1$ and $y = \mp 1$

Therefore, stationary points are $(1, -1)$ and $(-1, 1)$.

Q. 14

Changing the equation $x\frac{dy}{dx} = x^2 + 3y$ in the form $\frac{dy}{dx} + P(x).y = Q(x)$.

$$\frac{dy}{dx} - \frac{3}{x}y = x$$

It is a standard form of Leibnitz linear equation.

Integrating Factor I.F. = $e^{-\int Q(x)dx}$

Solution is I.F. = $e^{\int \frac{-3}{x}dx} = x^{-3}$

$$y(I.F.) = \int Q(I.F.)dx + C$$

$$yx^{-3} = \int x.x^{-3}\,dx + C$$

$$yx^{-3} = -x^{-1} + C$$

Putting, x = 1 and y = 2

$$C = 3$$

So, the solution is $y = -x^2 + 3x^3$

Q. 17

Changing the equation $(x+1)\frac{dy}{dx} - y = \sin x$ in the form $\frac{dy}{dx} + P(x).y = Q(x)$

$$\frac{dy}{dx} + \left(-\frac{y}{1+x}\right) = \sin x$$

It is a standard form of Leibnitz linear equation.

Integrating Factor I.F. = $e^{-\int Q(x)dx}$

Solution is I.F. = $e^{\int \frac{-3}{x}dx} = x^{-3}$

Thus, I.F. = $e^{-\int \frac{1}{1+x}dx} = e^{\ln(1/(1+x))} = \frac{1}{1+x}$

Q. 18

Slope of tangent at point (x_0, y_0) is $(m) = f'(x)$

Consider, Equation of tangent is $y = mx + c$

At point, (x_0, y_0) $y_0 = f'(x)\,x_0 + c$

$\Rightarrow$ $c = y_0 - f'(x)\,x_0$

Hence, Equation of line is $y = f'(x)\,x + y_0 - f'(x)\,x_0$

This line intersect at x axis, therefore put y = 0.

$f'(x)\,x + y_0 - f'(x)\,x_0 = 0$

$$x = x_0 - \frac{y_0}{f'(x_0)}$$

Point of intersection = $(x_0 - \frac{y_0}{f'(x_0)}, 0)$

Q. 19

Interval (h) = π/4

X	0	π/4	π/2
y = cos x	1	$1/\sqrt{2}$	0

Trapezoidal rule gives

$$I = \frac{h}{2}\,[\,y_0 + 2(y_1 + y^2)\,]$$

$$= \frac{3.14}{2 \times 4}\,[\,1 + 2(1\ \sqrt{2}\ + 0)]$$

$$= 0.947$$

Q. 20

According to mean value theorem

$$\frac{f(b) - f(a)}{b - a} = f'(x)$$

$$\Rightarrow \quad \frac{f(1) - f(0)}{1 - 0} = e^1$$

$$\Rightarrow \quad \frac{f(1) - 5}{1} = e^1$$

$$\Rightarrow \quad f(1) = e + 5$$

And $f(0) = 6$

Hence, value of f (1) lies between 6 and e + 5.

Q. 21

General equation is $m^2 - 3m - 4 = 0$

$$m = 4, -1$$

Thus solution of differential equation is $y = Ae^{-x} + Be^{4x}$

Q. 22

x	$f(x) = e^{-x^2}$
1.1	0.298
1.2	0.237
1.3	0.184
1.4	0.141
1.5	0.105

Applying Simpson's formula

$$(I) = \int_{1.1}^{1.5} f(x)dx = \frac{0.1}{3} = [(0.298 + 0.105) + 4(0.237 + 0.141)$$

$$+ 2(0.184)] = 0.08$$

Q. 24

$f(x,y) = 3x^2 - 4xy + 2y^2$

$$\frac{\partial f}{\partial x} = 6x - 4y$$

$$\frac{\partial f}{\partial y} = -4x + 4y$$

Now,

$$\frac{\partial f}{\partial x} + \frac{\partial f}{\partial y} = 6x - 4y - 4x + 4y$$

Putting, x = 1 and y = 2

$$\frac{\partial f}{\partial x} + \frac{\partial f}{\partial y} = 6 - 8 - 4 + 8 = 2$$

Q. 25

Given

$$I = \int_0^{\pi/2} \frac{\cos^2 x\, dx}{1 + \sin x}$$

$$= \int_0^{\pi/2} \frac{(1 - \sin^2 x)dx}{1 + \sin x} \qquad (\because 1 + \sin^2 x = \cos^2 x)$$

$$= \int_0^{\pi/2} (1 + \sin x)dx$$

$$= [x + \cos x]_0^{\pi/2}$$

$$= \frac{\pi}{2} - 1$$

Q. 26

Given

$I = \int_c \{(x^2 y)\, dx + (x-z)\, dy + (xyz)dz\}$

Also

$y = x^2 \Rightarrow dy = 2x\, dx$

And $z = 0 \Rightarrow dz = 0$

Now

$I = \int_0^1 \{(x^2x^2)\, dx + (x-2)2x\, dx + (x.x^2.2)\times 0\}$

$= \left[\frac{x^5}{5} + \frac{2x^3}{3} - 2x^2\right]_0^1$

$= -\frac{17}{15}$

Q. 27

We have $a_n = (n^2 - 2n)/(3n^2 + n) = (n-2)/(3n+1)$ and

$a_{n+1} = \{(n+1)^2 - 2(n+1)\}/\{3(n+1)^2 + (n+1)\} = (n-1)/(3n+4)$

Apply D' Alembert's ratio test

$\therefore\ lim_{n\to\infty} \frac{a_{n+1}}{a_n} = lim_{n\to\infty} \frac{(n-1)(3n+1)}{(n-2)(3n+4)} = lim_{n\to\infty} \frac{(1-1/n)\,(3+1/n)}{(1-2/n)\,(3+4/n)}\ 1$

Thus, D' Alembert's ratio test fails.

Apply Limit form of Comparison test

Let, $V_n = n/(1+n)$

(V_n is a convergent series and $a_n \leq V_n$ for all values of n.)

Then,

$lim_{n\to\infty} \frac{V_n}{a_n} = lim_{n\to\infty} \frac{(n-2)(1+n)}{(3n+1)(n)} = lim_{n\to\infty} \frac{(1-2/n)(1/n+1)}{(3+1/n)(1/n)}\ 1/3$

Which is < 1. Hence, the given series is convergent.

Q. 28

Given that $f(x) = 1/(1+x^2)$

Interval is [0, 6]

X	$f(x) = 1/(1+x^2)$
0	1
1	0.5
2	0.2
3	0.1
4	0.0588
5	0.0385
6	0.027

According to Simpson's 1/3rd rule

$$\int_{x_1}^{x_n} f(x)\,dx = \frac{h}{3}[(y_0 + y_n) + 4[(y_1 + y_3 + y_5 \ldots\ldots) + 2[(y_2 + y_4 + y_6 \ldots\ldots)]$$

$$\int_0^1 [\frac{1}{1+x2}]dx = \frac{1}{3}[(1 + 0.027) + 4[(0.5 + 0.1 + 0.0385)$$

$$+ 2[(0.2 + 0.0588)] = 1.37$$

Q. 29

Since $\frac{dy}{dx} = x - e^{-x}$

Hence, $y = \int(x - e^{-x})dx$

i.e. $y = x^2/2 + e^{-x} + C$, which is the general solution.

Substituting the boundary conditions y = – 1 and x = 0, to evaluate C gives:

$0 + 1 + C = -1 \Rightarrow C = -2$

Hence the particular solution is $y = x^2/2 + e^{-x} - 2$

Q. 30

We have $I = \int_0^{\pi/2} \sin^3 x\,dx$

We know $\sin^3 x = (3\sin x - \sin 3x)/4$

Hence, $I = \int_0^{\pi/2} \left(\frac{3\sin x - \sin 3x}{4}\right)dx$

$$= \left[-\frac{3\cos x}{4} + \frac{\cos 3x}{12}\right]_0^{\pi/2}$$

$= 2/3$

Q. 31

Given that

$dy/dx = e^{3x-2y} + x^2 e^{-2y}$

or $dy/dx = (e^{3x} + x^2)/e^{2y}$

Multiply both side with e^{2y}

$$\int e^{2y}\frac{dy}{dx}dx = \int(e^{3x} + x^2)\ dx$$

Evaluate the integrals

$\frac{1}{2}e^{2y} = e^{3x}/3 + x^3/3 + C$

or $C = \frac{1}{2} e^{2y} - (e^{3x} + x^3)/3$

Q. 32

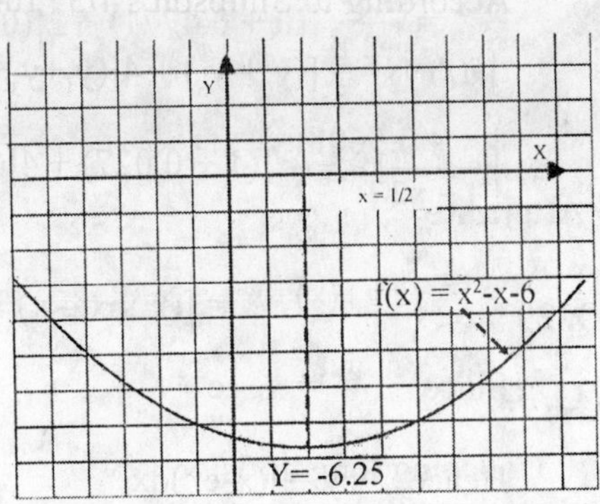

Given that $f(x) = x^2 - x - 6$

Put $dy/dx = 0$

$dy/dx = 2x - 1 = 0$

$\Rightarrow \quad x = 1/2$

Now put $x = 1/2$ in d^2y/dx^2

$d^2y/dx^2 = 2 > 0,$

Hence $f(x) = x^2 - x - 6$ has a minima at $x = 1/2$.

Alternate

Graphical representaion of $f(x) = x^2 - x - 6$

shows that function

f(x) has a minima

at $x = 1/2$.

Q. 33

Given that $f(x) = \log_{10}(x)$

Interval is [1, 3]

X	$f(x) = \log_{10}(x)$
1	0
1.2857	0.1091
1.5714	0.1963
1.8571	0.2688
2.1428	0.3310
2.4285	0.3853
2.7142	0.4336
3	0.4771

According to trapezoidal rule

$$\int_{x_1}^{x_n} f(x)dx = \frac{h}{2}\ [(y_0 + y_n) + 2[(y_1 + y_2 + y_3 \ldots\ldots)]$$

$= \frac{0.33}{2}$ [(0 + 0.4771) + 2[(0.1091 + 0.1963 + 0.2688 + 0.3310 + 0.3853 + 0.4336)]

= 0.5608

Alternate

$$\int_1^3 \log_{10}(x)\,dx = [(x.\ln x - x)/\ln(10)]_1^3 = 0.5608$$

Q. 34

Doing long division

$$I = \int_2^4 \left(-\frac{1}{x+1} + \frac{1}{x-1} + 1\right) dx$$

$$= -\int_2^4 \left(\frac{1}{x+1}\right) dx + \int_2^4 \left(\frac{1}{x-1}\right) dx + \int_2^4 1\, dx$$

For the integrand $\frac{1}{x+1}$ substitute $u = x + 1$ and $du = dx$.

This gives a new lower bound u = 1+2 = 3 and upper bound u = 1 + 4 =5.

Also, For the integrand $\frac{1}{x-1}$ substitute $s = x - 1$ and $ds = dx$.

This gives a new lower bound s = 2 – 1 = 1 and upper bound u =4 – 1 = 3.

$$I = -\int_3^5 \left(\frac{1}{u}\right) dx + \int_1^3 \left(\frac{1}{s}\right) dx + \int_2^4 1\, dx$$

The antidarative of 1/u is log(u).

$$= (-log(u))_3^5 + (-log(s))_1^3 + (x)_2^4$$

= – log(5/3) + log(3) + 2 = 2.25

Q. 36

Volume of water in the reservoir obtained by the Sympson's formula

$$V = \frac{h}{3} [(A_0 + A_n) + 4(A_1 + A_3 + \ldots\ldots) + 2(A_2 + A_4 + \ldots\ldots)]$$

$$= \frac{9}{3} [(210 + 170) + 4(250 + 350 + 230) + 2(320 + 290)]$$

= 14760 m³

Q. 37

We know, for a function(F) = ui + vj + zk, divergence is = $\frac{\delta u}{\delta x}+\frac{\delta v}{\delta y}+\frac{\delta z}{\delta z}$

Hence, $\text{Div(F)} = \frac{\delta(x^2y)}{\delta x}+\frac{\delta(-z^3+3x)}{\delta y}+\frac{\delta(4y^2)}{\delta z}$

$= 2xy - 0 - 0$

At, x = 1 and y = 2

Div(F) = 4

Q. 38

Differentiation of function (f) = $d(1+x^2)^{1/2}/dx$

$= \{(2x)(1+x^2)^{-1/2}\}/2$

$= x/(1+x^2)^{1/2}$

Q. 39

Let x = a sinθ then dx/dθ = a cosθ and dx = a cosθ dθ

Hence, $\int\sqrt{a^2-x^2}\ dx = \int\sqrt{a^2-a^2sin^2\theta}(a\,cos\theta\,d\theta)$

$= \int\sqrt{a^2(1-sin^2\theta)}(a\,cos\theta\,d\theta)$

$= \int\sqrt{a^2cos^2\theta}(a\,cos\theta\,d\theta)$

$= \int a^2cos^2\theta d\theta$

$= a^2\int\frac{1+cos2\theta}{2}d\theta$

(since, cos2θ = 2 cos²θ – 1)

$= \frac{a^2}{2}\left(\theta+\frac{sin2\theta}{2}\right)++C$

$= \frac{a^2}{2}(\theta+\frac{2sin\theta\,cos\theta}{2})+C$

(since, sin 2θ = 2 sinθ cosθ)

$= \frac{a^2}{2}(\theta+sin\theta\,cos\theta)+C$

Since, x = a sinθ then sinθ = x/a and θ = $\sin^{-1}(x/a)$

Also, $\cos^2\theta + \sin^2\theta = 1$, from which

$\cos\theta = (1-\sin^2\theta)^{1/2} = (1-(x/a)^2)^{1/2} = \frac{\sqrt{a^2-x^2}}{a}$

Thus, $\int\sqrt{a^2 - x^2}\ dx \quad = \frac{a^2}{2}\ (\theta + sin\theta\, cos\theta) + C$

$$= \frac{a^2}{2}\left(sin^{-1}\left(\frac{x}{a}\right) + (\frac{x}{a})\frac{\sqrt{a^2 - x^2}}{a}\right) + C$$

$$= \frac{a^2}{2} sin^{-1}\left(\frac{x}{a}\right) + (\frac{x}{2})\frac{\sqrt{a^2 - x^2}}{a} + C$$

Q. 40

Characteristic Equation

$A^2 - 6A + 25 = 0$

Hence, roots of quadratic equation is

$$A = \frac{-(-6) \pm \sqrt{(-6)^2 - 4 \times (1) \times (25)}}{2 \times 1} = 3 \pm 4i$$

Here, $3 \pm 4i = (\alpha \pm \beta i) \Rightarrow \alpha = 3$ and $\beta = 4$

General solution is $y = e^{\alpha x}(a \cos \beta x + b \sin(\beta x))$

$$= e^{3x}(a \cos 4x + b \sin(4x))$$

Q. 41

The Newton-Rapson method uses an iterative process to approach one root of a function

$x_{n+1} = x_n - f(x_n)/f'(x_n)$

$f(x) = x^4 + 2x^3 - 4x^2 + 3x - 1$

and $f'(x) = 4x^3 + 6x^2 - 8x + 3$

x_n	f(x)	f'(x)	X_{n+1}
0	-1	3	0.33333
0.33333	-0.35802	1.14815	0.64516
0.64516	0.01912	1.41026	0.65872
0.65872	0.00045	1.47703	0.65842
0.65842	0.0000002	1.47548	0.65842

Hence, Solution of equation f(x) = 0.65842

Q. 42

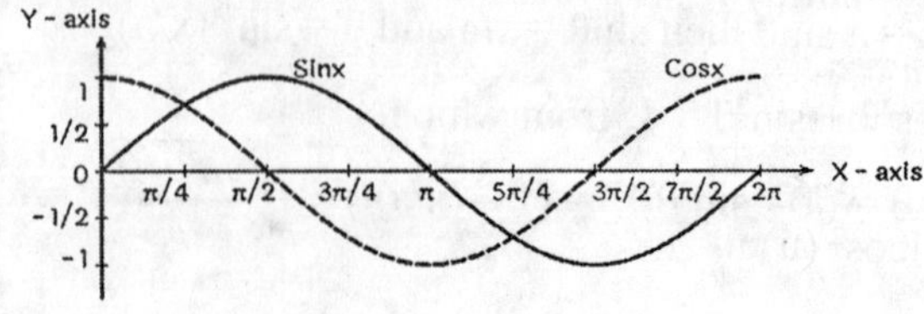

Q. 43

Given that $I = \int_0^{D/2} \frac{2R^2 x dx}{(R^2+x^2)^2}$

$= 2R^2 \int_0^{D/2} \frac{x dx}{(R^2+x^2)^2}$

Apply $u = R^2 + x^2$

$du = 2x\, dx$

$dx = du/2$

Again $I = R^2 \int_0^{D/2} \frac{du}{u^2}$

$= \left[-R^2/u\right]_0^{D/2}$

Substitute back $u = R^2 + x^2$

$I = \left[-R^2/(R^2+x^2)\right]_0^{D/2} = \frac{D^2}{4R^2 + D^2}$

Q. 44

According to Simpson's 1/3rd rule

$\int_{x_1}^{x_n} f(x)\, dx$

$= h/3[(y_0 + y_n) + 4[(y_1 + y_3 + y_5 \ldots\ldots) + 2[(y_2 + y_4 + y_6 \ldots\ldots)]$

$= \frac{2}{3}\ [(0+0) + 4[(0.8 + 2.1 + 2.7 + 0.9 + 0.2) + 2(1.5 + 2.4 + 1.7 + 0.4)]$

$= 25.86$ m

Q. 45

According to trapezoidal rule

$\int_{x_1}^{x_n} f(x)dx - \frac{h}{2}\ [(y_0 + y_n) + 2[(y_1 + y_2 + y_3 \ldots\ldots)]$

$\int_{4.0}^{5.2} ln(x)dx = \frac{0.2}{2}\ [(1.386 + 1.648) + 2[(1.435 + 1.482 + 1.526 + 1.569 + 1.609)]$

$= 1.83$

Q. 46

Put $x = \tan(u)$

$I = \int \frac{1}{\left(\frac{1}{cos(u)}\right)^4} cos^2(u)$

$= \int \cos^2(u)du$

$$= \int \frac{1 + cos2u}{2} du$$

$$= \frac{1}{2}(u + \frac{1}{2} sin(2u))$$

Putting u = arctan(x)

$$= \frac{1}{2}\left(arctan(x) + \frac{1}{2} sin(2arctan(x))\right) + C$$

Computing the boundaries for $x = 0 \rightarrow \infty$

$I = \pi/4 = 0.785$

Q. 47

Given

$D^2 + 4D + 5 = 0$

Solving the above equation, roots are

$\alpha = -2 + i$ and $= -2 - i$

We know

For two complex roots $\alpha \neq \beta$, Where $\alpha = u + iv$ and $\beta = u - iv$

The general solution will be $y = e^{ux}(a\cos(vx) + (b\sin(vx))$

Hence, General solution = $e^{-2x}(a \cos x + b \sin x)$

Q. 48

$f'(x) = 4(x^3 - 3x^2 + 3x - 1) = 4(x-1)^3 > 0$ for $x > 1$

Hence, f(x) has a minimum at x = 1

Q. 49

The given equation can be written as $(D^3 - 6D^2 + 11D - 6)y = 0$

The A.E is $m^3 - 6m^2 + 11m - 6 = 0$ Here, m = 1 is one of the roots.

m = 1	1	-6	11	-6
	0	1	-5	6
	1	-5	6	0

Therefore, $m^2 - 5m + 6 = 0$

m = 3, 2

m = 1, 2, 3 are the roots of AE.

Hence, C.F. = $C_1e^x + C_2e^{2x} + C_3e^{3x}$

Thus, $y = C_1e^x + C_2e^{2x} + C_3e^{3x}$ is the general solution.

Q. 50

$$lim_{x\to 0^\cdot}\left(\frac{e^x - e^{-x} - 2x}{x - sinx}\right) = \left(\frac{e^0 - e^{-0} - 2(0)}{0 - sin0}\right) = \frac{0}{0}$$

This is an indeterminate type so use L-Hopital's Rule which is the limit of the quotient of the derivative of the top and the derivative of the bottom as x goes to 0.

$$lim_{x\to 0^\cdot}\left(\frac{e^x + e^{-x} - 2}{1 - cosx}\right) = \left(\frac{e^0 + e^{-0} - 2}{1 - cos0}\right) = \frac{0}{0}$$

Still an indeterminate form so apply L-Hopital's rule again.

$$lim_{x\to 0^\cdot}\left(\frac{e^x - e^{-x}}{sinx}\right) = \left(\frac{e^0 - e^{-0}}{sin0}\right) = \frac{0}{0}$$

Still an indeterminate form so apply L-Hopital's rule again.

$$lim_{x\to 0^\cdot}\left(\frac{e^x + e^{-x}}{cosx}\right) = \left(\frac{e^0 + e^{-0}}{cos0}\right) = \frac{2}{1} = 2$$

Q. 51

x	0	π/12	π/6	π/4	π/3	5π/12	π/2
y = $\sqrt{cosx}$	1	0.9828	0.9306	0.8409	0.7071	0.5087	0

$$\int_0^{\pi/2}\sqrt{cosx}\,dx = \frac{h}{3}\,[y_1 + y_7 + 4(y_2 + y_4 + y_6) + 2(y_3 + y_5)]$$

$$= \frac{\pi}{36}\,[1 + 0 + 4(0.9828 + 0.8409 + 0.5087) + 2(0.9328 + 0.7071)]$$

$$= 1.1870$$

Q. 52

$$\lim_{x\to 0}\left(\frac{1}{x^2} + \frac{2}{x^2} + \frac{3}{x^2}\ldots\ldots\ldots\frac{x-1}{x^2} + \frac{1}{x}\right)$$

$$\Rightarrow \quad \lim_{x\to 0}\frac{1}{x^2}\{1 + 2 + 3 + \ldots\ldots(x-1)\} + \lim_{x\to 0}\left(\frac{1}{x}\right)$$

$$\Rightarrow \quad \lim_{x\to 0}\frac{1}{x^2}\left\{\frac{(x-1)x}{2}\right\} + \lim_{x\to 0}\left(\frac{1}{x}\right) \quad \{\because \sum\ (1 + 2 + 3 + \ldots\ldots n) = \frac{(n+1)n}{2}\}$$

$$\Rightarrow \quad \lim_{x\to 0}\frac{(x-1)}{2x} + \lim_{x\to 0}\left(\frac{1}{x}\right)$$

$$\Rightarrow \quad \lim_{x\to 0}\frac{(x+1)}{2x}$$

Differentiating numerator and denominator, and putting x = 0

$\Rightarrow \quad \lim_{x \to 0}\left(\frac{1}{2}\right) = \frac{1}{2}$

Q. 53

It's a second order linear ordinary differential equation.

General equation is $m^2 + m + 0.25 = 0 \Rightarrow m = -1/2$

Hence,

General solution $y(x) = e^{-1/2x}(C_1 + C_2 x)$

Sustitute $y(x) = 3 \Rightarrow C_1 = 3$

Again

$y'(x) = -\frac{1}{2} e^{-1/2x} C_1 - \frac{1}{2} e^{-1/2x} C_2 x + e^{-1/2x} C_2$

Sustitute $y'(x) = -3.5$ and $C_1 = 3 \Rightarrow C_2 = -2$

Hence, General solution $y(x) = e^{-1/2x}(3 - 2x)$

3

Vector Calculus

Graduate Aptitude Test in Engineering - 2008

Q. 1 The cross product of $\vec{x} = 2i + j$ and $\vec{y} = i + 2j + k$ is

(A) i - j + 2k
(B) i - 2j + 5k
(C) i - 2j - 5k
(D) 2i - 4j

Graduate Aptitude Test in Engineering - 2010

Q. 2 The curl of the vector A= xyi + yzj + zxk (i, j and k represents unit vectors along the three orthogonal axis) is

(A) xi + yj + zk
(B) – xi – yj – zk
(C) yi + zj + xk
(D) – yi - zj - xk

Q. 3 The angle of intersection between the planes and is

(A) 30°
(B) 60°
(C) 75°
(D) 90°

Graduate Aptitude Test in Engineering - 2011

Q. 4 The divergence of the vector F = sin(xy) i + y cos(z) j + xz cos(z)k (i, j and k represent unit vectors along the three orthogonal axes) at x = y = 0 is

(A) sin(z)
(B) cos(z)
(C) – sin(z)
(D) – cos(z)

Q. 5 A plane contains the following three points: P(2,1,5), Q(−1,3,4) and R(3,0,6). The vector perpendicular to the above plane can be represented as

(A) 2i – j + k
(B) 1 + 2j + 2k
(C) 2i + 3j + k
(D) i + 2j + k

Graduate Aptitude Test in Engineering - 2012

Q. 6 The line $y = x - 1$ can be expressed in polar coordinates (r,) as

(A) $r = \cos\theta$ (B) $r = \sin\theta$

(C) $r(\cos\theta + \sin\theta) = 1$ (D) $r(\cos\theta - \sin\theta) = 1$

Graduate Aptitude Test in Engineering - 2013

Q. 7 If $C = A \times B$, where $A = 2i - j + 3k$ and $B = i + 2j$; then C is

(A) $-6i + 3j + 5k$ (B) $6i + 3j - 5k$

(C) $-6i + 2j + 3k$ (D) $6i + 2j + 5k$

Graduate Aptitude Test in Engineering – 2015

Q. 8 The distance PQ between two position vectors $\overline{P} = -2i + 3j + 4k$ and $\vec{Q} = 4i + 5j + 7k$ is ....

Graduate Aptitude Test in Engineering – 2017

Q. 9 Direction cosines of vector $3i - 2j + 6k$ are

(A) (3/7, – 2/7, 6/7) (B) (– 3/7, 2/7, – 6/7)

(C) (– 7/3, 7/2, – 7/6) (D) (7/3, –7/2, 7/6)

Graduate Aptitude Test in Engineering – 2018

Q. 10 A particle is subjected to forces $P = 2i - 5j + 6k$ and $Q = -i + 2j - k$. The positional vectors are $\overrightarrow{OA} = 4i - 3j - 2k$ and $\overrightarrow{OB} = 6i + j - 3k$. Work done by the resultant force (F = P + Q) to move the particle from A to B is

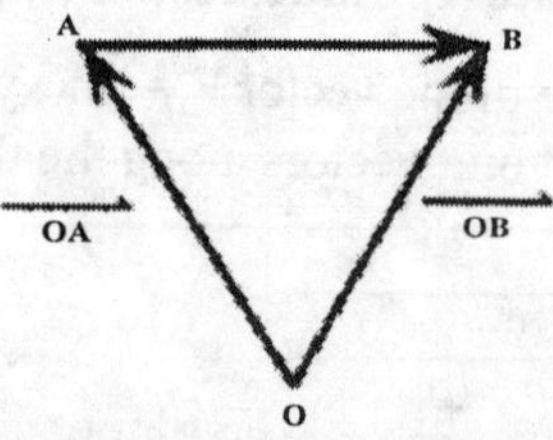

(A) 2 (B) –5

(C) 12 (D) –15

Graduate Aptitude Test in Engineering – 2019

Q. 11 For vectors $\vec{F} = 3xyi - y^2j$ and $\vec{R} = xi + yj$, the value of $\int_C \vec{F} \cdot d\vec{R}$ on the curve $C(y = 2x^2)$ in the x-y plane from (0, 0) to (1, 2) is

(A) −1.17 (B) 1.50

(C) −2.67 (D) 2.67

Q. 12 Directional derivative of $f(x, y, z) = xy^2 + yz^3$ at the point (2,-1, 1) in the direction of vector i + 2j + 2k is ....

Graduate Aptitude Test in Engineering – 2020

Q. 13 A particle moves along the curve $\vec{R} = (2t^3 + 4t^2)\, i + (3t^2 - 5t)\, j + (7t^2 + 6t)\, k$, where t is the time. The velocity component of the particle in the direction 3 i + 2 j + k at time t = 2 is

(A) $122/\sqrt{14}$ (B) 122

(C) $168/\sqrt{14}$ (B) 168

Graduate Aptitude Test in Engineering – 2021

Q.14 Let the vector $\vec{v} = v_1 i + v_2 j + v_3 k$ be a differentiable vector function of Cartesian coordinates x, y and z. The curl of the vector is $\vec{v}$ given by $curl\ \vec{v} =$

(A) $\left(\frac{\partial v_2}{\partial z} - \frac{\partial v_3}{\partial y}\right) i + \left(\frac{\partial v_3}{\partial x} - \frac{\partial v_1}{\partial z}\right) j + \left(\frac{\partial v_1}{\partial y} - \frac{\partial v_2}{\partial x}\right) k$

(B) $\left(\frac{\partial v_3}{\partial z} - \frac{\partial v_2}{\partial y}\right) i + \left(\frac{\partial v_1}{\partial x} - \frac{\partial v_3}{\partial z}\right) j + \left(\frac{\partial v_2}{\partial y} - \frac{\partial v_1}{\partial x}\right) k$

(C) $\left(\frac{\partial v_3}{\partial y} - \frac{\partial v_2}{\partial z}\right) i + \left(\frac{\partial v_1}{\partial z} - \frac{\partial v_3}{\partial x}\right) j + \left(\frac{\partial v_2}{\partial x} - \frac{\partial v_1}{\partial y}\right) k$

(D) $\left(\frac{\partial v_2}{\partial y} - \frac{\partial v_3}{\partial z}\right) i + \left(\frac{\partial v_3}{\partial z} - \frac{\partial v_1}{\partial x}\right) j + \left(\frac{\partial v_1}{\partial x} - \frac{\partial v_2}{\partial y}\right) k$

Answers Key

1	2	3	4	5	6	7	8	9	10
C	D	D	B	D	D	A	7	A	D
11	12	13	14						
A	-3.67	C	C						

Explanations

Q. 1

Required cross product = $\begin{bmatrix} i & j & k \\ 2 & 1 & 0 \\ 1 & -2 & 1 \end{bmatrix}$ = i - 2j - 5k

Q. 2

Given vector A = xyi + yzj + zxk

$$\text{Curl of given vector is} = \begin{bmatrix} i & j & k \\ \frac{\partial}{\partial x} & \frac{\partial}{\partial y} & \frac{\partial}{\partial z} \\ xy & yz & zx \end{bmatrix} = -yi - zj - xk$$

Q. 3

$$\text{Angle of intersection} = \cos^{-1}\left(\frac{a_1a_2 + b_1b_2 + c_1c_2}{\sqrt{a_1^2 + b_1^2 + c_1^2}\sqrt{a_2^2 + b_2^2 + c_2^2}}\right)$$

$$= \cos^{-1}\left(\frac{1\times 2 - 3\times 4 + 2\times 5}{\sqrt{1^2 + 3^2 + 2^2}\sqrt{2^2 + 4^2 + 5^2}}\right) = 90$$

Q. 4

Vector F = sin(xy)i + y cos(z)j + xz cos(z)k

We know, Divergence is $\frac{\delta u}{\delta x} + \frac{\delta v}{\delta y} + \frac{\delta z}{\delta z}$.

$$\text{Thus,}\quad div(F) = \frac{\delta(sin(xy))}{\delta x} + \frac{\delta(y\,cos(z))}{\delta y} + \frac{\delta(xz\,cos(z))}{\delta z}$$

$$= x\cos(xy) + \cos(z) + xz\sin(z)$$

At point, x = y = 0

div(F) = cos(z)

Q. 5

Plane passing through points P(2,1,5), Q(-1,3,4) and R(3,0,6) will be

$$f(x,y,z) = \begin{bmatrix} x-2 & y-1 & z-5 \\ -1-2 & 3-1 & 4-5 \\ 3-2 & 0-1 & 6-5 \end{bmatrix} = \begin{bmatrix} x-2 & y-1 & z-5 \\ -3 & 2 & -1 \\ 1 & -1 & 1 \end{bmatrix} = 0$$

$$= x + 2y + z - 9 = 0$$

As we know gradient of any function gives the perpendicular vector to that function.

Hence,

Perpendicular vector of f(x,y,z) = i + 2j + k

Q. 6

The polar form of y= x − 1 will be

Thus co-ordinates are (x, x − 1)

Thus, $\tan\theta = \frac{x-1}{x}$

And r $= \sqrt{2x^2 - 2x + 1}$

Thus, $\sin\theta = \frac{x-1}{r}$ (i)

And $\cos\theta = \frac{x}{r}$ (ii)

thus by equation (i) and (ii), we get $r(\cos\theta - \sin\theta) = 1$

Polar form is $r(\cos\theta - \sin\theta) = 1$

Q. 7

$$C = A \times B = \begin{bmatrix} i & j & k \\ 2 & -1 & 3 \\ 1 & 2 & 0 \end{bmatrix}$$

$$= -6i + 3j + 5k$$

Q. 8

We have $\vec{P} = -2i + 3j + 4k$ and $\vec{Q} = 4i + 5j + 7k$

We know

Distance between two position vectors $= [(4-2)^2 + (5-3)^2 + (7-4)^2]^{1/2}$

$= 7$

Q. 9

Vector given $= 3i - 2j + 6k$

Resultant of vector $= (3^2 + (-2)^2 + 6^2)^{1/2} = 7$

Hence, Direction cosines of the vector are $= (3/7, -2/7, 6/7)$

Q. 10

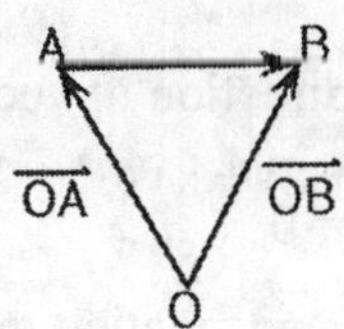

Refering the figure

$\overline{OA} - \overline{AB} = \overline{OB}$

$\Rightarrow \quad \overline{AB} = 6i + j - 3j - (4i - 3j - 2k) = 2i + 4j - k$

Force ($\vec{F}$) $= 2i - 5j + 6k + (-i + 2j - k) = i - 3j + 5k$

$\therefore$ Work done (W) $= \vec{F} \cdot \overline{AB}$

$= (i - 3j + 5k).(2i + 4j - k) = -15$ unit

Q. 11

Given

$y = 2x^2 \quad \Rightarrow dy = 4x\, dx$

Putting these values in $\vec{F}$ and $\vec{R}$

$\vec{F} = 6x^3i - 4x^4j$ and $\vec{R} = xi + 2x^2j \Rightarrow dxi + 4xdxj$

$$\oint \vec{F}.d\vec{R} = \int_0^1 \left(6x^3 i - 4x^4 j\right).(dxi + 4xdxj)$$

$$= \int_0^1 6x^3 dx - 16x^5 dx$$

$$= \left[\frac{6x^4}{4} - \frac{16x^6}{6}\right]_0^1 = -1.17$$

Q. 12

The gradient of the function and evaluate it at the given point:

$\nabla (xy^2 + yz^3)$ at the point (2,-1, 1) = (1, -3, -3)

Normalized or unit vector for given vector = (1/3, 2/3, 2/3)

Finally, the directional derivative is the dot product of the gradient and the normalized vector:

(1, -3, -3).(1/3, 2/3, 2/3) = -11/3 = -3.67

Q. 13

Differentiating the $\vec{R}$ gives velocity $\vec{v} = (6t^2 + 8t)i + (3t^2 - 5t)j + (7t^2 + 6t)k$

Putting, t = 2

$\vec{v} = 40i + 7j + 34\,k$

Velocity component in direction of vector given

$= (40i + 7j + 34\,k).\,(3i + 2j + k)/(3^2 + 2^2 + 1^2)^{1/2} = 168/14^{1/2}$

Q. 14

Given vector $A = v_1 i + v_2 j + v_3 k$

Curl of given vector is = $\begin{bmatrix} i & j & k \\ \frac{\partial}{\partial x} & \frac{\partial}{\partial y} & \frac{\partial}{\partial z} \\ v_1 & v_2 & v_3 \end{bmatrix}$

$$= \left(\frac{\partial v_3}{\partial y} - \frac{\partial v_2}{\partial z}\right)i + \left(\frac{\partial v_1}{\partial z} - \frac{\partial v_3}{\partial x}\right)j + \left(\frac{\partial v_2}{\partial x} - \frac{\partial v_1}{\partial y}\right)k$$

4

Statistics and Probability

Graduate Aptitude Test in Engineering - 2007

Q. 1 For station X, the maximum one day rainfall with 25 years return period is 100 mm. The probability of a on day rainfall equal to or greater than 100 mm at station X occurring at least once in 15 successive years is

(A) 0.458 (B) 0.500

(C) 0.637 (D) 0.990

Graduate Aptitude Test in Engineering - 2008

Q. 2 If one bucket contains 8 red balls and 2 black balls and another bucket contains 7 red balls and 3 black balls, the probability of having at least one red ball from drawings of one ball from each of the two buckets is

(A) 0.94 (B) 0.84

(C) 0.56 (D) 0.38

Q. 3 In a factory 30% of the machines are assembled by robots and 70% are assembled by manual labour. Reliability of the first type of machines is 0.9 and that of the second type of machines is 0.8. One piece of machine was found to be reliable. The probability of the machine having been assembled by robot is

(A) 0.325 (B) 0.565

(C) 0.675 (D) 0.835

Q. 4 If f(x) is a perfect normal distribution with mean and standard deviation of 5 and 1 respectively, then the value of f(x) for x = 6 is

(A) 0.124 (B) 0.242

(C) 0.482 (D) 0.524

Graduate Aptitude Test in Engineering - 2009

Q. 5 The probability function value [f(x)] at x = 3 for Poisson distribution with mean of 2 is

(A) 0.12 (B) 0.18

(C) 0.24 (D) 0.30

Graduate Aptitude Test in Engineering - 2011

Q. 6 A statistical measure of the variability of a distribution around its mean is referred to as

(A) Coefficient of determination (B) Standard error

(C) Coefficient of variation (D) Standard deviation

Graduate Aptitude Test in Engineering - 2012

Q. 7 Approximate percentage of scores that fall within $\pm \sigma$ (standard deviation) of the mean in a normal distribution is

(A) 34 (B) 68

(C) 95 (D) 99

Graduate Aptitude Test in Engineering - 2013

Q. 8 Poisson distribution having a mean of 5 will have $\sqrt{5}$ as

(A) Median (B) Mode

(C) Standard deviation (D) Variance

Q. 9 Box 1 contains 15 balls out of which 3 are red. Box 2 contains 12 balls out of which 4 are red. If one ball is drawn at random from each box simultaneously, the probability of getting at least one red ball is

(A) 0.07 (B) 0.47

(C) 0.53 (D) 0.75

Graduate Aptitude Test in Engineering - 2014

Q. 10 If f(x) is a normal distribution with mean 8 and standard deviation 1, the value of f(x) for x = 10 is

(A) 0.05 (B) 0.14

(C) 0.25 (D) 0.73

Q. 11 If two independent variables X and Y are uncorrelated then

(A) Cov(X,Y) = 0 (B) Cov(X,Y) > 0

(C) Cov(X,Y) < 0 (D) − 1 < Cov(X,Y) <1

Graduate Aptitude Test in Engineering - 2015

Q. 12 Daily sales figures for a week are shown in the bar chart given below. The mean daily sales and the standard deviation of the sample mean (integer value) are .... and .... respectively.

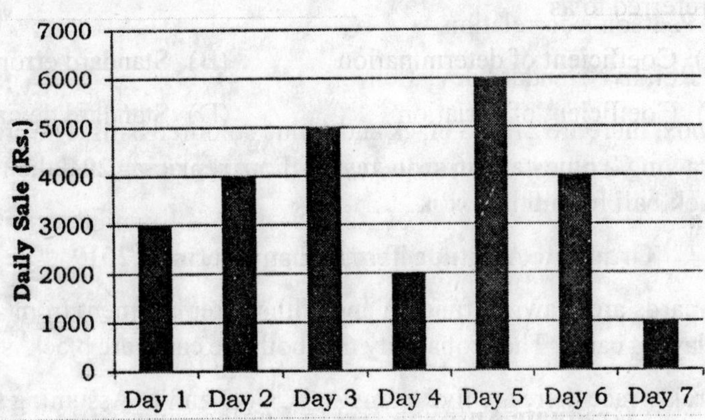

(A) 3571, 1718 (B) 3715, 1915

(C) 3571, 1178 (D) 3715, 1591

Q. 13 A coin is tossed ten times. The probability of getting five heads and five tails is ....

Graduate Aptitude Test in Engineering - 2016

Q. 14 The function f(x) represents a normal distribution whose standard deviation and mean are 1 and 5, respectively. The value of f(x) at x = 5 is

(A) 0.0 (B) 0.159

(C) 0.282 (D) 0.398

Graduate Aptitude Test in Engineering - 2017

Q. 15 A box contains three white and four red balls. Two balls are drawn randomly in sequence. If the first draw resulted in a red ball, the probability of getting a second red ball in next draw is

(A) 0.33 (B) 0.50

(C) 0.67 (D) 0.75

Q. 16 The probability of getting two heads and two tails from four tosses of the same coin is ....

Graduate Aptitude Test in Engineering - 2018

Q. 17 For accepting or rejecting a null hypothesis, which one of the following is NOT used as a significance test method in statistics ?

(A) Z-test

(B) Student's t-test

(C) Pearson's correlation

(D) Relative standard deviation

Q. 18 In a box, there are 2 red, 3 black and 4 blue coloured balls. The probability of drawing 2 blue balls in sequence without replacing, and then drawing 1 black ball from this box is ....%.

Graduate Aptitude Test in Engineering - 2019

Q. 19 Two cards are drawn at random and without replacement from a pack of 52 playing cards. The probability that both the cards are black is ....

Q. 20 In a relay race there are five teams A, B, C, D and E. Assuming that each team has an equal chance of securing any position (first, second, third, fourth or fifth) in the race, the probability that A, B and C finish first, second and third, respectively is

(A) 1/60 (B) 1/20

(C) 1/10 (D) 3/10

Q. 21 The mean absolute deviation about the median for the data 3, 9, 5, 3, 12, 10, 18, 4, 7, 19, 21 is ....

Graduate Aptitude Test in Engineering - 2020

Q. 22 Two playing cards are drawn at random, but in succession from a pack of red cards (26 in number) without replacement. The probability of drawing a king first, followed by drawing a queen is $Px10^{-3}$. The value of P is ....

Graduate Aptitude Test in Engineering - 2021

Q. 23 The probabilities of A and B are given by P(A) = 0.35 and P(B) = 0.25, respectively. If A and B are mutually exclusive so that P(AUB) = P(A) + P(B), then the value of P(A/AUB) is ....

Q. 24 A nine-member committee of an Agricultural University consists of 4 B. Tech., 3 M. Tech., and 2 Ph. D. students. It is decided to remove three students from the committee at random. The probability of removing 2 students from the same category and the third one from any other category is ....

Answers Key

1	2	3	4	5	6	7	8	9	10
A	A	A	B	B	D	B	C	B	A
11	12	13	14	15	16	17	18	19	20
A	A	0.246	D	B	0.375	D	7.14	0.245	A
21	22	23	24						
5.27	6.15	0.5833	0.655						

Explanations

Q. 1

Probability of occurrence of 100 mm rainfall in 25 years = $\frac{1}{25} = 0.04$

Probability of non-occurrence in 25 years = 1 – 0.04 = 0.96

Probability of non occurrence in 15 successive years = $0.96^{15} = 0.542$

Probability of rainfall occurrence in 15 successive years = 1 – 0.542 = 0.458

Q. 2

For bucket 1

Probability of drawing one black ball = ${}^{2}C_{1} / {}^{10}C_{1} = 0.2$

For bucket 2

Probability of drawing one black ball = ${}^{3}C_{1} / {}^{10}C_{1} = 0.3$

Hence,

Probability of having both balls black= 0.3×0.2 = 0.06

And Probability of having at least one red ball = 1 – 0.06 = 0.94

Q. 3

Required probability = $\frac{0.3 \times 0.9}{0.3 \times 0.9 + 0.7 \times 0.8} = 0.325$

Q. 4

Given that

Mean(μ) = 5

Standard Deviation(σ) =1

Applying normal distribution formula

$$f(x) = \frac{1}{\sqrt{2\pi\sigma}} e^{\frac{-z^2}{2}}, \text{ where } z = \frac{x-\mu}{\sigma} = \frac{6-5}{1} = 1$$

$$f(x) = \frac{1}{\sqrt{2\pi \times 1}} e^{\frac{-1^2}{2}} = 0.242$$

Q. 5

Given that

Mean (m) = 2

Applying Poisson distribution;

$$f(x) = \frac{e^{-m}.m^x}{x!}$$

$$f(3) = \frac{e^{-2}.2^3}{3!} = 0.18$$

Q. 7

About 68% of values drawn from a normal distribution are within one standard deviation σ away from the mean; about 95% of the values lie within two standard deviations; and about 99.7% are within three standard deviations. This fact is known as the **68-95-99.7** (empirical) rule, or the **3-sigma rule.**

Q. 8

We know,

In Poisson distribution,

Standard deviation = √Mean

Q. 9

Probability of getting both ball not red = $\frac{12}{15}.\frac{8}{12} = \frac{8}{15}$

Thus, probability to get atleast one red ball = $1 - \frac{8}{15} = \frac{7}{15} = 0.47$

Q. 10

Given that

Mean(μ) = 8

Standard Deviation(σ) =1

Applying normal distribution formula

$$f(x) = \frac{1}{\sqrt{2\pi\sigma}} e^{\frac{-z^2}{2}}, \text{ where } z = \frac{x-\mu}{\sigma} = \frac{10-8}{1} = 2$$

$$f(x) = \frac{1}{\sqrt{2\pi \times 1}} e^{\frac{-(2)^2}{2}} = 0.054$$

Q. 12

Mean of daily sales ($\overline{x}$) = (3000 + 4000 + 5000 + 2000 + 6000 + 4000 + 1000)/7

= 3571.43 Rs. = 3571 Rs. (Integer value)

$$\text{Standard deviation } (\sigma) = \sqrt{\frac{\sum(x_i - \overline{x})^2}{n-1}}$$

Here, $\sum(x_i - \overline{x})^2$ = [(3000 – 3571)^2 + (4000 – 3571)^2 + (5000 – 3571)^2 + (2000 – 3571)^2 + (6000 – 3571)^2 + (4000 – 3571)^2 + (1000 – 3571)^2 = 17714287

$$\text{Hence, } (\sigma) = \sqrt{\frac{17714287}{7-1}} \; 1718.25 \text{ Rs.} = 1718 \text{ Rs. (Integer value)}$$

Q. 13

Probability of getting heads (p) = 1/2

Probability of not getting heads or getting tails (q) = 1/2

Applying, Binomial theorem

$P(k) = P(k \text{ success}) = {}^{n}C_k\, p^k\, q^{n-k}$

$= {}^{10}C_5\, (1/2)^5\, (1/2)^5$

$= 0.246$

Q. 14

Given that

Mean(μ) = 5

Standard Deviation(σ) =1

Applying normal distribution formula

$$f(x) = \frac{1}{\sqrt{2\pi\sigma}} e^{\frac{-z^2}{2}}, \text{ where } z = \frac{x-\mu}{\sigma} = \frac{5-5}{1} = 0$$

$$f(x) = \frac{1}{\sqrt{2\pi \times 1}} e^{\frac{-(0)^2}{2}} = 0.398$$

Q. 15

Total ball remaining after first draw = 6 (3 red and 3 white balls)

Required probability = $({}^3C_1 \times {}^3C_0)/{}^6C_1$

= 3/6 = 0.5

Q. 16

Total outcomes are = 16, such as

H,H,H,H	H,H,H,T	H,H,T,H	H,T,H,H	H,T,T,H	H,T,H,T	H,T,T,T	H,H,T,T
T,T,T,T	T,T,T,H	T,T,H,T	T,H,T,T	T,H,H,T	T,H,T,H	T,H,H,H	T,T,H,H

Favorable conditions/outcomes = 6 (Underlined conditions)

Hence, required probability = 6/16 = 0.375

Q. 18

Non combinations approach:-

First, the probability of drawing a blue ball at the first draw, again a blue at the second and a black at the third. The first probability is 4/9, the second, conditional on the first is 3/8 and the third conditional on the other two is 3/7.

∴ Required Probability = (4/9)×(3/8)×(3/7) = 0.0714 or 7.14 %

Q. 19

Probability that first card will be black = 26/52

Probability that second card will be black = 25/51

Required probability = (26/52)× (25/51) = 0.245

Q. 20

Probability for A to be on first position = 1/5

Probability for B to be on second position = 1/4

Probability for C to be on third position = 1/3

Required probability = (1/5)(1/4)(1/3) = 1/60

Q. 21

Arrange the data in asscending order to find out the median

3, 3, 4, 5, 7, 9, 10, 12, 18, 19, 21

Here, Median = 9

Mean Absolute Deviation = $\frac{\sum|x - median|}{n}$

$= \frac{|3-9|+|3-9|+|4-9|+|5-9|+|7-9|+|9-9|+|10-9|+|12-9|+|18-9|+|19-9|+|21-9|}{11}$

= 5.27

Q. 22

Total conditions of drawing a king from 2 red kings = $^2C_1$

Total conditions of drawing a queen from 2 red queens = $^2C_1$

Required Probability = $^2C_1 \times {}^2C_1 / {}^{26}C_2$

= 6.15×10^{-3}

Q. 23

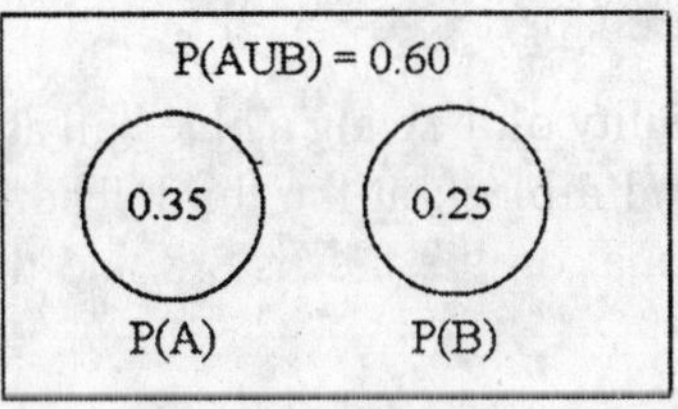

P{A∩(AUB)} = P(A)

P(A) = 0.35 ; P(B) = 0.25 ; P(AUB) = 0.60

$$P(A/AUB) - \frac{P\{A \cap (AUB)\}}{P(AUB)}$$

$$= \frac{P(A)}{P(AUB)} = \frac{0.35}{0.60} = 0.5833$$

Q. 24

Required probability

$= ({}^4C_2 \times {}^3C_1 + {}^4C_2 \times {}^2C_1 + {}^3C_2 \times {}^4C_1 + {}^3C_2 \times {}^2C_1 + {}^2C_2 \times {}^3C_1 + {}^2C_2 \times {}^4C_1)/({}^9C_3)$

= 0.655

5

Laplace Transforms

Graduate Aptitude Test in Engineering - 2007

Q. 1 $\frac{\omega}{(s+a)^2+\omega^2}$ is the Laplace Transform of

(A) exp(-at) sinωt (B) exp(-at) sinhωt

(C) at sinωt (D) at sinhωt

Graduate Aptitude Test in Engineering - 2008

Q. 2 Inverse Laplace Transform of $1/(s-2)^2$ is

(A) exp(2t) (B) t exp(2t)

(C) 2t exp(2t) (D) t^2 exp(2t)

Graduate Aptitude Test in Engineering - 2009

Q. 3 $\frac{1}{s^2-a^2}$ is the Laplace Transform of

(A) sin at (B) t sin at

(C) sinh at (D) a^{-1} sinh at

Graduate Aptitude Test in Engineering - 2010

Q. 4 The Laplace Transform of $t^3 e^{4t}$ is

(A) $\frac{4!}{(s-3)^4}$ (B) $\frac{3!}{(s-4)^4}$

(C) $\frac{3!}{(s-4)^3}$ (D) $\frac{4!}{(s-4)^3}$

Graduate Aptitude Test in Engineering - 2012

Q. 5 The inverse Laplace Transform of $\frac{s^2}{(s-3)^3}$ can be written as $\frac{e^{3t}}{2}$ (At^2+Bt+C). The values of A, B and C, respectively are

(A) 3, 5 and 7 (B) 2, 10 and 12

(C) 10, 12 and 4 (D) 9, 12 and 2

Graduate Aptitude Test in Engineering - 2020

Q. 6 Let a function f(t) = 4 cos 2t + 6 e^{-8t}. The Laplace transform of the given function f(t), L{f(t)} = F(s) = $\int_0^\infty f(t)e^{-st}\,dt$ is

(A) $\frac{4s}{s^2+4}+\frac{6}{s+8}$ (B) $\frac{4s}{s^2-4}+\frac{6}{s-8}$

(C) $\frac{8}{s^2+4}+\frac{6}{s+8}$ (D) $\frac{8}{s^2-4}+\frac{6}{s-8}$

Answers Key

1	2	3	4	5	6
A	B	D	B	D	A

Explanations

Q. 1

We know

$$L^{-1}\left\{\frac{\omega}{s^2+\omega^2}\right\} = \sin \omega t$$

Applying shifting property

$$L^{-1}\left\{\frac{\omega}{s^2+\omega^2}\right\} = e^{-at} \sin \omega t$$

Q. 2

We know

$$L^{-1}\left\{\frac{1}{s^2}\right\} = t$$

Applying shifting property

$\therefore$ $$L^{-1}\left\{\frac{1}{(s-a)^{n+1}}\right\} = \frac{1}{n!}e^{at}t^n$$

Therefore, $$L^{-1}\left\{\frac{1}{(s-2)^2}\right\} = t\exp(2t)$$

Q. 3

$$L^{-1}\left\{\frac{1}{s^2-a^2}\right\} = \frac{\sinh at}{at}$$

Q. 4

We know

$$L\{t^n e^{at}\} = \frac{n!}{(s-a)^{n+1}}$$

$$\therefore \quad L\{t^3 e^{4t}\} = \frac{3!}{(s-4)^{3+1}} = \frac{3!}{(s-4)^{3+1}}$$

Q. 5

$$\frac{s^2}{(s-3)^3} = \frac{P}{s-3} + \frac{Q}{(s-3)^2} + \frac{R}{(s-3)^3}$$

Hence $Ps^2 - 6Ps + 9P + 3Qs - 3Q + R = s^2$

Equating s^2, s terms

$\Rightarrow \quad P = 1, \quad Q = 6, \quad R = 9$

Hence

$$L^{-1}\left\{\frac{s^2}{(s-3)^3}\right\} = L^{-1}\left\{\frac{P}{s-3} + \frac{Q}{(s-3)^2} + \frac{R}{(s-3)^3}\right\}$$

$$= e^{3t} + 6te^{3t} + \frac{9}{2}\, t^2e^{3t}$$

$$= \frac{e^{3t}}{2}\,(9t^2 + 12t + 2)$$

Given

$$L^{-1}\left\{\frac{s^2}{(s-3)^3}\right\} = \frac{e^{3t}}{2}(At^2 + Bt + C) = \frac{e^{3t}}{2}\left(9t^2 + 12t + 2\right)$$

$$= (At^2 + Bt + C) = (9t^2 + 12t + 2)$$

Therefore, $A = 9, B = 12, C = 2$

Q. 6

In the question paper the term $F(s) = \int_0^\infty f(t)e^{-s}dt$

should be $F(s) = \int_0^\infty f(t)e^{-st}dt$

$\mathscr{L}\{f(t)\} = F(s) =$

We know,

$f(t)=e^{at}, t \geq 0$	$F(s)=\frac{1}{s-a}, S>a$
$f(t)=\cos(kt), t \geq 0$	$F(s)=\frac{s}{s^2+k^2}$

⇒ $\mathscr{L}\{\cos 2t\} = s/s^2+4$ and $\mathscr{L}\{e^{-8t}\} = 1/s+8$

Hence, $\mathscr{L}\{f(t)\} = 4s/(s^2+4) + 6/(s+8)$

II. Farm Power

6

Farm Power Sources

APPSC AEES Agriculture Engineering Exam 2016

Q. 1 Sun's radiation corresponding to that of a black body temperature is about

(A) 50000 K (B) 60000 K

(C) 70000 K (D) 140000 K

Q. 2 The percentage of bran received from paddy

(A) 5 (B) 10

(C) 15 (D) 20

Q. 3 The velocity required to operate a wind mill is more than

(A) 5 km/h (B) 10 km/h

(C) 5 miles per hour (D) 10 miles per hour

Q. 4 Main constituent of biogas is

(A) Methane (B) Ethane

(C) Butane (D) Carbon dioxide

Q. 5 In floating type biogas plant, the gasholder is made of

(A) Metal (B) Wood

(C) Cement and Bricks (D) All of the above

Q. 6 The gas production per unit volume of digester capacity will be maximum when the diameter to depth ratio of digester will be in the range of

(A) 0.66 to 1.00 (B) 0.75 to 1.20

(C) 0.85 to 1.30 (D) 0.90 to 1.40

Q. 7 The wet process of biomass digestion involve

(A) Pyrolysis and Gasification

(B) Hydrogenation and liquefaction

(C) Liquefaction and Pyrolysis

(D) Anaerobic digestion and fermentation

MPPSC Assistant Agricultural Engineer 2013

Q. 8 In order that the plants in every corner should receive maximum light, the green house axis is placed in .... direction.

(A) North – South (B) East - West

(C) Any direction (D) None of these is correct

Q. 9 Green houses can be broadly classified as wooden framed, pipe framed and truss framed structures based on ....

(A) Size (B) Shape

(C) Construction (D) Utility

Q. 10 The green house structures in India are usually designed to withstand wind speeds of .... km/h and wind pressure of .... kg/sq.m.

(A) 50 and 50 (B) 50 and 110

(C) 110 and 50 (D) 110 and 110

Q. 11 The rate of air movement should range between .... cum/h/sq.m. of green house floor area in Horizontal Air Flow (HAF) cooling system.

(A) More than 90 (B) 72 to 90

(C) 54 to 72 (D) 36 to 54

Q. 12 .... gas is present in biogas in larger proportion.

(A) Acetylene (B) Carbon monoxide

(C) Ethylene (D) Methane

Q. 13 On an average sunny days in India, the solar collector having area of 1 Sq.m. can provide .... litre of hot water at about 600 °C.

(A) 75 (B) 100

(C) 125 (D) 150

Q. 14 The solar energy is converted into .... by the solar cells in solar panel.

(A) Chemical energy (B) Electrical energy

(C) Mechanical energy (D) None of these is correct

Q. 15 The minimum wind speed required for working of a wind mill is .... m/sec.

(A) 10 (B) 5.56

(C) 5 (D) 2.78

Q. 16 Biomass can be converted into

(A) Liquid fuel (B) Hydrogen

(C) Producer gas (D) All options are correct

Q. 17 Remote sensing and GIS techniques can be employed for .... purposes.

(A) Improvement of natural resources management

(B) Land use planning according to capability

(C) Planning of watershed development work

(D) All options are correct

UKPSC Combined Junior Engineering Exam 2013 Agricultural Engineering Paper-I

Q. 18 The first tractor plant in India was set up by

(A) Hindustan (B) Ford

(C) HMT (D) Eicher

Q. 19 Which of the following has most crucial role in promoting farm mechanization ?

(A) Electric motor (B) Power tiller

(C) Diesel engine (D) Tractor

Q. 20 The power generated in a windmill

(A) Depends on the height of tower.

(B) Depends on wind velocity.

(C) Can be increased by planting tall trees close to tower.

(D) Is more in rainy season.

Q. 21 Minimum wind velocity for working of a wind mill is

(A) 5 kmph (B) 12 kmph

(C) 25 kmph (D) 50 kmph

Q. 22 Optimum solid-water ratio for highest gas production is

(A) 1 : 10 (B) 1 : 15

(C) 1 : 20 (D) 1 : 25

Q. 23 Which of the following is a non-renewable source of energy ?

(A) Fossil fuels (B) Wood

(C) Sun (D) Wind

Q. 24 Where is the largest solar plant of India located ?

(A) Odisha (B) Rajasthan

(C) Gujarat (D) Maharashtra

Q. 25 Main combustible constituent of biogas is

(A) Carbon dioxide (B) Ethane

(C) Butane (D) Methane

Q. 26 The main component of a biogas plant is

(A) Digester (B) Gas holder

(C) Slurry mixing tank (D) All of above

Q. 27 Which is the ultimate source of energy ?

(A) Water (B) Sun

(C) Fossils fuels (D) Uranium

Q. 28 The rate of doing work is known as

(A) Energy (B) Power

(C) Horse power (D) None of the above

Q. 29 A normal human being develops about

(A) 1 HP (B) 0.50 HP

(C) 0.30 HP (D) 0.10 HP

Q. 30 Steam engines are

(A) Internal combustion engines

(B) Semi-internal combustion engines

(C) External combustion engines

(D) None of the above

Q. 31 Wind mills can be used for

(A) Water pumping (B) Electricity generation

(C) Both of (A) & (B) (D) None of these

Q. 32 Which one of the following energy sources is not renewable ?

(A) Solar energy (B) Wind energy

(C) Coal energy (D) Nuclear energy

Q. 33 Biogas plants which are generally used for gas production are generally made of

(A) Bricks (B) Steel

(C) Both of (A) & (B) (D) None of these

Q. 34 Solar energy can be converted into

(A) Thermal energy (B) Electrical energy

(C) Mechanical energy (D) All of the above

Q. 35 In houses active system of solar energy

(A) Uses solar collectors for warming the space

(B) Heating is done either by water or air

(C) Works on natural convection

(D) Both of the (A) & (B)

Q. 36 Electricity from solar energy can be produced by

(A) Solar power production system

(B) Solar photo-voltaic cells

(C) Both of the (A) and (B)

(D) None of these

Q. 37 Wind energy is available in the form of

(A) Kinetic energy (B) Potential energy

(C) Both of the above (A) and (B) (D) None of these

Q. 38 Which of the following do not apply in case of wind energy ?

(A) Wind energy is variable (B) Wind energy is steady

(C) Wind energy is irregular (D) Wind energy is intermittent

Q. 39 Advantages of electrical power over other sources are

(A) Highest efficiency

(B) Adaptive to varying speeds

(C) Adaptable to all farm operations

(D) Both (A) and (B)

Q. 40 Manufacture of Massey Fergusson Tractor is

(A) HMT, Pinjore (B) Sonalika, Hoshiarpur

(C) TAFE, Chennai (D) PTL, Mohali

Q. 41 Main advantages of mechanical power are

(A) Requires frequent rest periods

(B) Available in different sizes

(C) Not affected by hot or cold weather

(D) Both (B) & (C)

Q. 42 The solar photo-voltaic system include

(A) Solar cell array (B) Battery Bank

(C) Inverter and charge controller (D) All above three

Q. 43 A domestic solar water heater of 100 litres per day capacity can save electricity annually equal to

(A) 500 kWh (B) 1000 kWh

(C) 1500 kWh (D) 2000 kWh

Q. 44 Biogas in a biogas digester is produced by chemical process, known as

(A) Pyrolysis (B) Fermentation

(C) Radiation (D) Hydration

Q. 45 Production of biogas through anaerobic digestion of biomass depends on

(A) Temperature (B) C : N Ratio

(C) Pressure (D) All above three

Q. 46 The useful sun-shine days per year in most part of India are

(A) 250 to 300 (B) 175 to 230

(C) 200 to 250 (D) None of the above

Q. 47 Gas production in biogas plant is retarded when temperature goes below

(A) 10 °C (B) 15 °C

(C) 20 °C (D) 25 °C

Q. 48 The best sites for wind are found in

(A) Mountains (B) Plains

(C) Offshores and seacoast (D) None of the above

Q. 49 Photo-voltaic solar cells are made of

(A) Silicon (B) Carbon

(C) Gun metal (D) Silver

Q. 50 Biogas is a mixture of

(A) CO_2 & H_2 (B) CO_2 & H_2O

(C) CO_2 & CH_4 (D) All three

Q. 51 To feed even the smallest size of biogas plant, the minimum number of cattle requirement is

(A) 2 to 4 (B) 4 to 6

(C) 6 to 8 (D) 8 to 10

Q. 52 Tidal energy is most popular and feasible method of producing power by

(A) River (B) Well

(C) Sea (D) Tank

Q. 53 On an average, one ton of paddy contains bran about

(A) 30 kg (B) 40 kg

(C) 45 kg (D) 50 kg

Q. 54 Darrieus type rotor mill is example of which type of wind mill ?

(A) Vertical axis high velocity (B) Vertical axis low velocity

(C) Horizontal axis high velocity (D) Horizontal axis low velocity

Q. 55 Biogas is a flammable gas and its heating value is approximately

(A) 17,925 J/m^3 (B) 29,874 J/m^3

(C) 19,725 J/m^2 (D) 92,874 J/m^2

Q. 56 Solar energy can be utilized by the application of solar cells to produce

(A) DC electricity (B) AC electricity

(C) DC & AC electricity (D) None of the above

Q. 57 Sun drying is an example of which type of drying ?

(A) Convective (B) Conductive

(C) Radiation (D) None of the above

Q. 58 The first practical solar cell was made in the year

(A) 1954 (B) 1860

(C) 1760 (D) 1800

Q. 59 A wind mill cannot be used for the purpose of

(A) Water lifting (B) Ploughing

(C) Power generation (D) Chaff cutting

Q. 60 Wind speed is measured by

(A) Anemometer (B) Wind vane

(C) Photometer (D) Rotavator

Q. 61 Solar energy reaches Earth's surface in about

(A) 4 minutes (B) 8 minutes

(C) 10 minutes (D) 20 minutes

Q. 62 Which of the following is not a type of biogas plant ?

(A) KVIC Model (B) Deenbandhu Model

(C) Gandhi Model (D) Fixed Dome Model

Q. 63 Biogas contains .... % Methane.

(A) 92 – 94 (B) 25 – 50

(C) 80 – 90 (D) 50 – 75

Q. 64 Anaerobic digestion of organic matter to produce biogas is completed in

(A) 2-stages (B) 3-stages

(C) 4-stages (D) 6-stages

Q. 65 Which of the following is a part of the cooling system for an engine ?

(A) Radiator (B) Fly wheel

(C) Injector (D) Clutch

Q. 66 Biogas mixture have lowest amount of

(A) Hydrogen sulfide (B) Hydrogen

(C) Nitrogen (D) Oxygen

Q. 67 Which of the following is not a source of non-conventional energy ?

(A) Geothermal (B) Biomass

(C) Wind energy (D) None of the above

Q. 68 In India primary source of energy is

(A) Fossil (B) Coal

(C) Petrol (D) Diesel

Punjab Mechanic Agricultural Machinery Instructor 2014

Q. 69 The latitude of a place is expressed by its angular distance in relation to

(A) Equator (B) South Pole

(C) Axis of the Earth (D) North Pole

RPSC AEn Pre Exam 2013 (Agricultural Engineering)

Q. 70 Now a days farmers in India are widely adopting small tractors on their fields. What is common range of Horse Power of these tractors ?

(A) Below 5 HP (B) Up to 20 HP

(C) 25 to 50 HP (D) None of the above

Q. 71 Solar energy is distributed over entire surface of earth facing the sun and it seldom exceeds

(A) 1.0 MW/m^2 (B) 1.0 kW/m^2

(C) 1.35 MW/m^2 (D) 1.20 MW/m^2

Q. 72 Solar energy reaching per square meter of the earth's atmosphere called solar constant is equal to

(A) 1.63 kW (B) 3.63 kW

(C) 1.36 kW (D) 3.36 kW

Q. 73 In photovoltaic devices, sun light falls on special semiconductor material, which converts

(A) More than 90 % of the sunlight energy directly into DC electricity

(B) At least 75 % of the sunlight energy directly into DC electricity

(C) It can convert 100 % sunlight energy directly into DC electricity

(D) About 15 % of the sunlight energy directly into DC electricity

Q. 74 Solar cooker can have

(A) Only a single build in reflector

(B) Can have either single or even the double sided reflectors

(C) At least two reflectors

(D) None of the above is absolutely correct

Q. 75 Renewable energy sources are essentially

(A) Flows of energy (B) Stocks of energy

(C) Motion of energy (D) None of the above

Q. 76 For the production of biogas, the temperature range 24 °C to 45 °C belongs to

(A) Thermophilic zone (B) Psychrophilic zone

(C) Mesophilic zone (D) All of the above

Q. 77 The gasifier which gives lowest amount of tar is known as

(A) Imbert gasifier (B) Stratified gasifier

(C) Fluidized bed gasifier (D) None of the above

Q. 78 Calorific value of producer gas when air is used as a gasification medium

(A) 4.0 – 6.0 MJ/Nm^3 (B) 4.0 – 6.0 kJ/Nm^3

(C) 5.0 – 8.0 kJ/Nm^3 (D) None of the above

Q. 79 Blade surface of sail type windmill can be made from

(A) Cloth (B) Nylon

(C) Plastic (D) All of the above

Q. 80 Digest slurry is a good source of

(A) Microorganisms (B) Micronutrients

(C) Macronutrients (D) Fungi

OCS AEn Exam 2011 (Pre) – II (Agricultural Engineering)

Q. 81 Major constituents of solid biomass is

(A) Cellulose (B) Hemicellulose

(C) Lignin (D) Carotene

Q. 82 In wind power generation, the maximum fraction of wind power that can be extracted (power coefficient) is

(A) 0.32 (B) 0.45

(C) 0.59 (D) 1

Q. 83 Fixed dome biogas plant with cylindrical digester is called

(A) KVIC model (B) Deenbandhu model

(C) Janata model (D) CIAE model

Q. 84 Major by-products of biodiesel production is

(A) Alcohol (B) Ethanol

(C) Glycerol (D) Methanol

Q. 85 Power Tillers used in Indian agriculture are in power range of

(A) 1 to 3.5 kW (B) 3.5 to 9 kW

(C) 9 to 12 kW (D) 12 to 15 kW

CGPSC State Engineering Services Exam (Agriculture Engineering) Part - II

Q. 86 The first use of polythene as a greenhouse cover was introduced in the year

(A) 1956 (B) 1948

(C) 1936 (D) 1957

(E) 1967

Q. 87 The greenhouse constructed on leveled ground with an intention to intercept more light is known as

(A) Even Span (B) Ridge and furrow

(C) Modified Quonset (D) Gothic

(E) Both (A) and (B)

Q. 88 If the greenhouse span is equal to greater than 15 m, the frames used are

(A) Strut (B) Truss

(C) Foundation (D) Purline

(D) Both (B) and (C)

Q. 89 The type of ventilation suitable for tropical greenhouse

(A) Top (B) Top and side

(C) Lateral (D) Side

(E) Opposite

Q. 90 Greenhouse crops are subjected to light intensities with maximum range on clear summer days

(A) 3.2 K lux (B) 62 K lux

(C) 129.6 K lux (D) 32.2 K lux

(E) 140.2 K lux

Q. 91 Visible radiation in the greenhouse has the wave length of

(A) 400 – 700 mm (B) 500 – 720 mm

(C) 300 – 400 mm (D) 200 – 300 mm

(E) 100 – 200 mm

Q. 92 For most crops, the acceptable range of relative humidity is between

(A) 50 – 80 % (B) 10 – 50 %

(C) More than 80 % (D) 80 – 90 %

(E) None of these

Q. 93 The fan and pad system can lower the temperature of incoming air by about .... percent between dry and wet bulb temperature.

(A) 50 (B) 80

(C) 90 (D) 100

(E) 110

Q. 94 The plant become taller than DIF become

(A) Zero (B) Highly positive

(C) Highly Negative (D) Both (A) and (B)

(E) Both (B) and (C)

Q. 95 Most crops will responded favorable to CO_2 at .... PPM

(A) 1000 – 1200 (B) 900 – 800

(C) 700 – 600 (D) 500 – 300

(E) 200 – 100

Q. 96 The east-west orientation of greenhouse maintain the better light level in

(A) Summer (B) Winter

(C) Rainy season (D) Both (A) and (B)

(E) Both (B) and (C)

Q. 97 In general percent of CO_2 in the atmosphere is .... %

(A) 0.0566
(B) 0.0345
(C) 0.0897
(D) 0.001
(E) 0.656

Q. 98 The most common and least expensive heating system is the

(A) Unit heater
(B) Central heater
(C) Radiation heater
(D) Axial Flow
(E) None of these

Q. 99 Greenhouse structures should be designed to resist a wind velocity .... km/h.

(A) 130
(B) 250
(C) 170
(D) 200
(E) None of these

Q. 100 The structural member that is running throughout the length of greenhouse is

(A) Side posts
(B) Purlins
(C) Hoop
(D) Column
(E) Apron

Q. 101 The major disadvantage of rigid panel as covering material is

(A) Short Life
(B) Breakage prone
(C) Non uniform lighting
(D) Collecting dust and algae
(E) None of these

Q. 102 Most of heat loss by conduction in greenhouse occur through

(A) Covering material
(B) Floor
(C) Wall
(D) Air leak
(E) Doors

Q. 103 For a 30m×4m size pipe framed greenhouse, if hoops are spaced at 1.25 m, the number of hoops required are

(A) 23 (B) 24

(C) 25 (D) 26

(D) 29

Q. 104 The structural design that is suitable for growing nursery stock

(A) Gothic roof (B) Hoop type roof

(C) Gable roof (D) Gutter connected roof

(E) Both (B) and (C)

Q. 105 The CO_2 enrichment method that is advantageous during winter is ....

(A) Solid CO_2 (B) Liquid CO_2

(C) Combustion (D) Aerated CO_2

(E) Both (A) and (B)

Q. 106 The off season use of greenhouse is for

(A) Storage of crops (B) Bird rearing

(C) Drying of crops (D) Mushroom cultivation

(E) None of these

CGPSC State Engineering Services Exam Paper – I

Q. 107 The average farm power availability in India by 2010 was about

(A) 1.55 kW/ha (B) 1.65 kW/ha

(C) 1.75 kW/ha (D) 1.85 kW/ha

(E) 1.95 kW/ha

Q. 108 The word tractor appeared first on record in a patent issued on a tractor engine invented by George H. Harris in the year

(A) 1860 (B) 1870

(C) 1880 (D) 1890

(E) 1915

Q. 109 The rate at which solar energy arrives at the top of the atmosphere is called the

(A) Solar radiation (B) Solar constant

(C) Solar reflection (D) Solar diffusion

(E) Solar absorption

Q. 110 The instrument which measures the total hemispherical solar radiation is

(A) Pyranometer (B) Pyrheliometer

(C) Pycnometer (D) Sine wave

(E) Thermometer

Q. 111 The average wind speed desirable to have (WECS) wind mills generators start turning

(A) 5 - 8 km/hr (B) 8 - 10 km/hr

(C) 12 - 16 km/hr (D) 20 - 25 km/hr

(E) 27 - 30 km/hr

Q. 112 The Instrument used for measurement wind speed from a height of 10 m is

(A) Hygrometer (B) Anemometer

(C) Odometer (D) Dynamometer

(E) Pycnometer

Q. 113 Bio gas a mixture of methane, carbon dioxide and some other trace impurities contains a major portion of methane of

(A) 30 – 40 % (B) 40 – 50 %

(C) 50 – 75 % (D) > 75 %

(E) < 75 %

Q. 114 The calorific value of biogas is

(A) 3000 kcal/kg (B) 4000 kcal/kg

(C) 5000 - 5500 kcal/kg (D) 6000 - 6500 kcal/kg

(E) 2800 kcal/kg

Q. 115 The pH in the biogas digester should be maintained at the following level, that the microorganisms will be very active and bio digestion will be very efficient

(A) 9 - 10 (B) 4 - 6

(C) 6.5 - 7.5 (D) 8.2

(E) 9

Q. 116 What will be the charcoal yield from 1000 kg of dry wood in the process of pyrolysis

(A) 200 kg (B) 250 kg

(C) 300 kg (D) 500 kg

(E) 650 kg

Q. 117 The Stephen Boltzmann law is applicable for heat transfer by

(A) Convection

(B) Conduction

(C) Radiation

(D) Conduction and radiation combined

(E) Convection and conduction combined

Q. 118 Optical pyrometer is used to measure

(A) Low temperature (B) High temperature

(C) High intensity (D) High velocity

(E) High intensity and velocity

Q. 119 Producer gas has the following impurity, which needs removal before use in engine

(A) Nitrogen (B) Particulate material and tar

(C) Methane (D) Corbin dioxide

(E) Helium

Q. 120 Which of the following dryer comes under mixing type

(A) Baffle dryer (B) Circular metal bin dryer

(C) Rectangular metal bin dryer (D) Deep bed dryer

(E) Batch dryer

Q. 121 The covered floor area recommended for Boar as per IS : 3916 - 1966

(A) 6.25 to 7.50 (B) 7.50 to 9.00

(C) 0.96 to 1.80 (D) 1.80 to 2.70

(E) 1.95 to 2.95

Q. 122 IS code recommended for farm cattle housing for large dairy farms

(A) IS: 6027 – 1970 (B) IS: 2981 - 1964

(C) IS: 1166 – 1986 (D) IS: 5284 - 1969

(E) IS: 14433 - 2007

Q. 123 Sludge from septic tank should be removed in

(A) 1 or 2 years (B) 4 years

(C) 3 years (D) 5 years

(E) 6 years

UKPSC A.En. Exam 2012 Paper - II

Q. 124 The power developed by an average pair of bullocks is about

(A) 1.75 HP (B) 1.0 HP

(C) 0.75 HP (D) 2.0 HP

Q. 125 Central Farm Machinery Training and Testing Institute (CFMTTI) is located at

(A) Delhi (B) Hissar, Haryana

(C) Budni, Madhya Pardesh (D) Both (B) and (C)

Q. 126 Which of the following sources of energy is the renewable energy ?

(A) Coal (B) Natural gas

(C) Petrol (D) Wind

Kerla AAO Exam 2017

Q. 127 The maximum power that can be developed by an average man for doing farm work is

(A) 0.075 HP (B) 0.075 kW

(C) 0.750 HP (D) 0.0750 W

Graduate Aptitude Test in Engineering - 2007

Q. 128 A flat plate solar collector with an absorber area for 1.0×1.5 m receives a solar flux of 850 W m^{-2} on the top cover. The indicated solar flux absorbed in the absorber plate is 600 W m^{-2}. The ambient temperature is 297 K. The heat loss coefficient of the collector at the side, bottom and top are 0.35, 0.65 and 3.50 W m^{-2} K^{-1} respectively with a collector heat-removal factor of 0.85. The collector fluid temperature is 333 K.

(i) Useful heat gain rate for the collector is

(A) 558.45 W (B) 604.35 W

(C) 657.01 W (D) 711.02 W

(ii) Instantaneous collector efficiency is

(A) 43.80 % (B) 47.40 %

(C) 51.53 % (D) 55.76 %

Q. 129 A solar photovoltaic system comprising solar photovoltaic array, inverter and a motor-pump unit is installed for supplying drinking water in a village. There are 24 modules in the array and each module contains 36 number of cells of size 104×104 mm with a conversion efficiency of 12.8%. the global solar radiation incident normally on the cells is 945 W m^{-2}. The power consumed in lifting the water is found to be 435 W. If the pump-motor unit efficiency is 45%, the efficiency of the inverter is

(A) 56.21 % (B) 69.42 %

(C) 80.25 % (D) 85.52 %

Q. 130 A single phase 230 V electric motor while running at 1400 rpm develops a torque of 3.1 Nm. If the phase angle between the voltage and current is 38 and the power efficiency of the motor is 80%, the amount of electric current drawn by the electric motor is

(A) 2.470 A (B) 3.135 A

(C) 4.810 A (D) 5.512 A

Graduate Aptitude Test in Engineering - 2008

Q. 131 A farmer wishes to construct a 5 m^3 capacity biogas plant with a cylindrical digester. The depth of the digester below the ground level is restricted to 5 m. Assume that 1.0 kg of cow dung produces 0.04 m^3 of gas per day and that the bulk density of wet cow dung is 1100 kg m^{-3}. If equal amount of water on volume basis is added to the dung for slurry preparation and the retention period is taken as 40 days, the diameter of the digester tank will be

(A) 0.24 m (B) 1.08 m

(C) 1.52 m (D) 2.31 m

Graduate Aptitude Test in Engineering - 2009

Q. 132 The type of gasifier which produces nearly tar free producer gas is

(A) Counter current gasifier (B) Co-current gasifier

(C) Cross-draught gasifier (D) Fluidized bed gasifier

Q. 133 A load cell employing Wheatstone bridge circuit has two fixed resistors and two strain gauges all of which have a value of 120 ohm. the gauge factor is 2.1. The strain in each of the two strain gauges, one is tension and the other is compression, is 1.65×10^{-4}. If the battery current in the initial balanced condition of the bridge is 50 mA, the sensitivity of the load cell in v/strain will be

(A) 0.001 (B) 0.002

(C) 6.3 (D) 12.6

Graduate Aptitude Test in Engineering - 2010

Q. 134 A farmer constructed a 2 m^3 40 days HRT (hydraulic retention time) Deenbandhu model biogas plant. the gas will be solely used for cooking in a stove with a burner efficiency of 45 %. If the density of biogas is 0.94 kg m^{-3} with a heating value of 21 MJ kg^{-1}, the total effective energy available per day is MJ will be

(A) 17.77 (B) 18.91

(C) 24.47 (D) 39.48

Q. 135 The Local Apparent Time (LAT) corresponding to 14 h 30' Indian Standard Time (IST) at a place in India (19 07'N, 70 51'E) in the month of April with a time correction of zero min will be

(A) 13 h 51' 24" (B) 14 h 9' 39"

(C) 14 h 30' (D) 15 h 8' 36"

Graduate Aptitude Test in Engineering - 2011

Q. 136 A horizontal axis windmill having 8 blades is used for pumping water. Each blade has a tip radius of 1.0 m and a mean chord of 0.1 m. Assuming blade length equal to tip radius, solidity of the windmill is

(A) 0.03 (B) 0.10

(C) 0.25 (D) 0.80

Graduate Aptitude Test in Engineering - 2012

Q. 137 The constituent of producer gas which occupies the highest percentage by volume and helps in increasing its overall calorific value is

(A) CO (B) CO_2

(C) H_2 (D) CH_4

Q. 138 A horizontal axis drag type wind mill with square blades and a horizontal axis lift type wind mill with airfoil section blades having same rotor size are installed at a height of 10 m above the ground. The average wind speed is 25 km h^{-1}. The maximum power coefficient for drag type and lift type wind mills is 0.148 and 0.593, respectively. If the maximum power extracted by drag type wind mill is 5 kW, the corresponding power extracted by lift type wind mill, in kW is

(A) 8.43 (B) 12.63

(C) 18.03 (D) 20.03

Graduate Aptitude Test in Engineering - 2013

Q. 139 A piston pump is driven by a 5 m diameter horizontal axis wind turbine for supplying water from a borehole with a total pump head of 10 m. The mean velocity of air is 18 km/hand the density of air is 1.29 kg/m^3. The actual power coefficient of the wind turbine is 0.30 and the overall pump efficiency is 60%. Neglecting the transmission losses, the expected pump discharge in L/s will be

(A) 2.90 (B) 5.80

(C) 28.50 (D) 32.27

Graduate Aptitude Test in Engineering - 2014

Q. 140 Match the processes given in Group-I with the derived products given in Group-II.

Group-I	Group-II
i. Transestrification	a. Producer gas
ii. Pyrolysis	b. Ethanol
iii. Yeast fermentation	c. Biogas
iv. Anaerobic digestion	d. Biodiesel

(A) i-a; ii-c; iii-b; iv-d (B) i-d; ii-c; iii-b; iv-a

(C) i-b; ii-d; iii-a; iv-c (D) i-d; ii-a; iii-b; iv-c

Q. 141 The day length (sunshine hours) on 31st May 2014 at a place in India (26° 18' N, 73° 01' E) will be………………

Graduate Aptitude Test in Engineering – 2015

Q. 142 In the design of an agitator vessel with model volume V_1 and prototype volume V_2, the scale-up ratio is given by

(A) V_2/V_1 (B) $(V_2/V_1)^{1/2}$

(C) $(V_2/V_1)^{1/4}$ (D) $(V_2/V_1)^{1/3}$

Q. 143 The conversion efficiency of a solar cell is 12 %. For a maximum power output of 9×10^{-3} W at an incident solar radiation of 250 W/m^2, required surface area of the solar cell in mm^2 will be ………

Q. 144 The non-combustible constituents of the producer gas are

(A) Carbon monoxide and Hydrogen

(B) Hydrogen and Methane

(C) Nitrogen and Carbon dioxide

(D) Carbon monoxide and Nitrogen

Q. 145 A propeller type wind turbine of 8 m diameter generates 4 kW electrical power. If the overall efficiency of power generation system is 32% and the density of air is 1.2 kg/m^3, the average wind speed in m/s ………

Graduate Aptitude Test in Engineering – 2016

Q. 146 A gasifier uses rice husk as fuel and generates producer gas containing CO–23%, CO_2–4.4%, O_2–2.6% and N_2 – 70%; all expressed in mole%. Atomic mass of C, O and N are 12, 16 and 14, respectively. Average molecular weight of the producer gas in kg kgmol^{-1} is ………

Graduate Aptitude Test in Engineering – 2017

Q. 147 A cylindrical parabolic solar collector is designed to heat a fluid that enters the absorber at 140° C at a flow rate of 5 kg min^{-1}. The specific heat capacity of the fluid is 1.5 kJ kg^{-1} C^{-1} and its outlet temperature is 180° C. If the incident beam radiation on the plane of aperture is 3000 kJ h^{-1} m^{-2} and useful projected area of the reflectors 2 m × 10 m, the efficiency of the collector in percentage will be ……….

Graduate Aptitude Test in Engineering – 2018

Q. 148 Match the following items of Column I with the corresponding items of Column II:

Column I	Column II
P. Cetane number	1. Beam radiation
Q. Pyrheliometer	2. Anti-knock quality
R. Octane number	3. Total solar radiation
S. Pyranometer	4. Ignition quality

(A) P-2, Q-3, R-1, S-4 (B) P-4, Q-1, R-2, S-3

(C) P-4, Q-3, R-2, S-1 (D) P-2, Q-4, R-1, S-3

Q. 149 Constituents of the producer gas contributing to its heating value are

(A) CO and CO_2 (B) CH_4 and CO_2

(C) CO and H_2 (D) CO_2 and H_2

Q. 150 A 2 m^3 biogas plant is to be operated using cow-dung as feedstock. The cow-dung is mixed with water in proportion of 1:1 (by weight) to form a slurry of density 1090 kg m^{-3}. The yield of biogas is 0.036 m^3 kg^{-1} of cow-dung. If the hydraulic retention time is 40 days, volume of the digester of the biogas plant will be ………m^3.

Graduate Aptitude Test in Engineering – 2019

Q. 151 The amount of biogas required to run a diesel engine is 0.65 m^3/(kW h). The minimum size of the Deenbandhu model biogas plant in m^3 required to run a 1 kW (brake power) diesel engine daily for one hour is

(A) 1 (B) 2

(C) 3 (D) 4

Q. 152 A horizontal axis drag type wind rotor, fitted with 4 thin rectangular blades having drag coefficient 1.29, is used to extract power when the average wind velocity in the rotor plane is 10 km/h. The maximum power coefficient is

(A) 0.148 (B) 0.191

(C) 0.393 (D) 0.593

Graduate Aptitude Test in Engineering – 2020

Q. 153 At the maximum power output of a solar panel, the voltage and current are 18 V and 5.56 A, respectively. If the open circuit voltage and short circuit current of the same solar panel are 21.6 V and 6.11 A, respectively, the fill factor of the panel is ....

Graduate Aptitude Test in Engineering – 2021

Q. 154 In an ordinary chimney, the draught is 12 mm of water column. Assuming density of water to be 1000 kg/m^3, the pressure difference between the outside air and gas at the base of the chimney in Pa is.........

Q. 155 Rushton turbine having an impeller diameter of 20 cm and operating at a stirrer speed of 200 rpm is used in a mixing tank. If the tank receives air at a volumetric flow rate of 0.2 m^3/min, the non-dimensional Froude Number, N_{Fr} is. ....

Q. 156 A solar panel has length of 1.3 m and width of 0.65 m. The solar cells cover 90% of the panel area and its conversion efficiency is 13.7%. For a total solar radiation of 750 W/m^2, the panel output voltage is 18 V at its maximum power output. If two such panels are connected in series to supply power to run a thresher, the current in A that can be supplied by the two panels at the maximum power output is

(A) 2.17 (B) 3.01

(C) 4.34 (D) 8.68

Q. 157 A horizontal axis lift type wind rotor of diameter 4 m is used to run a pump at a wind velocity of 15 km/h at standard atmospheric pressure and temperature (density of air is 1.23 kg/m^3). If velocity of wind leaving the rotor blade is reduced to one-third of the approaching wind velocity, the thrust acting on the blade of the wind rotor in N is..........

Answers Key

1	2	3	4	5	6	7	8	9	10
B	A	B	A	A	A	D	B	C	C
11	12	13	14	15	16	17	18	19	20
D	D	B	B	D	D	D	D	D	B
21	22	23	24	25	26	27	28	29	30
B	C	A	BONUS	D	D	B	B	D	C
31	32	33	34	35	36	37	38	39	40
C	C	C	D	D	C	A	B	D	C
41	42	43	44	45	46	47	48	49	50
D	D	C	B	D	A	C	C	A	C
51	52	53	54	55	56	57	58	59	60
A	C	D	A	A	A	C	A	B	A
61	62	63	64	65	66	67	68	69	70
B	C	D	C	A	D	D	B	A	B
71	72	73	74	75	76	77	78	79	80
A	C	D	B	A	C	A	A	D	B
81	82	83	84	85	86	87	88	89	90
A	C	C	C	B	B	E	B	B	C
91	92	93	94	95	96	97	98	99	100
A	A	B	A	A	B	B	A	A	B
101	102	103	104	105	106	107	108	109	110
D	A	C	B	B	C	B	D	B	A
111	112	113	114	115	116	117	118	119	120
C	B	C	C	C	C	C	B	B	A
121	122	123	124	125	126	127	128	129	130
A	A	A	B	C	D	B	A, A	D	B
131	132	133	134	135	136	137	138	139	140
C	B	D	B	A	C	A	D	A	D
141	142	143	144	145	146	147	148	149	150
13.54	D	300	C	7.46	28.81	30%	B	C	4.077
151	152	153	154	155	156	157			
B	B	0.76	117.58	0.23	C	119.32			

Explanations

Q. 21

As per standards, wind speed to run wind mill should be 10 kmph. In the options lowest value after 10 kmph is 12 kmph.

Q. 24

The Solar power plant in Kamuthi, Tamil Nadu is not only India's but also the world's largest plant. And now, India's biggest solar power plant in Bhadla, Rajasthan is being set up.

Q. 25

Typical composition of biogas is such as

Components	Household waste	Wastewater treatment plants sludge	Agricultural wastes	Waste of agri-food industry
CH_4 % vol.	50-60	60-75	60-75	68
CO_2 % vol.	38-34	33-19	33-19	26
N_2 % vol.	5-0	1-0	1-0	-
O_2 % vol.	1-0	< 0,5	< 0,5	-

Q. 53

It is considered that paddy contains 20% husk, 5−12% bran depending on the milling degree and 68−75 % milled rice or white rice depending on the variety.

Here, 5 % - 12 % of 1 ton paddy will be = 50 kg – 120 kg

Q. 128 (i)

Collector absorbing area (A) = $1.0\times1.5\ m^2 = 1.5\ m^2$

Total solar incident flux rate (R) = $800\ W/m^2 \Rightarrow 800\times1.5 = 1275\ W$

Indicated solar flux absorbing rate by flat plate (R_a) = $600\ W/m^2$

$\Rightarrow 600\times1.5 = 900\ W$

Total heat loss coefficients(C) = $0.35 + 0.65 + 3.5 = 4.5\ W/m^2\ K$

Temperature difference (ΔT) = 333.297 = 36 K

$\therefore$ Heat loss rate from collector (H_l) = $A\times C\times \Delta T$

$= 1.5\times4.5\times36 = 243\ W$

Useful heat gain rate of the collector = $R_a - H_l = 900 - 243 = 657\ W$

$\because$ Collector heat removal factor = 0.85

Thus, Heat gain rate of the collector = $657\times0.85 = 558.45\ W$

(ii)

$$\text{Collector efficiency} = \frac{\text{Heat gain rate of the collector}}{\text{Total solar incident flux rate}} = \frac{558.45}{1275}$$

$= \times100 = 43.80\ \%$

Q. 129

Number of modules = 24

Number of cells in each module = 36

Size of a cell = 104×104 mm^2 = 0.0108 m^2

Total cell area = $24\times36\times0.0108 = 9.33$ m^2

Global radiation incident flux = 945 W/m^2

Conversion efficiency (η_c) = 12.8%

Power produced in array (P) = Global radiation incident flux × cell area

$= 945\times9.33 = 8816.85$ W

Power output from array (P) = Power produced in array × Conversion efficiency

$= 8816.85\times0.128 = 1128.57$ W

Consider, Inverter efficiency (η_i) = y%

Pump motor efficiency (η_p) = 45%

Power consumed in water lifting = 435 W

Power consumed in water lifting = P× η_i × η_p

$435 = 1128.57\times y\times0.45$

$\Rightarrow$ $y = 85.65\ \% \approx 85.52\ \%$

Q. 130

Voltage (V) = 230 V

Phase angle between voltage and current (θ) = 38

Speed of motor (N) = 1400 rpm = 23.33 rps

Torque developed by motor(T) = 3.1 Nm

$\therefore$ Power developed by motor (P) = $2\ \pi NT$

$= 2\times3.14\times23.33\times3.1 = 454.19$ W

Here power efficiency (η) = 80%

Thus, Actual power developed by motor (P) = 454.19/0.80 =567.74 W

We know that

$$P = V\,I\cos\theta$$

$$\Rightarrow \quad 567.74 = 230 \times I \times \cos 38$$

Hence, Electric current drawn by motor (I) = 3.134 A

Q. 131

Capacity of biogas plant = 5 m^3

0.04 m^3 of gas is produced from 1 kg cow dung/day

$$\therefore\ 5\ m^3 \text{ of gas is produced from } = \frac{1\times 5}{0.04} = 125 \text{ kg cow dung/day}$$

$$\text{Volume of digester} = \frac{\text{Total volume of slurry}}{\text{Bulk density}} = \frac{2\times 125}{1140} = 0.2193\ m^3/d$$

Here retention period = 40 days

$$\therefore \text{ Total gas produced in digester} = 40\times 0.2193 = 8.772\ m^3$$

We know that

$$\text{Volume of digester} = \frac{\pi}{4} d^2 h \qquad (\because h = 5 \text{ m})$$

$$8.772 = \frac{3.14}{4} \times d^2 \times 5$$

$$\Rightarrow \quad d = 1.495 \text{ m} \approx 1.52 \text{ m}$$

Q. 133

We know

$$\text{Gauge factor } (G_f) = \frac{\Delta R / R}{\in}$$

$$R = G_f \times \in \times R$$

$$\Delta R = 2.1 \times \in \times \frac{120}{2}$$

$$= 252 \in \text{ohm}$$

Hence,

$$\text{Required sensitivity of the load cell in v/strain} = \frac{\Delta V}{\in} = \frac{I\Delta R}{\in}$$

$$= \frac{50\times 10^{-3} \times 252 \in}{\in}$$

$$= 12.6$$

Q. 134

Capacity of biogas plant (V) = 2 m^3 /day

Density of biogas (ρ) = 0.94 kg/ m^3

$\therefore$ Mass of biogas = $0.94 \times 2 = 1.88$ kg

Heating value of biogas (CV) = 21 MJ/kg

Here, Burner efficiency = 45%

$\therefore$ Effective energy available per day = $21 \times 1.88 \times 0.45 = 18.91$ MJ

Q. 135

We know

Local apparent time (LAT) = IST ± 4(Longitude of IST − Longitude of place) min + time correction

(Negative sign is taken if longitude of location is eastward)

Local apparent time (LAT) = 14:30 − 4(82.5 − 72.51) min + 0

(Longitude of Indian standard Allahabad is 82.5°)

LAT = 13 h 51 min 24 seconds

Q. 136

$$\text{Solidity of windmill} = \frac{nC}{\pi r} = \frac{8 \times 0.1}{3.14 \times 1}$$

$$- 0.25$$

Q. 138

We know that

$$\text{Power generated by windmill (P)} = \frac{1}{2} \times C_p \rho A V^3$$

Here, all the parameters are same for both the wind mills excluding power coefficient (C_p).

Therefore $P \propto C_p$

$$\text{Or} \quad \frac{P_2}{P_1} = \frac{(C_p)_2}{(C_p)_1}$$

$$\Rightarrow \quad P_2 = 5 \times \frac{0.593}{0.148} = 20.03 \text{ W}$$

Q. 139

$$\text{Theoretical power of windmill} = \frac{\rho.A.v^3.Cp}{2}$$

$$= \frac{1.29 \times 19.625 \times 5^3 \times 0.30}{2}$$

$$= 474.68 \text{ kg.m}^2\text{/sec}$$

We know

Hydraulic power (P_h) = ρ.g.h.Q

$$474.68 \times 0.60 = 1000 \times 9.81 \times 10 \times Q$$

$$\Rightarrow \quad Q = 2.90 \text{ L/s}$$

Q. 141

No. of days from 1 January 2014 to 31 May 2014 (N) = 152

$$\text{Declination } (\delta) = 23.45 \sin[\frac{360}{365}(284 + N)]$$

$$= 23.45 \sin[\frac{360}{365}(284 + 152)] = 22.04^o$$

Given that, $\varnothing = 26°18' = 26.30^o$

$$\text{Day length} = \frac{2}{15}\cos^{-1}(-\tan\varnothing\tan\delta)$$

$$= \frac{2}{15}\cos^{-1}(-\tan 26.30 \tan 22.04) = 13.54 \text{ hours}$$

Q. 142

For both laminar and turbulent equations

Agitator vessel volume (V) = kT^3

Where, k = proportionality constant.

T = top width or diameter of vessel.

Therefore, Scale up ratio (R)= $(V_2/V_1)^{1/3}$

Q. 143

Incident solar radiation = 250 W/m²

$= 250\times10^{-6}$ W/mm²

Effective incident solar radiation = $250\times10^{-6}\times 0.12 = 30\times10^{-6}$ W/mm²

Maximum power output = 9×10^{-3} W

We know

Incident solar radiation × Cell surface area = Power output

$$30\times10^{-6}\times A = 9\times10^{-3}$$

$$A = 300 \text{ mm}^2$$

Q. 145

We know that

Power of windmill (P) = $\frac{1}{2} \times \rho A V^3$

∵ Overall efficiency of power generation = 0.32,

Therefore, power of windmill = 4000/0.32 W

$$\Rightarrow \quad \frac{4000}{0.32} = \frac{1}{2}\times1.2\times\frac{3.14\times8^2}{4}\times V^3$$

$$\Rightarrow \quad V = 7.46 \text{ m/s}$$

Q. 146

We know

Average Molecular Weight

= ∑(mass of gas componenets×mole%)

= (12+16)×0.23 + (12+32)×0.044 + 32×0.026 + 28×0.70

= 28.81 kg kg mol^{-1}

Q. 147

We know,

Rate of heat used through collector = $mC_p\Delta T$

= 5×1.5×(180 – 140)/60 = 5 kW

And, Total solar incident flux rate = 3000×2×10/3600 = 16.67 kW

Therefore, Efficiency of collector = 5/16.67 = 0.30 or 30 %

Q. 150

Given that, Hydraulic retention time = 40 days

We know

0.036 m3 of bio-gas is produced from 1 kg cow dung/day

Hence,

Cow dung required for 2 m³ of bio-gas over 40 days = $\frac{2}{0.036} \times 402$

= 2222.22 kg

∵ Cow-dung is mixed with water in proportion of 1:1 (by weight)

Therefore total slurry needed = 2×2222.22 = 4444.44 kg

Volume of digester = Total weight of slurry /Bulk density

= 4444.44/1090 = 4.077 m³/d

Q. 151

Biogass required to run 1 kW diesel engine for 1 hour = 0.65 m³

As per recent researches,

The average yield of biogas digester = 0.35 m³/ m³ digester capacity

Digester volume = 0.65/0.35 = 1.86 m³

Hence, Plant capacity will be greater than 1.86 m³ and minimum will be 2 m³ in this situation.

Q. 152

Max. Power coefficient of a drag based wind mill = $4C_d/27$

= 4×1.29/27 = 0.191

Q. 153

Fill Factor = Actual maximum obtained power/(Open circuit voltage × Short circuit current)

= 18×5.56/(21.6×6.11) = 0.76

Q. 154

Density of air (ρ_a) = 1.225 kg/m³

Net pressure difference = $(\rho_w - \rho_a)gh$

= (1000 – 1.225)×9.81×0.012 = 117.58 Pa

Q. 155

Impeller Diameter(D) = 0.2 m

Operating speed of turbine impeller (N) = 200 rpm = 3.33 rps

Froude number = $\frac{N^2 D}{g}$

$= \frac{(3.33)^2 \, 0.2}{9.81} = 0.226$ or 0.23

Q. 156

Power generated by both the panels = Panel area x Solar radiation

= 1.3x0.65x750x0.90 = 570.375 W

$\because$ Conversion efficiency = 13.7%

$\therefore$ Actual Power generated = 570.375x13.7/100 = 78.14 W

We know

Power = VI

$\Rightarrow$ 78.14 = 18xI

$\Rightarrow$ I = 4.34 A

We know, current remains same in series connection of circuit. Hence, current value will be 4.34 A.

Q. 157

Upstream side wind velocity (V) = 15 km/h = 4.17 m/s

Downstream side wind velocity (U) = 4.17/3 = 1.39 m/s

Pressure Difference (ΔP) = $\frac{1}{2}$ $\rho(V^2 - U^2)$

$\frac{1}{2}$ = 1.23× (4.17^2 - 1.39^2) × = 9.50 Pa

Thrust acting on rotor blade = Rotor cross sectional area× ΔP

$\frac{\pi \times (4)^2}{4} \times 9.50 = 119.32$ N

7

Thermodynamic Principles of I.C. Engine

7

Thermodynamic Principles of I.C. Engine

UKPSC AE Agricultural Engineering Paper - II 2013

Q. 1 Direction of rotation can be reversed in

(A) 2- stroke engine (B) 4- stroke engine

(C) Both (A) and (B) (D) None of these

Q. 2 High speed engines have

(A) 1000 or less r.p.m (B) 1000 or more r.p.m.

(C) 500 r.p.m. (D) 300-500 r.p.m.

Q. 3 The oil pressure in lubricating system of tractor engine is around

(A) 1.0 kg/cm^2 (B) 1.5 kg/cm^2

(C) 2.0 kg/cm^2 (D) 3.0 kg/cm^2

Q. 4 The average firing interval for 6 cylinder, 4 cycle engine is

(A) 120° (B) 180°

(C) 240° (D) 360°

Q. 5 The hydrostatic hydraulic system of a tractor is based on

(A) Bernoulli's theorem (B) Archimede's principle

(C) Pascal's law (D) None of these

Q. 6 The oil to be used in engine for lubrication purpose is

(A) SAE 90 (B) SAE 50

(C) SAE 30 (D) None of these

Q. 7 Which is the correct firing order of six cylinder engine?

(A) 1-2-3-4-5-6 (B) 1-5-2-4-3-6

(C) 6-5-4-3-2-1 (D) 1-5-3-6-2-4

Q. 8 Which of the following describes the best meaning of Ergonomics?

(A) Crop sciences

(B) Machine economics

(C) Worker and working environment

(D) None of these

Q. 9 To raise the temperature of a unit mass of a substance by one degree, the energy required is known as

(A) Specific gravity (B) Specific heat

(C) Latent heat (D) Specific temperature

Q. 10 Thermal efficiency of a diesel engine varies from

(A) 25 to 32% (B) 32 to 38%

(C) 40 to 45% (D) 45 to 50%

Q. 11 Combustion is which type of process?

(A) Isothermic (B) Endothermic

(C) Isopleth (D) Exothermic

Q. 12 The first law of thermodynamics is a statement about

(A) Conservation of energy (B) Conservation of momentum

(C) Conservation of heat (D) Conservation of work

Q. 13 Diesel cycle is also known as

(A) Combination cycle (B) Constant volume cycle

(C) Constant pressure cycle (D) (A) and (B) both

Q. 14 A petrol engine is easily identified by seeing

(A) Spark plug (B) Oil cleaner

(C) Fly wheel (D) Injector

Q. 15 The constant volume cycle is also called

(A) Carnot cycle (B) Otto cycle

(C) Diesel cycle (D) Rankine cycle

Q. 16 The efficiency of a Carnot engine depends on

(A) Design of engine (B) Temperature difference

(C) Type of material used (D) Capacity of engine

Q. 17 The air standard efficiency of a Otto cycle compared to diesel cycle for the given compression ratio is

(A) Same (B) Less

(C) More (D) Unpredictable

Q. 18 The pressure and temperature at the end of compression stroke in a petrol engine are of the order of

(A) 4-6 Kg/cm^2 and 200-250°C

(B) 6-12 Kg/cm^2 and 250-350°C

(C) 12-20 Kg/cm^2 and 350-450°C

(D) 20-30 Kg/cm^2 and 450-500°C

Q. 19 The temperature gauge on the tractor dashboard indicates the temperature of

(A) Coolant circulating in the engine

(B) Engine running at standard speed

(C) Lubricating oil in the sump

(D) Ideal engine running temperature

Q. 20 Internal energy of ideal gas is a function of

(A) Temperature and volume (B) Pressure and volume

(C) Pressure and temperature (D) Temperature alone

Q. 21 Characteristic gas constant of any perfect gas

(A) Increases with increase in temperature

(B) Increases with increase in pressure

(C) Is a function of pressure and temperature

(D) Is constant

Q. 22 Zeroth law of thermodynamics defines

(A) Internal energy (B) Temperature

(C) Enthalpy (D) Pressure

Q. 23 Ignition quality of diesel fuel is indicated by

(A) Octane number (B) Calorific value

(C) Centane number (D) Ignition temperature

Q. 24 The use of pressurized radiator cap in forced circulation water cooling system in tractor engines helps in

(A) Reducing the evaporation losses

(B) Increasing the engine operating temperature

(C) Maintaining the appropriate range of temperature of water

(D) Increasing the radiator capacity

Q. 25 The second law of thermodynamics describes the statement about

(A) Enthalpy (B) Work

(C) Entropy (D) Heat

Q. 26 Thermodynamic equilibrium of a system does not require

(A) Mechanical equilibrium (B) Thermal equilibrium

(C) Optical equilibrium (D) Chemical equilibrium

Punjab Mechanic Agricultural Machinery Instructor 2014

Q. 27 Automobile engines are usually designed as multi-cylinder engine because of

(A) Economy reasons

(B) Higher efficiency

(C) Better balance, uniform torque output

(D) Lower fuel consumption

Q. 28 Which of the following processes is not associated with diesel cycle

(A) Constant volume (B) Constant pressure

(C) Isothermal (D) Adiabatic

Q. 29 In I.C. engines, the power developed inside the cylinder is called

(A) Break horse power (B) Pumping power

(C) Indicated horse power (D) None of these

Q. 30 Diesel engine as compared to petrol engine is

(A) More difficult to ignite (B) Less difficult to ignite

(C) Highly ignitable (D) None of above

Q. 31 The petrol engine works on

(A) Diesel cycle (B) Brayton cycle

(C) Ericsson cycle (D) Otto cycle

Q. 32 The thermal efficiency of two-stroke cycle engine as compared to four-stroke engine is

(A) More (B) less

(C) Same (D) None of the above

Q. 33 The minimum number of rings in a piston are

(A) Two (B) Three

(C) Four (D) Six

Q. 34 The ignition coil in tractor acts as a

(A) Rectifier (B) Switch

(C) Step-up transformer (D) DC to AC converter

OCS AEn Exam 2011 (Pre) – II (Agricultural Engineering)

Q. 35 The brake thermal efficiency of diesel engine at full load and rated speed is about

(A) 40 % (B) 60 %

(C) 75 % (D) 90 %

Q. 36 Maximum engine power in tractor engine is generated

(A) At rated speed

(B) At maximum engine speed

(C) At 50 % of maximum engine speed

(D) At 1200 engine rpm

UKPSC A.En. Exam 2007 Paper – II

Q. 37 For an ideal gas, the universal gas constant(R) is related to molecular heat capacity at constant pressure(C_p) and at constant volume(C_v) as

(A) $C_p + C_v = R$ (B) $C_p - C_v = R$

(C) $C_p \times C_v = R$ (D) $C_p \div C_v = R$

Q. 38 The theoretical cycle on which steam engine works is

(A) Carnot cycle (B) Joule's cycle

(C) Rankine's cycle (D) Dual cycle

Q. 39 The Otto cycle is the theoretical cycle of a

(A) Petrol engine (B) Gas engine

(C) High speed diesel engine (D) All of the above

Q. 40 Compression ratio in diesel engine varies between

(A) 15 – 22 : 1 (B) 25 – 30 : 1

(C) 10 – 15 : 1 (D) None of the above

Q. 41 The first law of thermodynamic is a statement about

(A) Conservation of heat (B) Conservation of momentum

(C) Conservation of work (D) Conservation of energy

Q. 42 Thermal efficiency of diesel engine is about

(A) 22 % (B) 34 %

(C) 44 % (D) 54 %

Q. 43 Heat energy added during a process indicates

(A) Negative entropy (B) Zero entropy

(C) Positive entropy (D) Both (A) and (B)

Q. 44 A boiler is also known as

(A) Water heater (B) Steam generator

(C) Gas producer (D) None of the above

Q. 45 Second law of thermodynamics defines

(A) Entropy (B) Enthalpy

(C) Heat (D) Work

Q. 46 The unit of Temperature in SI system is

(A) Centigrade (B) Celsius

(C) Fahrenheit (D) Kelvin

Q. 47 The thermal efficiency of the engine is given by the equation

$$\eta = 1 - \left(\frac{1}{r}\right)^{\gamma-1}\left(\frac{\sigma^{\gamma}-1}{\gamma(\sigma-1)}\right)$$

where "σ" is the

(A) Compression ratio (B) Stroke bore ratio

(C) Air constant (D) Cut off ratio

Q. 48 Joule's statement establishes that during a cycle

(A) Heat transfer is equal to work transfer

(B) Work transfer is only a fraction of heat transfer

(C) Heat transfer is only a fraction of work transfer

(D) There is no relationship between work and heat transfer

Q. 49 Power developed by an average pair of bullocks is about

(A) 250 watts (B) 500 watts

(C) 750 watts (D) 1000 watts

Q. 50 The BHP of an engine increases or decreases, if the diameter of the piston is increased by 20 % and the stroke length is reduced by 20 %, all other factors remaining the same

(A) Does not changes (B) 40 % increases

(C) 15 % decreases (D) 15 % increases

Q. 51 The mechanical efficiency of an engine is expressed as

(A) BHP×100/IHP (B) IHP×100/BHP

(C) BHP×IHP×100 (D) BHP×IHP/100

Q. 52 During compression stroke in a four stroke diesel engine

(A) The inlet valve is open

(B) Exhaust valve is open

(C) Both (A) and (B) valves are open

(D) Both inlet and exhaust valves are closed

Q. 53 The Direct Injection system in a tractor engine is preferred because of

(A) Lesser injection pressure
(B) Lower speed of tractor engine
(C) Better fuel efficiency
(D) Better tractor sound

Q. 54 Best operating temperature of an I.C. engine lies between

(A) 140^0 to 200 °C (B) 100^0 to 120 °C

(C) 140^0 to 200 °F (D) 100^0 to 120 °F

Q. 55 The tappet clearance is adjusted in an engine, when

(A) Inlet valve is in shut position
(B) Exhaust valve is in shut position
(C) Inlet valve is in open position
(D) Both (A) and (B)

Q. 56 Firing interval between successive power strokes in different cylinders of the two strokes engine is determined as

(A) FI = 720^0/No. of cylinders (B) FI = 540^0/No. of cylinders

(C) FI = 360^0/No. of cylinders (D) FI = 180^0/No. of cylinders

UKPSC A.En. Exam 2012 Paper - II

Q. 57 At STP, one gram of hydrogen occupies a volume of

(A) 9.4 litre (B) 10.5 litre

(C) 11.2 litre (D) None of the above

Q. 58 In an adiabatic process which of the following is correct ?

(A) $q = \Delta E + W$ (B) $\Delta E = q$

(C) $p\Delta V = 0$ (D) $q = 0$

Q. 59 The change of state of a gas with respect to pressure and volume, when temperature remains constant, is called

(A) Charle's law (B) Boyle's law

(C) Otto cycle (D) Diesel cycle

Q. 60 A diesel engine works on the principle of

(A) Variable pressure cycle (B) Variable volume cycle

(C) Constant pressure cycle (D) None of the above

Q. 61 The specific heat at constant pressure of a real gas

(A) Increases with increase in temperature

(B) Decreases with increase in temperature

(C) Remains constant with increase in temperature

(D) None of the above

Q. 62 A steam engine is an

(A) I.C. engine

(B) Compression ignition engine

(C) Spark ignition engine

(D) External combustion engine

Q. 63 An adiabatic process takes place at

(A) Constant heat (B) Constant enthalpy

(C) Constant temperature (D) Constant pressure

Q. 64 A process in which volume of a system remains constant is called

(A) Isobaric process (B) Adiabatic process

(C) Isochoric process (D) Isothermal process

Q. 65 For the same compression ratio efficiency of Otto cycle is

(A) Same as diesel cycle (B) More than diesel engine

(C) Less than diesel engine (D) None of the above

Q. 66 A perfect gas at 27 °C is heated at constant pressure till its volume is doubled. The final temperature of the gas will be

(A) 54 °C (B) 327 °C

(C) 600 °C (D) 654 °C

Q. 67 A closed thermodynamic system is a system of fixed

(A) Energy (B) Mass

(C) Temperature (D) None of the above

Q. 68 The latest heat is the amount of heat transfer required to cause a phase change in unit mass of a substance at a

(A) Constant pressure (B) Constant temperature

(C) Both of the above (D) None of the above

Q. 69 The second law of thermodynamics leads to a property called

(A) Entropy (B) Enthalpy

(C) Heat (D) Work

Q. 70 In an Isothermal process

(A) Temperature increases gradually

(B) Pressure remains constant

(C) No change in internal energy

(D) Volume remains constant

Q. 71 The function of following fittings in boiler is to regulate the flow of water

(A) Stop valve (B) Feed check valve

(C) Safety valve (D) Blow off cock

Q. 72 Thermal efficiency of a diesel engine is about

(A) 44 % (B) 54 %

(C) 60 % (D) 34 %

Q. 73 The cetane number is the percentage of cetane in the mixture of cetane and Alphamethyl nepthalene by

(A) Weight (B) Mass

(C) Volume (D) None of the above

Q. 74 The thermal efficiency of a theoretical Otto cycle

(A) Increases with increase in compression ratio

(B) Increases with increase in isentropic index

(C) Does not depend on pressure ratio

(D) Follows all the above

Q. 75 Work done in a free expansion process is

(A) Zero (B) Minimum

(C) Maximum (D) Positive

Q. 76 The value of universal gas constant in SI unit is

(A) 0.8134 J/mole/K (B) 8.314 J/mole/K

(C) 8314 J/mole/K (D) 83.14 J/mole/K

Q. 77 Theoretical cycle on which a petrol engine works, is

(A) Carnot cycle (B) Otto cycle

(C) Rankine cycle (D) Dual cycle

Q. 78 In a 4 stroke diesel engine during the compression stroke

(A) The inlet valve is open. (B) The exhaust valve is open.

(C) Both valves are open. (D) Both valves are closed.

Q. 79 Tappet is a part of

(A) Engine lubrication system (B) Engine igniation system

(C) Engine valve operating system (D) None of the above

Q. 80 The common firing order for 4 cylinder engine is

(A) 1 – 3 – 4 – 2 (B) 1 – 2 – 4 – 3

(C) 1 – 2 – 3 – 4 (D) Both (A) and (B)

Q. 81 A two-stroke cycle engine can be operated in

(A) Clockwise direction (B) Anti-clockwise direction

(C) In both the directions (D) None of the above

Graduate Aptitude Test in Engineering - 2008

Q. 82 One kilogram of air is subjected to polytropic compression from a volume of 28m^3 and a pressure of 101 kPa to a volume of 2m^3 and pressure of 2 MPa. The external work required to make this compression possible

(A) 1.66 MJ (B) 2.93 MJ

(C) 3.04 MJ (D) 8.92 MJ

Graduate Aptitude Test in Engineering – 2009

Q. 83 In a diesel engine with variable compression ratio, the initial compression ratio is 16:1.The ratio of specific heats is 1.4. for the same cut-off ratio of 4.0, if the compression ratio is increased by 25%, the air standard thermal efficiency of the engine will be

(A) Increased by 1.0% (B) Increased by 2.8%

(C) Increased by 3.5% (D) Increased by 4.0%

Graduate Aptitude Test in Engineering - 2012

Q. 84 A diesel engine running in dual fuel mode with diesel as pilot fuel and producer gas as primary fuel produces 3.5 kW at rated engine speed and is coupled directly to a generator for producing electricity. The amount of diesel and producer gas consumed per hour is 460 ml and 12.5 m^3, respectively.

(i) Assuming calorific value of diesel and producer gas as 35280 and 3.97 MJ m^{-3}, respectively, the brake thermal efficiency of the engine in percentage is

(A) 17.19 (B) 19.13

(C) 22.79 (D) 25.32

(ii) If generator efficiency is 90%, the maximum electricity produced, in kW is

(A) 2.85 (B) 3.00

(C) 3.15 (D) 3.50

Graduate Aptitude Test in Engineering - 2013

Q. 85 A 4-cylinder, 4-stroke compression ignition engine has piston stroke of 10.5 cm and cylinder bore of 11 cm. At a mean piston speed of 7 m s^{-1}, the developed brake mean effective pressure is 650 kPa. The brake power in kW developed by the engine is

(A) 39.40 (B) 43.24

(C) 86.48 (D) 172.96

Graduate Aptitude Test in Engineering - 2014

Q. 86 A tractor tyre contains 31 L of air at a pressure of 190 kPa and a temperature of 30°C Using R = 8.314 J $(gmol)^{-1}$ K^{-1} and molecular mass of air = 29 g $(gmol)^{-1}$, the mass of air contained in the tyre is $M \times 10^{-3}$ kg. The value of M is ....

Graduate Aptitude Test in Engineering - 2016

Q. 87 The intake pressure of a diesel engine is 1 bar and pressure at the end of the compression is 34 bar. The adiabatic exponent is 1.3 and the expansion ratio is 7. The diesel cycle efficiency in percentage is ....

Graduate Aptitude Test in Engineering - 2017

Q. 88 Air temperature at beginning of adiabatic compression in an IC engine is 27 °C and the engine compression ratio is 16:1. The temperature of the air at the end of the compression in °C will be ....

Graduate Aptitude Test in Engineering - 2019

Q. 89 A 3-cylinder, 4-stroke CI engine coupled with a turbocharger has a bore and stroke length of 120 mm and 130 mm, respectively. The engine is running at 1600 rpm with a volumetric efficiency of 150%. The air to fuel ratio for complete combustion on weight basis is 14.9 : 1 and the density of air entering the cylinder is 1.2 kg m^{-3} The fuel consumption in kg h^{-1} is

(A) 17.05 (B) 25.57

(C) 33.33 (D) 51.14

Graduate Aptitude Test in Engineering - 2020

Q. 90 The cooling system of a tractor fitted with diesel engine rejects 0.58 kW of heat per kW of brake power. It requires 0.16 L/s of water per kW of heat rejection from the engine to maintain a temperature drop of 6 °C of water as it moves from the top of radiator to its bottom. If the engine develops 45 kW brake power, the required water flow rate in the radiator in L/s is ....

Answers Key

1	2	3	4	5	6	7	8	9	10
A	B	D	A	C	C	B	C	B	B
11	12	13	14	15	16	17	18	19	20
D	A	C	A	B	B	C	B	A	B
21	22	23	24	25	26	27	28	29	30
D	B	C	C	C	C	C	C	C	A
31	32	33	34	35	36	37	38	39	40
D	B	B	C	A	A	B	C	A	A
41	42	43	44	45	46	47	48	49	50
D	B	C	B	A	D	D	A	C	D
51	52	53	54	55	56	57	58	59	60
A	D	C	C	D	C	C	D	B	C
61	62	63	64	65	66	67	68	69	70
A	D	B	C	B	A	B	C	A	C
71	72	73	74	75	76	77	78	79	80
B	D	C	D	A	C	B	D	C	D
81	82	83	84	85	86	87	88	89	90
C	D	D	B,C	B	0.0678	48.62	636.43	B	4.17

Explanations

Q. 7

The firing order is the sequence of power delivery of each cylinder in a multi-cylinder reciprocating engine.

The firing order theory is considered to maximize the power transferred to crank, based on one by one pushing of the piston. Choice of order is a consideration for vibration, power, efficiency, fuel consumption etc.

Possible firing orders of engine are such as

Type of engine	Possible firing orders
3 Cylinder engine	1-3-2, 1-2-3
4 Cylinder engine	1-3-4-2, 1-2-4-3
6 Cylinder engine	1-5-3-6-2-4, 1-5-4-6-2-3, 1-2-4-6-5-3, 1-2-3-6-5-4
8 Cylinder in-line engine	1-6-2-5-8-3-7-4
8 Cylinder V shape engine	1-5-4-8-6-3-7-2, 1-8-4-3-6-5-7-2, 1-6-2-5-8-3-7-4, 1-8-7-3-6-5-4-2, 1-5-4-2-6-3-7-8

Q. 10

Thermal Efficiencies of engines are such as

Engine Type	Thermal efficiencies
Diesel engine	32 to 38 per cent
Petrol engine	25 to 32 per cent

Q. 66

We know

For Isobaric process

$T_1/V_1 = T_2/V_2$

$\Rightarrow \quad T_2 = (2V/V)\times 27 = 54\ ^0C$

Q. 82

Given that

$P_1 = 101$ kPa, $P_2 = 2000$ kPa

$V_1 = 28$ m^3, $V_2 = 2$ m^3

For polytropic compression

External work required (W) = $\frac{P_2V_2 - P_1V_1}{1-n}$

Where, $\quad n = \frac{\log\frac{P_1}{P_2}}{\log\frac{V2}{V1}}$

$= \frac{\log\frac{101}{2000}}{\log\frac{2}{28}} = 1.1314$

Therefore, $W = \frac{2000\times 2 - 101\times 28}{1-1.1314} = -8919$ kJ = - 8.92 MJ

Here work is done on system so it is negative.

Q. 83

Given that

Compression ratio(r) = 16; Cutoff ratio (σ) = 4

Ratio of specific heat (γ) = 1.4

$\text{SAE} = 1 - \left(\frac{1}{r}\right)^{\gamma-1}\left(\frac{\sigma^{\gamma}-1}{\gamma(\sigma-1)}\right)$

$$= 1-\left(\frac{1}{16}\right)^{1.4-1}\left(\frac{4^{1.4}-1}{1.4(4-1)}\right) = 53.15\ \%$$

If, compression ratio is increased by 25%.

Compression ratio(r) = 20

$$\text{SAE} = 1-\left(\frac{1}{r}\right)^{\gamma-1}\left(\frac{\sigma^{\gamma}-1}{\gamma(\sigma-1)}\right)$$

$$= 1-\left(\frac{1}{20}\right)^{1.4-1}\left(\frac{4^{1.4}-1}{1.4(4-1)}\right) = 57.15\ \%$$

$\therefore$ Increase in SAE = 57.15 – 53.15 = 4 %

Q. 84 (i)

Power generated by consumption of diesel and producer gas

$= 35280\times10^{6}\times0.46\times10^{-3} + 3.97\times10^{6}\times12.5 = 65853800$ J/h

$= 18.29$ kW

$$\text{Brake thermal efficiency} = \frac{\text{Brake power}}{\text{Power generated by fuel consumption}}$$

$$= \frac{3.5}{18.29} = 19.13\%$$

(ii)

Generator efficiency = 90%

$\therefore$ Power produced = $3.5\times0.90 = 3.15$ kW

Q. 85

Cylinder cross sectional area (A) = $\pi\ d^2/4$

$= 3.14\times0.11\times0.11/4 = 9.4985\times10^{-3}\ m^2$

Discharge (Q) = A.v

$= 9.4985\times10^{-3}\times7 = 0.0665\ m^3/\sec$

Mean effective Pressure (P_e)= 650 kPa

$\because$ Break Power = Q× P_e

$= 0.0665\times 650$

$= 43.23$ kW

Q. 86

We know, Gas equation

$$PV = \frac{W}{m} RT$$

$$W = \frac{PVm}{RT}$$

$$= \frac{190 \times 10^3 \times 31 \times 10^{-3} \times 29}{8.314 \times 303}$$

$= 67.8$ g $= 67.8\times 10^{-3}$ kg

Q. 87

$$SAE = 1-\left(\frac{1}{r}\right)^{\gamma-1}\left(\frac{\sigma^{\gamma}-1}{\gamma(\sigma-1)}\right)$$

$$= 1-\left(\frac{1}{34}\right)^{1.3-1}\left(\frac{7^{1.3}-1}{1.3(7-1)}\right)$$

$= 48.62$ %

Q. 88

For adiabatic process

$$T_1V_1^{\gamma-1} = T_2V_2^{\gamma-1}$$

$$\Rightarrow \quad T_2 = T_1\frac{V_1^{\gamma-1}}{V_2^{\gamma-1}} = T_1\left(\frac{V_1}{V_2}\right)^{\gamma-1}$$

$$\Rightarrow \quad T_2 = 300 \times \left(\frac{16}{1}\right)^{1.4-1} = 909.43 \text{ K or } 636.43\ ^0\text{C}$$

Q. 89

For 4 stroke cycle engine

Engine dispalcement (Q) = AL(N/2)n

$= (3.14\times0.12^2/4)\times0.13\times(1600/2)\times3$

$= 3.527$ m^3/min or 211.62 m^3/h

Quantity of air taken in = 211.62×1.2×1.50 = 380.92 kg/h

Therefore, Fuel consumption = 380.92/14.9 = 25.57 kg/h

Q. 90

Water flow rate = 0.58×45×0.16 = 4.17 L/s

8

I.C. Engine Fuel System

8

I.C. Engine Fuel System

MPPSC Assistant Agricultural Engineer 2013

Q. 1 A hunting governor is ....

(A) Less sensitive
(B) More sensitive
(C) More stable
(D) None of these is correct

Q. 2 Specific fuel consumption of petrol engine is .... specific fuel consumption of diesel engine.

(A) Lesser than
(B) More than
(C) Equal to
(D) None of these is correct

UKPSC Combined Junior Engineering Exam 2013 Agricultural Engineering Paper-I

Q. 3 The specific gravity of light diesel oil is

(A) 0.82
(B) 0.85
(C) 0.92
(D) 0.95

Punjab Mechanic Agricultural Machinery Instructor 2014

Q. 4 Fuel pump is part of

(A) Lubrication system
(B) Fuel supply system
(C) Cooling system
(D) None of the above

RPSC AEn Pre Exam 2013 (Agricultural Engineering)

Q. 5 In I.C. engine, the air or air-fuel mixture is drawn into the cylinder during

(A) Suction stroke
(B) Combustion stroke
(C) Ignition
(D) Combustion or compression

Q. 6 "Octane number" is associated to

(A) Ignition quality of fuel (B) Fuel consumption

(C) Fuel supply system (D) All of the above

Q. 7 Fuel injection pressure in diesel engine rises more than

(A) 50 kg/cm^2 (B) 70 kg/cm^2

(C) 90 kg/cm^2 (D) 120 kg/cm^2

Q. 8 The function of Governor in I.C. engines, is to

(A) Control engine speed (B) Control fuel supply rate, only

(C) Control air-fuel mixture (D) Control detention

OCS AEn Exam 2011 (Pre) – II (Agricultural Engineering)

Q. 9 The variable speed governors are used on

(A) Stationary diesel engine (B) Petrol engine

(C) Tractor engine (D) None of the above

CGPSC State Engineering Services Exam Paper – I

Q. 10 Specific fuel consumption of diesel engine ranges between

(A) 100 - 130 g/hp-h (B) 130 - 160 g/hp-h

(C) 160 - 200 g/hp-h (D) 200 - 230 g/hp-h

(E) 200 - 290 g/hp-h

Q. 11 Stroke – bore ratio of tractor engine is about

(A) 1.25 (B) 1

(C) 1.5 (D) 1.45

(E) 2.15

Q. 12 Calorific value of High Speed Diesel (HSD) oil is

(A) 10300 kcal/kg (B) 10500 kcal/kg

(C) 10850 kcal/kg (D) 11100 kcal/kg

(E) 9600 kcal/kg

Q. 13 Instrument which is used to measure the specific gravity of fuel is

(A) Hygrometer (B) Thermometer

(C) Hydrometer (D) Anemometer

(E) Ammeter

Q. 14 Internal Combustion engine cylinder is made with

(A) Copper (B) Cast iron

(C) Aluminum (D) Steel

(E) Tin

Q. 15 Of the total heat generated during operation of I.C engine, the quantity of heat removed by cooling system is about

(A) 20 % (B) 30 %

(C) 40 % (D) 50 %

(E) 70 %

Q. 16 Hit and miss system of governor is mostly used in

(A) Scooter (B) Gas engines

(C) Tractors (D) Moto car

(D) Diesel engine

UKPSC A.En. Exam 2007 Paper – II

Q. 17 In an internal combustion engine, the turbocharger is driven by

(A) Intake air (B) Exhaust gas

(C) Flywheel of the engine (D) Gasoline

Q. 18 The specific fuel consumption of diesel engine as compared to that for petrol engine is

(A) Higher (B) Lower

(C) Same for the same output (D) Same for the same speed

Q. 19 In a petrol engine, the function of a "Choke" is to

(A) Heat the engine

(B) Provide rich air fuel mixture

(C) Increase the efficiency of the cylinder

(D) Remove the cylinder blockage

Q. 20 Ignition quality of diesel fuel is indicated by

(A) Octane number (B) Cetane number

(C) Calorific value (D) Ignition temperature

Q. 21 The commercial diesel fuels have got cetane rating varying from

(A) 10 to 20 (B) 22 to 28

(C) 30 to 60 (D) 70 to 80

Q. 22 In a diesel engine, fuel is drawn from the fuel tank by the

(A) Feed pump (B) Fuel injection pump

(C) Oil pump (D) None of the above

Q. 23 The functions of fuel injection system are

(A) To measure the correct amount of fuel

(B) To maintain correct timing of injection

(C) To atomize the fuel for quick ignition

(D) All above functions

Q. 24 Which of the following is a component of the diesel engine ?

(A) Spark plug (B) Carburetor

(C) Injector (D) Ignition system(High Tension)

Q. 25 The proper air fuel ratio for a diesel engine is

(A) 8 : 1 (B) 15 : 1

(C) 20 : 1 (D) Variable

Q. 26 Governor hunting in a tractor is due to

(A) Large variation in speed (B) Vibration in tractor

(C) Higher noise in tractor (D) None of the above

Q. 27 A breather of an engine connects the space above the oil level with the

(A) Fuel pump (B) Oil pump

(C) Outside atmosphere (D) Water pump

UKPSC A.En. Exam 2012 Paper - II

Q. 28 Which of the following is a component of petrol engine ?

(A) Injector (B) Carburettor

(C) Fuel injection pump (D) All of the above

Q. 29 Specific fuel consumption of petrol engine is

(A) 200 gm/bhp hr (B) 290 gm/bhp hr

(C) 350 gm/bhp hr (D) 150 gm/bhp hr

Q. 30 Governor hunting occurs due to

(A) Deficiency of fuel

(B) Inefficient working of governor

(C) Combustion of excess fuel in engine

(D) All of the above

Q. 31 Blue smoke is produced in engine due to

(A) Excess fuel consumption

(B) Overheating of engine

(C) Combustion of lubricating oil with fuel

(D) None of the above

Q. 32 Choke is used in carburetor engine

(A) Before starting the engine (B) Before closing the engine

(C) Both (A) and (B) (D) None of the above

Q. 33 Injection pressure of the fuel nozzle in diesel engine is

(A) 120 – 200 kg/cm^2 (B) 50 – 100 kg/cm^2

(C) 35 – 45 kg/cm^2 (D) > 200 kg/cm^2

Q. 34 The tetra ethyl lead is added to the fuel to improve

(A) Viscosity of fuel (B) Cetane number

(C) Flash point (D) Octane number

Q. 35 Oil pump used in forced feed lubrication system is driven by

(A) Camshaft (B) Crankshaft

(C) Flywheel (D) None of the above

Q. 36 The antiknock quality of a petrol is determined by

(A) Cetane number (B) Octane number

(C) Knocking index (D) None of the above

Kerla AAO Exam 2017

Q. 37 The major pollutant present in diesel engine emission is

(A) Carbon dust (B) Oxides of nitrogen

(C) Carbon monoxide (D) Both (A) and (B)

Graduate Aptitude Test in Engineering - 2007

Q. 38 An oil engine works on the ideal diesel cycle with a compression ratio of 18:1. The constant pressure energy addition ceases at 10% of the stroke. The intake pressure and temperature are 100 kPa and 300 K respectively. The hourly air consumption is 100 m^3. If the ratio of specific heats is 1.4, the maximum temperature in the cycle is

(A) 953.3 K (B) 1334.6 K

(C) 2154.5 K (D) 2573.9 K

Graduate Aptitude Test in Engineering - 2008

Q. 39 A 20 kW four stroke cycle diesel engine is running at 2400 rpm and maintaining an ignition delay of 18° during combustion. When the engine speed is reduced by 25 percent, the ignition delay increases by 4°. If the specific fuel consumption is 0.20 kg/(kW.h), the percent change in the fuel consumption during the above conditions of combustion is

(A) 37.0 (B) 38.64

(C) 61.36 (D) 62.96

Graduate Aptitude Test in Engineering – 2011

Q. 40 A two cylinder four stroke diesel engine develops 15 kW power at 2400 rpm. Its brake specific fuel consumption is 0.268 kg/(kW.h). If the specific gravity of fuel is 0.85, quantity of fuel injected per cylinder per cycle in m^3 is

(A) 0.028×10^{-6} (B) 0.033×10^{-6}

(C) 0.056×10^{-6} (D) 0.065×10^{-6}

Q. 41 An engine is to be run in dual fuel mode using diesel and producer gas with diesel as pilot fuel. Operation NOT required while running the engine is

(A) Cooling of the producer gas

(B) Cleaning of the producer gas

(C) Preheating of the producer gas

(D) Mixing of producer gas with air

Graduate Aptitude Test in Engineering - 2012

Q. 42 The power developed and the exhaust gas temperature of a diesel engine compared to a spark ignition engine of the same size and running at the same speed respectively, are

(A) Higher and lower (B) Higher and higher

(C) Lower and higher (D) Lower and lower

Graduate Aptitude Test in Engineering - 2014

Q. 43 A four-stroke, four cylinder diesel engine running at 2000 rpm develops brake power of 60 kW and the fuel consumption is 0.30 kg/(kW.h). The engine has a bore of 120 mm and stroke of 100 mm. If air-fuel ratio is 15:1 and air density is 1.15 kg m^{-3}, the volumetric efficiency of the engine in percent is

(A) 43.25 (B) 66.32

(C) 75.22 (D) 86.50

Graduate Aptitude Test in Engineering – 2015

Q. 44 The brake power of a six-cylinder, four stroke diesel engine running at 3000 rpm is 125 kW. Its brake specific fuel consumption is 200 g kW^{-1} h^{-1}. Assuming specific gravity of fuel as 0.85, the volume of fuel to be injected per cycle in *ml* is ....

Q. 45 The connecting rod of an internal combustion engine is subjected to

(A) Compression only (B) Tension only

(C) Both compression and tension (D) Torsion only

Graduate Aptitude Test in Engineering – 2017

Q. 46 A diesel fuel has same ignition delay as that of blend of two reference fuels namely 56% n-cetane and 44% hepta-methylnonane. The cetane number of the diesel fuel is ....

Q. 47 In fuel property determination, Reid vapour pressure test is used for measuring

(A) Volatility (B) Viscocity

(C) Sulphur content (D) Carbon residue

Graduate Aptitude Test in Engineering – 2018

Q. 48 A power tiller with diesel engine produces 12 kW of brake power with a brake thermal efficiency of 30%. If the air-fuel ratio in the combustion process is 14:1 and heating value of the fuel is 45 MJ kg^{-1}, air consumption rate of the engine will be .... kg h^{-1}.

Q. 49 Air enters into an engine cylinder at a pressure of 100 kPa and temperature of 300 K in the suction stroke of an ideal Otto cycle. The cylinder clearance volume is 600 cm^3 and compression ratio is 8:1. The expansion stroke of the cycle is polytropic in nature. The pressure at the beginning of the expansion stroke is 10 MPa and the temperature at the end of the expansion stroke is 1800 K. The polytropic exponent of the expansion stroke is ....

Graduate Aptitude Test in Engineering – 2020

Q. 50 A diesel engine when operates with biodiesel blend B20 (20% biodiesel and 80% diesel by volume) develops a brake power of 10 kW with a brake specific fuel consumption of 0.26 kg /(kW· h). If the density of biodiesel is 880 kg/m^3 and that of diesel is 850 kg/m^3, the amount of biodiesel required to run the engine for 3 hours in L is ....

Graduate Aptitude Test in Engineering – 2021

Q. 51 Stoichiometric air-fuel ratio of an SI engine is 14.7:1. If equivalence ratio (λ) is 0.92, the actual air-fuel ratio maintained during the engine operation is ....

Answers Key

1	2	3	4	5	6	7	8	9	10
B	B	C	B	A	A	D	A	C	C
11	12	13	14	15	16	17	18	19	20
A	B	C	B	B	B	B	B	B	B
21	22	23	24	25	26	27	28	29	30
C	A	D	C	B	A	C	B	B	B
31	32	33	34	35	36	37	38	39	40
C	A	A	D	A	B	D	D	D	B
41	42	43	44	45	46	47	48'	49	50
C	D	D	0.054	C	62.6	A	44.80	1.353	1.82
51									
13.52									

Explanations

Q. 3

Specific gravities of fuel oil are such as

Fuel oil	Specific gravity	Temperature(^{0}C)
Diesel Fuel Oil 2D/3D/4D/5D	0.81 - 0.97	
1. Diesel Oil (Gas Oil)	0.81 - 0.87	
2. Light Fuel Oil (LFO)	0.88 - 0.93	15.6
3. Medium Fuel Oil (MFO)	0.94 - 0.95	
4. Heavy Fuel Oil (HFO)	0.96 - 0.97	
Petrol Oil	0.739	15.6
Gas oils	0.89	15.6
Petroleum oil	0.82	15.6

Q. 10

The specific fuel consumption of diesel engine is 160 to 200 g/hp/hr.

The specific fuel consumption of petrol engine is 250 to 290 m/hp/hr.

Q. 12

Calorific values of different fuels

Name of fuel	Calorific value (kcal/kg)
Wood	4000
Coal	7000
Diesel	10500
Kerosene	11200
Petrol	11700
Biogas	8200 – 9400
Natural Gas	7700 - 11700

Q. 16

Hit and Miss - Controls the engines speed by cutting off the engines ignition system when it runs too fast. This is the "miss" part. When the engine slows too much, it allows it to "hit" again and pick speed back up.

Q. 31

White smoke indicates – Presence of water in fuel

Blue smoke indicates – Burning of lubricant in cylinder

Black smoke indicates – Engine is overloaded

Q. 32

A choke valve is sometimes installed in the carburetor of internal combustion engines. Its purpose is to restrict the flow of air, thereby enriching the fuel-air mixture while starting the engine. When doing a cold start, the choke should be closed to limit the amount of air going in. This increases the amount of fuel in the cylinder and helps to keep the engine running, while it is trying to warm up.

Q. 38

For Diesel Cycle

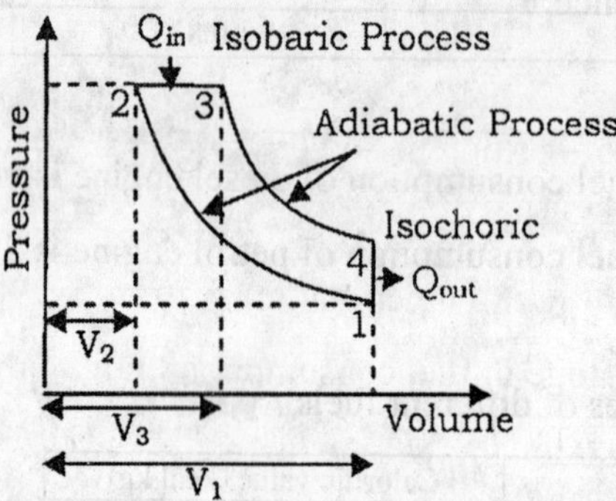

Intake pressure (P_1) = 100 kPa; Intake temperature (T_1) = 300 K

$$\text{Compression ratio}(r) = \frac{\text{Clearance Volvme} + \text{Swept Volume}}{\text{Clearance Volume}}$$

$$= \frac{V_1}{V_2} = 18:1 \Rightarrow V_1 = 18V_2$$

And $\quad V_3 = V_2 + 0.10(V_1 - V_2)$

$\Rightarrow \quad V_3 = V_2 + 0.10(18V_2 - V_2) = 2.7V_2$

Now, for adiabatic process 1-2

$$T_1V_1^{\gamma-1} = T_2V_2^{\gamma-1}$$

$$\Rightarrow \quad T_2 = T_1\frac{V_1^{\gamma-1}}{V_2^{\gamma-1}} = T_1\left(\frac{V_1}{V_2}\right)^{\gamma-1}$$

$$\Rightarrow \quad T_2 = 300 \times \left(\frac{18}{1}\right)^{1.4-1} = 953.3 \text{ K}$$

Again, For Isobaric process 2-3

$$T_3/V_3 = T_2/V_2$$

$$\Rightarrow \quad T_3 = (V_3/V_2)\times T_2$$

Maximum temperature in diesel cycle will be at point 3, thus

$$T_3 = 2.7\times 953.3 = 2573.91 \text{ K}$$

Q. 39

In the given problem, fuel consumption is depending only on time period of ignition delay. All other parameters are not changing for both the cases. Hence,

CASE I

Time period of ignition period $(T_1) = \frac{18}{2400\times 360}$ min

CASE II

Time period of ignition period $(T_2) = \frac{22}{1800\times 360}$ min

∴ Percent change in fuel consumption

$$= \frac{T_2 - T_1}{T_1}\times 100$$

$$= \frac{\frac{22}{1800\times 360} - \frac{18}{2400\times 360}}{\frac{18}{2400\times 360}}\times 100 = 62.96\ \%$$

Q. 40

Power developed = 15 kW

Total revolutions = 2400 rpm

Thus, no. of cycles = 1200 cycles/min = 1200×60 cycles/h

∵ Density of oil = 0.85×1000 = 850 kg/m^3

Thus quantity of fuel injected per cycle per cylinder

$$= \frac{15\times 0.268}{1200\times 60\times 850\times 2}$$

$$= 0.033\times 10^{-6} \text{ m}^3$$

Q. 43

Given

Fuel consumption (F) = 0.30×60 = 18 kg / h = 0.3 kg/ min

Air fuel ratio (A/F) = 15;

N = 2000/2 = 1000 (∵ For four stroke engine, rpm = N/2)

Number of cylinder (n) = 4; Stroke length (L) = 0.1 m; Bore (D) = 0.12 m

We know

$$\text{Volumetric efficiency } (\eta_v) = \frac{\left(\frac{A}{F}\right)\frac{F}{\rho_a}}{LANn} \times 100$$

$$\eta_v = \frac{(15)\times\frac{0.30}{1.25}}{0.1\times\frac{3.14}{4}\times 0.12\times 0.12\times(2000/2)\times 4} = \times 100$$

$$\eta_v = 86.50\ \%$$

Q. 44

Power developed = 125 kW

Specific fuel consumption = 200 g kW^{-1} h^{-1} = 200×10^{-3} kg kW^{-1} h^{-1}

Total revolutions = 3000 rpm

Thus, no. of cycles = 3000/2 = 1500 cycles/min = 1500×60 cycles/h

∵ Density of oil = 0.85×1000 = 850 kg/m^3

Thus quantity of fuel injected per cycle per cylinder

$$= \frac{125\times10^{-3}\times 200}{1500\times 60\times 850\times 6}$$

$$= 5.45\times10^{-8}\ m^3$$

$$= 0.054\ ml$$

Q. 46

We know

CN of fuel = Percentage of n-cetane + 0.15×(Percentage of iso-cetane)

= 56 + 0.15×44 = 62.6

Q. 48

Given,

Brake Power = 12 kW; Break thermal efficiency = 30 %

Therefore, Fuel power required = 12/0.3 = 40 kW = 40 kJ/s = 144000 kJ/h

Again, Heating value of fuel is = 45 MJ/kg = 45000 kJ/kg

Hence, Fuel required to generate the above fuel power = 144000/45000

= 3.2 kg/h

∵ Air-fuel ratio = 14 : 1

∴ Air consumption rate of the engine = 14×3.2 = 44.80 kg/h

Q. 49

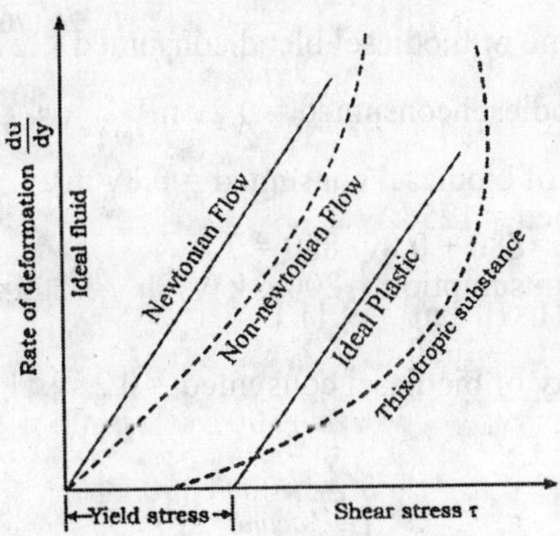

Given that,

Suction stroke (1-2)

P_1 = 100 kPa; T_1 = 300 K

Compression Stroke (2-3)

V_2 = 600 cm^3;

Expansion Stroke (3-4)

P_3 = 10 Mpa = 10000 kPa; V_3 = 600 cm^3;

Exhaust Stroke (4-1)

T_4 = 300 K

Compression ratio = V_1 / V2 = 8 : 1

⇒ $V_1 = 8 \times 600 = 4800\ cm^3$

For Constant volume rejection (4-1)

$P_4/T_4 = P_1/T_1$

⇒ $P_4 = 100 \times 1800/300 = 600$ kPa

Again, For adiabatic expansion or Polytropic Process (3-4)

$P_3/P_4 = (V_4/V_3)^\gamma$

⇒ $\gamma = \log(P_3/P_4)/\log(V_4/V_3) = \{\log(10000/600)\}/\{\log(4800/600)$

$= 1.353$

Q. 50

Biodiesel blend consumed = $0.26 \times 10 \times 3 = 7.8$ kg

Consider volume of biodiesel blend consumed = v m^3

Quantity of biodiesel consumed = 0.2v m^3

And, Quantity of biodiesel consumed = 0.8v m^3

Therefore, $0.2v \times 880 + 0.8v \times 850 = 7.8$

⇒ $v = 9.11 \times 10^{-3}\ m^3 = 9.11$ L

Hence, Quantity of biodiesel consumed = $0.2 \times 9.11 = 1.82$ L

Q. 51

$$\text{Equivalence ratio}(\varphi) = \frac{\left(\frac{F}{A}\right)actual}{\left(\frac{F}{A}\right)stoichiometric}$$

$$\Rightarrow \quad 0.91 = \frac{\left(\frac{F}{A}\right)actual}{(1/14.7)}$$

$$\Rightarrow \quad \left(\frac{F}{A}\right) \text{actual} = 0.91/14.7$$

$$\Rightarrow \quad \left(\frac{A}{F}\right) \text{actual} = 13.52$$

9

I.C. Engine Auxiliary Systems Lubrication, Ignition and Cooling System

MPPSC Assistant Agricultural Engineer 2013

Q. 1 Which option about the statements made below is correct?

(1) The cooling system maintains engine temperature at 71 to 82°C for petrol engines.

(2) The thermostat valve begins to open at about 60°C and opens widely at about 71°C.

(A) Statement (1) is true, but statement (2) is false.

(B) Statement (2) is true, but statement (1) is false.

(C) Both statements (1) and (2) are true.

(D) Both Statements (1) and (2) are false.

UKPSC Combined Junior Engineering Exam 2013 Agricultural Engineering Paper-I

Q. 2 Which of the following tractor has the air cooled engine ?

(A) Eicher (B) HMT

(C) Mahindra (D) Ford

Q. 3 The common firing order of a 4 stroke 4 cylinder engine is given as

(A) 1 – 2 – 3 – 4 (B) 1 – 3 – 2 – 4

(C) 1 – 3 – 4 – 2 (D) 1 – 4 – 3 – 2

Q. 4 Compression ratio of diesel engine is

(A) 5 : 1 (B) 5 : 1 to 11 : 1

(C) 11 : 1 to 14 : 1 (D) 14 : 1 to 21 : 1

Q. 5 A cylinder block of tractor engine is normally made of

(A) Cast Iron (B) Cast steel

(C) Heat treated steel alloy (D) Spring steel

Q. 6 Engine oil is changed in tractor after

(A) 80 hours (B) 120 hours

(C) 200 hours (D) 300 hours

Q. 7 A connecting rod of a tractor engine is made of

(A) Forged steel (B) Mild steel

(C) Cast steel (D) Cast iron

Q. 8 Stroke bore ratio for under square or long-stroke engine is

(A) 1.25 (B) < 1

(C) 1.75 (D) 2.00

Q. 9 Piston position in the engine when it reaches the bottom is known as

(A) Crank end dead centre (B) Bottom dead centre

(C) Both of the above (A) & (B) (D) None of the above

Q. 10 In diesel engine, the heat is taken in at

(A) Constant pressure (B) Constant volume

(C) Both (A) & (B) (D) None of the above

Q. 11 When the engine cycle is completed in two revolutions of the crankshaft, the engine is known as

(A) Two stroke cycle engine (B) Four stroke cycle engine

(C) Eight stroke cycle engine (D) None of the above

Q. 12 Hot spark plugs are used in

(A) Slow speed engine (B) Low compression engine

(C) Two stroke petrol engine (D) All of the above

Q. 13 Carburettor is found in

(A) Petrol engine (B) Diesel engine

(C) Both of these (D) None of these

Q. 14 In IC engine, the diameter of cylinder is known as

(A) Bore (B) Stroke

(C) Piston displacement (D) None of the above

Q. 15 Mostly the type of piston pin used is

(A) Semi-floating (B) Floating

(C) Fixed (D) All above three

Q. 16 Types of rings used in engine are

(A) Compression rings (B) Oil rings

(C) Both (A) & (B) (D) None of the above

Q. 17 Air cleaner is a part of

(A) Engine lubricating system (B) Engine cooling system

(C) Engine air fuel supply system (D) Hydraulic system

Q. 18 In diesel cycle maximum efficiency is obtained with

(A) Early fuel cut-off (B) High compression ratio

(C) Both (A) & (B) (D) None of the above

Q. 19 Primary function of lubrication is to reduce

(A) Friction (B) Wear

(C) Power loss (D) All these

Q. 20 Farm tractors used in India do not have

(A) Carburetor (B) Gear box

(C) Differential (D) Clutch

Punjab Mechanic Agricultural Machinery Instructor 2014

Q. 21 The purpose of engine lubrication is

(A) Reduce frictional effects (B) Cooling effect

(C) Sealing effect (D) All of the above

RPSC AEn Pre Exam 2013 (Agricultural Engineering)

Q. 22 Which of the following is not the part of air-cooled engine ?

(A) Radiator (B) Spark plug

(C) Fins (D) Carburetor

Q. 23 The grade of oil used in tractor engine during summer should be

(A) SAE 30 (B) SAE 40

(C) SAE 60 (D) SAE 90

CGPSC State Engineering Services Exam Paper – I

Q. 24 In spark ignition engines the spark plug gap setting usually kept between

(A) 0.4 - 0.6 mm (B) 0.5 - 0.7 mm

(C) 0.5 - 0.85 mm (D) 0.6 - 0.9 mm

(E) 0.5 - 0.95 mm

Q. 25 In petrol engine, thermostat valve open fully at the temperature of

(A) 70 ºC (B) 75 ºC

(C) 82 ºC (D) 88 ºC

(E) 90 ºC

Q. 26 Specific gravity of fully charged electrolyte solution of lead-acid battery is

(A) 1.280 (B) 1.225

(C) 1.550 (D) 2.280

(E) 0.280

UKPSC A.En. Exam 2007 Paper – II

Q. 27 In an engine the oil is used for

(A) Lubrication only (B) Cooling only

(C) For both (A) and (B) (D) None of the above

UKPSC A.En. Exam 2012 Paper - II

Q. 28 Thermostat valve is used in

(A) Lubrication system (B) Steering system

(C) Braking system (D) Cooling system

Q. 29 Spark plug is used in

(A) Petrol engine (B) Biodiesel fuelled engine

(C) Diesel engine (D) All of the above

Q. 30 The most common type of cooling system found in general purpose tractors is

(A) Air cooling

(B) Thermo-siphon system

(C) Open jacket system

(D) Force feed water circulation system

Graduate Aptitude Test in Engineering - 2007

Q. 31 A lubricating oil with high viscosity index is desirable for tractor engine due to

(A) More variation of viscosity with temperature

(B) Less variation of viscosity with temperature

(C) High pour point

(D) High cloud point

Q. 32 As compared to diesel, the heating value and exhaust emissions such as CO, CO_2 and smoke density of biodiesel when used in compression ignition engine are

(A) Lower and higher respectively

(B) Higher and Lower respectively

(C) Lower and Lower respectively

(D) Higher and higher respectively

Q. 33 A tractor engine developing 30 KW rejects heat at the rate of 0.58 kW per kW of engine output. A water cooling system is to be installed in the tractor. The expected temperature rise as air moves through the radiator is 20 K. the frontal area of the radiator is limited to 0.16 m^2. If density of air is 1.29 kg m^{-3} and specific heat of air is 1.0 kJ kg^{-1} K^{-1}, the amount of air to be blown per unit time through the radiator frontal area is

(A) 0.674 m^3s^{-1}

(B) 0.870 m^3s^{-1}

(C) 1.162 m^3s^{-1}

(D) 1.502 m^3s^{-1}

Graduate Aptitude Test in Engineering - 2012

Q. 34 The type of pump used in forced water cooling system of a tractor engine is

(A) Piston (B) Centrifugal

(C) Gear (D) Vane

Answers Key

1	2	3	4	5	6	7	8	9	10
A	A	C	D	A	B	A	B	C	A
11	12	13	14	15	16	17	18	19	20
B	D	A	A	B	C	C	C	D	A
21	22	23	24	25	26	27	28	29	30
D	A	B	C	C	A	C	D	A	D
31	32	33	34						
B	C	A	B						

Explanations

Q. 1

Thermostat valve starts opening at 70 ^{0}C and fully opens at 82 ^{0}C for petrol engine. In diesel engine it starts opening at 70 ^{0}C and fully opens at 90^{0}C.

Q. 23

The Society of Automotive Engineers (SAE) developed a scale for both engine (motor oil grades) and transmission oils. SAE30 has a viscosity rating of 30 at a temperature of 100 degrees Celsius. Now, multi grade engine oil like 5W-30, 10W-30 is being used instead of SAE 30. Where W denotes "Winter" and second number after "-" denotes viscosity of oil at 100 ^{0}C. A 5W-30 motor oil performs like a SAE 5 engine oil would perform at the cold temperature specified, but still has the SAE 30 viscosity at 100 °C which is engine operating temperature.

Q. 26

Specific gravity of fully charged electrolyte solution of lead-acid battery is about 1.28 and in discharge condition, it falls below 1.12 (it is considered to be 1).

Q. 34

Given

Density of air (ρ) = 1.29 kg/ m^3; C_p = 1.0 kJ kg^{-1} K^{-1}; ΔT = 20 K

Volume of air blown per unit time = V m^3/ s (Consider)

Heat rejected by tractor engine = 30×0.58 = 17.4 kW = 17.4 kJ/ s

Also,

Heat rejected = $mC_p\Delta T = V\rho C_p\Delta T$

$= V\times1.29\times1\times20 = 25.8$ V kJ/ s

Now,

$25.8\ V = 17.4$

$\Rightarrow \quad V = 0.674$ m^3/ s

10

Power Transmission Mechanism and Drive System

UKPSC AE Agricultural Engineering Paper - II 2013

Q. 1 In which of the following tractor components, the speed reduction takes place?

(A) Gear box (B) Differential

(C) Final drive (D) All of the above

Q. 2 Brake lining is mounted on :-

(A) Wheel cylinder (B) Brake drum

(C) Brake shoe (D) Master cylinder

Q. 3 For an engine running at 1500 r.p.m. the cam shaft speed will be :-

(A) 750 r.p.m (B) 1500 r.p.m

(C) 2250 r.p.m (D) 3000 r.p.m

Q. 4 A tractor with 21 splines P.T.O. shaft will have standard speed of:-

(A) 540 r.p.m. (B) 1000 r.p.m.

(C) 1540 r.p.m. (D) 1800 r.p.m.

Q. 5 For a tractor P.T.O. shaft rotates in

(A) Clockwise direction as viewed from end of shaft

(B) Anti-clockwise direction as viewed from end of shaft

(C) Both the direction

(D) None of the above

Q. 6 Oil pump in an engine is driven by:-

(A) Cam shaft (B) Crank shaft

(C) Timing gear (D) Rocker shaft

APPSC AEES Agriculture Engineering Exam 2016

Q. 7 The most used and least efficient power outlet of a tractor is

(A) Power take-off shaft in the front

(B) Drawbar in the rear

(C) P.T.O. in the rear

(D) None of the above

UKPSC Combined Junior Engineering Exam 2013 Agricultural Engineering Paper-I

Q. 8 Gears are used for

(A) Changing the speed of rotation(B)Changing direction of shafting

(C) Changing direction of rotation(D) All of the above

Q. 9 Connecting rod generally have

(A) I – shape (B) T-shape

(C) Angle shape (D) Flat shape

Q. 10 B.H.P. of an engine indicates

(A) Power at PTO pulley (B) Power in cylinder

(C) Power on fly wheel (D) Frictional power

Q. 11 Variable speed governors are used on

(A) Tractor engine (Diesel) (B) Stationary engine (Diesel)

(C) Petrol engine (D) None of above

Q. 12 Speed reduction in a tractor at

(A) Differential (B) Final drive

(C) Gear box (D) None of above

Q. 13 Power transmission from the source of power to the machines is done by the following system :

(A) Pullies and belts (B) Flat belts

(C) V-belts (D) All of the above

Q. 14 Fly wheel of the engine helps

(A) To run the engine smoothly

(B) Stores energy during power stroke

(C) Returns the energy during idle stroke

(D) All of the above

Q. 15 To maintain a constant speed of engine under varying load conditions, the engines are provided with

(A) Nozzle (B) Governor

(C) Gear (D) Thermostat

Q. 16 Which of the following part is not connected to the piston of an engine ?

(A) Rings (B) Gudgeon pin

(C) Connecting Rod (D) Fly wheel

Q. 17 To measure the tappet clearance, which gauge is used

(A) Go gauge (B) Not go gauge

(C) Feeler gauge (D) All of the above

Q. 18 The clutch is located between the engine and

(A) Gear box (B) Differential gear

(C) Rear axle (D) Universal joint

Q. 19 The ratio of break horse power to indicated horse power is called

(A) Volumetric efficiency (B) Overall efficiency

(C) Indicated thermal efficiency (D) Mechanical efficiency

Q. 20 The drive from gear box to the rear axle is taken by

(A) Propeller shaft (B) Clutch

(C) Differential gear (D) Universal joint

Q. 21 In four wheel drive, which type of axle is used

(A) Front live (B) Rear live

(C) Any one (D) Both axles

Q. 22 Braking is produced by the frictional effect between the brake drum and the

(A) Brake shoes (B) Wheel cylinder pistons

(C) Wheel rim (D) Wheel studs

RPSC AEn Pre Exam 2013 (Agricultural Engineering)

Q. 23 Function of crank shaft, is to

(A) Turn the wheel (B) Power the piston

(C) Rotate piston (D) Stop the engine

Q. 24 The tie rod of steering mechanism of a tractor is actuated by

(A) Tie rod (B) Wheel spindle

(C) Drag line (D) Toe in

Q. 25 The standard PTO speed in a tractor is

(A) 540 (B) 1000

(C) Both (A) and (B) (D) None of the above

Q. 26 The crankshaft and rear axle of tractor are attached at

(A) Right angle to each other (B) 30°

(C) 60° (D) 120°

OCS AEn Exam 2011 (Pre) – II (Agricultural Engineering)

Q. 27 Maximum engine torque in Tractor engine is generated

(A) At rated speed

(B) At about 400 rpm less than rated speed

(C) At about 400 rpm more than rated speed

(D) At 1200 engine rpm

CGPSC State Engineering Services Exam Paper – I

Q. 28 In case of drive, the grip between pulley and belt is obtained by

(A) Compression (B) Friction

(C) Cohesion force (D) Tension

(E) Adhesion force

Q. 29 As per ASAE standard, the width of V-belt with designation B is

(A) 16.7 mm (B) 13.6 mm
(C) 25.2 mm (D) 32.5 mm
(E) 19.8 mm

Q. 30 Which type of clutch is mostly used in Power Tiller

(A) Friction Clutch (B) Single disc clutch
(C) Dog clutch (D) Cone clutch
(E) Multi plate clutch

Q. 31 Fluid Coupling works on the principle of

(A) Hydrokinetics (B) Hydrodynamics
(C) Hydrostatics (D) Pascal's law
(E) Rittingers law

Q. 32 Gear in which the teeth are cut obliquely across the perimeter of gear is

(A) Herring bone gear (B) Double helical gear
(C) Helical spur gear (D) Bull gear
(E) Crown wheel

Q. 33 In internal expanding shoe type brake system of a tractor, the moment of the cam is caused by

(A) Brake drum (B) Brake pedal
(C) Brake lining (D) Brake shoe
(E) Fulcrum

Q. 34 In differential unit of a tractor, the bevel pinion is in mesh a large bevel wheel is known as

(A) Heringbone gear (B) Crown wheel
(C) Spur gear (D) Helical gear
(E) Differential lock

Q. 35 According to ASAE standard, the six spline PTO speed is about

(A) 536 ± 10 rpm (B) 540 ± 10 rpm
(C) 550 ± 10 rpm (D) 560 ± 10 rpm
(E) 565 ± 10 rpm

Q. 36 The space that supports the crankshaft in the cylinder blocks is called as

(A) Camshaft (B) Main journal

(C) Crank journal (D) Power train

(E) Piston

UKPSC A.En. Exam 2007 Paper - II

Q. 37 In a tractor, if one wheel is locked, the speed of the other one is increased by

(A) 4 times (B) 2 times

(C) 3 times (D) 1.50 times

Q. 38 A pulley 30 cm diameter is driving, another pulley of 20 cm diameter using a belt drive without any slippage. The velocity ratio will be

(A) 1.00 (B) 1.50

(C) 0.50 (D) 2.00

Q. 39 A differential in a tractor is provided to work as

(A) Compensating devices in turns (B) Gear reduction unit

(C) Torque multiplier (D) None of the above

Q. 40 A two wheel small tractor (Power Tiller) is preferably recommended for the best use as

(A) Primary tillage operation with mould board plough

(B) Rota-tilling of hard field

(C) Puddling of paddy field with the rota-tiller

(D) None of the above

Q. 41 The following type of clutch is used in power tiller

(A) Dog clutch (B) Friction clutch

(C) Fluid coupling (D) Multiple plate clutch

Q. 42 In an engine, the ratio of total volume of the cylinder to the clearance volume is known as

(A) Stroke ratio (B) Bore ratio

(C) Compression ratio (D) Swept volume

Q. 43 Piston rings of a tractor are made of

(A) Cast iron (B) Cast steel

(C) High carbon steel (D) Drop forge steel

Q. 44 Indicated Horse Power of a four stroke single cylinder engine is determined by the following formula

(A) IHP = PLAN/4500 (B) IHP = PLAN/(4500×2)

(C) IHP = PLAN/(75×4) (D) IHP = PLAN/(4500×4)

Q. 45 In an IC engine, indicated mean effective pressure is

(A) Maximum pressure in compression stroke

(B) Maximum pressure in power stroke

(C) Average pressure in power stroke

(D) Minimum pressure in power stroke

Q. 46 Differential in a tractor is used

(A) To change the speed of a tractor

(B) To vary the spacing between rear wheels

(C) Both (A) and (B)

(D) To permit the tractor to take turn without overturning

Q. 47 An IC engine converts the reciprocating motion of piston in to rotary motion of the crankshaft by means of

(A) Connecting rod (B) Push rod

(C) Camshaft (D) Gudgeon pin

Q. 48 The power available at the crankshaft in a tractor is

(A) B.M.E.P. (B) Electric power

(C) I.H.P. (D) B.H.P.

Q. 49 In a power transmission system, the "final drive" stands for

(A) Gear reduction unit of the power train

(B) Flywheel

(C) Rear wheel

(D) P.T.O.

Q. 50 The transmission system of modern tractors is mostly

(A) Selective sliding gear type (B) Fixed sliding gear type

(C) Hand Lever type (D) Friction type

UKPSC A.En. Exam 2012 Paper - II

Q. 51 Power available at the crank shaft of the engine is called as

(A) BHP (B) IHP

(C) FHP (D) None of the above

Q. 52 In a diesel engine, the duration between time of injection and time of ignition is called

(A) Period of ignition (B) Delay period

(C) Explosion period (D) Burning period

Q. 53 ASAE standard PTO speed (rpm) is

(A) 536 (B) 1000

(C) 1440 (D) 2000

Q. 54 The most used and least efficient power outlet of a tractor is

(A) Power take off shaft in the front

(B) Power take off shaft in the rear

(C) Drawbar in the rear

(D) None of the above

Q. 55 The reciprocating motion of piston is converted to rotary motion of flywheel by

(A) Crankshaft (B) Piston

(C) Connecting rod (D) All the above

Q. 56 The power outlet from a tractor for agricultural operations are

(A) By PTO shaft and power lift (B) By belt pulley arrangement

(C) By hitches and drawbar (D) All of the above

Q. 57 Differential in tractor increases the wheel speed while turning of

(A) Both the wheels (B) Inner wheel

(C) Outer wheel (D) None of the above

Q. 58 If one wheel of a tractor is locked by differential lock, the speed of the other wheel is increased by

(A) Increased (B) Decreased

(C) Remains same (D) Nothing can be said

Q. 59 A tractor with 20 splines PTO shaft will have standard speed of

(A) 1000 rpm (B) 540 rpm

(C) 1080 rpm (D) 1500 rpm

Q. 60 The differential unit of a tractor has a set of

(A) Rack and pinion gear (B) Spur gears

(C) Bevel gears (D) Worm and gear

Q. 61 In tractor the driving wheel requires power supply at

(A) Low rpm (B) High torque

(C) Low rpm and high torque (D) High rpm and low torque

Q. 62 Standard speed of P.T.O. is

(A) 2000 rpm (B) 1440 rpm

(C) 1000 rpm (D) 540 rpm

Q. 63 Tractors are mostly equipped with high speed engines running at about

(A) 2000 rpm (B) 4000 rpm

(C) 500 rpm (D) 6000 rpm

Kerla AAO Exam 2017

Q. 64 Final drive provided in tractors for

(A) Increasing the speed (B) Final speed reduction

(C) Reducing traction (D) Reducing torque

Q. 65 Tractor pulling power is transmitted to implement through the

(A) Hydraulic system (B) Three point linkage

(C) Lubricating system (D) Steering system

Graduate Aptitude Test in Engineering - 2007

Q. 66 A flat leather belt with 9×250 mm cross section is used to drive a cast iron pulley of diameter 0.90 m running at 336 rpm. The active arc of contact on the smaller pulley is 120°. The belt weighs 980 kg m^{-3}. Coefficient of friction between the leather and cast iron is 0.35. Centrifugal tension experienced by the belt is

(A) 5.5 N (B) 56.4 N

(C) 552.8 N (D) 2211.2 N

Q. 67 A flange mounted shear pin is used on a shaft as a safety device. The steel shear pin has a diameter of 2.38 mm and is to be mounted on the flange of a shaft rotating at 650 rpm. The maximum power transmitted by the shaft is 4.5 kW. If the shear strength of the material of pin is 310 MPa, The radial distance of its mounting is

(A) 5.02 mm (B) 11.98 mm

(C) 47.94 mm (D) 301.20 mm

Q. 68 As per ASABE standards, the three-point hitch of a two-wheel drive tractor with a maximum drawbar power of 45 kW comes under the category

(A) I (B) II

(C) III (D) IV

Q. 69 The mechanical efficiency of a power tiller engine developing 7.5 kW is 80%. The calorific value of diesel is 45 MJ kg^{-1}. If the indicated thermal efficiency is 35%, the brake specific fuel consumption of the engine is

(A) 0.135 kg $kW^{-1}h^{-1}$ (B) 0.228 kg $kW^{-1}h^{-1}$

(C) 0.245 kg $kW^{-1}h^{-1}$ (D) 0.286 kg $kW^{-1}h^{-1}$

Graduate Aptitude Test in Engineering - 2008

Q. 70 The function of a differential lock used in a rear wheel driven tractor is

(A) To operate both the rear wheels at the same speed

(B) To operate both the rear wheels at differential speeds

(C) To operate both the rear wheels at the same torque

(D) To evenly distribute the power to both the wheels

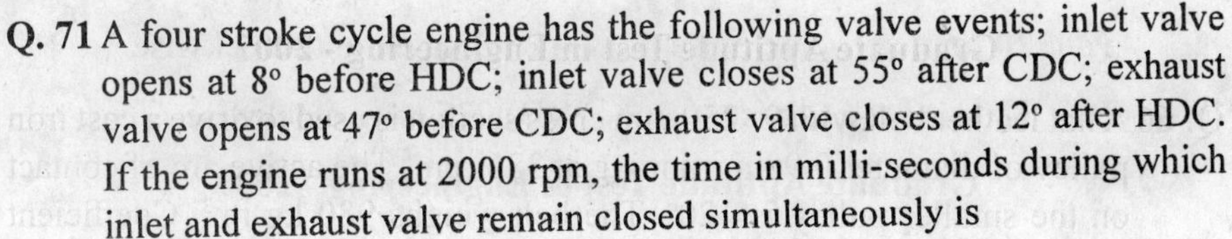

Q. 71 A four stroke cycle engine has the following valve events; inlet valve opens at 8° before HDC; inlet valve closes at 55° after CDC; exhaust valve opens at 47° before CDC; exhaust valve closes at 12° after HDC. If the engine runs at 2000 rpm, the time in milli-seconds during which inlet and exhaust valve remain closed simultaneously is

(A) 19.67 (B) 21.50

(C) 40.58 (D) 80.67

Q. 72 At an engine throttle position of 75%, the high idle speed of the engine is shifted by 200 rpm towards the maximum torque position. If the engine is maintaining a uniform speed of 2475 rpm at a given load, the governor regulation is

(A) 8.42 % (B) 8.10 %

(C) 7.77 % (D) 3.88 %

Q. 73 The torque exerted on the crankshaft of a two stroke engine is given by the equation

$T(N\ m) = 450 + 30 \sin 2\theta - 90 \cos 2\theta$

Where is the crank angle displacement from the inner dead centre. If the resisting torque is constant, the power developed by the engine at a speed of 1500 rpm is

(A) 20.50 kW (B) 35.30 kW

(C) 70.69 kW (D) 135.00 kW

Q. 74 A single plate dry type clutch is to be designed for a tractor engine to transmit its maximum torque with the following data.

The torque developed by the engine at governor's maximum = 125 N m; the engine torque reserve capacity = 20 percent; coefficient of friction = 0.3; maximum facing pressure = 0.1 MPa. Considering uniform pressure, if the outer diameter of the plate is 1.5 times the inner diameter, the outer diameter of the plate will be

(A) 165.38 mm (B) 224.46 mm

(C) 238.50 mm (D) 300.52 mm

Q. 75 In an epicyclic gear train, an arm carries two wheels A and B having 24 teeth and 30 teeth respectively. If the arm rotates at 100 rpm in the clockwise direction about the centre of the wheel. A which is fixed, then the speed of wheel B on its own axis is

(A) 20 rpm, anti-clockwise (B) 25 rpm, anti-clockwise

(C) 180 rpm, clockwise (D) 225 rpm, clockwise

Graduate Aptitude Test in Engineering - 2009

Q. 76 As per BIS standard, the power tests for tractor PTO includes

(A) Maximum power, varying load and varying speed tests

(B) Varying speed and maximum power tests

(C) Varying load and varying speed tests

(D) Varying load and maximum power tests

Q. 77 The high idle speed of an engine is 2240 rpm and the governor regulation is 11.5%. The peak torque of 180 N m occurs at 1450 engine rpm. If lugging ability is 28 Nm, the engine power is kW at governor's maximum position will be

(A) 31.8 (B) 35.7

(C) 37.6 (D) 42.2

Q. 78 The final drive system of a tractor comprised a planetary gear drive. The ring gear has 70 teeth and is held stationary. Power comes into the sun gear which has 34 teeth and rotates clockwise at 100 rpm. Power comes out of the gear set on the planet carrier which drives the rear axle shaft. The speed of the rear axle in rpm is

(A) 32.7 (B) 47.0

(C) 54.4 (D) 71.4

Q. 79 In a single V-belt drive having 30° groove angle, the cross-sectional area of the belt is 250 mm^2 and the mass density of the belt material is 1200 kg m^{-3}. The maximum stress bearing capacity of the belt material is 6000 kN m^{-2}. Coefficient of friction between the belt and the pulley is 0.2 and the angle of lap on the driving pulley is 150°. The maximum power transmitted through the V-belt in kW is.

(A) 5.6 (B) 10.5

(C) 17.7 (D) 35.4

Graduate Aptitude Test in Engineering - 2010

Q. 80 In a tractor differential, the pinion on the propeller shaft has 12 teeth and the crown gear has 60 teeth. The propeller shaft rotates at 1000 rpm

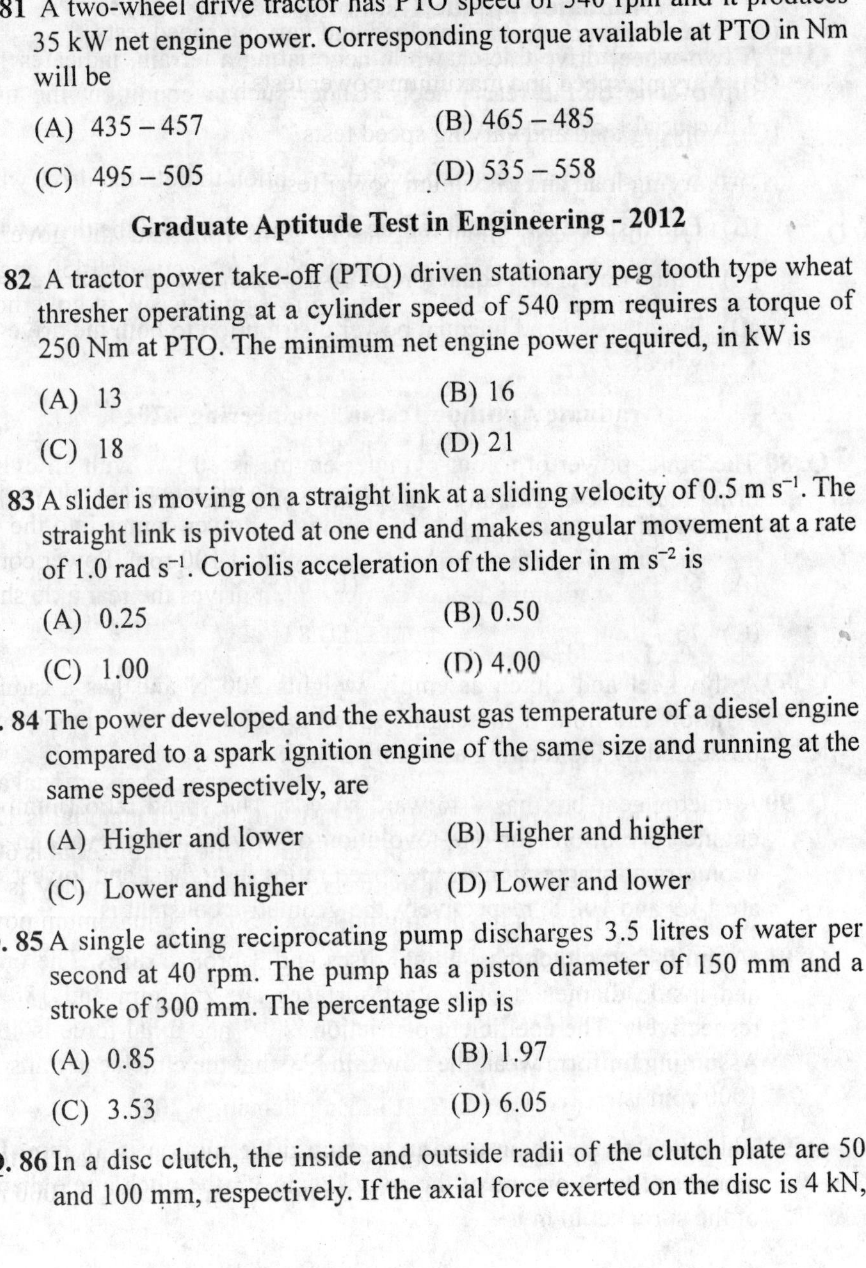

and the right rear axle rotates at 210 rpm while taking a left turn. The rotation of the left rear axle is rpm will be

(A) 170 (B) 180

(C) 190 (D) 200

Q. 81 A two-wheel drive tractor has PTO speed of 540 rpm and it produces 35 kW net engine power. Corresponding torque available at PTO in Nm will be

(A) 435 – 457 (B) 465 – 485

(C) 495 – 505 (D) 535 – 558

Graduate Aptitude Test in Engineering - 2012

Q. 82 A tractor power take-off (PTO) driven stationary peg tooth type wheat thresher operating at a cylinder speed of 540 rpm requires a torque of 250 Nm at PTO. The minimum net engine power required, in kW is

(A) 13 (B) 16

(C) 18 (D) 21

Q. 83 A slider is moving on a straight link at a sliding velocity of 0.5 m s^{-1}. The straight link is pivoted at one end and makes angular movement at a rate of 1.0 rad s^{-1}. Coriolis acceleration of the slider in m s^{-2} is

(A) 0.25 (B) 0.50

(C) 1.00 (D) 4.00

Q. 84 The power developed and the exhaust gas temperature of a diesel engine compared to a spark ignition engine of the same size and running at the same speed respectively, are

(A) Higher and lower (B) Higher and higher

(C) Lower and higher (D) Lower and lower

Q. 85 A single acting reciprocating pump discharges 3.5 litres of water per second at 40 rpm. The pump has a piston diameter of 150 mm and a stroke of 300 mm. The percentage slip is

(A) 0.85 (B) 1.97

(C) 3.53 (D) 6.05

Q. 86 In a disc clutch, the inside and outside radii of the clutch plate are 50 and 100 mm, respectively. If the axial force exerted on the disc is 4 kN, the maximum pressure experienced by the clutch plate in N mm^{-2} under uniform wear conditions is

(A) 0.13 (B) 0.17

(C) 0.25 (D) 0.51

Graduate Aptitude Test in Engineering - 2013

Q. 87 A two-wheel drive tractor, while negotiating a terrain, indicates 100% slip of one of the rear wheels. Under such a condition, the use of differential lock causes

(A) Equal speed and equal power distribution to both the drive wheels

(B) Equal speed and equal torque distribution to both the drive wheels

(C) Equal power and equal torque distribution to both the drive wheels

(D) Equal speed and unequal power distribution to both the drive wheels

Graduate Aptitude Test in Engineering - 2014

Q. 88 The brake power of a four-cylinder engine is 30 kW with all cylinder firing and 20 kW with any one cylinder cut. The mechanical efficiency of the engine in percent is

(A) 60 (B) 67

(C) 75 (D) 83

Q. 89 A flywheel and clutch assembly weights 200 N and has a radius of gyration 150 mm. If the engine speed is 3000 rpm, the kinetic energy possessed by the rotating assembly in kJ is ....

Q. 90 A tractor gear box has 8 forward speeds. The speed ratio (number of engine revolutions for one revolution of driving wheel) vary in exact geometrical progression. If the speed ratios in highest and lowest gears are 14.9 and 108.8, respectively, the geometric constant is ....

Q. 91 A multi-disc clutch has 4 steel discs and 3 bronze discs. The outside and inside diameters of contact surfaces are 250 mm and 180 mm, respectively. The coefficient of friction is 0.3 and axial force is 400 N. Assuming uniform wear, the power in kW that the clutch can transmit at 1000 rpm is ....

Q. 92 The pitch of the chain used in a chain drive motion is 38 mm. If the number of teeth on one of the sprockets is 35, the pitch circle diameter of the sprocket in m is ....

Graduate Aptitude Test in Engineering – 2015

Q. 93 A multiple disc clutch is to transmit 15 kW at 750 rpm. The inner and outer radii of the friction surfaces are 60 mm and 100 mm, respectively. The coefficient of friction is 0.1 and the maximum allowable pressure is 350 kPa. Assuming uniform wear, the number of pair of contact surfaces required is ....

Q. 94 The speed reductions in the 1st low gear of a tractor gearbox and differential with final drive are 5 : 1 and 40 : 1, respectively. For the tractor developing 24 kW power at an engine rpm of 2000 with an overall power transmission efficiency of 80%, the total torque in kN m available at the wheel axle will be

(A) 22.92 (B) 28.64

(C) 16.54 (D) 18.33

Graduate Aptitude Test in Engineering – 2016

Q. 95 A force of 8.0 kN is applied perpendicularly to the axis of a crankpin having circular cross- sectional area. The allowable shear stress of the crankpin material is 40.0 N mm^{-2}. If the crankpin fails under double shear, the design diameter of the crankpin in mm is ....

Q. 96 The ASAE-SAE standard for tractor 3-point hitches has been categorized as Category I to IV on the basis of

(A) Maximum drawbar power (B) Maximum drawbar pull

(C) Brake power of tractor engine (D) Maximum PTO power

Q. 97 As per ASAE standards, the diameter of 1000 rpm-PTO shaft with 20 splines is

(A) 30 mm (B) 35 mm

(C) 40 mm (D) 45 mm

Q. 98 In restrained link operation of three point hitches, the line of pull passes

(A) Through the virtual hitch point and bending force exists on lower links

(B) Above the virtual hitch point and bending force exists on lower links

(C) Below the virtual hitch point and no bending force exists on lower links

(D) Through the virtual hitch point and tensile force exists on lower links

Q. 99 While testing a wheat thresher at the recommended throughput, 80 N m torque is recorded at 750 rpm at the main shaft of the threshing cylinder, which is operated by a 200 mm diameter v-pulley. The overload factor and unit mass of the v-belt are 1.2 and 0.9 kg m^{-1}, respectively. At the condition of maximum power transmission, the maximum tension in the v-belt in N is ....

Q. 100 A tractor PTO operated 4-disc rotary mower is harvesting with a forward speed of 3.5 km h^{-1}. The cutting circle diameter and rpm of each disc are 60 cm and 1400, respectively. The peak cutting force experienced by each rotary disc is 110 N. The peak overall motion resistance is found to be 3.6 kN. If the overall power conversion efficiency of the tractor is 82%, required peak engine brake power in kW will be ....

Q. 101 In a tractor rear axle differential, the bevel pinion has 12 teeth and bevel/crown gear has 42 teeth. The input speed and torque of the bevel pinion are 520 rpm and 1200 N m, respectively. There is a planetary gear between the differential unit and each half axle with 4 : 1 speed reduction. The left wheel encounters poor traction when the tractor is moving in straight path that causes 15% drop in the left axle torque. If the differential efficiency is 0.98, the right axle torque under locked differential condition in N m will be ....

Graduate Aptitude Test in Engineering – 2017

Q. 102 In a tractor power transmission, the input pinion (24 teeth) is in mesh with a gear(46 teeth) on counter shaft and another gear (20 teeth) of counter shaft is in mesh with main shaft gear (50 teeth). The engine is running at 1800 rpm, differential gear ratio is 3.5:1 and final drive ratiois 4:1. If the tractor is fitted with 1.2 m diameter rear wheels, the forward speed of the tractor in km h^{-1} is ....

Graduate Aptitude Test in Engineering – 2018

Q. 103 For a two-wheel drive tractor, use of differential lock in adverse field condition ensures

(A) Equal distribution of torque to both the drive wheels.

(B) Equal distribution of power to both the drive wheels.

(C) Distribution of higher amount of torque to the wheel under poor traction.

(D) Distribution of higher amount of torque to the wheel under better traction.

Q. 104 An epicyclic gear train consists of a sun gear, four planet gears pinned on a planet carrier (which is mounted on the sun gear) and a fixed ring gear (in which the entire sun-planet gear assembly is inserted). If the ring gear has 56 teeth and each planet gear has 18 teeth, the number of teeth on the sun gear is ....

Q. 105 An open belt drive system transmits 5 kW power using a flat belt of width and thickness as 80 mm and 5 mm, respectively. The driving shaft speed is 1500 rpm and the driven shaft speed is 500 rpm. The coefficient of friction between the belt and pulley is 0.20, and the wrap angle of the belt is 168°. If diameter of the smaller pulley is 200 mm, maximum stress induced in the belt will be .... MPa.

Q. 106 A single plate dry clutch transmits 15 kW power at 1200 rpm. The clutch sustains a maximum axial load of 2.65 kN. The ratio of outer to inner diameter of the frictional surface is 1.25:1. Considering uniform wear with a coefficient of friction of 0.3 on both the frictional surfaces, the outer diameter of the clutch plate is .... mm.

Graduate Aptitude Test in Engineering – 2019

Q. 107 A pair of straight teeth spur gears is transmitting power at 500 rpm. The pinion has 16 standard full depth involute teeth of module 8 mm. The pitch line velocity of the pinion in m/s is ....

Q. 108 Head pulley of a bucket elevator has an effective radius of 150 mm. In order to obtain the most satisfactory discharge from this elevator, the speed of the head pulley in rpm is

(A) 36 (B) 44

(C) 50 (D) 77

Graduate Aptitude Test in Engineering – 2020

Q. 109 A tractor PTO operated rotary disc mower has 4 rotating discs and the width of cut of each disc is 60 cm without any overlap. The specific power losses to air, stubble and gear-train friction is 2.5 kW per meter of cutting width. The specific cutting energy requirement is 2.0 kJ /m^2. If the machine is operated at a forward speed of 6 km/h, the PTO power requirement in kW is ....

Q. 110 A V-belt drive transmits 10 kW power at a belt velocity of 8 m/s. The angle of contact on the smaller pulley is 170° and groove angle of the pulley is 38°. The coefficient of friction between the pulley and the belt is 0.28 and the maximum permissible stress of the belt is limited to 4 MPa. Neglecting centrifugal effect of the belt, the minimum cross-sectional area of the V-belt in mm^2 is ....

Graduate Aptitude Test in Engineering – 2021

Q.111 A tractor PTO driven rotavator with a rotor radius 30 cm has 20 L-shaped blades each of width 12 cm. These blades are fixed at a radial distance of 7 cm from the center of the rotor shaft to the brackets attached to the rotor shaft. When this rotavator is operated at a forward speed of 4.5 km/h and at a depth of 12 cm, the resultant soil force of 150 N tangential to the rotor circumference acts at the middle of the blade width. The torsional moment acting on the blade in N/m is...........

Q. 112 Two meshed involute gears transmit 1.0 kW power. The pressure angle is 20° and the pitch circle diameter of the large gear rotating at a speed of 600 rpm is 20 cm. If only a pair of teeth meshes at a time, the normal force acting between the meshed teeth in N will be ....

(Take $\pi = 3.14$)

Answers Key

1	2	3	4	5	6	7	8	9	10
D	C	A	B	A	C	C	D	A	C
11	12	13	14	15	16	17	18	19	20
A	A	D	D	B	D	C	A	D	A
21	22	23	24	25	26	27	28	29	30
A	A	A	C	C	A	B	B	A	C
31	32	33	34	35	36	37	38	39	40
B	C	B	B	B	B	B	C	A	C
41	42	43	44	45	46	47	48	49	50
A	C	A	B	C	D	A	D	A	A
51	52	53	54	55	56	57	58	59	60
A	B	A	B	A	D	C	A	A	C
61	62	63	64	65	66	67	68	69	70
C	D	A	B	B	C	C	B	D	A
71	72	73	74	75	76	77	78	79	80
B	A	C	B	C	A	A	A	D	C
81	82	83	84	85	86	87	88	89	90
D	B	C	D	A	C	D	C	22.34	0.752
91	92	93	94	95	96	97	98	99	100

8.1	0.424	3	D	11.29	A	D	B.	166.38	27.73
101	102	103	104	105	106	107	108	109	110
9325	6.07	D	20	1.795	166.92	3.35	D	14	339
111	112								
9	169.46								

Explanations

Q. 5

Standard types of Power Take-Off (P.T.O.) are such as

PTO Type	Type 1	Type 2	Type 3
Diameter	1 3/8" (35 mm)	1 3/8" (35 mm)	1 3/4" (45 mm)
Speed	540 RPM	1000 RPM	1000 RPM
Gear teeth	6	21	20
Rotation	Clockwise, as viewed from end of shaft(behind the tractor)		

The first industry standard for PTO design was adopted by ASAE (the American Society of Agricultural Engineers). The PTO rotational speed was specified as 536 ± 10 rpm; the direction was clockwise. The speed was later changed to 540 rpm.

Q. 38

Velocity ratio = diameter of driven pulley/ diameter of driving pulley

$= 20/30 = 0.67$

Q. 66

Belt cross section (A) $= 9\times250 = 2250\ mm^2 = 2.25\times10^{-3}\ m^2$

Linear speed of belt (V) $= \pi DN$

$= 3.14\times0.90\times336 = 949.536\ m/min = 15.82\ m/s$

Mass of the belt per unit length of belt (m) $= 980\times2.25\times10^{-3} = 2.21$ kg

We know

Centrifugal tension $(T_c) = mV^2$

$T_c = 2.21\times(15.82)^2 = 553$ N 552.8 N

Q. 67

Shear strength of shear pin material $(\sigma) = 310$ MPa $= 310\times10^6$ Pa

Shear pin diameter (D) = 2.38 mm = 0.00238 m

Speed of shaft (N) = 650 rpm = 10.83 rps

Power transmitted by shaft (P) = 4.5 kW = 4500 W

Consider radial distance of shear pin centre from centre of shaft = R m

We know that,

Power transmitted by shaft(P) = $4.936 \times d^2 \times \sigma \times R \times N$

$\Rightarrow \quad 4500 = 4.936 \times (0.00238)^2 \times 310 \times 10^6 \times R \times 10.83$

$\Rightarrow \quad R = 0.04794 \text{ m} = 47.94 \text{ mm}$

Q. 68

According to ASBAE, Tractor categorization is given as:

Category	Tractor drawbar power
0	<15 kW (<20 hp)
1	15-35 kW (20-45 hp)
2	30-75 kW (40-100 hp)
3	60-168 kW (80-225 hp)
4	135-300 kW (180-400 hp)

Q. 69

Break horse power (BHP) = 7.5×0.80 = 6 kW

Break thermal efficiency (BTE) = Mechanical efficiency × Indicated thermal efficiency

$$BTE = 0.80 \times 0.35 = 0.28$$

We know

$$BTE = \frac{\text{Break Horse Power}}{\text{Fuel consumption} \times \text{Calorific Value}}$$

$$\Rightarrow \quad 0.28 = \frac{6 \times 1000}{\text{Fuel consumption} \times 45000000}$$

$\Rightarrow \quad$ Fuel consumption = 4.76×10^{-4} kg/ s = 1.714 kg/ h

$\therefore \quad$ Specific fuel consumption = Fuel consumption/BHP

$\Rightarrow \quad SFC = 1.714/6 = 0.286 \text{ kg kW}^{-1} \text{ h}^{-1}$

Q. 71

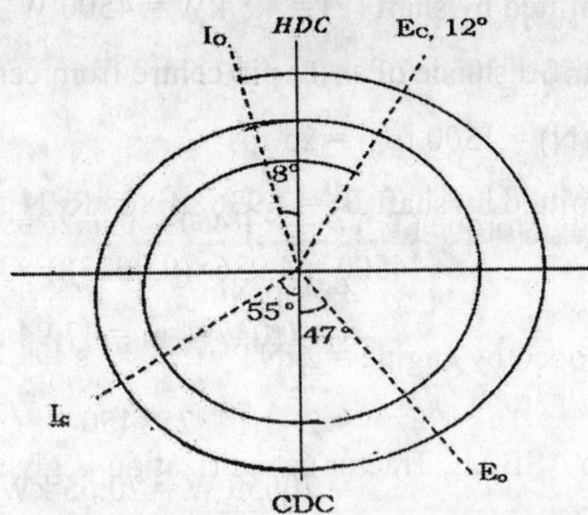

Inlet valve closes at 55° after CDC (I_C) and remains close till I_O.

∴ Inlet valve remains close for 35 + 360 + 82 = 477° (I_C to I_O)

Exhaust valve remain close for 78 + 43 = 121° (E_C to E_O)

But both valves remains close simultaneously for

(I_C to E_C) + (E_C to E_O) = 35 + 82 + 8 + 12 + 121 = 258

We know that

Time for both valve closing

$$= \frac{\theta}{360} \times \frac{1}{N}$$

$$= \frac{258}{360} \times \frac{60}{2000} = 0.0215 \text{ seconds} = 21.5 \text{ milli-seconds}$$

Q. 72

Engine minimum or no load speed (N_0) = 2475 rpm

In maximum torque condition, engine speed is decreased by 200 rpm.

Hence, Engine speed (minimum) at full load(N) = 2475 – 200 = 2275 rpm

$$\therefore \quad \text{Governor regulation(R)} = \frac{N - N_0}{\bar{N}} \times 100$$

$$= \frac{N - N_0}{(N + N_0)/2} \times 100$$

$$= \frac{2475 - 2275}{(2475 + 2275)/2} \times 100$$

$$= 8.42 \%$$

Q. 73

Given that

Torque (T) = 450 + 30 sin2θ - 90 cos2θ Nm

Engine speed (N) = 1500 rpm = 25 rps

$$\therefore \quad \text{Average torque } (T_{av}) = \frac{1}{\pi} \times \int_0^{\pi} (450 + 30\sin 2\theta - 90\cos 2\theta)\, d\theta$$

$$= 450 \text{ Nm}$$

Power developed by engine = 2πNT

$$= 2 \times 3.14 \times 25 \times 450$$

$$= 70650 \text{ W} = 70.65 \text{ kW}$$

Q. 74

Inner radius = r; Outer radius (R) = 1.5 r

n = 2 (∵ For single plate clutch, both sides of the clutch are effective.)

Torque (T) = 125×0.8 = 100 N-m (20 % torque reserved.)

$$\text{Mean radius } (r_m) = \frac{2}{3}\left(\frac{R^3 - r^3}{R^2 - r^2}\right)$$

$$= \frac{2}{3}\left(\frac{(1.5r)^3 - r^3}{(1.5r)^2 - r^2}\right) = 1.27 \text{ r}$$

We know

Torque (T) = μ.n.W.r_w

$$= \mu \times 2 \times 2\pi P_{max} \times r \times (R - r) \times r_w \quad (\because W = 2\pi P_{max}.r.(R-r))$$

Now,

$$\Rightarrow \quad 100 = 0.3 \times 2 \times 2 \times 3.14 \times 100000 \times r \times 0.5r \times 1.27r$$

$$\Rightarrow \quad r = 0.0748 \text{ m}$$

Hence,

$$D = 2R$$

$$= 2 \times 1.5 \times r = 0.2244 \text{ m} = 224.4 \text{ mm}$$

Q. 75

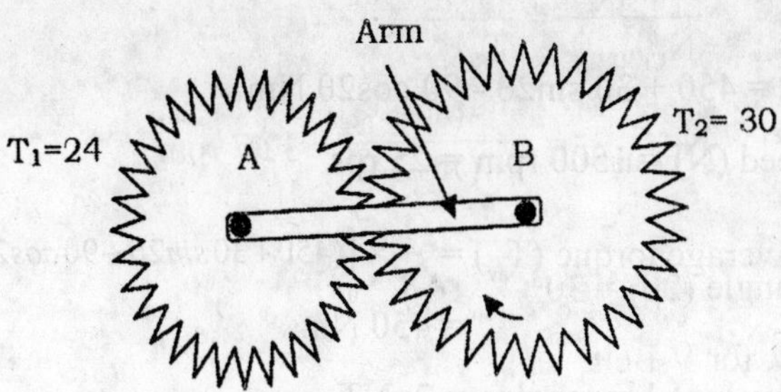

In this type of gear mechanism gear B rotates in the direction of rotating arm.

Rpm of gear $(N_B) = N_{ARM}(1+T_A/T_B)$

$= 100(1 + 24/30)$

$= 180$ rpm, clockwise

Q. 77

We know,

$$\text{Governor regulation} = \frac{2(N_{max} - N_{min})}{(N_{max} + N_{min})} = 0.115$$

$$\Rightarrow \quad \frac{2(2240 - N_{min})}{(2240 + N_{min})} = 0.115$$

$\Rightarrow \quad N_{min} = 1996.41$ rpm $= 33.274$ rps

$\because$ Effective torque $= 180 - 28 = 152$ N-m

$\therefore$ Power developed $= 2\,\pi\,NT$

$= 2\times3.14\times33.274\times152 = 31762$ W $= 31.8$ kW

Q. 78

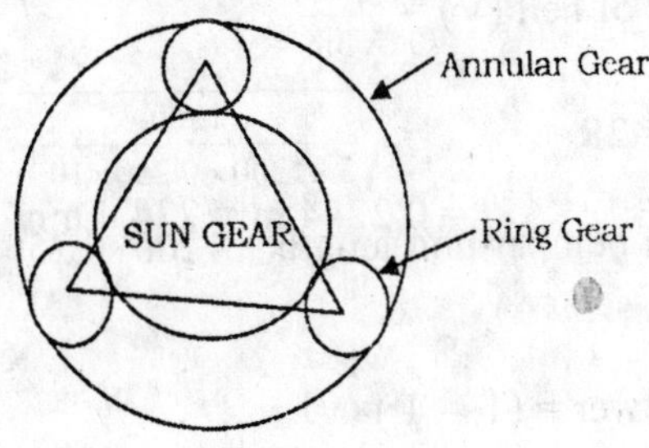

We know

$$\frac{N_{SUN}}{N_{CARRIER}} = \frac{T_{RING}}{T_{SUN}}$$

$$N_{CARRIER} = \frac{100}{1+(70/34)} = 32.7 \text{ rpm}$$

Q. 79

Groove angle (2β) = 30°

We know, for V-Belt

$$\ln\left(\frac{T_t}{T_s}\right) = \mu\theta \text{ cosec } \beta$$

$$\Rightarrow \ln\left(\frac{T_t}{T_s}\right) = 0.2\times2.62\times\text{cosec}15 \quad (\because 150^o = 2.62 \text{ rad})$$

$$\Rightarrow \ln\left(\frac{T_t}{T_s}\right) = 2.205$$

$$\Rightarrow \frac{T_t}{T_s} = 7.6 \qquad \text{(i)}$$

Maximum tension in belt (T) = $6000\times10^3\times200\times10^{-6}$ N

$$T = 1500 \text{ N}$$

$\because$ Centrifugal tension (T_c) = T/3

$$\Rightarrow \quad T_c = 500 \text{ N}$$

Also $\quad T_t + T_c = T$

$$\Rightarrow \quad T_t = 1000 \text{ N}$$

From equation (i)

$$T_s = T_t/7.6 = 1000/7.6 = 131.58 \text{ N}$$

We also know

Linear velocity of belt (V) = $\sqrt{\frac{T}{3m}}$

$$= \sqrt{\frac{1500}{3\times1200\times0.250\times10^{-6}}} = 40.825 \text{ m/s}$$

($\because$ m = mass of belt per unit length = $1200\times0.250\times10^{-6}$ kg)

Therefore,

$$\text{Power} = (T_t - T_s)\times V$$

$$= (1000 - 131.58)\times40.825 = 35.45 \text{ kW}$$

Q. 80

As we know

$$N_1T_1 = N_2T_2$$

Thus, $T_2 = 12\times1000/60 = 200$ rpm

Here, right axle rotates at 210 rpm, therefore left axle will rotate at 190 rpm.

Q. 81

PTO speed (N) = 540 rpm; Engine power produced (P) = 35 kW

We know PTO efficiency of a tractor is 88.3 % (Approx.).

Thus power available at PTO will be = 3.5×0.883 = 30.905 kW

We know

$$P = 2\pi NT/60$$

$$\Rightarrow \quad 30.905\times1000 = 2\times3.14\times540\times N/60$$

$$\Rightarrow \quad N = 546.8 \text{ Nm}$$

Thus torque available at PTO will be in between 535 – 558.

Q. 82

Power required to operate the thresher

$$(P) - 2\,\pi\,NT$$

$$= 2\times3.14\times\frac{540}{60}\times250 = 14130 \text{ W} = 14.13 \text{ kW}$$

$\because$ PTO efficiency = 88.3%

$\therefore$ Engine power required = 14.13/0.883 = 16 kW

Q. 83

Given

Sliding velocity (v) = 0.5 m/ s; Angular velocity (ω) = 1.0 rad/ s

Applying formula

$$\text{Coriolis acceleration} = 2\,\omega v$$

$$= 2\times1\times0.5 = 1 \text{ m/ s}^2$$

Q. 85

Actual discharge (Q_a) = 3.5 L/s

Theoretical discharge (Q_t) = $\frac{\pi}{4}$ d^2hN

$$= \frac{3.14}{4} \times (0.15)^2 \times 0.3 \times \frac{40}{60}$$

$$= 0.0035325 \text{ m}^3/\text{s} = 3.53 \text{ L/s}$$

∴ Percentage slip $\frac{Q_t - Q_a}{Qt} = \frac{3.53 - 3.5}{3.53} \times 100 = \times 100 = 0.85\%$

Q. 86

We know

Axial force (W) = $2\pi P_{max}.r.(R-r)$

⇒ $4000 = 2 \times 3.14 \times P_{max} \times 50 \times (100-50)$

⇒ $P_{max} = 0.25 \text{ N/mm}^2$

Q. 88

Indicated horse power(IHP), during 1st cylinder cut (I_1) = 30 – 20 = 10 kW

Similarly, I_2 = 30-20 = 10 kW

And I_3 = 30-20 = 10 kW

And I_4 = 30-20 = 10 kW

Thus, total IHP = 10 + 10 + 10 + 10 = 40 kW

Mechanical Efficiency = $\frac{BHP}{IHP} \times 100$

$$= \frac{30}{40} \times 100 = 75\ \%$$

Q. 89

Mass (m) = $\frac{\text{Weight}}{g} = \frac{200}{9.8} = 20.41$ Kg

Kinetic Energy(K.E.) = $\frac{1}{2} I\omega^2 = \frac{1}{2} mK^2 \left(\frac{2\pi N}{60}\right)^2$

⇒ $\frac{1}{2} \times 20.41 \times (0.15)^2 \times \left(\frac{2 \times 3.14 \times 3000}{60}\right)^2 = 22.34$ kJ

Q. 90

n^{th} term of a Geometric Progression $(T_n) = a(r)^{n-1}$

$$14.9 = 108.8 \times (r)^7$$

$$\Rightarrow \quad r = 0.752$$

Q. 91

$Z = (3+4)-1 = 6$, $D_o = 250$ mm $= 0.25$ m, $D_i = 0.18$m

$\mu = 0.3$, $W = 400$ N, $N = 1000$ rpm

$$\text{Uniform wear} \Rightarrow r_m = \frac{D_0 + D_i}{4} = 0.1075 \text{ m}$$

$$\text{Torque (T)} = n\mu W r_m$$

$$= 6 \times 0.3 \times 400 \times 0.1075 = 77.4 \text{ Nm}$$

$$\therefore \quad \text{Power (P)} = \frac{2\pi NT}{60}$$

$$= \frac{2 \times 3.14 \times 77.4 \times 1000}{60} = 8101.2 \text{ W} = 8.1 \text{ kW}$$

Q. 92

$$\text{Pitch Diameter (PD)} = \frac{P}{sin(180/N)}$$

$$= \frac{30}{sin(180/35)} = 424 \text{ mm} = 0.424 \text{ m}$$

Q. 93

$D_o = 200$ mm $= 0.20$ m; $D_i = 0.12$ m; $\mu = 0.1$, $N = 750$ rpm

$$\text{Uniform wear} \Rightarrow r_m = \frac{D_0 + D_i}{4} = \frac{0.20 + 0.12}{4} = 0.08 \text{ m}$$

$$\therefore \quad \text{Power (P)} = \frac{2\pi NT}{60}$$

$$15 = \frac{2 \times 3.14 \times T \times 750}{60}$$

$$\Rightarrow \quad T = 0.1911 \text{ kN m}$$

Again we know,

Torque (T) $= n\, \mu\, W r_m$

$$\because \quad \text{Axial load (W)} = 2\pi p_{max} (R_o - R_i) R_i$$

$\therefore$ Torque (T) = $n\,2\mu p_{max}(R_o - R_i)R_i r_m$

$0.1911 = n\times0.1\times2\times3.14\times350\times(0.1-0.06)\times0.06\times0.08$

$n = 4.51 \approx 4$

$\because$ For n number of discs or plates, there must be n – 1 pairs of contact surfaces.

Hence, Number of contact surfaces is = n – 1 = 4 – 1 = 3

Q. 94

Engine speed (N) = 2000 rpm

Hence,

Speed at 1st low gear = 2000×(1/5) = 400 rpm ($\because$ Speed reduction of gearbox.)

Speed of wheel axle = 400×(1/40) = 10 rpm ($\because$ Speed reduction due to final drive.)

Power available at wheel axle (P) = 24×0.80 = 19.2 kW

($\because$ Overall power transmission efficiency = 80%)

We know,

$$\text{Power (P)} = \frac{2\pi NT}{60}$$

$$\Rightarrow \quad 19.2 = \frac{2\times3.14\times10\times T}{60}$$

$$\Rightarrow \quad T = 18.34 \text{ kN-m}$$

Q. 95

Consider, diameter of the crankpin = d m

Here, crankpin fails under double shear hence effective stress will be 2×40 = 80 N/mm²

We know,

Axial Force = shear stress × Cross sectional area

$$\Rightarrow \quad 8000 = 80\times3.14\times(d/2)^2$$

$$\Rightarrow \quad d = 11.29 \text{ mm}$$

Q. 97

Basic Specifications of rear PTO, as standardized by ASAE-SAE.

PTO TYPE	1	2	3
Direction of rotation	Clockwise viewed from behind		
Nominal speed, rpm	540	1000	1000
Nom. diameter, mm	35	35	45
Number of splines	6	21	20
Spline profile	Straight	Involute	Involute

Q. 99

Linear speed of belt (V) = πDN

$= 3.14 \times 0.2 \times (750/60) = 7.85$ m/sec

For maximum power transmission

Maximum tension (T_{max}) = $3mV^2$

$= 3 \times 0.9 \times (7.85)^2$

$= 166.38$ N

Q. 100

Torque developed in rotory disc shaft (T) = Force×Radius

$= 110 \times 0.30 \times 4 = 132$ N-m

Power required to drive rotary discs = $2\pi NT$

$= 2 \times 3.15 \times (1400/60) \times 132$

$= 19.34$ Kw

Power consumed against motion resistance = Speed×Motion Resistence

$= (3500/3600) \times 3.5 = 3.40$ kW

Hence, Total Power consumed = 19.34 + 3.40

= 22.74 kW

$\because$ Overall efficiency = 0.82

Hence, Engine brake power = 22.74/0.82

= 27.73 kW

Q. 102

Speed of input shaft of differential gear = 1800×(24/46)×(20/50) = 375.65 rpm

Rear wheel axle speed = 375.65/(3.5×4) = 26.83 rpm

(∵ Differential gear ratio = 3.5:1 and Final Drive gear ratio = 4:1)

Hence, Forward speed of rear wheel

$= \pi DN$

$= 3.14 \times 1.2 \times (26.83 \times 60)$

= 6065.73 m/h or 6.07 km/h

Q. 104

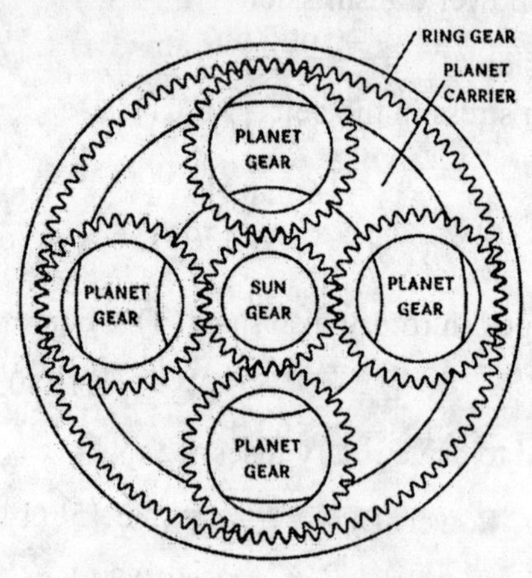

For epyciclic gear train

$T_a = T_s + 2T_p$

Where,

T_a = number of teeth on ring gear

Ts = number of teeth on sun or centre gear

T_p = number of teeth on planet gear

Hence, $56 = T_s + 2 \times 18$

$\Rightarrow$ $T_s = 20$

Q. 105

We know, for flat-Belt

$\ln (Tt/Ts) = \mu\theta$

$\Rightarrow$ $\ln (Tt/Ts) = 0.2\times2.931$ ($\because$ $168° = 2.931$ rad)

$\Rightarrow$ $\ln (Tt/Ts) = 0.5862$

$\Rightarrow$ $(Tt/Ts) = 1.797 \Rightarrow Tt = 1.797\ Ts$ (i)

Velocity of belt (v) = $3.14\times0.2\times1500/60 = 15.7$ m/s

$\because$ Power = $(Tt - Ts)\times V$

$\Rightarrow$ $5000 = (Tt - Ts)\times15.7$

$\Rightarrow$ $1.797\ Ts - Ts = 318.47$

$\Rightarrow$ $Ts = 399.59$ N and $Tt = 1.797\times399.59 = 718.06$ N

$\therefore$ Maximum stress in the belt = Tt/ Cross section area of belt

$= 718.06/(0.080\times0.005)$

$= 1795150\ N/m^2$ or 1.795 MPa

Q. 106

$\mu = 0.3$, Power = 15 kW = 15000 W, N = 1200 rpm;

Axial Load = 2.65 kN = 2650 N

$\because$ $\text{Power (P)} = \dfrac{2\pi NT}{60}$

$\Rightarrow$ $15000 = \dfrac{2\times3.14\times T\times1200}{60}$

$\Rightarrow$ NT = 119.43 Nm

We know, for Uniform wear

$r_m = (R_0+R_i)2$

Here, $Ro/Ri = 1.25 \Rightarrow r_m = (R_0+R_0/1.25)/2 = 0.9\ R_0$

Again

Torque (T) = $n\mu Wr_m$ ($\because$ n = 2 for single plate clutch)

$\Rightarrow$ $119.43 = 2\times0.3\times2650\times0.9\ R_0$

$\Rightarrow$ $R_0 = 0.08346$ m = 83.46 mm

Hence, Outer Diameter = $2\times83.46 = 166.92$ mm

Q. 107

Module = 8 = Pitch dia./ No. of teeths

⇒ Pitch dia. = 8×16 = 128 mm or 0.128 m

Pitch line velocity = πDN/60

= 3.14×0.128×500/60 = 3.35 m/s

Q. 108

Speed of head pulley = V/(2πr) = $\sqrt{gr}$ /(2πr)

=√(9.81×0.15)/(2×3.14×0.15)

= 1.28 rps or 77.26 rpm

Q. 109

The power loss due to air, stubble and gear trains = 2.5×4×0.60 = 6 kW

Power used for cutting = 2×(6000/3600) ×4×0.60 = 8 kW

Total PTO power required = 6 + 8 = 14 kW

Q. 110

Groove angle (2β) = 38°

We know, for V-Belt

$$ln\left(\frac{T_t}{T_s}\right) = \mu\theta = \text{cosec}$$

$$\Rightarrow \quad ln\left(\frac{T_t}{T_s}\right) = 0.28\times2.97\times\text{cosec}19 (\because 170^\circ = 2.97 \text{ rad})$$

$$\Rightarrow \quad ln\left(\frac{T_t}{T_s}\right) = 2.554$$

$$\Rightarrow \quad \frac{T_t}{T_s} = 12.86 \qquad \text{(i)}$$

Maximum tension in belt (T) = 4000A kN

(A = cross sectional area of the v belt.)

∵ Centrifugal tension (T_c) = 0 kN

Also $T_t + T_c = T$

⇒ T_t = 4000A kN

From equation (i)

$$T_s = T_t/12.86 = 311.04A \text{ kN}$$

We also know

$$\text{Power} = (T_t - T_s)\times V$$

$$\Rightarrow \quad 10 = (4000A - 311.04A)\times 8$$

$$\Rightarrow \quad A = 3.39\times 10^{-4} \text{ m}^2 \text{ or } 339 \text{ mm}^2$$

Q. 111

Here,

Rotor diameter (r) = 30 cm or 0.30 m

Tangential force (K_s) = 150 N

Total Number of blades (N) = 20

The number of blades which action jointly on the soil (n_e) = 0.25 (It is assumed that 1/4 of total number of blades cut soil simultaneously).

We know

Soil force acting on each blade (K_e) $= \dfrac{K_s}{Nn_e}$

$$= \frac{150}{20\times 0.25} = 30 \text{ N}$$

$\Rightarrow$ Torsional Moment acting on each blade = F×r

$$= 30\times 0.30 = 9 \text{ N-m}$$

Q. 112

We know,

$$\text{Power (P)} = \frac{2\pi NT}{60}$$

$$\Rightarrow \quad 1000 = \frac{2\times 3.14\times 600\times T}{60}$$

$$\Rightarrow \quad T = 15.9236 \text{ N-m}$$

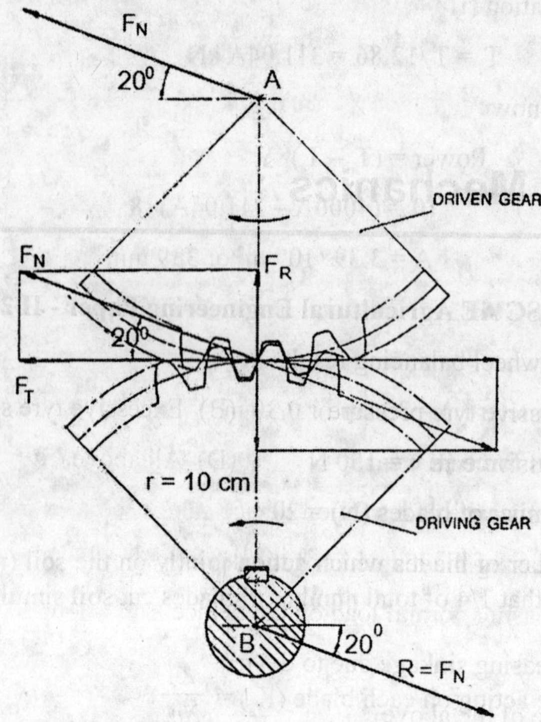

Referring above figure

Torque transmitted (T) = $F_T \times r$

$\Rightarrow \quad 15.92 = F_T \times 0.10$

$\Rightarrow \quad F_T = 159.236$ N

And

Force acting between two gears (F_N) = $F_T/\cos 20$

$\Rightarrow \quad F_N = 159.236/\cos 20 = 169.46$ N

11

Traction Mechanics

UKPSC AE Agricultural Engineering Paper - II 2013

Q. 1 Improper wheel balancing results in :-

(A) Excessive tyre pressure (B) Excessive tyre speed

(C) Excessive tyre wear (D) All above

Q. 2 Rolling resistance of a traction device can be reduced by:-

(A) Increasing the diameter of the device

(B) Increasing normal load on the device

(C) Increasing sinkage due to device

(D) None of the above

MPPSC Assistant Agricultural Engineer 2013

Q. 3 Shifting the centre of gravity of a tractor towards its front wheel may create the problem of ....

(A) Instability (B) Overturning

(C) Steering (D) None of these is correct

Q. 4 Pneumatic wheels have the advantage of .... over steel wheels.

(A) Less fuel consumption

(B) Less power required for same load

(C) Decreased rolling resistance

(D) None of these is correct

UKPSC Combined Junior Engineering Exam 2013 Agricultural Engineering Paper-I

Q. 5 Inflation pressure of front wheels of tractor (2 wheel drive)

(A) 0.8 – 1.2 kg/cm^2 (B) 1.2 – 2.0 kg/cm^2

(C) 2.0 – 2.5 kg/cm^2 (D) > 2.5 kg/cm^2

OCS AEn Exam 2011 (Pre) – II (Agricultural Engineering)

Q.6 The maximum drawbar power of two-wheel drive tractor is achieved at the wheel slip of

(A) 0 to 10 % (B) 10 to 20 %

(C) 20 to 30 % (D) 30 to 40 %

Q. 7 Tractive performance of a tractor may be improved by

(A) Ballasting front wheels (B) Ballasting rear wheels

(C) Reducing inflation pressure of traction wheels

(D) All of the above

CGPSC State Engineering Services Exam Paper – I

Q. 8 Ratio of the total force output of the traction device in the direction of travel to the dynamic weight on the traction device is known as

(A) Tractive efficiency (B) Draft

(C) Rolling resistance (D) Coefficient of traction

(E) Rim pull

UKPSC A.En. Exam 2012 Paper - II

Q. 9 In wet field the traction of tractor can be improved by using

(A) Bigger size of tyre (B) Additional weight

(C) Cage wheel (D) None of the above

Q. 10 In general the weight distribution on rear and front axle of an Indian tractor is

(A) 40 : 60 (B) 60 : 40

(C) 70 : 30 (D) 80 : 20

Q. 11 In tyre size 12.4 × 28, the number 12.4 represents

(A) Sectional diameter of tyre (B) Tyre pressure

(C) Ply rating of tyre (D) Rim diameter

Q. 12 The inflation pressure (kg/cm^2) in the rear wheels of the tractor for field operations vary between

(A) 1.3 – 1.7 (B) 2.2 – 2.6

(C) 1.8 – 2.1 (D) 0.8 – 1.2

Graduate Aptitude Test in Engineering - 2007

Q. 13 A 35 kW two-wheel drive tractor weighing 20 kN is fitted with 6-16 8PR tyre at the front axle and 13.6-28 12PR tyre at the rear axle. The ratio of section height and section width for all tyres is 0.75. The tractor has a wheel base of 2.1 m and the center of gravity is located 0.7 m ahead of the rear axle center on a horizontal plane. The tractor is to be towed on a level ground having sandy clay loam soil at 10% moisture content with a come index of 1200 kPa.

(i) The wheel numeric for each of the rear wheels is

(A) 39.50 (B) 58.17

(C) 79.01 (D) 116.37

(ii) Rolling resistance of each of the front wheels is

(A) 0.244 kN (B) 0.354 kN

(C) 0.575 kN (D) 0.707 kN

(iii) If the same tractor is to be towed on a level ground with compacted dry clay soil, the force required for towing is

(A) 0.27 kN (B) 0.40 kN

(C) 0.53 kN (D) 0.80 kN

Graduate Aptitude Test in Engineering - 2008

Q. 14 A rear wheel driven tractor weighing 20 kN has 40 percent of its weight supported by the front wheels. The tractor is pulling a trailed plough with a forward speed of 5 km/h on a flat land. The plough exerts a drawbar pull of 8.0 kN with the line of pull making an angle of 15 with the horizontal in the vertical plane. The drawbar hitch height is 500 mm.

(i) The coefficient of traction developed by the tractor for this operation is

(A) 0.15 (B) 0.49

(C) 0.50 (D) 0.56

(ii) If the wheel slip is 20 percent and the coefficient of rolling resistance is 0.04, the tractive efficiency of the tractor is

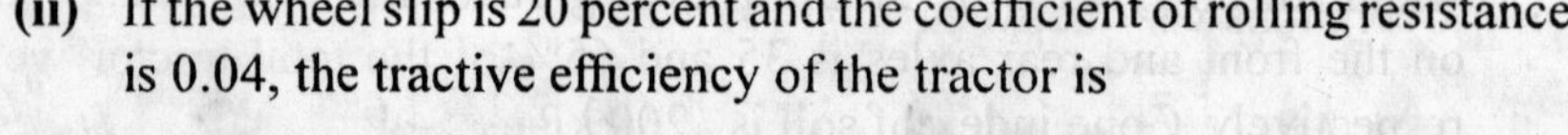

(A) 92.39 % (B) 87.17 %

(C) 73.96 % (D) 21.79 %

Q. 15 The nature of variation of tractive efficiency (TE) with wheel slip (S) in a rear wheel driven tractor is

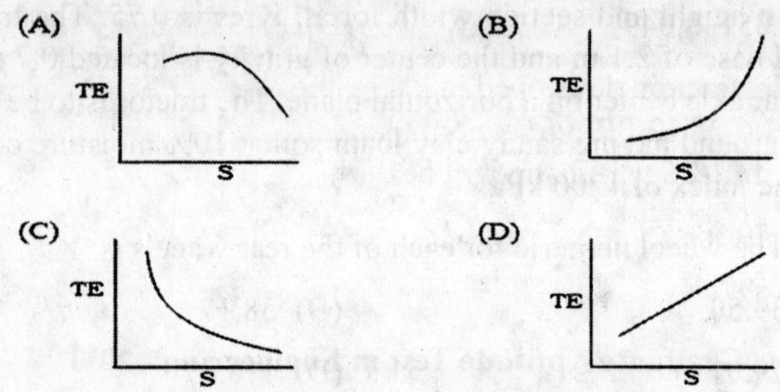

Graduate Aptitude Test in Engineering - 2009

Q. 16 The diameter of an undeflected tractor wheel fitted with 13.6-28, 12 PR tyre with an aspect ratio of 0.75 is

(A) 0.99 m (B) 1.05 m

(C) 1.23 m (D) 1.40 m

Q. 17 Two-wheel drive tractor is pulling a load of 12 kN horizontally on a leveled surface at a forward speed of 5.0 km h^{-1}. The rolling radius of the traction wheel and wheel slip are 0.65 m and 20% respectively. If the rear axle torque is 9 kN m, the tractive efficiency is

(A) 56.7 % (B) 62.1 %

(C) 69.3 % (D) 78.5 %

Graduate Aptitude Test in Engineering - 2010

Q. 18 A 37 kW two-wheel drive tractor weighing 20 kN with a wheel base of 2.1 m is having the option to be fitted with either 12.4-28 12PR or 13.6-28 12 PR at the rear axle. The ratio of section height and section width for all tyres is 0.75. On a level ground, the weight distribution on the front and rear axles is 35 and 65% of the total tractor weight, respectively. Cone index of soil is 1200 kPa.

(i) The motion resistance ratio of each of the rear wheels when fitted with the above-mentioned tyres at normal tyre inflation pressure while moving on a level ground will be

(A) 0.04, 0.04 (B) 0.047, 0.055

(C) 0.051, 0.049 (D) 0.057, 0.055

(ii) Net traction developed in kN by the rear wheels when fitted with 13.6-28 12 PR tyre at normal inflation pressure on a level ground with 15% wheel slip will be

(A) 8.79 (B) 9.18

(C) 9.78 (D) 10.32

Graduate Aptitude Test in Engineering - 2011

Q. 19 A towed pneumatic wheel (width to diameter ratio of 0.3) is to be rolled on a leveled concrete surface. Total wheel load is 2000 N.

(i) The force in N required to roll the wheel on the horizontal concrete surface would be

(A) 80 (B) 91

(C) 800 (D) 912

(ii) The minimum slope in degrees of the concrete surface with respect to the horizontal at which the wheel itself will start rolling downward is

(A) 2.29 (B) 5.28

(C) 23.58 (D) 27.13

Q. 20 Which one of the following is NOT a towed wheel ?

(A) Wheels of power tiller

(B) Front wheels of two wheel drive tractor

(C) Wheels of bullock cart

(D) Wheels of trailer

Graduate Aptitude Test in Engineering - 2012

Q. 21 Which one of the following statements is NOT appropriate regarding cone index

(A) It reflects strength of soil

(B) It is a composite parameter

(C) It is dimensionless

(D) It is measured at a constant penetration rate of 30 mm/s

Q. 22 A two wheel drive tractor, weighing 15.84 kN with a wheel base of 2160 mm, has the static weight divided between the front and rear axles in the ratio of 30 : 70 on a horizontal level surface. The hitch point is at a height of 700 mm from the ground and at a horizontal distance of 120 mm to the rear side from the center of the rear axle. Pull acts at an angle of 12° downwards from the horizontal. The maximum pull in kN, when the front wheels would just start rising from the ground is

(A) 1.48 (B) 14.46

(C) 39.04 (D) 85.54

Graduate Aptitude Test in Engineering - 2013

Q. 23 A two-wheel drive tractor pulls an implement that has a draft force of 11.5 kN. The total motion resistance of the tractor is 2.5 kN. Under these circumstances, the slip of the drive wheels is 20%. If the power loss in transmission is 20%, the percentage of power lost in converting engine power into drawbar power is ....

Graduate Aptitude Test in Engineering - 2014

Q. 24 A four-wheel-drive tractor has a static weight of 50 kN with 40% weight on rear axle and 60% on front axle. The wheel base is 2 m. The tractor is pulling a disc harrow that exerts a level drawbar pull at a hitch height of 0.5 m from the ground. During the operation, when the dynamic reaction on each axle is same, the dynamic traction ratio developed by the tractor is

(A) 0.2 (B) 0.4

(C) 0.5 (D) 0.8

Graduate Aptitude Test in Engineering – 2016

Q. 25 A tractor weighing 21 kN has 70% static weight on rear axle and its wheel base is 1.8 m. The drawbar hitch is located 25 cm behind the rear axle centre and 35 cm above the ground level. To overcome longitudinal instability, the front end loading is provided at a distance of 20 cm ahead of the front axle centre. It is observed that, there is front-end instability in the tractor due to a pull of 30 kN inclined at 20° downward from the horizontal. A minimum front-end load required to overcome the instability in N is ....

Graduate Aptitude Test in Engineering – 2017

Q. 26 A tractor pulls 8 kN drawbar load against 4 kN rolling resistance. If the tractor develops 57%, tractive efficiency, the slip experienced by the tractor in percentage will be ....

Q. 27 A load of 3 kN is acting on a tyre having 150 mm nominal width. The effective friction coefficient of tyre and ground interaction is 0.6 and the kingpin offset is 10 mm. Assuming the tyre impression on ground as circle of diameter same as the tyre nominal width, the kingpin torque of the tyre in N m will be ....

Q. 28 A tractor of 19.5 kN weight and 1.8 m wheel base has 70% static weight on the rear axle. It pulls 8 kN drawbar load parallel to the ground through a hitch point located 450 mm above the ground. The dynamic weight on the rear axle of the tractor under operating condition in kN will be ......

Graduate Aptitude Test in Engineering – 2018

Q. 29 A two-wheel drive tractor weighing 18 kN has a wheel base 1.8 m. Its centre of gravity is located 600 mm ahead of the rear axle centre, under static condition, on a level ground. When this tractor is used to pull a disc plough hitched at a height of 390 mm from the ground, the draft observed is 6 kN. The change in reaction on rear wheels of the tractor due to pull in kN is

(A) 1.30 (B) 1.95

(C) 3.90 (D) 5.85

Q. 30 A two-wheel drive tractor with rear wheel rolling radius of 600 mm develops a tractive force of 18.5 kN at a wheel slip of 12%. The engine speed is 2400 rpm and the transmission ratio from the engine to the rear wheels is 120:1. If the tractor experiences a motion resistance of 1.75 kN, the drawbar power developed by the tractor will be .... kW.

Graduate Aptitude Test in Engineering – 2019

Q. 31 A two-wheel drive tractor while operating a plough at a forward speed of 5 km/h experiences a wheel slip of 15%. If the angular speed of the rear axle is 2.4 rad/s, the rolling radius of the traction wheel in meter will be

(A) 0.49 (B) 0.58

(C) 0.68 (D) 0.75

Q. 32 During operation, a two-wheel drive tractor with a total weight of 2000 kg has a weight distribution of 35% and 65% in front and rear axles, respectively. The width and diameter of the tyres fitted to the front axle are 0.18 m and 0.56 m, and those of the rear axle are 0.34 m and 1.10 m, respectively. If tyre deflection is 20%, then rolling resistance (in kN) of the tractor in a soil with average cone index 1000 kPa at a wheel slip of 15% will be ....

Q. 33 A two-wheel drive tractor is taking a turn with a radius of curvature 5.0 m. The minimum horizontal distance between the tipping axis and line of action of the CG is 800 mm. The angle between the line of action of centrifugal force and perpendicular direction to the tipping plane is 15°. If the vertical distance of the CG from the ground level is 900 mm, the limiting speed (in km/h) of the tractor to prevent overturning is ....

Graduate Aptitude Test in Engineering – 2020

Q. 34 A self-propelled wheel does not have

(A) Wheel torque (B) Tractive power

(C) Rolling resistance (D) Drawbar pull

Q. 35 A towed pneumatic wheel with an unloaded radius of 330 mm covers a distance of 9.9 m in 5 revolutions without any skid. Assuming the rolling radius to be same as the static loaded radius of the wheel, the deflection of the wheel in mm is ....

Q. 36 A two-wheel drive tractor with a total weight of 25 kN has a wheelbase of 2.2 m and its centre of gravity lies 0.7 m ahead of the centre of rear axle. A steady horizontal pull is applied at a drawbar hitch height of 0.5 m on a concrete surface such that the weight on front axle becomes 20% of static weight of the tractor. The coefficient of net traction is ....

Graduate Aptitude Test in Engineering – 2021

Q. 37 During operation of a two-wheel drive tractor with a total weight of 20 kN in pure sandy soil (angle of internal friction is 26.5°), the weight distribution at the front and rear axles are found to be 35% and 65%, respectively. If extra weight of 2.5 kN is added to each of the rear wheels, the change in maximum thrust developed by the tractor in per cent will be ....

Q. 38 In a tyre axis system as defined by Society of Automotive Engineers, the moment acting about z-axis is called

(A) Aligning torque (B) Over turning torque

(C) Rolling resistance moment (D) Lateral moment

Answers Key

1	2	3	4	5	6	7	8	9	10
C	A	C	C	C	B	D	D	C	B
11	12	13	14	15	16	17	18	19	20
A	D	C, A, D	B, C	A	C	C	D, C	A, A	A
21	22	23	24	25	26	27	28	29	30
C	B	47.428	B	751.50	14	97	15.65	A	18.26
31	32	33	34	35	36	37	38		
C	1.3263	24.20	D	15	0.65	38.42	A		

Explanations

Q. 13 (i)

Given

Tractor weight (W) = 20 kN

Cone index of soil (CI) = 1200 kPa

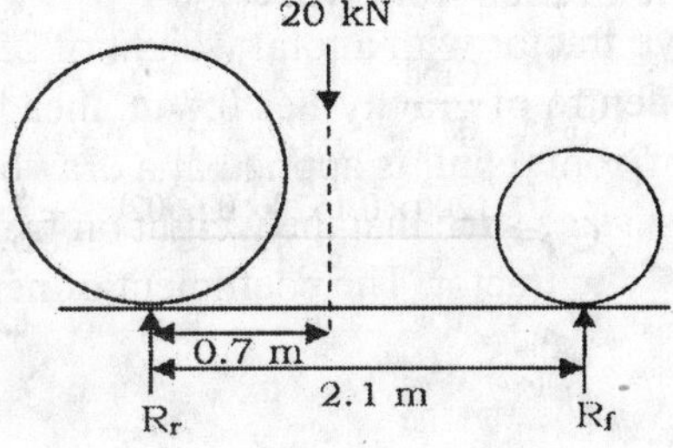

Taking moment with respect to front wheel

$$R_r \times 2.1 = 20 \times (2.1 - 0.7)$$

$$R_r = 13.33 \text{ kN}$$

Therefore, dynamic weight on each rear wheel (R_{r1}) = 13.33/2 = 6.67 kN

Rear axle of tractor is fitted with 13.6-28 12R tyres.

Section width (b) = 13.6 inches = 13.6×0.0254 = 0.3454 m

Section height (h) = 0.75×0.3454 = 0.2591 m

Rim diameter = 28 inches = 28×0.0254 = 0.7112 m

∴ Overall diameter of rear wheel (d) = Rim diameter×1.062 + 2×Section height

(Here, 1.062 is multiplied with rim diameter because in deflected condition, rim diameter is 1.062 times greater than actual.)

$$d = 0.7112 \times 1.062 + 2 \times 0.2591 = 1.273 \text{ m}$$

∴ Wheel numeric of each rear wheel $(C_n) = \dfrac{Clbd}{R_{r1}}$

$$C_n = \frac{1200 \times 0.3454 \times 1.273}{6.67} = 79.11$$

(ii)

Dynamic weight on front wheels (R_f) = 20 – 13.33 = 6.67 kN

Dynamic weight on each front wheel (R_{f1}) = 6.67/2 = 3.34 kN

Front axle of tractor is fitted with 6-16 8R tyres.

Overall diameter of front wheel (d)

$$= 2 \times 0.75 \times 6 \times 0.0254 + 16 \times 0.0254 \times 1.062$$

$$= 0.6602 \text{ m}$$

∴ Wheel numeric of each front wheel

$$(C_n) = \frac{Clbd}{R_{f1}}$$

$$C_n = \frac{1200 \times 0.1524 \times 0.6602}{3.34} = 36.1489$$

We know that

Rolling resistance coefficient (ρ) = $\frac{1.2}{C_n} + 0.04$

$= \frac{1.2}{36.1489} + 0.04 = 0.0732$

And Rolling resistance coefficient (ρ) = $\frac{\text{Towed or rolling resistance}}{\text{Dynamic load}}$

$\Rightarrow \quad TF = \rho \times R_{fl}$

$TF = 0.0732 \times 3.34 = 0.2445$ kN

(iii)

For level ground with compacted dry clay soil, rolling resistance coefficient is = 0.04

$\therefore$ Towed force = 0.04×20 = 0.80

Q. 14 (i)

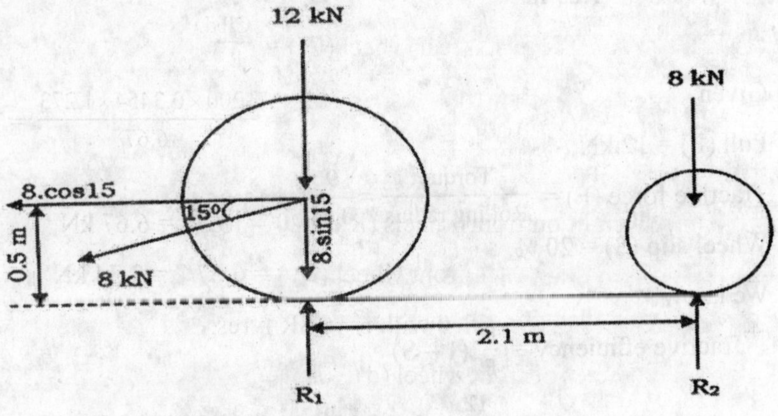

Generally, wheel base of tractor is 2.1 m.

Thus weight transfer on rear wheel is = $\frac{8.\cos 15}{2.1} \times 0.5 = 1.84$ kN

Total dynamic load on rear wheel = 1.84 + 12 + 8.sin15 = 15.91 kN

Coefficient of traction (μ) = $\frac{\text{Traction force}}{\text{Normal load}} = \frac{8.\cos 15}{15.91} = \frac{7.73}{15.91} = 0.49$

(ii)

Given that

Wheel slip(S) = 20%

Coefficient of rolling resistance (ρ) = 0.04

$$\text{Tractive efficiency} = \frac{\mu}{\mu + \rho}(1 - S) \times 100$$

$$= \frac{0.49}{0.49 + .04}(1 - 0.20) \times 100 = 73.96\%$$

Q. 16

∵ Tractor wheel is fitted with 13.6-28 12R tyres.

Section width (b) = 13.6 inches = 13.6×0.0254 = 0.3454 m

Section height (h) = 0.75×0.3454 = 0.2591 m

Rim diameter = 28 inches = 28×0.0254 = 0.7112 m

∴ Overall diameter of wheel (d) = Rim diameter + 2×Section height

(Here 1.062 is not multiplied with rim diameter because in undeflected condition, rim diameter is taken as it is.)

$$d = 0.7112 + 2 \times 0.2591$$

$$= 1.23 \text{ m}$$

Q. 17

Given

Pull (P) = 12 kN;

$$\text{Tractive force (F)} = \frac{\text{Torque}}{\text{Rolling radius}} = \frac{9}{0.65} = 13.85 \text{ kN}$$

Wheel slip (S) = 20 %

We know

$$\text{Tractive efficiency} = \frac{P}{F}(1 - S)$$

$$= \frac{12}{13.85}(1 - 0.20)$$

$$= 69.31\ \%$$

Q. 18 (i)

Dynamic weight on rear wheels (R_r) = 13 kN

Therefore, dynamic weight on each rear wheel (R_{r1}) = 13/2 = 6.5 kN

For 12.4-28 12PR tyres

Rear axle of tractor is fitted with 12.4-28 12R tyres.

Section width (b) = 12.4 inches = 12.4×0.0254 = 0.315 m

Section height (h) = 0.75×0.315 = 0.236 m

Rim diameter = 28 inches = 28×0.0254 = 0.7112 m

∴ Overall diameter of rear wheel (d) = Rim diameter×1.062 + 2×Section height

(Here 1.062 is multiplied with rim diameter because in deflected condition, rim diameter is 1.062 times greater than actual)

$$d = 0.7112 \times 1.062 + 2 \times 0.236 = 1.227 \text{ m}$$

∴ Wheel numeric of each rear wheel $(C_n) = \dfrac{Clbd}{R_{rl}}$

$$C_n = \frac{1200 \times 0.315 \times 1.227}{6.5} = 71.355$$

And Motion resistance ratio $(\rho) = \dfrac{1.2}{C_n} + 0.04$

$$= \frac{1.2}{71.355} + 0.04 = 0.057$$

For 13.6-28 12PR tyres

Section width (b) = 13.6 inches = 13.6×0.0254 = 0.3454 m

Section height (h) = 0.75×0.3454 = 0.2591 m

Rim diameter = 28 inches = 28×0.0254 = 0.7112 m

∴ Overall diameter of rear wheel

(d) = Rim diameter×1.062 + 2×Section height

$$d = 0.7112 \times 1.062 + 2 \times 0.2591 = 1.273 \text{ m}$$

∴ Wheel numeric of each rear wheel

$$(C_n) = \frac{Clbd}{R_{rl}}$$

$$C_n = \frac{1200 \times 0.3454 \times 1.273}{6.5} = 81.174$$

And Motion resistance ratio $(\rho) = \dfrac{1.2}{C_n} + 0.04$

$$= \frac{1.2}{81.174} + 0.04 = 0.055$$

(ii)

Net traction ratio $(\mu) = 0.75(1 - e^{-0.3C_n S})$

$$= 0.75\times(1-e^{-0.3\times81.174\times0.15})$$

$$= 0.7306$$

$\therefore$ Net traction developed = 13×0.7306 = 9.5 kN

Q. 19 (i)

We know, the force required to roll the pneumatic wheel on horizontal surface

= 0.04 × Wheel load

Therefore, Force = 0.04×2000 = 80 N

(ii)

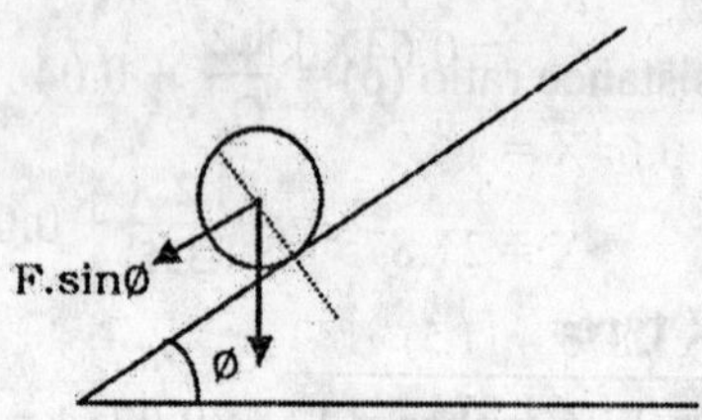

As shown in figure

F sin∅ = 80

Thus, ∅ = 2.29

Q. 22

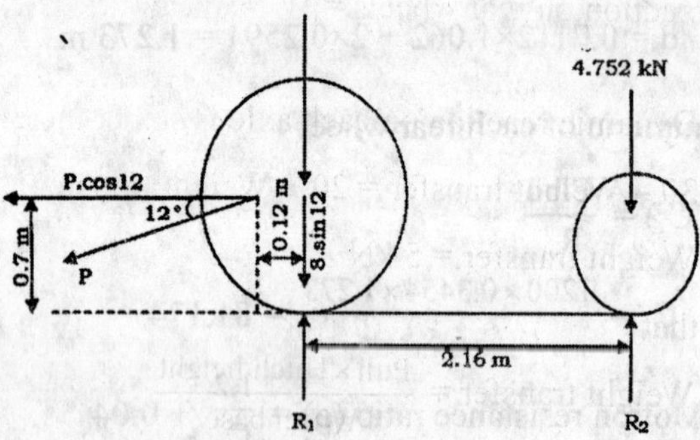

Static weight of front wheels (R_f) = 15.84×0.30 = 4.752 kN

Static weight of rear wheels (R_r) = 15.84×0.70 = 11.088 kN

Referring figure

Taking moment around point A

$$4.752\times2.16 = (P\sin 12)\times0.12 + (P\cos 12)\times0.7$$

$$\Rightarrow \quad P = 14.46 \text{ kN}$$

Q. 23

Consider total engine power = X kN

Total draft = 11.5 + 2.5 = 14 kN

Due to 20% wheel slippage power loss = 20X/100

$\therefore$ Power used in transmission = X – X×20/100 = 4X/5

Due to 20% power loss, power consumed in draft

$$= 4X/5 - (4X/5)\times(20/100)$$

$$= 0.64X \text{ kN}$$

Thus, $0.64X = 14$

$$\Rightarrow \quad X = 21.875 \text{ kN}$$

$$\text{Power Loss} = \frac{(21.875 - 11.5)\times100}{21.875} = 47.428\ \%$$

(Because useful pull is 11.5 kN.)

Q. 24

Dynamic reaction on front wheels = W_f – Weight transfer = 30 – Weight transfer

Dynamic reaction on rear wheels = W_r + Weight transfer = 20 + Weight transfer

$\because$ Dynamic reaction on each axle is same, therefore

30 – Weight transfer = 20 + Weight transfer

$\Rightarrow$ Weight transfer = 5 kN

We know that

$$\text{Weight transfer} = \frac{\text{Pull}\times\text{Hitch height}}{\text{Wheel base}}$$

$$\Rightarrow \quad 5 = \frac{\text{Pull}\times0.5}{2}$$

$$\Rightarrow \quad \text{Pull} = 20 \text{ kN}$$

$$\therefore \text{Dynamic traction ratio} = \frac{\text{Pull}}{\text{Weight of Tractor}} = \frac{20}{50} = 0.4$$

Q. 25

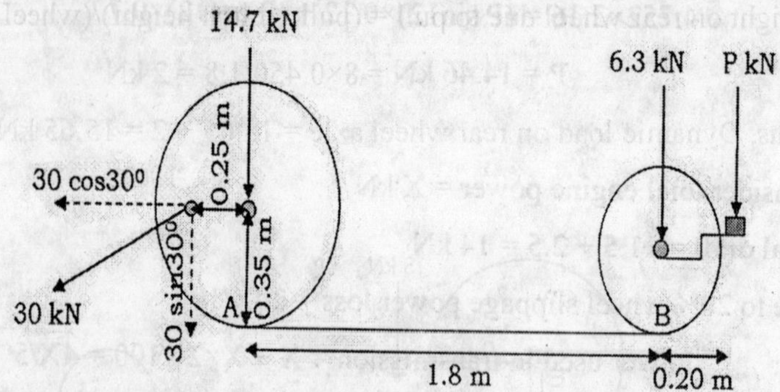

Taking moment around point A

$$6.3\times1.8 + P\times2 = 30\cos30^0\times(0.35) + 30\sin30^0\times(0.25)$$

$$\Rightarrow \quad P = 0.7515 \text{ kN or } 751.50 \text{ N}$$

Q. 26

We know

$$\text{Tractive efficiency} = \text{Pull}\times(1-s)\times100/\text{Drawbar load}$$

$$\Rightarrow \quad 0.57 = 4\times(1-s)/8$$

$$\Rightarrow \quad s = 14\ \%$$

Q. 27

We know,

Torque required at the kingpin for steering axle (T) = $W.u(B^2/8 + E^2)^{1/2}$

Where, W = Weight on steering axle = 3×2 = 6 kN

u = Coefficient of tyre and ground interaction = 0.6

B = Nominal width of tyre = 0.15 m

E = Kingpin offset = 0.01 m

Therefore $\quad T = 6\times0.6\times\{(0.15)^2/8 + (0.01)^2\}^{1/2}$

$= 0.194$ kN-m or 194 N-m

Hence, Kingpin torque for the tyre= 194/2 = 97 N-m

Q. 28

Weight of tractor = 19.50 kN

and Static weight on rear wheel axle = 19.50×0.70 = 13.65 kN

Weight on rear wheel due to pull = (pull × hitch height)/(wheel base)

= 8×0.450/1.8 = 2 kN

Thus, Dynamic load on rear wheel axle = 13.65 + 2 = 15.65 kN

Q. 29

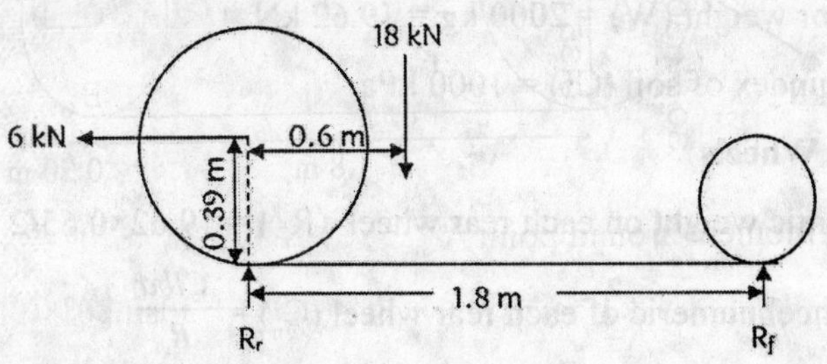

In static condition

Rear wheel reaction = 18×1.2/1.8 = 12 kN

When pull of 6 kN is applied

Taking moment arround front wheel

18×1.2 + 6×0.39 = Rr×1.8

⇒ Rr = 13.3 kN

∴ Change in reaction on rear wheel = 13.3 – 12

= 1.3 kN

Q. 30

Drawbar pull = 18.5 – 1.75 = 16.75 kN

Rear wheel speed (N) = 2400/120 – 20 rpm or 0.33 rps

(∵ Transmission ratio is 120 : 1) Tractor speed

(v) = 2πrN

= 2×3.14×0.6×0.33 = 1.24 m/s

∵ Actual velocity of tractor = v(1 – s) = 1.24×(1 – 0.12) = 1.09 m/s

Therefore, Drawbar power = Pull×Velocity = 16.75×1.09 = 18.26 kW

Q. 31 (C)

We know

$V_a/V_t = 1 - s$

$V_a / \omega r = 1 - s$

$\Rightarrow \quad r = 5000/[2.4\times3600\times(1\text{-}0.15)] = 0.68$ m

Q. 32

Given

Tractor weight (W) = 2000 kg = 19.62 kN

Cone index of soil (CI) = 1000 kPa

Rear Wheels

Dynamic weight on each rear wheel $(R_{r1}) = 19.62\times0.65/2 = 6.38$ kN

$\therefore$ Wheel numeric of each rear wheel $(C_n) = \dfrac{CIbd}{R_{r1}}$

$$C_n = \frac{1000\times0.34\times1.10}{6.38} = 58.62$$

Rolling resistance coefficient $(\rho) = \dfrac{1.2}{58.62} + 0.04 = 0.0605$

Rolling resistance $= 2\times0.0605\times6.38 = 0.7720$ kN

Front Wheels

Dynamic weight on each front wheel $(R_{f1}) = 19.62\times0.35/2 = 3.43$ kN

$\therefore$ Wheel numeric of each front wheel $(C_n) = \dfrac{CIbd}{R_{f1}}$

$$C_n = \frac{1000\times0.34\times0.56}{3.43} = 29.39$$

Rolling resistance coefficient $(\rho) = \dfrac{1.2}{29.39} + 0.04 = 0.0808$

Rolling resistance $= 2\times0.0808\times3.43 = 0.5543$ kN

Hence, Total rolling resistance of tractor = 0.5543 + 0.7720 = 1.3263 kN

Q. 33

Limiting speed $= \sqrt{\dfrac{gAR}{Z_{cg}cos\varnothing}}$

$$= \sqrt{\frac{9.81\times0.8\times5}{0.9\times cos15}} = 6.72 \text{ m/s or } 24.20 \text{ km/h}$$

Q. 35

Wheel radius in unloaded condition = 330 mm

In loaded condition

Distance covered in one revolution = Perimeter of wheel

$\Rightarrow \quad 9.9/5 = 2\times3.14\times R$

$\Rightarrow \quad R = 0.315$ m or 315 mm

Deflection of the wheel = 315 – 300 = 15 mm

Q. 36

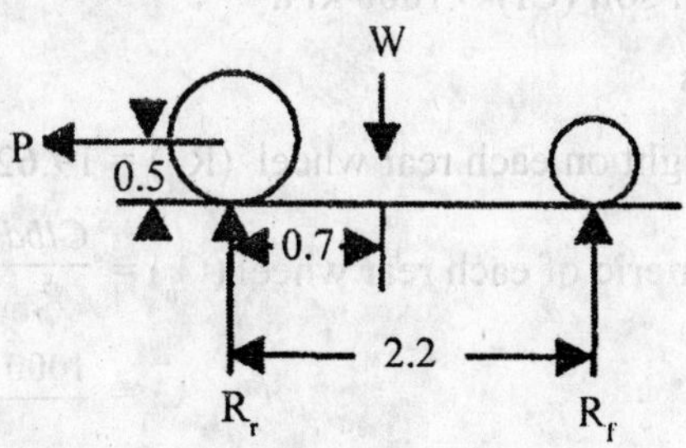

$R_f = 25\times0.2 = 5$ kN and $R_r = 20$ kN

Referring above figure, Taking moment about rear wheel.

$$5\times2.2 + P\times0.5 = 25\times0.7$$

$\Rightarrow \quad P = 13$ kN

Hence,

Coefficient of net traction = 13/20 = 0.65

Q. 37

Weight on rear tyres = 20×0.65 = 13 kN

Thrust/Tractive force developed = 13×tan26.5 = 6.48 kN

If extra weight of 2.5 kN is added on each wheel.

Total weight on rear tyres = 2.5×2 + 13 = 18 kN

Thrust/Tractive force developed = 18×tan26.5 = 8.97 kN

$$\text{Percentage change} = \frac{8.97-6.48}{6.48}\times100 = 38.42\ \%$$

12

Tractor Chassis Mechanism

APPSC AEES Agriculture Engineering Exam 2016

Q. 1 Toe-in and toe-out are related with

(A) Front axle of tractor (B) Rear axle of tractor

(C) Tractor steering (D) Transmission system

Q. 2 Angle between centre line of kingpin of tractor and vertical line is called

(A) Tilt angle (B) Camber Angle

(C) Caster angle (D) All of the above

Punjab Mechanic Agricultural Machinery Instructor 2014

Q. 3 The basic tractor structure consists of frame, axles, wheels and

(A) Steering (B) Suspension

(C) Brakes (D) Differential gear

OCS AEn Exam 2011 (Pre) – II (Agricultural Engineering)

Q. 4 The longitudinal stability of the tractor can be improved by

(A) Putting dead weight on the front axle

(B) Putting dead weight on rear axle

(C) Decreasing air pressure of rear wheel

(D) Decreasing air pressure of front wheel

CGPSC State Engineering Services Exam Paper – I

Q. 5 The angle between the centre line of the tractor front tyre and the vertical line is known as

(A) Caster angle (B) Toe-in
(C) Tit angle (D) Gang angle
(E) Camber angle

Q. 6 Steering shaft of tractor steering system is made with
(A) Copper (B) Cast iron
(C) Wood (D) Steel
(E) Bronze

UKPSC A.En. Exam 2007 Paper – II

Q. 7 Tractors having pneumatic wheels are called
(A) Power tractors (B) Crawler tractors
(C) Industrial tractors (D) Wheel tractors

Q. 8 The angle between plane of front wheel of tractor and vertical is called as
(A) Clearance angle (B) Camber angle
(C) Toe-in (D) Toe-out

UKPSC A.En. Exam 2007 Paper – II

Q. 9 The tyre pressure in a tractor should be checked
(A) Just after returning from work (B) When the tire is cold
(C) When the tractor is working (D) Any one of these

Q. 10 In tractors, the wheel alignment can be adjusted by
(A) Drag link (B) Radius link
(C) Both (A) and (B) (D) Varying the length of the rod

Q. 11 The kingpin inclination, in steering system of a tractor ranges from
(A) 15 – 20 degrees (B) 10 – 15 degrees
(C) 6 – 9 degrees (D) 0 – 5 degree

Kerala AAO Exam 2017

Q. 12 The inflation pressure of the tractor rear wheel varies between
(A) 0.8 to 1.5 kg/cm^2 (B) 1.5 to 2.5 kg/cm^2
(C) 2.5 to 3.5 kg/cm^2 (D) 3.5 to 4.5 kg/cm^2

Graduate Aptitude Test in Engineering - 2007

Q. 13 A two-wheel drive tractor weighing 20 kN has a wheel base of 2.1 m with a static weight distribution of 35 % and 65 % at the front and rear axles respectively. On a level ground, the tractor moves at a speed of 4 km h^{-1}. Considering small steer angle, the cornering force acting on each of the front tyre for a tuning radius of 1.8 m is

(A) 0.244 kN (B) 0.454 kN

(C) 0.489 kN (D) 0.907 kN

Graduate Aptitude Test in Engineering - 2010

Q. 14 Match all items in Group I with correct options from those in Group II.

	Group I		Group II
i.	Slider crane mechanism	a.	Tractor steering
ii.	four bar linkage mechanism	b.	Attachment of pitman to knife head
iii.	Ball and socket joint	c.	Planting unit of rice transplanter
iv.	Worm and roller type unit	d.	Vertical conveyor reaper

(A) i - d, ii - c, iii - b, iv – a (B) i - b, ii - c, iii - a, iv - d

(C) i - d, ii - c, iii - a, iv – b (D) i - b, ii - d, iii - a, iv - c

Graduate Aptitude Test in Engineering - 2014

Q. 15 A tractor is provided with an Ackerman steering gear mechanism. If b is the wheel base, c the distance between the front wheel pivot points, θ the steering angle of the inner wheel and ∅ the steering angle of the outer wheel, the fundamental relationship to be satisfied to avoid skidding of the two front wheels during a turn is given by

(A) $\tan\varnothing - \tan\theta = c/b$ (B) $\cot\varnothing - \cot\theta = c/b$

(C) $\tan\varnothing - \tan\theta = b/c$ (D) $\cot\varnothing - \cot\theta = b/c$

Graduate Aptitude Test in Engineering - 2016

Q. 16 The effective temperature (ET) scale developed in 1972 on the basis of a human model is

(A) Heart rate (B) Blood pressure

(C) Psychological response (D) Physiological response

Q. 17 Natural frequency of an undamped operator seat is 5 Hz, and combined weight of the seat and the operator is 880 N. If there are four springs fitted in parallel below the operator seat, the spring rate (or stiffness) of each spring in kN m^{-1} is ....

Graduate Aptitude Test in Engineering - 2019

Q. 18 The combined mass of a tractor seat and operator is 75 kg and the undamped natural frequency of the operator seat is 10 rad/s. If the seat suspension damping rate is 600N/(m s), the damping ratio is

(A) 0.2 (B) 0.4

(C) 0.6 (D) 0.8

Graduate Aptitude Test in Engineering - 2020

Q. 19 In a tractor steering system, the angle made by the kingpin axis projected on the longitudinal plane of the tractor with the vertical axis is known as

(A) Kingpin inclination (B) Caster angle

(C) Camber angle (D) Steering angle

Q. 20 The sound pressure level on the operator's seat of a tractor is 80 dB. If the reference sound pressure is 2x 10^{-5} N/m^2, the root mean square (RMS) sound pressure in N/m^2 is ....

Answers Key

1	2	3	4	5	6	7	8	9	10
A	C	B	A	E	D	D	C	B	D
11	12	13	14	15	16	17	18	19	20
C	A	A	B	B	D	Bonus	B	B	0.2

Explanations

Q. 13

Given

Weight on front tyres (W_f) = 20×0.35 = 7 kN

Turning radius (R) = 1.8 m; Acceleration due to gravity (g) = 9.81 m/ s^2

Speed of tractor (u) = 4 km/ h = 1.11 m/ s

We know

$$\text{Cornering force} = \frac{W_f.u^2}{gR}$$

$$= \frac{7 \times (1.11)^2}{9.81 \times 1.8} = 0.488 \text{ kN}$$

Hence, cornering force on each of the front tyre = 0.488/2

= 0.244 kN

Q. 17

Equivalent spring constant (k_{eq}) = k/4

Where k = spring constant of each one

Total Mass of the system (m) = 880 N = 89.70 kg

We know

$$\text{Natural Frequency of seat} = \frac{1}{2\pi}\sqrt{\frac{k_{eq}}{m}}$$

$$5 = \frac{1}{2\pi}\sqrt{\frac{k}{4\times 89.70}}$$

$\Rightarrow$ k = 88444 N/m or 88.44 kN/m

Q. 18

Seat suspension damping ratio(C) = 600 N/(m s)

Natural Frequency(C_c) = 2mω = 2×75×10 = 1500 N/(m s)

Hence,

Damping ratio = C/C_c

= 600/1500 = 0.4

Q. 20

We know that

$$\text{Sound pressure level (SPL)} = 20\log\left(\frac{P_{rms}}{P_0}\right)$$

$$\Rightarrow \quad 80 = 20.\log\left(\frac{P_{rms}}{2\times 10^{-5}}\right)$$

$$\Rightarrow \quad P_{rms} = 0.2\ \text{N/m}^2$$

13

Tractor Hydraulics and Implement Control

UKPSC Combined Junior Engineering Exam 2013 Agricultural Engineering Paper-I

Q. 1 A device which is used to store energy in hydraulic system is known as

(A) Motor (B) Cylinder

(C) Accumulator (D) Pump

Punjab Mechanic Agricultural Machinery Instructor 2014

Q. 2 In hydraulic system, the power used is

(A) Mechanical (B) Hydrostatic

(C) Electrical (D) Frictional

CGPSC State Engineering Services Exam Paper – I

Q. 3 Tractor hydraulic system works based on the principle of

(A) Newton's law (B) Charles law

(C) Rittinger's law (D) I[st] law of thermodynamics

(E) Pascal's law

UKPSC A.En. Exam 2012 Paper - II

Q. 4 In a tractor the three point hitch is operated by

(A) PTO (B) Gear system

(C) Hydraulic control (D) Differential

Graduate Aptitude Test in Engineering - 2007

Q. 5 A piston with 50 mm diameter and length 50 mm is to be moved at a velocity of 0.25 m s^{-1} in a hydraulic cylinder with 50.2 mm diameter. The cylinder is full of oil with a kinematic viscosity of 9×10^{-4} m^2 s^{-1}

and a density of 880 kg m^{-3}. Assuming pressure difference between inside and outside of the cylinder as zero, the force required to move the piston is

(A) 7.772 N (B) 15.543 N

(C) 76.243 N (D) 152.476 N

Graduate Aptitude Test in Engineering - 2008

Q. 6 A double acting hydraulic cylinder has a piston diameter of 40 mm and the rod diameter equal to one-half the piston diameter. For a constant pressure of 4 MPa, the difference in load carrying capacity between extension and retraction is

(A) 0 kN (B) 1.26 kN

(C) 3.77 kN (D) 6.29 kN

Q. 7 A hydraulic motor receives a flow rate of 72 L min^{-1} at a pressure of 12 MPa. The motor speed is 800 rpm. If the motor has a power loss of 3 kW, the actual torque delivered by the motor is

(A) 136.08 N m (B) 171.89 N m

(C) 204.62 N m (D) 262.84 N m

Graduate Aptitude Test in Engineering - 2011

Q. 8 A vertical hydraulic cylinder having an inside diameter of 100 mm is used in a hoist for lifting load. A constant flow rate pump capable of delivering 12 litre per minute at a maximum pressure of 18 MPa is used for supplying the fluid. Volumetric and mechanical efficiencies of the pump are 91% and 85% respectively.

(i) The rate of lifting of load in m s^{-1} and maximum amount of load that can be lifted in kN are

(A) 0.020, 180 (B) 0.025, 141

(C) 0.064, 45 (D) 0.080, 565

(ii) The size of motor required for operating the pump in kW is

(A) 3.60 (B) 3.96

(C) 4.24 (D) 4.65

Graduate Aptitude Test in Engineering - 2013

Q. 9 A double acting hydraulic cylinder has a rod diameter equal to one-half the piston diameter. If the system pressure is maintained constant, the ratio of load carrying capacity of extension stroke to that of retraction stroke is

(A) 0.75 (B) 1.00

(C) 1.33 (D) 4.00

Graduate Aptitude Test in Engineering - 2014

Q. 10 A hydraulic circuit uses a pump having a fixed displacement volume of 12.5 $cm^3\ rev^{-1}$ driven at 1500 rpm. The pump has a volumetric efficiency of 85% and an overall efficiency of 75%. If the system pressure is set at 15 MPa by the relief valve, the power required to drive the pump in kW will be

(A) 2.99 (B) 4.53

(C) 5.31 (D) 7.53

Graduate Aptitude Test in Engineering – 2015

Q. 11 A double acting hydraulic cylinder has a bore of 200 mm with a piston rod of 140 mm diameter. The extend speed of the piston is 80 mm/s. If the flow rate of oil during retraction is same as that of the extending, the retract speed of the piston in mm/s is ....

Answers Key

1	2	3	4	5	6	7	8	9	10
C	B	E	C	BONUS	B	A	B, D	C	C
11									
156.86									

Explanations

Q. 5

Cross sectional area of piston (A) = $\frac{\pi D^2}{4}$

$$= \frac{3.14\times(0.050)^2}{4} = 1.9625\times10^{-3}\ m^2$$

Dynamic viscosity of oil (μ) = ν.ρ

$$= 9\times10^{-4}\times880 = 0.792\ kg/m.s = 7.77\ N/m.s$$

∵ Inside and outside pressure of the cylinder is zero, Hence, piston will move/ work against the viscous force of hydraulic oil.

We know

$$\text{Viscous force } (F) = \mu.A.\frac{dv}{dx}$$

$$F = 7.77\times1.9625\times10^{-3}\times\frac{0.25}{0.050} = 76.243\times10^{-3}\,N$$

Q. 6

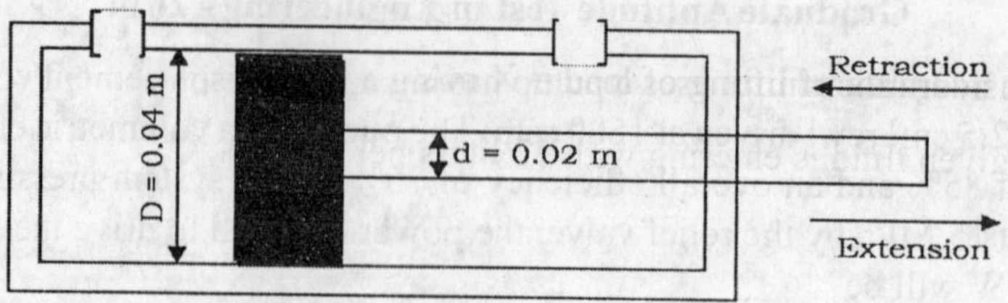

Piston diameter (D) = 0.04 m

Piston rod diameter (d) = 0.02 m

Pressure (P) = 4 MPa = 4×10^6 Pa

Piston extending force (F_e) = P×A = $P\times\frac{\pi}{4}D^2$

Piston retracting force (F_r) = P×A = $P\times(\frac{\pi}{4}D^2-\frac{\pi}{4}d^2)$

∴ Difference in load carrying capacity between extension and retraction

$$= P\times\frac{\pi}{4}D^2 - P\times(\frac{\pi}{4}D^2-\frac{\pi}{4}d^2) = P\times\frac{\pi}{4}d^2$$

$$= 4\times10^6\times\frac{3.14}{4}\times(0.02)^2 = 1.26\text{ kN}$$

Q. 7

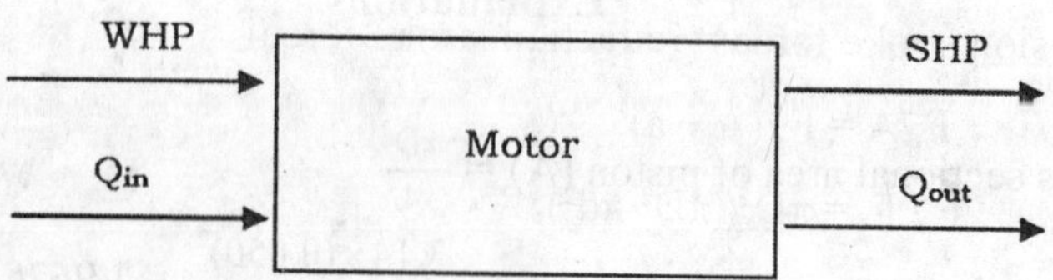

Hydraulic motor flow rate (Q) = 72 L/min = 1.2×10^{-3} m³/s

Pressure developed (P) = 1.2 MPa = 12×10^6 Pa; Motor speed (N) = 800 rpm = 13.33 rps

$\therefore$ Power developed = PQ = $1.2\times10^{-3}\times12\times10^{6}$ = 1440 W = 14.4 kW

Here power loss of motor is 3 kW.

Thus, Actual power delivered by motor = 14.4 − 3 = 11.4 kW

We know that

$$\text{Power} = 2\pi NT$$

$$\Rightarrow T = 11.4\times 1000/(2\times3.14\times13.33) = 136.08 \text{ Nm}$$

Q. 8 (i)

Consider rate of lifting of load = X m/s

As given fluid is entering with 12 liters per minute.

Thus

Cross sectional area of cylinder × X = $12\times10^{-3}/60$

$$= (3.14\times(0.1)^2/4) \times X = 12\times10^{-3}/60$$

Solving this we get X = 0.025 m/sec

As we know F = P×A

Thus, amount of load that can be lifted will be

$$F = 18\times10^{6}\times(3.14\times(0.1)^2/4) = 141.3 \text{ kN}$$

(ii)

Power required to lift the fluid = Q×P

$$= 18\times10^{6}\times(12\times10^{-3}/60) = 3.6 \text{ kW}$$

$\because$ Volumetric and mechanical efficiencies are 91% and 85% respectively, thus size of motor to run the pump will be = 3.6/(0.91)×(0.85) = 4.65 kW

Q. 9

Extension stroke force = retraction stroke force

$$F_1/A = F_2/(A - a)$$

$$F_1/F_2 = \pi D^2/(\pi D^2 - \pi d^2)$$

Putting. d = D/2

$$F_1/F_2 = 4/3 = 1.33$$

Q. 10

Theoretical volumetric flow rate = $12.5 \times 1500 = 18750 \ cm^3/min$

$= 3.125 \times 10^{-4} \ m^3/s$

∵ Volumetric efficiency = 85 %

Actual flow rate (Q) = $0.85 \times 3.125 \times 10^{-4} = 2.66 \times 10^{-4} \ m^3/s$

System pressure (P) = 15 MPa = 15×10^6 Pa

∵ Output power pump = P×Q

$= 15 \times 10^6 \times 2.66 \times 10^{-4} = 3990$ W

∵ Overall efficiency = 75 %

Therefore,

Power required to drive the pump = 3990/0.75 = 5320 W = 5.32 kW

Q. 11

Piston diameter/ bore (D) = 200 mm; Piston rod diameter (d) = 140 mm

Piston extending cross sectional area

$$(A) = \frac{\pi}{4} D^2$$

$$= \frac{3.14}{4}(200)^2 = 31400 \ mm^2$$

Piston retracting cross sectional area

$$(a) = (\frac{\pi}{4} D^2 - \frac{\pi}{4} d^2)$$

$$= \frac{3.14}{4}(200)^2 - \frac{3.14}{4}(140)^2 = 16014 \ mm^2$$

∵ Flow rate of oil during retraction is same as that of the extending.

Hence, $Q_{extending} = Q_{retracting}$

$$AV_e = aV_r$$

$$\Rightarrow \quad 31400 \times 80 = 16014 \times V_r$$

$$\Rightarrow \quad V_r = 156.86 \ mm/s$$

14

Tractor Design: Role of Comfort, Health and Safety of Human

MPPSC Assistant Agricultural Engineer 2013

Q. 1 In agricultural tractors, the vibration level is in the range of .... Hz.

(A) 0 – 2 (B) 2 - 8

(C) 8 – 20 (D) More than 20

Q. 2 The safe level of noise for human being ranges between .... db.

(A) 100 – 110 (B) 90 - 100

(C) 80 – 90 (D) 70 – 80

Q. 3 In proper seat suspension design, the transmissibility should be ....

(A) Less than unity (B) Between 1 and 2

(C) More than 2 (D) None of these is correct

Q. 4 The temperature range of ....^{0}C is most practical comfort zone for a human operator.

(A) 5 to 10 (B) 10 to 20

(C) 20 to 30 (D) More than 30

OCS AEn Exam 2011 (Pre) – II (Agricultural Engineering)

Q. 5 The undamped natural frequency of the wheeled tractor lies in the range of

(A) 1 to 2 Hz (B) 3 to 10 Hz

(C) 11 to 15 Hz (D) 16 to 25 Hz

Q. 6 The sound level of an agricultural tractor should not exceed

(A) 110 dB (B) 100 dB

(C) 90 dB (D) 80 dB

CGPSC State Engineering Services Exam Paper – I

Q. 7 In tractor centre of gravity measurement test, the driver should be replaced with the weight of

(A) 70 kg (B) 75 kg

(C) 65 kg (D) 80 kg

(E) 85 kg

Q. 8 Test code for agricultural tractor is

(A) IS: 5994-1998 (B) IS: 5994-2000

(C) IS: 5994-1970 (D) IS: 5994-1960

(E) IS: 5994-1995

Q. 9 Test code for Power Tiller

(A) IS: 9935-1948 (B) IS: 9935-1958

(C) IS: 9935-1988 (D) IS: 9935-1919

(D) IS: 9935-2002

UKPSC A.En. Exam 2012 Paper - II

Q. 10 "Tachometer" is used to measure

(A) Pressure (B) Power

(C) Rpm (D) All the above

Q. 11 The maximum human tolerance to vibration is

(A) 8 – 12 Hz (B) 6 – 10 Hz

(C) 4 – 8 Hz (D) None of the above

Graduate Aptitude Test in Engineering - 2007

Q. 12 While testing a tractor, the airborne sound intensity is increased so that the root mean square sound pressure is doubled. The corresponding increase in sound pressure level to the reference sound pressure of 2×10^{-5} Pa is

(A) 2 dB (B) 4 dB

(C) 6 dB (D) 8 dB

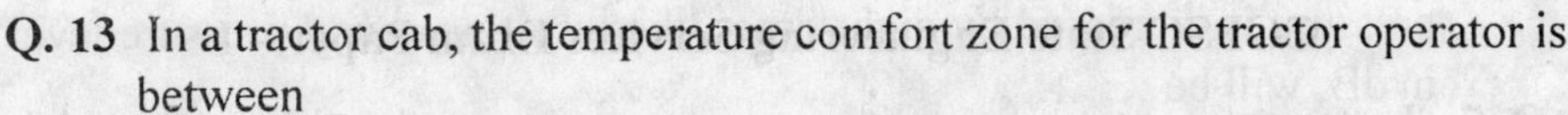

Q. 13 In a tractor cab, the temperature comfort zone for the tractor operator is between

(A) 287 and 291 K (B) 288 and 293 K

(C) 291 and 297 K (D) 295 and 301 K

Graduate Aptitude Test in Engineering - 2009

Q. 14 A tractor seat suspension system with a seat and operator mass of 90 kg has a seat suspension damping rate of 350 N s m^{-1}. If the spring rate of the system is 5 N mm^{-1}, the damping ratio of the system is

(A) 0.13 (B) 0.26

(C) 0.39 (D) 0.52

Graduate Aptitude Test in Engineering - 2010

Q. 15 The tractor seat vibrates with a frequency of 1 Hz when there is no damping. When damping is provided, the frequency of damped vibration is reduced by 10%. The damping factor is

(A) 0.21 (B) 0.39

(C) 0.44 (D) 0.93

Graduate Aptitude Test in Engineering - 2011

Q. 16 The range of frequency of vertical vibration of tractor most harmful to the operator's body at a root mean square acceleration of 1.0 m s^{-2} in Hertz is

(A) 0.4 – 0.8 (B) 4.0 – 8.0

(C) 400 – 800 (D) 4000 – 8000

Graduate Aptitude Test in Engineering - 2013

Q. 17 During a test, sound level was measured as 90 dB in the operator's cabin on a tractor. Taking reference sound pressure as 2×10^{-5} N m^{-2}, the measured RMS sound pressure in N m^{-2} is

(A) 6.32 (B) 6.32×10^{-1}

(C) 1.8×10^{-3} (D) 6.32×10^{-10}

Graduate Aptitude Test in Engineering – 2015

Q. 18 For a reference sound pressure of 2×10^{-5} N m^{-2}, the sound level measured at the operator's workspace of a tractor was 80 dB. If the RMS sound pressure is increased by eight times, the resulting sound pressure level in dB, will be ....

Answers Key

1	2	3	4	5	6	7	8	9	10
B	B	A	C	B	C	B	A	E	C
11	12	13	14	15	16	17	18		
C	C	C	B	C	B	B	98.06		

Explanations

Q. 12

Consider, $P_{rms} = P$

We know that

Sound pressure level $(SPL)_1 = 20 \log \left(\frac{P_{rms}}{P_0}\right)$

$$= 20.\log \left(\frac{P}{P_0}\right) \quad \text{(i)}$$

If root mean square sound pressure is doubled

Sound pressure level $(SPL)_2 = 20 \log \frac{2P}{P_0}$

$$= 20 \log(2) + 20 \log(\frac{P}{P_0}) \quad \text{(ii)}$$

Subtracting equation (ii) from (i)

$$SPL_2 - SPL_1 = 20 \log 2 = 6 \text{ db}$$

Q. 14

Given that

Seat suspension damping rate (c) = 350 N s m^{-1}

Spring rate (k) = 5 N/ mm or 5000 N/ m; Operator mass (m) = 90 kg

$$\therefore \quad \text{Damping ratio} = \frac{c}{2\sqrt{mk}}$$

$$= \frac{350}{2\sqrt{90 \times 5000}}$$

$$= 0.26$$

Q. 15

Tractor seat no damping frequency (ω_s) = 1 Hz

Tractor seat damping frequency (ω_d) = 0.9 Hz (∵ 10% reduction)

We know that

$$\omega_d = \omega_s\sqrt{1-\varepsilon^2}$$

$$\Rightarrow \quad 0.9 = 1\times \sqrt{1-\varepsilon^2}$$

$$\Rightarrow \quad \varepsilon = 0.44$$

Hence, Damping factor (ε) is 0.44.

Q. 17

Given that

Sound pressure level = 90 db;

Reference sound pressure (P_0) = 2×10^{-5} N/m^2

We know

Sound Pressure Level = $20.\log(\frac{P}{P_0})$

$$\Rightarrow \quad 90 = 20\times\log(\frac{P}{2\times10^{-5}})$$

$$\Rightarrow \quad P = 6.32\times10^{-1}\ \text{N/m}^2$$

Q. 18

Consider, $P_{rms} = P$

We know that

Sound pressure level $(SPL)_1 = 20\log(\frac{P_{rms}}{P_0})$

$$= 20.\log(\frac{P}{P_0}) = 80\ \text{dB}$$

If root mean square sound pressure is increased by eight times.

Sound pressure level $(SPL)_2 = 20\log\left(\left(\frac{8P}{P_0}\right)\right)$

$$= 20\log(8) + 20\log(\frac{P}{P_0})$$

$$= 18.06 + 80 = 98.06\ \text{dB}$$

III: Farm Machinery

III. Farm Machinery

15

Soil Tillage and Sowing Machinery Seeders and Planters

UKPSC AE Agricultural Engineering Paper - II 2013

Q. 1 If the forward speed of a tractor drawn seed drill is increased twice, then the seed rate delivered will be

(A) Reduced to half (B) Double

(C) Increased by four time (D) Remains the same

Q. 2 Which is the most pertinent variable in selecting a machine ?

(A) Height of machine (B) Colour of machine

(C) Size of machine (D) Ground clearance

Q. 3 The unit draft in silt loam (moist) soil varies from :-

(A) 0.21 to 0.24 Kg/cm^2 (B) 0.71 to 0.89 Kg/cm^2

(C) 0.64 to 0.71 Kg/cm^2 (D) 0.36 to 0.43 Kg/cm^2

Q. 4 The use of a power tiller is considered economical when it is utilized for

(A) 800 hrs/year (B) 1000 hrs/year

(C) 1200 hrs/year (D) 1400 hrs/year

Q. 5 Tilt angle in a disc plough varies from :-

(A) 15 to 25° (B) 5 to 10°

(C) 25 to 30° (D) 10 to 15°

Q. 6 The direction of rotation of the rotor shaft of a rotavator is:-

(A) Opposite to the rotation of tractor wheels

(B) Same as rotation of tractor wheels

(C) Transverse to the rotation of tractor wheels

(D) None of these

Q. 7 One hectare is equal to :-

(A) 10,000 sq.m. (B) 100,000 sq.m.

(C) 1000 sq.m. (D) 100 sq.m.

Q. 8 The unit of draft of implement is :-

(A) Horse power (B) mm

(C) Foot (D) Kilogram force

Q. 9 Bakhar is a type of harrow :-

(A) Disc harrow (B) Blade harrow

(C) Spring tooth harrow (D) Spike tooth harrow

Q. 10 Usually mould boards are made from :-

(A) Mild steel (B) Soft center steel

(C) Chilled steel (D) Hardened carbon steel

Q. 11 The most commonly used implement for clod crushing is :-

(A) Grubber (B) Blade harrow

(C) Pata / Patela / Planker (D) None of above

Q. 12 Frog is a part of :-

(A) M. B. Plough (B) Disc Plough

(C) Disc harrow (D) Indigeneous plough

Q. 13 The angle between the axis of gang and line perpendicular to the direction of motion of disc harrow is called :-

(A) General angle (B) Disc angle

(C) Gang angle (D) Tilt angle

Q. 14 In ploughs, rear furrow wheels are provided to :-

(A) Reduce side thrust on the plough (B) Control depth of cut

(C) Transport Machines (D) None of these

Q. 15 Power developed by an average pair of bullocks is about :-

(A) 750 kW (B) 75 kW

(C) 750 W (D) 75 W

Q. 16 In seed drill, kind of device used to transmit power from ground wheel to seed metering mechanism is:-

(A) Gear (B) Chain and sprocket

(C) Belt and pulley (D) Rope and wheel

Q. 17 Functions of a planter are:-

(A) Open the furrow (B) Meter the seed

(C) Place the seed in furrow (D) All of these

Q. 18 Ballasts are used in rear tyres of a tractor to :-

(A) Increase stability of tractor (B) Increase traction

(C) Decrease front wheel slippage (D) Decrease tractor vibration

APPSC AEES Agriculture Engineering Exam 2016

Q. 19 Increasing length of land side of a one bottom right hand mould board plow will:

(A) Increase the amount of offset on the tractor

(B) Decrease the amount of offset

(C) No affect

(D) None of the above

Q. 20 Disc type furrow openers are especially suited to

(A) Stoney and root infested soils

(B) Poorly prepared bed in trashy soils

(C) Hard and sticky ground with considerable amount of debris and mulch

(D) All of the above

Q. 21 A power tiller is most suited for rotary cultivation because

(A) It generates negative draft

(B) Its traction requirement is low

(C) It provides high degree of soil pulverization

(D) All of the above

Q. 22 If the travel speed of a tillage tool is increased. The draft will

(A) Increase linearly (B) Decrease linearly

(C) Increase quadratically (D) Decrease quadratically

Q. 23 The penetration of animal-drawn disc harrows can be increased by

(A) Increasing the disc angle (B) Decreasing the tilt angle

(C) Adding more dead weight (D) All of the above

Q. 24 Jointer and coulter are the parts of

(A) Disc plough (B) Harrow plough

(C) Indigenous plough (D) MB plough

MPPSC Assistant Agricultural Engineer 2013

Q. 25 .... is the implement recommended to break hard pan at some depth in soil.

(A) M B plough (B) Disc harrow

(C) Rotavator (D) Subsoiler

Q. 26 A small ribbon like furrow slice directly in front of main plough bottom is turned over by

(A) Coulter (B) Furrow wheel

(C) Jointer (D) Land side

Q. 27 A bullock exerts average force equal to .... of its body weight.

(A) One tenth (B) Half

(C) One fourth (D) None of these is correct

Q. 28 A point which lies at a distance of 3/4th size of plough from share wing, where the resultant of horizontal and vertical forces acts is known as ....

(A) Centre of resistance (B) Centre of power

(C) Centre of pull (D) None of these is correct

Q. 29 A seed drill having 1.2 m width operates at 4 kmph for 8 hours. If the field efficiency is 80% what would be the effective field capacity?

(A) 4.35 ha (B) 3.072 ha

(C) 3.84 ha (D) None of these is correct

UKPSC Combined Junior Engineering Exam 2013 Agricultural Engineering Paper-I

Q. 30 The size of an M.B. plough is expressed in terms of its

(A) Width of cut (B) Length of share

(C) Depth of cut (D) None of the above

Q. 31 Which one from the below is a multipurpose tool ?

(A) Indigenous plough (B) M.B. Plough

(C) Cultivator (D) Multipurpose tool bar carrier

Q. 32 Which type of seed drill is mostly used in our country ?

(A) Cup feed type (B) Variable orifice type

(C) Disc type (D) Fluted roller type

Q. 33 A power tiller is most suited for

(A) Transport work (B) Stationary operation

(C) Rotary cultivation (D) None of the above

Q. 34 Farming operation which requires maximum energy :

(A) Harvesting (B) Interculture

(C) Sowing (D) Tillage

Q. 35 Puddling is done to

(A) Reduce percolation of water

(B) Level the field

(C) Kill weeds

(D) Increase percolation of water

Q. 36 Vertical suction of a plough influences

(A) Pulverization (B) Depth of cut

(C) Width of cut (D) Direction of pull

Q. 37 Two primary tillage equipment are

(A) Mould board plough and disc harrow

(B) Disc plough and disc harrow

(C) Disc harrow and cultivator

(D) M.B. plough and sub-soiler

Q. 38 The size of a seed drill is expressed by

(A) Working capacity (B) No. of furrow opener

(C) Spacing between furrow opener (D) No. of furrow opener and spacing between them

Q. 39 An indigenous plough is

(A) A secondary tillage implement (B) A wet-land puddler

(C) A multi-purpose implement (D) A primary tillage implement

Q. 40 The work which requires pulling or drawing such as ploughing, seeding/planting etc. is known as

(A) Tractive work (B) Stationary work

(C) Both of the above (D) None of the above

Q. 41 The simplest form of dynamometer for measuring the draw bar power is

(A) Hydraulic dynamometer (B) Spring dynamometer

(C) Distortion type dynamometer (D) Strain gauge dynamometer

Q. 42 Indian farmers are adopting latest farm mechanization techniques in order to

(A) Achieve the timelines in farm operations

(B) Reduction in loss of crops and food products

(C) Reduction in drudgery and labour

(D) All of the above

Q. 43 Tillage is the preparation of soil to provide

(A) Conditions favourable to plant growth for flood

(B) Weed free field during plant growth during flood

(C) Both (A) & (B)

(D) None of the above

Q. 44 Bullock operated mould board ploughs are

(A) Single bottom (B) Double bottom

(C) Multiple bottom (D) None of these

Q. 45 Gunnel is a part of

(A) Indigenous plough (B) Mould board plough

(C) Cultivator (D) None of the above

Q. 46 Materials which are used for making shares of the plough are

(A) Plain steel (B) Soft centre steel

(C) Chilled cast iron (D) Both (A) and (C)

Q. 47 The manual rotary paddy weeder is used for

(A) Weeding the paddy field by uprooting the weeds

(B) Burying weeds in wet soils

(C) None of the above

(D) Both (A) & (B)

Q. 48 The working capacity of hydraulic power operated land leveller is about

(A) 60 – 70 m^3/day (B) 90 – 110 m^3/day

(C) 110 – 160 m^3/day (D) None of the above

Q. 49 The function of seed drills is to sow seeds

(A) In furrows (B) At proper depth

(C) At uniform rate (D) All above three

Q. 50 Bulldozer is used for

(A) Cleaning forests

(B) Pushing soil upto small distances

(C) Filling big ditches/channels

(D) All above three

Q. 51 Inclined plate planter is used for sowing of

(A) Cotton (B) Legumes

(C) Groundnut (D) All above three

Q. 52 Size of planter is equal to

(A) No. of rows × row spacing

(B) No. of rows × depth of planking

(C) Both (A) and (B)

(D) None of the above

Q. 53 In crop production the first operation performed by the farmers is

(A) Planting of seed (B) Tillage

(C) Intercultural (D) Fertilizer application

Q. 54 Why do we use intercropping ?

(A) Suppression of weed (B) Reduction in plant disease

(C) Stability in more yield (D) All of these

Q. 55 The draft force applied on the implement by a tractor is measured by using

(A) Dynamometer (B) Altimeter

(C) Differential (D) Draft meter

Punjab Mechanic Agricultural Machinery Instructor 2014

Q. 56 Placing of seeds in holes is called

(A) Drilling (B) Dibbling

(C) Broadcasting (D) Seed dropping

Q. 57 The farm machinery can be categorized by

(A) Operation wise (B) Crop wise

(C) Season wise (D) All of the above

Q. 58 Which of the following is not the type of cultivator

(A) Disc type (B) Rotary type

(C) Tine type (D) Centrifugal type

Q. 59 The selection of a tractor is based on

(A) Land holding (B) Cropping pattern

(C) Soil condition (D) All of the above

RPSC AEn Pre Exam 2013 (Agricultural Engineering)

Q. 60 A "Dibbler" is used for

(A) Sowing the seeds (B) Broadcasting the seed

(C) Sowing the seed at fixed spacing (D) None of the above

Q. 61 Which of the following, is not the secondary tillage implement ?

(A) Disc harrow (B) Disc plough

(C) Bakhar and patela (D) Both (A) and (C)

OCS AEn Exam 2011 (Pre) – II (Agricultural Engineering)

Q. 62 The depth of cut of sub-soiler normally lies in the range of

(A) 5 to 10 cm (B) 10 to 20 cm

(C) 20 to 30 cm (D) 30 to 60 cm

Q. 63 In mould board plough, vertical/down suction is provided for maintaining uniform

(A) Width of cut (B) Depth of cut

(C) Draft (D) Pulverization of soil

Q. 64 In disc plough, the depth of cut can be increased by

(A) Decreasing tilt angle

(B) Increasing operational speed

(C) Decreasing disc angle

(D) Increasing disc angle

Q. 65 The soil cutting tools are generally made of

(A) High carbon steel (B) Low carbon steel

(C) Cast iron (D) High speed steel

Q. 66 Ploughing with tractor mounted mould board plough is normally done at a speed of (km/h)

(A) 1 to 3 (B) 4 to 6

(C) 8 to 10 (D) 10 to 12

Q. 67 In dry land area, the plantation method generally practiced is called

(A) Furrow plantation (B) Ridge plantation

(C) Flat bed plantation (D) Ridge and furrow plantation

Q. 68 While puddling the soil, maximum puddling occurs at moisture content between

(A) Wilting point and field capacity

(B) Wilting point and critical moisture level

(C) Dew point and critical moisture level

(D) Field capacity and saturation point

Q. 69 The optimum depth of water for puddling the soil for rice cultivation is

(A) 0 to 5 cm (B) 5 to 10 cm

(C) 15 to 20 cm (D) 20 to 25 cm

Q. 70 The type of furrow opener recommended for use in hard or trashy ground and also in wet sticky soil is

(A) Single disc type (B) Stub runner type

(C) Full or covered runner type (D) Hoe type

Q. 71 Planters differ from a seed drill in respect of

(A) Kind of power transmission system

(B) Kind of metering mechanism

(C) Kind of furrow opener used

(D) All of the above

Q. 72 Most commonly used metering mechanism for fertilizer is

(A) Spur wheel (B) Serrated disc

(C) Star wheel (D) Auger type

Q. 73 Cut-off devices is not required with

(A) Inclined plate metering (B) Horizontal plate metering

(C) Vertical plate metering (D) None of these

CGPSC State Engineering Services Exam Paper – I

Q. 74 Draft per unit area of tilled cross-section is known as

(A) Side draft (B) Draft

(C) Unit draft (D) Gathering

(E) Casting

Q. 75 Tillage system in which only isolated bands of soil are tilled

(A) Mulch tillage (B) Minimum tillage

(C) Zero tillage (D) Primary tillage

(E) Strip tillage

Q. 76 Part of the mould board plough bottom to which the other components of the plough bottoms are attached

(A) Gunnel (B) Landside

(C) Share (D) Frog

(E) Gauge wheel

Q. 77 Disc thickness of a standard disc plough ranging between

(A) 4 - 7 mm (B) 5 - 10 mm

(C) 8 - 13 mm (D) 10 - 12 mm

(E) 12 - 15 mm

Q. 78 Puddling is done in a standing water depth of

(A) 5 - 10 cm (B) 10 - 15 cm

(C) 10 - 15 mm (D) 25 - 40 mm

(E) 15 - 20 cm

Q. 79 An imaginary straight line passing from the centre of resistance through the clevis to the centre of pull is known as

(A) Vertical clevis
(B) Horizontal clevis
(C) Suction
(D) Centre of power
(E) Line of pull

Q. 80 What percentage of inter variation in fertilizer rate is allowed for a seed-cum-fertilizer drill ?

(A) 7.5
(B) 10.5
(C) 12.5
(D) 15
(E) 10

Q. 81 Landside of mould board plough is fastened to the frog with the help of

(A) Plough bolt
(B) Stud bolt
(C) Square bolt
(D) Carriage bolt
(E) Hexagonal head bolt

Q. 82 Cone Penetrometer is used to measure

(A) Stress
(B) Tension
(C) Compression
(D) Depth of plough
(E) Soil strength

Q. 83 Width of furrow cut by an animal drawn medium mould board plough is

(A) 100 - 150 mm
(B) 150 - 200 mm
(C) 200 - 250 mm
(D) 250 - 300 mm
(E) 300 - 350 mm

Q. 84 Draft on mould board plough developing 100 kg pull at 30^0 to the horizontal is

(A) 50 kg
(B) 75 kg
(C) 86 kg
(D) 95 kg
(E) 110 kg

Q. 85 Row spacing of cotton crop is

(A) 40 - 48 inches (B) 30 - 38 inches

(C) 50 - 58 inches (D) 60 - 68 inches

(E) 35 - 40 inches

Q. 86 Back furrow is caused while ploughing is done

(A) Center to side (B) Side to centre

(C) One-way ploughing (D) Two-way plough

(E) Casting

Q. 87 A device used to cut the furrow slice vertically from a land ahead of the bottom is

(A) Share (B) Coulter

(C) Jointer (D) Gauge wheel

(E) Landside

Q. 88 A strip of unploughed land that left at each end of the field for tractor turning is

(A) Gathering (B) Casting

(C) Head land (D) Crown

(E) Furrow wall

UKPSC A.En. Exam 2007 Paper – II

Q. 89 The working principle of mould board plough is

(A) Sliding (B) Rolling

(C) Both Rolling and Sliding (D) None of the above

Q. 90 Minimum tillage system is also known as

(A) Optimum tillage (B) Zero tillage

(C) Plough-plant method (D) Till-plough method

Q. 91 Factors affecting the draft of a tillage implement are

(A) Size of the implement

(B) Forward speed of the implement

(C) Depth of the furrow

(D) All of the above factors

Q. 92 Vertical suction in a mould board plough is provided to

(A) Increase the width of cut (B) Decrease the draft

(C) Ease in penetration (D) Decreases the depth oı cut

Q. 93 The tandem type harrow is also called as

(A) Single action disc harrow (B) Double action disc harrow

(C) Offset harrow (D) None of the above

Q. 94 Which component of the following is not attached to the frog of a mould board plough ?

(A) Mouldboard (B) Share

(C) Landslide (D) Disc coulter

Q. 95 A Sub-soiler plough is best suited for

(A) Breaking hardpan (B) Shallow ploughing

(C) Inter-cultivation (D) Deep ploughing

Q. 96 If a three bottom 40 cm mouldboard plough has a working depth of 15 cm against a draft of 1800 kg, the unit draft would be

(A) 1 kg/cm^2 (B) 2 kg/cm^2

(C) 3 kg/cm^2 (D) None of the above

Q. 97 Digging of holes for tree-plantation is often done by

(A) Auger (B) Soil sampler

(C) Post hole digger (D) Hoe

Q. 98 The seed drill and planter

(A) Opens a furrow in the soil (B) Meters the seeds

(C) Provides soil cover to seeds (D) All of the above

Q. 99 The most common type of seed metering device in Indian seed drills is

(A) Internal double run type (B) Cup feed type

(C) Cell-plate type (D) Fluted roller type

Q. 100 Calibration of a seed drill certifies

(A) Row to row spacing

(B) Plant to plant spacing

(C) Testing of seed drill for the correct seed rate

(D) Speed and draft of seed drill

Q. 101 Seed rate of a planter is controlled by

(A) Increasing the forward speed of the machine

(B) Increasing/ decreasing the speed of cell-plate

(C) Decreasing the speed of plantar

(D) Increasing the speed of planter

Q. 102 The star feed type metering device is used in

(A) Fertilizer drills (B) Seed drills

(C) Planter drills (D) None of the above

Q. 103 In a seed drill, the seed tube delivers the seed directly to

(A) Boot (B) Furrow

(C) Furrow opener (D) Seed metering system

Q. 104 Seed drills are used for

(A) Broadcasting the seeds (B) Broadcasting of fertilizers

(C) Sowing of seed in a row (D) None of the above

UKPSC A.En. Exam 2012 Paper - II

Q. 105 Spike tooth harrows and spring tine harrows are classified as

(A) Disk harrows (B) Drag harrows

(C) Blade harrows (D) None of the above

Q. 106 Fluted roller of a seed drill is driven by

(A) Rectangular shaft (B) Square shaft

(C) Round shaft (D) None of the above

Q. 107 The subsurface tillage tools that work at a shallow depth and do not turn the soil are called as

(A) Disk plough (B) Ridger

(C) Sweeps (D) None of the above

Q. 108 Disc angle adjustment influences

(A) Depth of cut
(B) Width of cut
(C) Soil break up
(D) Direction of travel

Q. 109 Vertical suction of a plough influences

(A) Pulverization
(B) Depth of cut
(C) Width of cut
(D) Direction of pull

Q. 110 For maximum moisture conservation ploughing is done with

(A) MB plough (Mould Board Plough)
(B) Cultivator
(C) Chisel plough
(D) Disc harrow

Q. 111 The vertical distance between the point of share to the beam of plough is called

(A) The vertical suction
(B) Throat clearance
(C) Horizontal suction
(D) None of the above

Q. 112 Which one of the following is not a primary tillage equipment ?

(A) Meston plough
(B) Disc plough
(C) Desi plough
(D) Cultivator

Q. 113 The type of bearing generally used to support the disc on a standard disc plough is

(A) Ball bearing
(B) Tapered roller bearing
(C) Plain roller bearing
(D) Bush bearing

Q. 114 Material used for manufacture of cultivator shovel :

(A) Brass
(B) Gun metal
(C) Mild steel
(D) Chilled cast iron

Q. 115 The optimum operating speed of a rotor in a rotavater is ....

(A) More than 500 rpm
(B) 350 – 500 rpm
(C) 150 – 350 rpm
(D) 20 – 150 rpm

Q. 116 Angle of seed hole displacement for picking seed by seed plate in a pneumatic vegetable planter for sowing radish is

(A) 7.5° (B) 15°

(C) 22.5° (D) 30°

Q. 117 Concavity of disc in disc harrow causes effect on ....

(A) Penetration (B) Pulverization

(C) Depth of furrow (D) Both (A) and (B)

Q. 118 Tilt angle is associated with

(A) Cultivator (B) Disc plough

(C) Mould board plough (D) Rotavator

Q. 119 The planter which works on suction principle is

(A) Pneumatic planter (B) Inclined plate planter

(C) Seed planter (D) None of the above

Q. 120 The drawbar pull of a tractor is measured by

(A) Hour – meter (B) Hydrometer

(C) Dynamometer (D) Current meter

Q. 121 A disc harrow can be used for

(A) To control weeds (B) To mulch the soil

(C) To pulverize the soil (D) All of the above

Q. 122 Deep ploughing is done beyond depths of about

(A) 10 – 12 cm (B) 12 – 15 cm

(C) 15 – 20 cm (D) > 30 cm

Q. 123 Which of the following factors affects the penetration of disc harrow ?

(A) Diameter of discs (B) Spacing between discs

(C) Weight of the harrow (D) All of the above

Q. 124 Scouring ability of a tillage tool is affected by

(A) Coefficient of soil – metal friction

(B) Soil cohesion

(C) Soil adhesion

(D) All of the above

Q. 125 The rake angle of a sweep type cultivator is

(A) 60 degree (B) 70 degree

(C) 90 degree (D) None of the above

Graduate Aptitude Test in Engineering - 2007

Q. 126 A right hand offset disk harrow is operating with front and rear gang angles of 15 and 21 respectively. The centers of the two gangs are 2.45 m and 4.25 m behind a transverse line through the hitch point on the tractor drawbar. The horizontal soil force components are: $L_f = 3.1$ kN, $S_F = 2.65$ kN, $L_T = 3.35$ kN, $S_T = 2.65$ kN. The amount of offset of the center of cut with respect of the hitch point is

(A) 0.740 m (B) 0.795 m

(C) 0.968 m (D) 1.006 m

Q. 127 A tractor drawn rotary cultivator is concurrent revolution mode is to be operated at a depth of 150 mm and at a forward speed of 3.6 km h^{-1}. The radius of working set is 280 mm. the number of blades, which would cut identical path is 3. The working width of the cultivator is 1.8 m. the cultivator is to be powered from the tractor PTO running at 540 rpm through a suitable gearbox. For getting a tilling pitch of 74.1 mm, the suitable gear ratio is

(A) 1:2 (B) 1:1.5

(C) 1.5:1 (D) 2:1

Graduate Aptitude Test in Engineering - 2008

Q. 128 A tractor drawn seed broadcaster is operated at 10.8 km h^{-1}. The broadcaster has a horizontal seed plate located inside the hopper above the ground level. The diameter of the plate is 300 mm and its angular velocity is 80 rad s^{-1}. If the air resistance is neglected, the resultant velocity with which the seed mass is approaching the furrow 3 seconds after its release from the hopper is

(A) 29.40 m s^{-1} (B) 30.52 m s^{-1}

(C) 31.75 m s^{-1} (D) 41.40 m s^{-1}

Graduate Aptitude Test in Engineering - 2009

Q. 129 A tractor drawn vertical rotor planter is operated in the field at a forward speed of 5 km/h. the effective diameter of the ground wheel of the planter is 0.5 m and the transmission ratio between the ground wheel and the rotor shaft is 1:1.

(i) If the skid of the ground wheel of the planter is 10%, the speed of rotor in rpm will be

(A) 26 (B) 39

(C) 48 (D) 58

(ii) If the number of cells on the vertical rotor is 20, the plant to plant distance in a row in mm will be

(A) 87 (B) 128

(C) 157 (D) 174

Q. 130 The disc and tilt angles of a single bottom disc plough having 0.76 m disc diameter are 50° and 25°, respectively. if the depth of ploughing is 0.25 m, width of cut of the plough in m is

(A) 0.508 (B) 0.526

(C) 0.546 (D) 0.559

Q. 131 The standard of a cultivator experiences a maximum bending moment of 120 Nm. It has a rectangular cross-section with sides in the ratio of 3:1. if the permissible bending stress of the material is 8×10^{7} N m^{-2}, the cross-sectional area of the cultivator standard in square millimeter is

(A) 222 (B) 300

(C) 492 (D) 624

Graduate Aptitude Test in Engineering - 2010

Q. 132 A 2×0.3 m tractor drawn mould board plough while operating at a depth of 0.15 m has a draft of 2.5 kN at a forward speed of 3 km h^{-1} with a field efficiency of 75%. When the speed of operation is increased by 20%, draft increased by 10%. Assuming field efficiency, soil pulverization and soil inversion to be the same at both the speed, the performance index of the plough increases by

(A) 0 % (B) 9 %

(C) 20 % (D) 30 %

Graduate Aptitude Test in Engineering - 2011

Q. 133 The metering mechanism of a seed drill is driven by the ground wheels at a velocity ratio of 1:2. When forward speed of the seed drill is increased from 3.0 to 3.45 km h^{-1}, the seed rate would

(A) Increase by 15 % (B) Decrease by 13 %

(C) Decrease by 15 % (D) Remain the same

Q. 134 A maize planter drops seeds at 0.20 m interval. The seed weight is 200 g per 1000 seeds. If the row to row spacing is 0.25 m, the seed rate in kg ha^{-1} is

(A) 5 (B) 10

(C) 20 (D) 40

Q. 135 The resultant soil reaction force acting on a single bottom mould board plough has a component of 1200 N acting at an angle of 31° downwards in a vertical plane. This plane is along the direction of travel. Weight of the plough is 620 N. Neglecting the side forces, the magnitude of pull in N would be

(A) 618 (B) 1028

(C) 1350 (D) 1609

Graduate Aptitude Test in Engineering - 2012

Q. 136 The draft and total power requirement of a rotary cultivator operating in concurrent mode as compared to a spring tyne cultivator of equal cutting width under the same operating conditions, respectively are

(A) Higher and higher (B) Lower and lower

(C) Lower and higher (D) Higher and lower

Q. 137 During field operation, the shank of a tractor drawn rigid tyne sweep type cultivator is mainly subjected to

(A) Bending (B) Shear

(C) Torsion (D) Bending and torsion

Q. 138 Soybean is to be planted with a precision planter that meters 54 seeds per revolution of the metering disc powered from a ground wheel of diameter 490 mm. The desired plant population is 44800 per ha with a row to row spacing of 0.75 m. The germination percentage is 84. The planter is to be operated at 2.5 km h^{-1} with a 10% skid of ground wheel.

(i) The angular speed of ground wheel in rpm is

(A) 20.3 (B) 24.6

(C) 28.3 (D) 32.6

(ii) The angular speed ratio of metering disc to ground wheel for obtaining the desired plant population is

(A) 0.125:1 (B) 0.150:1

(C) 0.225:1 (D) 0.250:1

Graduate Aptitude Test in Engineering - 2013

Q. 139 The rear furrow wheel in a tractor mounted disc plough is provided to

(A) Reduce the frictional power loss

(B) Maintain the uniform depth of cut

(C) Reduce the side draft

(D) Improve the penetration of the plough

Graduate Aptitude Test in Engineering - 2014

Q. 140 A cultivator with a working width of 1.2 m utilizes 95% of its width due to overlapping while operating at a forward speed of 2 km/h. If the time lost in turning and other interruptions is 50 minutes per hectare, the field efficiency of the cultivator in percent is

(A) 79.84 (B) 83.34

(C) 86.97 (D) 90.20

Q. 141 A two-row horizontal potato planter with 0.6 m ridge spacing has 9 cups on each seed plate of 0.4 m diameter. For each revolution of the ground wheel, the seed plate makes half a revolution. The diameter of the ground wheel is 0.5 m. If the planter uses cut tubers each of 25 g mass, the seed rate in kg/ha is ....

Graduate Aptitude Test in Engineering – 2015

Q. 142 The horizontal component of soil forces acting on the front gang of a right hand offset disc harrow in the directions parallel and perpendicular to the direction of motion are 8 kN and 3 kN respectively. The corresponding forces on the rear gang are 6 kN and 7 kN, respectively. If the horizontal component of pull acts towards the left of the direction of motion, its magnitude in kN will be ....

Q. 143 A two-wheel drive tractor is operating a mould board plough at an average speed of 4 km/h. The draft and the rear axle torque are found to be 30 kN and 25 kN m, respectively. If the rolling radius of traction wheel is 0.7 m and the wheel slip is 20%, the tractive efficiency in percent will be

(A) 67.19 (B) 46.71

(C) 72.87 (D) 84.00

Graduate Aptitude Test in Engineering – 2016

Q. 144 A vertical rotor planter has 8 cells on each rotor. The rolling radius of the ground wheel is 200 mm. The ratio of rpm of the ground wheel to that of the rotor shaft is 2:3. If the planting is done at a forward speed of 3.5 km h^{-1}, the plant spacing in the rows in mm will be ....

Q. 145 An inclined blade cutting tool of 250 mm width is operating at 200 mm cutting depth. The normal load on the tool and the coefficient of soil-metal friction are 1000 N and 0.3, respectively. The soil cutting force per unit length of cutting edge is 20 N mm^{-1}. The tool lift angle is 40°. The required specific draft (or unit draft) in kPa is ....

Graduate Aptitude Test in Engineering – 2017

Q. 146 A subsoiler operating at 400 mm depth requires 15 kW peak drawbar power at 3 km h^{-1} speed. The standard is rigidly fixed vertically on the main frame. The resultant soil resistance acts horizontally at a vertical distance of 450 mm from the main frame. The standard has rectangular cross-section with width to thickness ratio of 4:1, and it fails due to bending. If the allowable bending stress is 90 N mm^{-2}, the width of the standard in mm will be ....

Q. 147 The horizontal component of resultant soil thrust (T) acting on each gang of a single acting disc harrow is 1650 N. The resultant downward load (W) acting on each gang is 2500 N. The perpendicular distance of T from the gang axis is 200 mm. In order to get a uniform depth of cut, the distance between the line of action of W and the centre of gang in mm will be ....

Q. 148 A 9-row fluted type seed drill with 400 mm ground wheel diameter is used for sowing wheat at 200 mm row spacing. Each fluted roller discharges 6500 mm^3 volume of seeds per revolution. The ratio of ground wheel rpm to fluted roller shaft rpm is 2:1. If the bulk density of wheat is ....

Graduate Aptitude Test in Engineering – 2018

Q. 149 A tractor drawn five-tine sweep type cultivator is set at an angle (load angle) of 45° with the soil surface during its operation. The sweep of each tine experiences a normal force of 960 N on its surface. If the coefficient of soil-metal friction is 0.3, the total draft requirement of the cultivator is .... kN.

Q. 150 A precision seed planter can plant 10 seeds per revolution of the metering plate with a row to row distance of 450 mm at a speed of 6 km h^{-1}. A plant population of 36 plants per square meter is desired for the crop. The germination percentage of the seed is 90%. If the planter ground wheel has a rolling radius of 350 mm, the speed ratio of metering plate to the ground wheel required to obtain the desired plant population is ....

Graduate Aptitude Test in Engineering – 2019

Q. 151 The farm machine/implement used only for preparing wetland is

(A) Rotavator (B) Disk harrow

(C) Hydro-tiller (D) Cultivator

Q. 152 The total width between the two extreme furrow openers in a tractor drawn 9-row wheat seed drill is 1.6 m. The average mass of wheat seeds dropped per meter of row length in each furrow opener is 2.15 g. Seed rate obtained with the seed drill in kg/ha is ....

Q. 153 A level field of 1.2 ha (120 m x 100 m) is ploughed using a reversible mould board plough with a total effective cutting width of 0.64 m. The average field overlap is 80 mm between two consecutive laps. The average time taken for each turn is 30 seconds and the mean operating speed is 5.0 km/h. The maximum effective field capacity in ha/h is

(A) 0.207 (B) 0.236

(C) 0.283 (D) 0.318

Graduate Aptitude Test in Engineering – 2020

Q. 154 A tractor drawn right-hand offset disk harrow experiences longitudinal and side soil reactions in the front gang as 3.0 kN and 2.5 kN, respectively as compared to 3.5 kN and 4.0 kN in the rear gang. The longitudinal distance of centers of front and rear gangs are located at 2.5 m and 4.0 m, respectively behind the tractor hitch point. The required amount of offset of the disk harrow in m is ....

Q. 155 A tractor operated 9-row precision planter has 16 cells on the metering plates. The speed ratio of the metering plates to the ground drive wheel is 1 :2 and the rolling diameter of the ground drive wheel is 40 cm. Assuming no skid, the plant to plant spacing in rows in mm is

(A) 39 (B) 50

(C) 157 (D) 314

Graduate Aptitude Test in Engineering – 2021

Q. 156 During ploughing with a tractor mounted mould board plough, the mast of three point hitch system would be

(A) Inclined 5 to 20º with horizontal

(B) Nearly vertical

(C) Parallel to the direction of travel of the tractor

(D) Parallel to the rear axle of the tractor

Q. 157 Useful soil reaction forces acting on a tractor drawn mould board plough during operation are 2.0 kN, 0.9 kN and 0.6 kN along longitudinal, transverse and vertical directions, respectively. The soil-metal friction angle is 25°. Neglecting the effects of weight of the implement and the vertical soil reaction, the estimated draft in N is ....

Q. 158 A bushy crop with stem cross-sectional diameter 6 mm is to be cut by impact force at a height of 50 mm above the soil surface. Based on the entire stem cross-section, the modulus of elasticity is 1500 N/mm^2 and ultimate tensile strength is 35 N/mm^2. The force in N that would cause failure of the stem due to bending is (Take $\pi = 3.14$)

(A) 14.84 (B) 23.52

(C) 29.69 (D) 44.53

Answers Key

1	2	3	4	5	6	7	8	9	10
D	A	D	A	A	B	A	D	B	B
11	12	13	14	15	16	17	18	19	20
C	A	C	A	C	B	D	B	C	C
21	22	23	24	25	26	27	28	29	30
A	C	D	D	D	C	A	A	B	A
31	32	33	34	35	36	37	38	39	40
D	D	C	D	A	B	D	D	D	A
41	42	43	44	45	46	47	48	49	50
B	D	C	A	B	D	D	C	D	D
51	52	53	54	55	56	57	58	59	60
D	A	B	D	A	B	D	D	D	C
61	62	63	64	65	66	67	68	69	70
B	D	B	D	A	B	A	D	B	A
71	72	73	74	75	76	77	78	79	80
B	C	A	C	E	D	B	A	E	A
81	82	83	84	85	86	87	88	89	90
A	E	B	C	E	A	B	C	A	C
91	92	93	94	95	96	97	98	99	100
D	C	B	D	A	A	C	D	D	C
101	102	103	104	105	106	107	108	109	110
B	A	A	C	B	B	C	B	B	A
111	112	113	114	115	116	117	118	119	120
B	D	B	D	C	C	D	B	A	C
121	122	123	124	125	126	127	128	129	130
D	C	D	D	B	A	D	D	D, A	D
131	132	133	134	135	136	137	138	139	140
B	B	D	D	D	C	A	B, A	C	A
141	142	143	144	145	146	147	148	149	150
1194.21	14.56	67.2	104.67	117.46	129.28	132	110.13	4.41	3.96
151	152	153	154	155	156	157	158		
C	107.50	A	1.50	C	B	2408	A		

Explanations

Q. 29

Actual Field capacity = Width×Speed

$= 1.2 \times 4000 = 3.84$ ha

Effective field capacity $= 0.80 \times 3.84 = 3.072$ ha

Q. 24

Draft = 100×cos30 = 86 kg

Q. 96

Total Cross sectional area of plough bottom = 3×40×15 = 1800 cm^2

Hence, Unit draft = Draft/Cross sectional area of furrow

= 1800/1800 = 1 kg/cm^2

Q. 126

Given that

$S_r = 2.65$ kN; $S_f = 2.65$ kN; $L_r = 3.35$ kN; $L_f = 3.1$ kN

d = 4.25 m; b = 2.45 m

We know that

$$\text{Amount of offset} = \frac{S_r.d - S_f.d}{L_f + L_r}$$

$$= \frac{2.65 \times 4.25 - 2.65 \times 2.45}{3.1 + 3.35} = 0.740 \text{ m}$$

Q. 127

Given that

Forward speed of cultivator (S) = 3.6 km/ h = 1 m/ sec

Tilling pitch (P) = 0.0741 m; Number of teeth or blades (T) = 3

$$\because \quad \text{Forward Speed (S)} = \frac{TPN}{60}$$

$$\Rightarrow \quad 1 = \frac{3 \times 0.0741 \times N}{60}$$

$$\Rightarrow \quad N = 270 \text{ rpm}$$

We know that

$$\text{Gear ratio} = \frac{N_{PTO}}{N}$$

$$\Rightarrow \quad \text{G.R.} = \frac{540}{270}$$

$$\Rightarrow \quad \text{G.R.} = 2 : 1$$

Q. 128

Angular velocity of plate (ω) = 80 rad/ s; Radius of plate (r) = 0.150 m

Planter speed (V_p) = 10.8 km/ h = 3 m/ s

We know that

Linear velocity of plate (u) = ω.r

$\Rightarrow$ $u = 0.15\times80 = 12$ m/ s

∵ Seed is falling under gravity, thus final velocity of seed

$$(V_s) = u + gt$$

$\Rightarrow$ $V_s = 12 + 9.8\times3$

$\Rightarrow$ $V_s = 41.4$ m/ s

∵ Planter speed is very low, therefore minor, effect on resultant velocity of seed.

Or we can further solve

$$\text{Resultant velocity } (V_r) = \sqrt{V_s^{\,2} + V_p^{\,2}}$$

$$V_r = \sqrt{(41.4)^2 + (3)^2}$$

$$V_r = 41.29 \text{ m/ sec}$$

Q. 129 (i)

Speed of planter (S) = 5 km/ h or 83.33 m/ min

Diameter of ground wheel (D) = 0.5 m

We know that

$$\pi.D.N = S$$

$\Rightarrow$ $N = \dfrac{83.33}{0.5\times3.14} = 53.08$ rpm

∵ Skid of ground wheel is 10%, therefore

Actual speed of ground wheel $= \dfrac{53.08}{0.9} = 58.97$ rpm ≈ 58 rpm

∵ Transmission ratio = 1:1 $\Rightarrow$ Speed of rotor = 58 rpm

(ii)

Distance traveled by ground wheel in one revolution = $3.14\times0.5\times1.1$

$= 1.73$ m

∵ Number of plant placed in one revolution of ground wheel is 20.

Thus

$$\text{Plant to plant distance} = \frac{1.73}{20} = 0.087 \text{ m} = 87 \text{ mm}$$

Q. 130

Given that

Disc angle (∅) = 50°; Tilt angle (θ) = 25°

Depth of ploughing (h) = PQ = 0.25 m; Diameter of disc (d) = 0.76 m

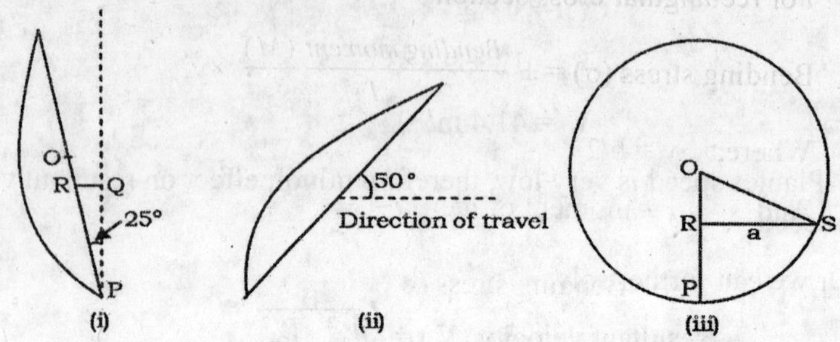

According to picture (i)

$$\cos\theta = \frac{PQ}{PR} = \frac{h}{PR}$$

$$\Rightarrow \quad PR = \frac{0.25}{\cos 25} = 0.276 \text{ m}$$

And

$$OR = OP - PR = d/2 - PR$$

$$\Rightarrow \quad OR = 0.76/2 - 0.276 = 0.104 \text{ m}$$

Also

$$RS = \sqrt{(0.38)^2 + (0.104)^2}$$

$$RS = a = 0.3655 \text{ m}$$

Referring picture (iii)

Width of cut of the plough = 2a sin ∅

$$= 2 \times 0.3655 \times \sin 50 = 0.5599 \text{ m}$$

Q. 131

Given that

Bending moment (M) = 120 Nm; Side ratio (b/d) = 3:1 ⇒ b = 3d

Permissible bending stress (σ) = 8×10^7 N/m^2

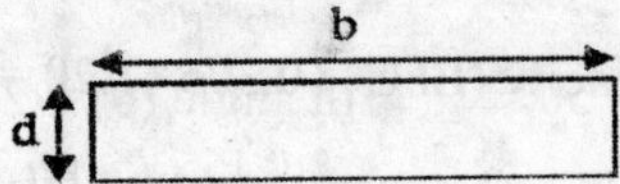

For rectangular cross section

$$\text{Bending stress } (\sigma) = \pm \frac{\textit{Bending moment } (M)}{I} \times y$$

Where; $y = b/2$

And I = moment of inertia = $\dfrac{db^3}{12}$

$$\therefore \quad \text{Bending stress } (\sigma) = \frac{6M}{db^2}$$

$$\Rightarrow \quad 8\times10^7 = \frac{6\times120}{9d^3}$$

$$\Rightarrow \quad d = 10 \text{ mm}$$

and $\quad b = 30$ mm

$\therefore$ Cross sectional area (b×d) = 10×30 = 300 mm^2

Q. 132

Case I

Depth of operation (d) = 0.15 m; Draft (D) = 2.5 kN

Effective field capacity (EFC)

$$= S \times W \times \text{Field efficiency}$$

$$= \frac{3000}{3600} \times 0.6 \times 0.75$$

$$= 0.375 \text{ m}^2\text{/h}$$

$\therefore$ Performance Index (PI_1)

$$= \frac{\textit{Depth} \times \textit{Pulverization} \times \textit{Inversion} \times \textit{Effective field capacity}}{\textit{Draft}}$$

$$= \frac{0.15 \times Pulverization \times Inversion \times 0.375}{2.5}$$

$= 0.0225 \times \text{Pulverization} \times \text{Inversion}$

Case II

Depth of operation (d) = 0.15 m; Draft (D) = 2.5×1.1 = 2.75 kN

Effective field capacity (EFC = 0.375×1.20 = 0.45 m²/h

$$\therefore \quad \text{Performance Index } (PI_2) = \frac{0.15 \times Pulverization \times Inversion \times 0.45}{2.75}$$

$= 0.0245 \times \text{Pulverization} \times \text{Inversion}$

$$\therefore \text{ Increase in performance index} = \left(\frac{PI_2 - PI_1}{PI_1}\right) \times 100$$

$$= \left(\frac{0.0245 - 0.0225}{0.0225}\right) \times 100 = 8.89\ \% \approx 9\%$$

Q. 133

Speed of operation doesn't affect the seed rate of seed drill or planter.

Q. 134

$\because$ Maize planter drops seeds at 0.20 m interval.

$\therefore$ One seed is dropped in = 0.20×0.25 = 0.05 m²

Hence, number of seeds dropped per hectare $= \frac{10000}{0.05} = 200000$ seeds

$\because$ Weight of one seed = 0.2×10^{-3} kg

Therefore,

Seed rate = $200000 \times 0.2 \times 10^{-3} = 40$ kg/ ha

Q. 135

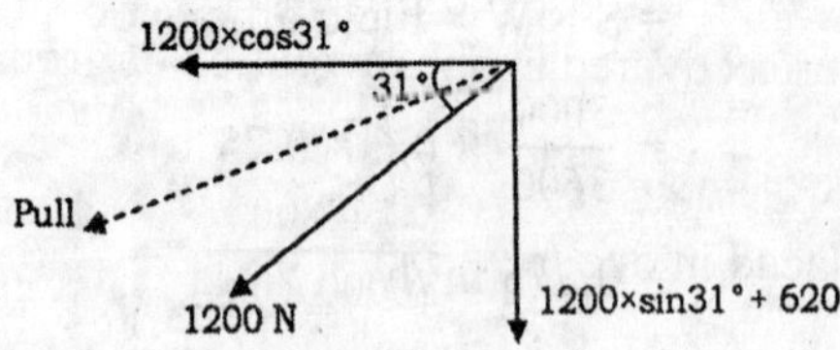

Horizontal component of soil reaction force acting plough bottom

$(H_f) = 1200 \times \cos 31$

$H_f = 1028.60$ N

Vertical component of soil reaction force acting plough bottom

$= 1200 \times \sin 31$

$= 618.04$ N

∵ Weight of plough = 620 N

∴ Net vertical force acting on plough bottom (V_f) = 620 + 618.04

= 1238.04 N

Hence,

Magnitude of pull = $\sqrt{H_f^2 + V_f^2}$

$\Rightarrow \quad = \sqrt{(1028.60)^2 + (1238.04)^2}$

Magnitude of pull = 1609.6 N

Q. 138 (i)

Speed of planter (S) = 2.5 km/ h = 41.67 m/ min

Diameter of ground wheel (D) = 0.490 m

We know that

$\pi.D.N = S$

$\Rightarrow \quad N = \frac{41.67}{0.490 \times 3.14} = 27.08$ rpm

∵ Skid of ground wheel is 10%, therefore

Actual speed of ground wheel = 27.08×0.9 = 24.37 rpm

(ii)

Distance covered in one revolution of ground wheel = 3.14×0.490×1.1

= 1.69 m

Hence, total area covered in one revolution of ground wheel = 16.9×0.75

= 1.27 m²

Total seed placed in one m² = $\frac{44800}{10000 \times 0.84}$ = 5.33 seeds/m²

∴ Seed placed in one revolution of ground wheel = 5.33×1.27 = 6.77 seeds

$\because$ Speed ratio $= \frac{N_p}{N_g} = \frac{\textit{Seed placed in one revolution of ground wheel}}{\textit{Seed placed in one revolution of metering plate}}$

Speed ratio $= \frac{6.77}{54} = 0.125$

Q. 140

Given that

Forward speed (S) = 2 km/h; Working width (W) = 1.2 m

Actual field capacity = S×W = $\frac{1.2 \times 2 \times 1000}{60}$ = 40 m²/min

Actual time taken per m² (T_A) $= \frac{1}{40}$ min/m²

Theoretical time taken per m² (T_T)

= Theoretical time + Turning loss

$$= \frac{60}{1.2 \times 0.95 \times 2 \times 1000} + \frac{50}{10000} \text{ min/m}^2$$

We know that

Field Efficiency $= \frac{T_A}{T_T} \times 100$

$$= \frac{\left(\frac{1}{40}\right)}{\left(\frac{60}{1.2 \times 0.95 \times 2 \times 1000} + \frac{50}{10000}\right)} \times 100 = 79.84\%$$

Q. 141

Distance traveled by ground wheel in one revolution = 3.14×0.5

= 1.57 m

Hence, total area covered in one revolution of ground wheel = 1.57×0.6

= 0.942 m²

$\therefore$ Revolutions made by ground wheel in one hectare of field

$= \frac{10000}{0.942}$

= 10615.71 revolutions

$\because$ Seed plate makes half revolutions to the ground wheel.

Hence, Number of revolutions of seed plate = 10615.71/2 = 5307.58 revolutions

Again,

Each revolution of seed plate drops = 9×0.025 = 0.225 kg mass of tubers

∴ Tuber dropped in 5307.58 revolutions = 5307.58×0.225 = 1194.21 kg

Hence, Seed rate = 1194.21 kg/ ha

Q 142

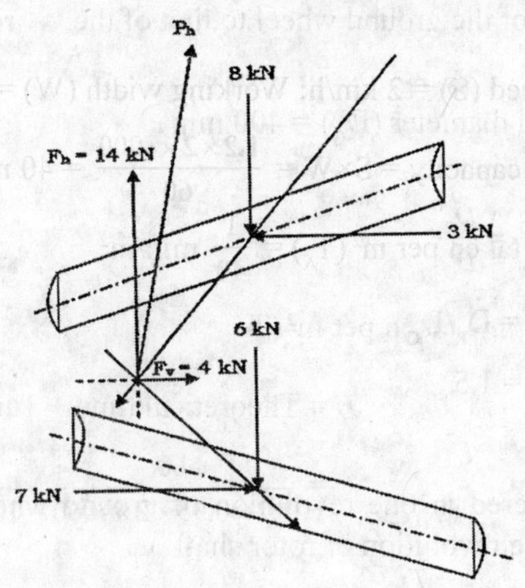

Referring above figure

Force component in direction parallel to direction of travel (F_h) = 8 + 6

= 14 kN

Force component in direction perpendicular to direction of travel

(F_v) = 7 – 3 = 4 kN

Therefore,

Magnitude of horizontal component of pull (P_h) = $4^2 + 14^2$

= 14.56 kN

Q. 143

Actual velocity (V_a) = 4×(1 – 0.2) = 3.2 km/h

We know

Tractive efficiency (TE) = Drawbar power/Axle power

$$= \frac{Pull \times V_a}{\frac{T}{r} \times V_t}$$

$$= \frac{30 \times 3.2}{(25/0.7) \times 4} = 0.672 \text{ or } 67.2\ \%$$

Q. 144

Ratio of rpm of the ground wheel to that of the rotor shaft (N_G/N_R) = 2:3

Ground wheel diameter (D_G) = 400 mm

We know

$$\frac{N_G}{N_R} = \frac{D_R}{D_G}$$

$\Rightarrow$ 2/3 = D_R/D_G

$\Rightarrow$ D_G/D_R = 1.5

It means,

Distance covered in one revolution of ground wheel = 1.5× distance covered in one revolution of rotor shaft

$\because$ Distance covered in one revolution of ground wheel = πD_G

= 3.14×400

= 1256 mm

and Distance covered in one revolution of rotor shaft = 8× Plant spacing

$\Rightarrow$ 1.5×8× Plant spacing = 1256

Hence, Plant spacing = 1258/(1.5×8)

= 104.67 mm

Q. 145

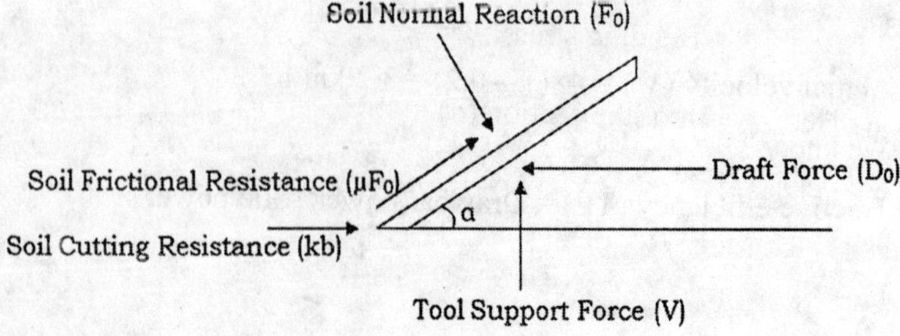

Equilizing forces of horizonatal Direction

$$D_0 = F_0 \sin\alpha + F_0 \cos\alpha + kb$$

$$= 1000\times\sin 40 + 1000\times 0.30\times\cos 40 + 500$$

$$= 5.873 \text{ kN}$$

Shear plane area (A) = b×d

$$= 0.25\times 0.2 = 0.05 \text{ m}^2$$

Hence, Specific draft (Unit Draft) $= D_0/ A$

$$= 5.873/0.05 = 117.46 \text{ kPa}$$

Q. 146

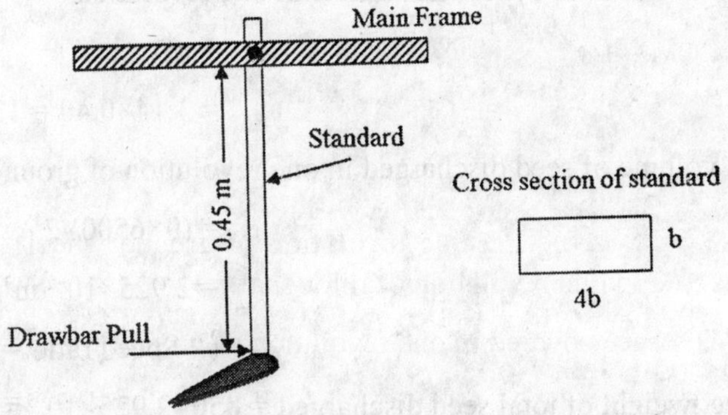

Drawbar Pull = Drawbar Power/ Forward Speed of tractor

$$= 15\times 1000/(3\times 1000/3600) - 18000 \text{ N}$$

Now taking bending moment around point from where standard attached with main frame

Bending Moment (M) = Drawbar Pull × Distance

$$= 18000\times 0.450 = 8100 \text{ N-m}$$

We know

$$\text{Bending stress} = \frac{6\times M}{bd^2}$$

Here, width of the section (d) = 4b

Hence, $$90\times 10^6 = \frac{6\times 8100}{b(4b)^2}$$

$\Rightarrow$ $$b = 0.03232 \text{ m Or } 32.32 \text{ mm}$$

∴ Width of section = 4d = 4×32.32 = 129.28 mm

Q. 147

In disk harrow

For uniform depth of penetration

Torque developed on gang axle due to soil reaction = Torque around gang axle due to weight or any other vertical load.

⇒ $W \times h = f \times T$

⇒ $h = 1650 \times 200/2500 = 132$ mm

Q. 148

Distance travelled by Ground wheel in one revolution

$= \pi D$

$= 3.14 \times 0.40 = 1.256$ m

Total volume of seed discharged in one revolution of ground wheel

$= (9 \times 6500)/2$

$= 2.925 \times 10^{-5}$ m³

(∵ Speed ratio = 2:1)

Hence, weight of total seed discharged = $850 \times 2.925 \times 10^{-5} = 0.0249$ kg

Area of field in which the seed discharged = $9 \times 1.256 \times 200$

$= 2.261$ m² or 2.261×10^{-4} ha

∴ Seed rate = $0.0249/ 2.261 \times 10^{-4} = 110.13$ kg/ha

Q. 149

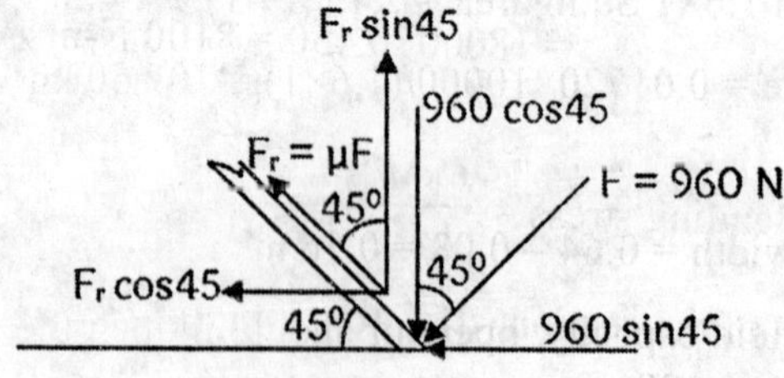

Referring above figure

Draft force on single tine $= F_r \cos45 + F \sin 45$

$= \mu F \cos45 + F \sin 45$

$= 0.2 \times 960 \times \cos45 + 960 \times \sin45$

$= 882.47$ N or 0.882 kN

Total Draft requirement of the cultivator = 5×0.882 = 4.41 kN

Q. 150

Diameter of ground wheel = 2×0.35 = 0.7 m

Distance covered by ground wheel in one revolution = $\pi D = 3.14 \times 0.7$

= 2.198 m

Row to row sapcing = 0.45 m

Area covered in one revolution of ground wheel = 0.45×2.198 = 0.9891 m^2

Given, Seeds to be sown = 36/0.90 = 40 seeds/ m^2

Total seeds sown in 0.9891 m^2 is

= 40×0.9891

= 39.56 seeds per revolution of ground wheel

Revolutions of ground wheel per seed = 1/39.56

Also, Revolutions of metering plate per seed = 1/10

∴ Required Speed ratio = 39.56/10 = 3.96

Q. 152

Seed dropped in 1.6×1 Sq.m area = 2.15×(9-1) = 17.20 g or 0.01720 kg

Hence, Seed rate = 0.01720×10000/(1.6×1) = 107.50 kg

Q. 153

Actual cutting width = 0.64 – 0.08 = 0.56 m

For maximum field capacity, operator should plough the land along the length of field (1.e. 120 m)

Hence, Number of laps = 100/0.56 = 178.57 or 179 laps

Therefore, Number of turns = 179 – 1 = 178

Turning loss = 178×30/3600 = 1.483 hours

Field ploughing time = 120×179/5000 = 4.296 hours

Hence, maximum field capacity = 1.2/(4.296+1.483) = 0.207 ha/h

Q. 154

Required amount of offset = (4×4 – 2.5×2.5)/(3+3.5) = 1.50 m

Q. 155

Distance traveled by ground wheel in one revolution of metering plate

= 3.14×40×2

= 251.2 cm or 2512 mm

Hence, Plant to plant spacing = 2512/16 = 157 mm

Q. 157

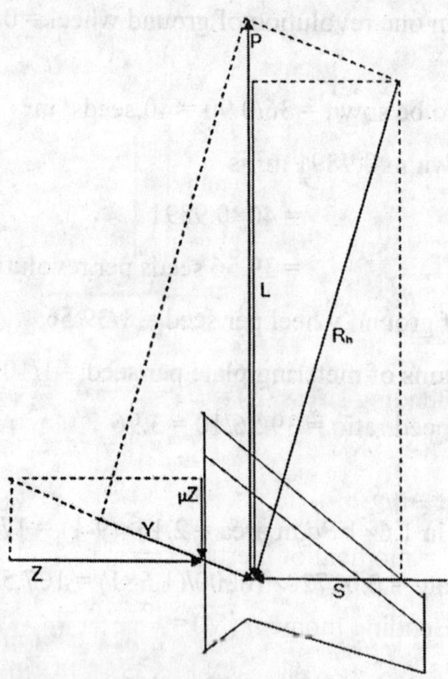

Given that

Horizontal component(L) = 2 kN

Side force(S=Z) = 0.9 kN (※When horizontal Pulling force is in the direction of travel, a Parasitic side force(Z) automatically introduced on side of the tool to counteract S.)

Frictional force(Y) = μZ = (tan25)*0.9 = 0.4197 kN

Referrring above figure

$$R_h = \sqrt{L^2 + S^2} = \sqrt{2^2 + 0.9^2} = 2.1932$$

Resultant of side force and accompany friction force on tool

$$Y = \sqrt{Z^2 + (\mu Z)^2} = \sqrt{0.9^2 + 0.4197^2} = 0.9930$$

Again $P = \sqrt{R^2 \quad Y^2} = \sqrt{2.1932^2 + 0.9930^2} = 2.4075 \text{ kN} = 2408 \text{ N}$

Q. 158

Given that

Bending stress (σ) = 35 N; Crop height (L) = 50 mm

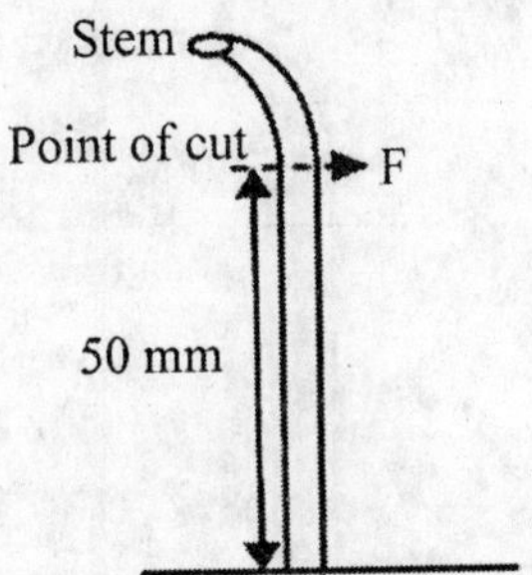

We know that

$$\text{Bending stress } (\sigma) = \pm \frac{\textit{Bending moment } (M)}{I} \times y$$

Where;

$y = d/2$

And $I = \text{moment of inertia} = \dfrac{\pi d^4}{64}$ (For circular sections)

$\therefore$ $\text{Bending moment (M)} = \dfrac{\sigma.\pi d^3}{32}$

$$M = \frac{35 \times 3.14 \times (6)^3}{32} = 741.83 \text{ N-mm}$$

$\because$ Crop stem is acting as a cantilever beam, therefore

$$\text{Horizontal force (F)} = \frac{M}{L} = \frac{741.83}{50} = 14.84 \text{ N}$$

16 Fertilizer Applications and Plant Protection Equipments

UKPSC AE Agricultural Engineering Paper - II 2013

Q. 1 Operating speed of power sprayer varies from

(B) Less inhaled by the operator

(C) Cause the chemical to break up

Q. 4 Application rate of sprayer is affected by

(A) Rate of discharge

Q. 5 Hand rotary duster

16

Fertilizer Applications and Plant Protection Equipments

UKPSC AE Agricultural Engineering Paper - II 2013

Q. 1 Operating speed of power sprayer varies from

(A) 2 - 5 km /hr (B) 4.8 - 8 km /hr

(C) 10 - 20 km /hr (D) 20 - 35 km /hr

Q. 2 Spraying of crops is usually avoided on windy days because

(A) Drift increases

(B) Gets inhaled by the operator

(C) Cause the chemical to break down

(D) None of the above

APPSC AEES Agriculture Engineering Exam 2016

Q. 3 The common type of fertilizer metering mechanism used on animal-drawn seed cum fertilizer drills is

(A) Star-wheel feed (B) Auger

(C) Stationary opening (D) Edge cell vertical rotor

UKPSC Combined Junior Engineering Exam 2013 Agricultural Engineering Paper-I

Q. 4 Application rate of sprayer is affected by

(A) Rate of discharge (B) Number of nozzle

(C) Forward speed (D) All of the above

Q. 5 Hand rotary dusters are more suitable for plant height upto

(A) 3 m (B) 5 m

(C) 7.5 m (D) 10 m

Punjab Mechanic Agricultural Machinery Instructor 2014

Q. 6 Airplane sprayers are used for

(A) Small areas (B) Large areas

(C) Battle fields (D) None of the above

Q. 7 To protect the agriculture from pests & diseases .... is used

(A) Ploughing (B) Spraying

(C) Cutting (D) None of the above

RPSC AEn Pre Exam 2013 (Agricultural Engineering)

Q. 8 What is the function of wheel hand hoe ?

(A) Weeding (B) Seeding

(C) Pulverization (D) Ditch making

Q. 9 In fluted feed type seed metering device, the groves on feed roller constructed at its

(A) Central area (B) Diagonal

(C) Periphery (D) Entire surface

Q. 10 The device used to spray the liquid is called

(A) Sprayer (B) Duster

(C) Pump (D) Both (A) and (C)

OCS AEn Exam 2011 (Pre) – II (Agricultural Engineering)

Q. 11 In the case of Ultra Low Volume (ULV) sprayer, the application rate lies in the range of

(A) 0.5 to 9 lit/ha (B) 10 to 20 lit/ha

(C) 20 to 30 lit/ha (D) 30 to 50 lit/ha

Q. 12 In hydraulic sprayers, the degree of atomization is primarily a function of

(A) Liquid pressure and the nozzle characteristics

(B) Air velocity

(C) Single of nozzle

(D) Single and shape of atomizer

Q. 13 Atomization of spray fluid can be achieved by

(A) Discharging the liquid through an orifice under pressure

(B) By breaking up the jet of liquid with a blast of air

(C) Both (A) and (B)

(D) None of the above

Q. 14 Hand operated sprayers are operated at a pressure of (kg/cm^2)

(A) < 10 (B) 10 to 15

(C) 15 to 20 (D) > 20

Q. 15 Application rates for dusters are (kg/ha)

(A) < 10 (B) 10 to 20

(C) 20 to 40 (D) > 40

Q. 16 Hollow cone nozzles employed on boom type of sprayers have spray angles between

(A) 30° to 60 ° (B) 40° to 60°

(C) 60° to 95° (D) 100° to 135°

Q. 17 In electrostatic spraying the total volume of liquid to be sprayed over one hectare

(A) < 1 litre (B) 15 litre

(C) 30 litre (D) > 30 litre

Q. 18 For a given spray sample

(A) The Volume Median Diameter(VMD) is equal to the Number Median Diameter(NMD)

(B) The VMD is less than NMD

(C) The VMD is larger than NMD

(D) None of the above

CGPSC State Engineering Services Exam Paper – I

Q. 19 Motorized knapsack sprayer hopper is made with

(A) Low density polyethylene (B) M.S sheet

(C) Copper (D) Plastic

(E) High density polyethylene

Q. 20 Pressure range built up in pressure chamber of Rocker Sprayer for conveniently spraying

(A) 10 - 14 kg/cm^2 (B) 14 - 18 kg/cm^2

(C) 15 - 19 kg/cm^2 (D) 7 - 10 kg/cm^2

(E) 17 - 21 kg/cm^2

UKPSC A.En. Exam 2012 Paper - II

Q. 21 The application rate of ULV sprayer per hectare is

(A) Up to 5 litres (B) 5 – 100 litres

(C) 400 litres (D) None of the above

Q. 22 Power sprayers are operated at a pressure

(A) 22 – 55 kg/m^2 (B) 10 – 20 kg/m^2

(C) 5 – 10 kg/m^2 (D) Less than 5 kg/m^2

Q. 23 The manually operated knapsack sprayer works on the principle of

(A) Hydraulic energy (B) Gaseous energy

(C) Centrifugal energy (D) Kinetic energy

Kerala AAO Exam 2017

Q. 24 Which one of the following is/are used as sprayable pesticide formulations ?

(A) Dusts (B) Wettable powders

(C) Emulsifiable concentrates (D) Both (B) and (C)

Q. 25 As per Fertilizer Control Order (FCO) standards, C : N ratio of compost should be

(A) Less than 20 : 1 (B) 20 : 1 to 30 : 1

(C) 30 : 1 to 40 : 1 (D) more than 40 : 1

Q. 26 Which weed is found in reclaim alkaline soil ?

(A) Cyprus rotandus (B) Parantheium hysterophorus

(C) Argemone mexicana (D) Amaranthus spp.

Q. 27 Which is the chemical formula for Urea ?

(A) $Ca(H_2PO_4)_2$ (B) NH_4NO_3

(C) $Co(NH_2)_2$ (D) $NaHCO_3$

Q. 28 The chemical name of Muriate of Potash is

(A) Potassium sulphate (B) Potassium nitrate

(C) Potassium phosphate (D) Potassium chloride

Graduate Aptitude Test in Engineering - 2007

Q. 29 A field sprayer having a boom with 20 nozzles spaced 0.46 m apart is to be designed for a maximum application rate of 750 liter ha^{-1} at 520 kPa pressure. The forward speed of travel is 6.5 km h^{-1}. Neglect field losses and assume that 10% of the pump output is bypassed.

(i) The required pump capacity is

(A) 67.95 litre min^{-1} (B) 74.75 litre min^{-1}

(C) 82.22 litre min^{-1} (D) 83.06 litre min^{-1}

(ii) If mechanical agitation requires 375 W input power and the pump efficiency is 70 %, the maximum power input required is

(A) 720 W (B) 879 W

(C) 1095 W (D) 1403 W

Graduate Aptitude Test in Engineering - 2008

Q. 30 A tractor sprayer boom is fitted with 20 hollow cone nozzles to achieve an application rate of 200 L ha^{-1}. During a calibration test the nozzle flow rate was found to be 1.25 L min^{-1}, whereas the rated nozzle flow rate of 0.473 L min^{-1} was available at 275 kPa.

(i) If the nozzle produces droplets with a volume median diameter of 200 micron at 1 MPa, the droplet size in micron at the desired flow rate is

(A) 104.17 (B) 160.91

(C) 248.60 (D) 594.89

(ii) If the forward speed of the tractor is 7.5 km h^{-1}, the field capacity of the sprayer in ha h^{-1} is

(A) 2.84 (B) 7.00

(C) 7.13 (D) 7.50

Graduate Aptitude Test in Engineering - 2010

Q. 31 A field sprayer having 16 fan type spray nozzles spaced 0.5 m apart is moving at a forward speed of 3.5 km h^{-1} with an application rate of 1 m^3 ha^{-1}. At a deposition level 430 mm below the tip of the nozzle, the discharge rate across a 0.2 m width at the centre of the sprayed tip is essentially constant at 15 ml min^{-1} per 10 mm of lateral distance. On each side of this 0.2 m centre strip, the discharge rate per mm of width decreases uniformly to zero at a lateral distance of 0.36 m from the nozzle centre line.

(i) The discharge rate per nozzle in m^3 h^{-1} will be

(A) 0.175 (B) 0.215

(C) 0.350 (D) 0.430

(ii) The nozzle tip height in mm above the deposition level that would give uniform coverage will be

(A) 602 (B) 546

(C) 501 (D) 477

Graduate Aptitude Test in Engineering - 2011

Q. 32 The Sauter mean diameter of liquid droplets of a hydraulic spray is

(A) Median diameter of droplets

(B) Surface area of droplets

(C) Mean diameter of droplets

(D) Volume to surface ratio of droplets

Q. 33 A hydraulic spray nozzle has a discharge of 450 ml min^{-1} at a pressure of 280 kPa. If the pressure is increased by 10%, the discharge will be

(A) Increased by 4.9% (B) Increased by 10.0%

(C) Increased by 21.0% (D) Decreased by 4.6%

Graduate Aptitude Test in Engineering - 2013

Q. 34 The cumulative discharge of a tractor mounted hydraulic sprayer having 7 nozzles is 2.0 L min^{-1} when the tractor is operated at 4 km h^{-1}. If the nozzle spacing is 0.3 m, the discharge in L ha^{-1} is

(A) 1000.02 (B) 285.72

(C) 166.67 (D) 142.86

Graduate Aptitude Test in Engineering - 2014

Q. 35 A knapsack sprayer is provided with a nozzle having rated delivery of 0.5 L min^{-1} at a pressure of 270 kPa. For a pressure setting of 210 kPa, the application rate per unit orifice area is 0.24 L $cm^{-2}s^{-1}$. If the coefficient of discharge of the nozzle is 0.75, the diameter of the nozzle orifice in mm is

(A) 1.71 (B) 1.76

(C) 2.28 (D) 2.59

Q. 36 Several identical sprinkler nozzles, each having discharge Q (litre per minute), are spaced in a grid of size L (metre) × S (metre). The application rate in mm h^{-1} is

(A) 60Q/(LS) (B) 3600Q/(LS)

(C) LS/(60Q) (D) LS/(3600Q)

Graduate Aptitude Test in Engineering – 2015

Q. 37 A power operated hydraulic sprayer operating at a pressure of 1.4 MPa has a boom with 11 nozzles spaced at 0.3 m interval. The input power of the pump is 1.5 kW and its mechanical efficiency is 75%. If the nozzle diameter and the discharge coefficient are 2.4 mm and 0.5, respectively, the average jet velocity of the nozzles in m/s is ....

Graduate Aptitude Test in Engineering – 2016

Q. 38 The rated nozzle flow rate and volume median diameter (VMD) of droplets of a hydraulic sprayer are 1.0 L min^{-1} and 200 µm, respectively at the rated pressure of 500 kPa. If the desired nozzle flow rate is 1.5 L min^{-1}, the droplet diameter in µm will be ....

Graduate Aptitude Test in Engineering – 2018

Q. 39 The discharge from a sprinkler nozzle depends on

(A) Operating pressure and nozzle geometry.

(B) Operating pressure and distribution pattern.

(C) Application rate and nozzle angle.

(D) Operating pressure and application rate.

Graduate Aptitude Test in Engineering – 2019

Q. 40 The type of typical spray distribution profile of a hollow cone nozzle is

(A) Steep sided slopes

(B) Gradual sloping sides

(C) Narrow topped with gradual slopes

(D) Narrow topped with steep sides

Q. 41 The application rate of an 18-nozzle hydraulic sprayer is 1120 L ha^{-1}. The nozzle spacing and forward speed are 400 mm and 3.4 km/h, respectively. The operating pressure is 2.1 MPa and the pump efficiency is 60%. If 10% of the pump output power is used for agitating the liquid, the power needed to operate the sprayer in kW is ....

Graduate Aptitude Test in Engineering – 2020

Q. 42 A hydraulic sprayer when operating at a speed of 10 km/h and working pressure of 420 kPa covers a width of 4.5 m. The power requirement and efficiency of the pump are 0.75 kW and 70%, respectively. Out of total pump discharge, 10% is bypassed for agitation purpose. The working pressure is increased to 500 kPa. Assuming no change in the width of coverage, the application rate of the sprayer in L/ha is

Graduate Aptitude Test in Engineering – 2021

Q. 43 A fertilizer drill with a row to row spacing of 40 cm, discharges 38 g of fertilizer per row per revolution of the metering wheel. The metering wheel is driven through a chain transmission system by ground wheel having 60 cm diameter. Neglecting skid of the ground wheel, for an application rate of 200 kg/ha, the speed ratio of ground wheel to metering wheel will be (Take $\pi = 3.14$)

(A) 1.40 : 1 (B) 2.52 : 1

(C) 3.64 : 1 (D) 4.76 : 1

Answers Key

1	2	3	4	5	6	7	8	9	10
B	A	A	D	A	B	B	A	C	A
11	12	13	14	15	16	17	18	19	20
A	A	C	A	A	C	A	C	E	B
21	22	23	24	25	26	27	28	29	30
A	A	A	D	A	C	C	D	D, D	C, D

31	32	33	34	35	36	37	38	39	40
A, D	D	A	D	C	A	32.31	152.63	A	A
41	42	43							
2.96	982.08	B							

Explanations

Q. 29 (i)

Given that

Width of field sprayer (W) = 20×0.46 = 9.2 m

Speed of travel (S) = 6.5 km/h = 108.33 m/min

Application rate (V) = 750 L/ha = 750×10^{-4} L/m^2

∴ Pump Capacity (Q) = S×W×V

$= 9.2\times108.33\times750\times10^{-4} = 74.75$ L/min

Here 10% of the pump output bypassed.

∴ Required pump capacity (Q) = 74.75/0.9 = 83.06 L/min

(ii)

Pump pressure (P) = 520 kPa = 520×10^3 Pa

Pump efficiency (η) = 70%

Mechanical agitation input power = 375 W

∴ Output power = P×Q

$$= 520\times10^3 \times \frac{83.06}{60\times1000} \times = 719.85 \text{ W}$$

∵ Pump efficiency (η) = 70%

Thus, Input power = 719.85/0.70 = 1028.36 W

∴ Maximum power input

= Pump input power + Mechanical agitation input power

= 1028.36 + 375 = 1403.36 W

Q. 30 (i)

Given that

Number of hollow cone (n) = 20; Application rate = 200 L/ha

Nozzle flow rate (Q) = 1.25 L/min g

Rated flow rate at 275 kPa = 0.473 L/min

We know that

$$\frac{Q_1}{Q_2} = \sqrt{\frac{P_1}{P_2}}$$

Therefore, $$\frac{1.25}{0.473} = \sqrt{\frac{P_1}{275}}$$

$\Rightarrow$ $P_1 = 1920.57$ kPa

We also know that $$\sqrt[3]{\frac{P_1}{P_2}} = \frac{D_2}{D_1}$$

$$\sqrt[3]{\frac{1000}{1920.67}} = \frac{D_2}{20}$$

$\Rightarrow$ $D_2 = 160.90$ micron

(ii)

Application rate (R) = 200 L/ ha

Total spray applied = 20×1.25 = 25 L/ min or 25×60 L/h = 1500 L/ h

$$\therefore \quad \text{Field capacity} = \frac{Total\ spray\ applied(L/h)}{Application\ rate(L/ha)}$$

$$= \frac{1500}{200} = 7.5 \text{ ha/ h}$$

Q. 31 (i)

Forward speed (V) = 3.5 km/h = 3500 m/h

Spacing between nozzles (S) = 0.5 m

Application rate (R) = 1 m^3/ ha = 1×10^{-4} m^3/m^2

Discharge rate = S.V.R

$$= 0.5\times3500\times1\times10^{-4} = 0.175 \text{ m}^3\text{/ h}$$

(ii)

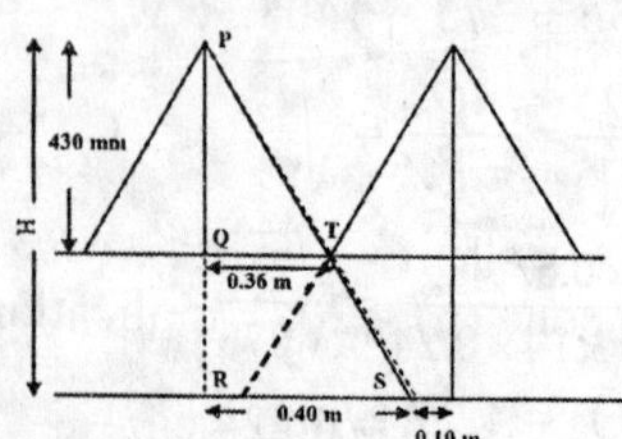

Spray radius in first condition (r) = 0.36 m;

Height of sprayer tip (h) = 430 mm

Spray radius in second condition (R) = 0.40m

Now, in equilateral ΔPQT and ΔPRS

$$\frac{PR}{PQ} = \frac{RS}{QT}$$

$$\Rightarrow \quad \frac{H}{430} = \frac{0.40}{0.36}$$

$$\Rightarrow \quad H = 477 \text{ mm}$$

Q. 33

Given that

$P_1 = 280$ kPa; $Q_1 = 450$ ml/min

$P_2 = 308$ kPa (10 % increased) ; $Q_2 = ?$

We know

$$Q_1/Q_2 = (P_1/P_2)^{1/2}$$

$$\Rightarrow \quad Q_2 = 450\times(308/280)$$

$$= 471.96 \text{ ml/min}$$

$\therefore$ Increase in discharge = (471.96 - 450)×100/450 = 4.9%

Q. 34

Area sprayed = Speed×Width

$= 4000\times0.3\times7 = 8400 \text{ m}^2\text{/hr} = 0.84$ Ha/hr

Thus, Discharge per hectare = Q/(S×W) = 2×60/0.84

$= 142.86 \text{ L ha}^{-1}$

Q. 35

We know that

$$\frac{Q_1}{Q_2} = \frac{Q_1}{AV} = \frac{4Q_1}{\pi d^2 V} =: \sqrt{\frac{P_2}{P_1}}$$

$$\Rightarrow \quad \frac{4\times0.5}{60\times d^2\times3.14\times0.24} = \sqrt{\frac{210}{270}}$$

$$\Rightarrow \quad d = 0.224 \text{ cm} = 2.24 \text{ mm}$$

Q. 36

Total Discharge through nozzles = Q L/min = 60 Q×10^6/h mm^3/h

We know that

Discharge = (Application rate)×(Area)

$$\Rightarrow \quad \text{Application rate} = \frac{60Q \times 10^6}{LS \times 1000 \times 1000}$$

$$= \frac{60Q}{LS}$$

Q. 37

Volumetric flow rate through nozzles (Q) = n A V C_d

$$= 11 \times (\pi \times 0.0024^2/4) \times V \times 0.5$$

$$= 2.4869 \times 10^{-5} V \text{ m}^3/\text{s}$$

We know

Power required to operate the pump

= Flow rate (Q)×Operating pressure(P)

$$\Rightarrow \quad 1.5 \times 1000 \times 0.75 = 2.4869 \times 10^{-5} V \times 1.4 \times 10^6$$

$$\Rightarrow \quad V = 32.31 \text{ m/s}$$

Q. 38

Given that

Nozzle flow rate (Q) = 1.5 L/min

Rated flow rate at 500 kPa = 1.0 L/min

We know that

$$\frac{Q_1}{Q_2} = \sqrt{\frac{P_1}{P_2}}$$

Therefore,

$$\frac{1.5}{1} = \sqrt{\frac{P_1}{P_2}}$$

$$\Rightarrow \quad P_1/P_2 = 2.25$$

We also know that $\sqrt[3]{\frac{P_1}{P_2}} = \frac{D_1}{D_2}$

$\Rightarrow \qquad (2.25)^{1/3} = \frac{200}{D_2}$

$\Rightarrow \qquad D_2 = 152.63$ micron

Q. 41

Discharge of sprayer = 1120×18×0.4×3.4×1000/(10000×1000×3600)

= 7.616×10^{-4} m³/s

Power consumed to pump and spray the liquid = $2.1\times10^{6}\times7.616\times10^{-4}$

= 1599.36 W or 1.599 kW

Consider, Power needed to operate the sprayer is = x kW

Hence,

x×0.60×0.90 = 1.599

(∵ Pump effeciency = 60% and 90% of total power is used to pump and spray the liquid)

$\Rightarrow \qquad$ x = 2.96 kW

Q. 42

We know

Power required = PQ

$\Rightarrow \qquad 0.75\times0.70 = 420\times Q_1$

$\Rightarrow \qquad Q_1 = 1.25\times10^{-3}$ m³/s

Again,

$$\frac{Q_1}{Q_2} = \sqrt{\frac{P_1}{P_2}}$$

$$\Rightarrow \qquad \frac{1.25\times10^{-3}}{Q_2} = \sqrt{\frac{420}{500}}$$

$\Rightarrow \qquad Q_2 = 1.364\times10^{-3}$ m³/s or 4910.4 L/h

10% by passed. Hence octual flow rate = 4910.4 x 0.90 = 4419.30 L/h

$$\text{Application rate} = \frac{4419.36\times10000}{10000\times4.5} = 982.08 \text{ L/h}$$

Q. 43

Area covered in one revolution of ground wheel = 3.14×0.06×0.40

= 0.7536 m^2

Revolutions of ground wheel for 1 ha area = 1/0.7536

= 13269.64 revolutions

Again, Revolutions of metering wheel for 1 ha area

= 200×1000/38

= 5263.16 revolutions

Hence, Required ratio = 13269.64/5263.16

= 2.52:1

17

Grain Harvesting Machinery Harvester and Thresher

UKPSC AE Agricultural Engineering Paper - II 2013

Q. 1 While operating a combine harvester, the grain not collected by the machine is called

(A) Field loss (B) Shattering loss

(C) Gathering loss (D) Reel loss

Q. 2 Multi crop thresher is equipped with

(A) Rasp-bar cylinder (B) Spike tooth cylinder

(C) Both (A) and (B) (D) Beater type cylinder

Q. 3 A reaper is used for

(A) Cutting crop (B) Cutting wood

(C) Cutting and threshing (D) Threshing

Q. 4 An aspirator uses the following principle to separate chaff from grain

(A) Difference in size

B) Difference in weight

(C) Difference in color

(D) Difference in terminal velocity

Q. 5 Blade diameter of gear type hand operated winnowing fan is

(A) 0.50 m (B) 0.25 m

(C) 0.75 m (D) 1.05 m

Q. 6 In a combine harvester, the reel index should normally be

(A) 0.75 to 1.0 (B) 1.50 to 1.75

(C) 1.75 to 2.0 (D) 1.25 to 1.50

Q. 7 A thresher causes more seed damage if:-

(A) Cylinder speed is increased

(B) Cylinder clearance is increased

(C) Aspirator speed is increased

(D) All of the above

Q. 8 A combine harvester discharges grains left on heads can be corrected by:-

1. Increasing the cylinder speed
2. Decreasing the cylinder concave clearance

(A) Only statement '1' is correct

(B) Only statement '2' is correct

(C) Statement '1' and '2' both are correct

(D) Neither statement '1' nor '2' is correct

Q. 9 A thresher sieve allows the following to pass through its perforations:-

(A) Clean grain (B) Chaff for blowing away

(C) Dust particles (D) Both grain and dust

APPSC AEES Agriculture Engineering Exam 2016

Q. 10 Increasing the cylinder speed results in

(A) Increase the threshing efficiency (B) Reduce cylinder loss

(C) Increase seed damage (D) All of the above

MPPSC Assistant Agricultural Engineer 2013

Q. 11 A 2.5 m wide combine is operating at a speed of 4 kmph. The combine operates for 8 hours in a day. How much area will be harvested considering overall efficiency of 60 per cent?

(A) 0.48 ha (B) 4.8 ha

(C) 8 ha (D) None of these is correct

Q. 12 Loop type threshing cylinders are generally used in ....

(A) Groundnut threshers (B) Maize threshers

(C) Paddy threshers (D) Wheat threshers

UKPSC Combined Junior Engineering Exam 2013 Agricultural Engineering Paper-I

Q. 13 An olpad thresher is operated

(A) By a tractor (B) By a pair of bullocks

(C) By an electric motor (D) Manually

Q. 14 Reaper is used for

(A) Land preparation (B) Sowing

(C) Harvesting (D) Threshing

Q. 15 Increasing the cylinder speed results in

(A) Increase threshing efficiency (B) Increase seed damage

(C) Reduce cylinder loss (D) All of the above

Q. 16 In combine, the length of rasp bar type cylinder varies from

(A) 2.0 – 2.5 m (B) 1.5 – 2.0 m

(C) 1.4 – 1.5 m (D) Less than 1.4 m

Q. 17 During the threshing of crops as far as possible ensure that

(A) Threshing yard should be away from the residential area

(B) It should be located near the railway track

(C) Adequate fire protection services must be provided

(D) Both (A) & (C)

Q. 18 The prime mover used in power thresher is

(A) Electric motor (B) Engine

(C) Tractor (D) All above three

Q. 19 Which of the following components are parts of threshing cylinder ?

(A) Concave (B) Rasp bar

(C) Spikes (D) Both (B) and (C)

Q. 20 A machine which cuts, thresh, clean and collect the grain simultaneously while moving through the field is known as

(A) Reaper (B) Combine

(C) Power thresher (D) None of the above

Punjab Mechanic Agricultural Machinery Instructor 2014

Q. 21 Huller is used for

(A) Hulling (B) Shelling

(C) Separation (D) Drying

Q. 22 Number of valves in combine harvester engine is

(A) One (B) two

(C) Three (D) four

Q. 23 Threshing of grain at higher moisture content, results in

(A) Breakage of grain

(B) Higher impact force requirement for grain detachment

(C) Both the above

(D) None of the above

RPSC AEn Pre Exam 2013 (Agricultural Engineering)

Q. 24 In thresher, the spikes are the part of

(A) Cleaning unit (B) Concave

(C) Cylinder or drum (D) Separator

Q. 26 The combine harvesters are generally fitted with the cylinder type

(A) Hammer (B) Spike tooth

(C) Raspbar (D) Wire-loop

OCS AEn Exam 2011 (Pre) – II (Agricultural Engineering)

Q. 26 Pods, heads or ears and free seeds lost during the cutting and conveying operations are called as

(A) Cylinder losses (B) Shoe losses

(C) Walker losses (D) Gathering losses

Q. 27 Speed of the cutter bar in case of combine (grain) is

(A) 400 to 550 rpm (B) 500 to 800 rpm

(C) 800 to 1000 rpm (D) 1000 to 1200 rpm

Q. 28 Most important factor affecting threshing efficiency is

(A) Moisture content of crop (B) Peripheral speed of the drum

(C) Cylinder concave clearance (D) Feed rate

Q. 29 Thresher blows broken grains with chaff due to

(A) Higher cylinder speed

(B) More feed rate

(C) More cylinder concave clearance

(D) Moisture content of crop is more

Q. 30 Most commonly used threshing mechanism on the grain combine is

(A) Peg tooth cylinder (B) Spike tooth cylinder

(C) Spring tooth cylinder (D) Rasp bar cylinder

CGPSC State Engineering Services Exam Paper – I

Q. 31 Proper registration of mower refers to the condition when knife section is stopped in the centre of

(A) Knife back (B) Its pitman

(C) Its guard (D) Its finger

(E) None of these

Q. 32 In tractor mounted vertical conveyer reaper, the crop is guided by

(A) Fly wheel (B) Guard

(C) Pressure spring (D) Cutter bar

(E) Star wheel

Q. 33 In cutter bar of mower, the knife is connected to

(A) Shoe (B) Ledger plate

(C) Cutter bar (D) Fly wheel

(E) Pitman

Q. 34 The cutter head of grain combine can be tilted to cut on slopes up to

(A) 55^0 (B) 75^0

(C) 95^0 (D) 65^0

(E) 80^0

Q. 35 For proper operation the reel index of combine should be between

(A) 1.50 to 1.5
(B) 1.25 to 1.5
(C) 1.75 to 1.5
(D) 2.5 to 3.0
(E) 1.0 to 1.5

Q. 36 Threshed grains passing out in the straw in the combine harvesting is known as

(A) Header loss
(B) Rack loss
(C) Processing loss
(D) Shoe loss
(E) Cleaning loss

Q. 37 Which cylinder speed is recommended for peanut thresher ?

(A) Slow speed
(B) Medium speed
(C) High speed
(D) 12-15 m/s
(E) 25-30 m/s

Q. 38 Machines may be tested at the request in writing of manufacturers/ accredited importers referred to as

(A) Author
(B) Report
(C) Applicant
(D) Manufacturer
(E) Customer

Q. 39 Overall width of flail shredder ranging between

(A) More than 100 cm
(B) 70 - 100 cm
(C) 50 - 70 cm
(D) 30 - 50 cm
(E) none of these

Q. 40 Effective cutting length of ISI specified sickle is about

(A) 150 mm
(B) 250 mm
(C) 300 mm
(D) 100 mm
(E) 125 mm

Q. 41 How many number of bars are spirally fixed on the rasp bar cylinder?

(A) 6 to 8 bars
(B) 6 to 10 bars
(C) 6 to 12 bars
(D) 6 to 14 bars
(E) 10 to 15 bars

UKPSC A.En. Exam 2007 Paper – II

Q.42 At a proper alignment of a mower, the cutter bar is at about

(A) 90 degree in the direction of motion

(B) 88 degree in the direction of motion

(C) 75 degree in the direction of motion

(D) 70 degree in the direction of motion

Q. 43 Registration is an adjustment in

(A) Thresher (B) Planter

(C) Digger (D) Reaper

Q. 44 Ledger plate is the part of

(A) Mould board plough (B) Planter

(C) Thresher (D) Reaper

Q. 45 A mower knife is said to be in proper registration when the knife section stops in the

(A) Centre (B) Left side of the guard

(C) Right side of the guard (D) None of the above

Q. 46 In general, a 1 m long mower cutter bar will have a lead of

(A) 20 mm (B) 30 mm

(C) 40 mm (D) 50 mm

Q. 47 A cutter bar of a tractor drawn mower ordinarily makes

(A) 800 – 1200 strokes/min (B) 1600 – 2000 strokes/min

(C) 2400 – 2800 strokes/min (D) 3200 – 3600 strokes/min

Q. 48 Pitman in a reaper is used to

(A) Provide rotary motion to cutter bar

(B) Convert rotary motion to reciprocatory motion

(C) Help in cutting crop

(D) None of the above

Q. 49 One of the objectives of chopping the fodder crop is

(A) To save moisture (B) To make easy in chewing

(C) To save storage space (D) Both (B) and (C)

Q. 50 A mower with 4 m width is moving with a speed of 5 km/hr. If it's field performance index is 75 %, the time taken by it to cover 1 hectare area of field would be

(A) 30 minutes (B) 40 minutes

(C) 60 minutes (D) None of the above

Q. 51 A self propelled machine is one which is

(A) Drawn by a tractor

(B) Drawn and operated by its own power source

(C) Operated by an external power source

(D) None of the above

Q. 52 In a combine, the function of header is to

(A) Cut and gather the crop

(B) Deliver the cut crop to cylinder

(C) Thresh the heads

(D) Both (A) and (B)

Q. 53 A thresher discharging broken grains, can be corrected by

(A) Increasing the drum speed

(B) Decreasing the drum speed

(C) Decreasing the concave clearance

(D) Decreasing the fan speed

Q. 54 Threshing drum used mostly in combine machine for wheat crop is

(A) Rasp bar type (B) Angle bar type

(C) Spike tooth type (D) Wire loop type

Q. 55 Threshing drum which are used in paddy thresher is

(A) Hammer type (B) Angle bar type

(C) Wire loop type (D) Spike tooth type

Q. 56 Size of a combine is expressed by

(A) Length of cutter bar
(B) Threshing capacity
(C) Area covered per unit time
(D) Working capacity

Q. 57 Ease in cutting crop can be achieved when cutting angle is

(A) 0^0
(B) $0^0 - 90^0$
(C) 90^0
(D) None of the above

Q. 58 A 2 m size combine is operating at 4 km/hr speed with an overall efficiency of 75 %. The area covered by this combine in hectare per hour will be

(A) 0.40
(B) 0.60
(C) 0.80
(D) 1.00

UKPSC A.En. Exam 2012 Paper - II

Q. 59 The process of separating grains from a mixture of grain and chaff in an air stream created artificially or naturally is called

(A) Winnowing
(B) Threshing
(C) Harvesting
(D) None of the above

Q. 60 The olpad thresher is pulled by

(A) Tractor
(B) Pair of bullocks
(C) Power tiller
(D) None of the above

Q. 61 Delayed harvesting of grains

(A) Increases field losses
(B) Reduction in milling yield
(C) Both of the above
(D) None of the above

Q. 62 The machine used for cutting grasses is known as

(A) Reaper
(B) Mower
(C) Windrower
(D) Combine harvester

Q. 63 The two types of cylinders used in wheat power thresher are

(A) Serrated disc and rasp bar
(B) Spike tooth and wire loop type
(C) Spike tooth and hammer mill
(D) Hammer mill and flails

Q. 64 A 2 m combine is operating at 4 kmph with an overall efficiency of .... %. The actual area covered by it is 0.60 ha/h.

(A) 75 % (B) 90 %

(C) 60 % (D) 80 %

Q. 65 A syndicator type thresher is a

(A) Hammer mill type thresher (B) Rasp bar type thresher

(C) Chaff cutter type thresher (D) None of these

Q. 66 The usual height of paddy stubbles after combine harvesting is

(A) 15 – 20 cm (B) 30 – 60 cm

(C) 60 – 70 cm (D) More than 75 cm

Q. 67 The power transmitting unit of a mower consists of

(A) Main axle (B) Gears

(C) Crankshaft and pitman (D) All of the above

Kerla AAO Exam 2017

Q. 68 Match the following

List-I	List-II
P. Mower	1. Machine to cut grain crops
Q. Sickle	2. Machine to harvest root crops
R. Reaper	3. Harvesting by manual power
S. Digger	4. Machine to cut grass and fodder crop

(A) P-4, Q-2, R-3, S-1 (B) P-4, Q-3, R-1, S-2

(C) P-3, Q-2, R-1, S-4 (D) P-1, Q-3, R-4, S-2

Graduate Aptitude Test in Engineering - 2008

Q. 69 A multi-crop thresher was tested utilizing power from a tractor PTO shaft and the fuel consumption recovered was 4.5 L h^{-1}. The brake thermal efficiency of the engine is 32 percent and density of the fuel having a heating value of 40 MJ kg^{-1} is 825 kg m^{-3}. If the transmission loss from the engine to PTO drive is 5 percent, the power consumed by the thresher is

(A) 8.66 kW (B) 12.54 kW

(C) 13.20 kW (D) 41.25 k

Graduate Aptitude Test in Engineering - 2009

Q. 70 A vertical conveyor reaper is to be used for harvesting wheat crop at a height of 30 mm above the ground. The ultimate tensile strength and diameter of the crop stem are 35 N mm^{-2} and 3 mm respectively. The friction coefficient of knife edge for wheat crop is 0.346 and the maximum oblique angle of the counter shear is 17°. The crop stem is of homogenous solid with a uniform circular section.

(i) The horizontal force in N that would cause bending failure of the crop stem is

(A) 1.55 (B) 3.09

(C) 4.64 (D) 6.19

(ii) The maximum clip angle in degrees between the knife and the counter shear is

(A) 2 (B) 17

(C) 19 (D) 36

Graduate Aptitude Test in Engineering - 2010

Q. 71 A disk type mower, operated by a tractor PTO, has six discs with a swath of 0.4 m per disk. The specific energy required for cutting is 2.1 kJ m^{-2} and specific power losses due to air, stubble and gear train friction are 2 kW m^{-1} of cutting width. if the mower with tractor requires a propelling force of 2 kN, the total power requirement for carrying out mowing in kW at a forward speed of 3 km h^{-1} is

(A) 6.47 (B) 7.57

(C) 8.33 (D) 10.67

Graduate Aptitude Test in Engineering - 2012

Q. 72 The thresher of a wheat combine harvester has an optimal throughput capacity of 2400 kg (crop) per hour. The harvester has a forward velocity of 4.5 km h^{-1}. Sample tests have revealed that the yield of crop in the field is 3000 kg (grain) per ha. Grain to straw ratio is 60 : 40. If the above throughput is to be maintained, the width of cut of the harvester in m, neglecting turning losses, is

(A) 0.71 (B) 1.07

(C) 1.78 (D) 2.96

Graduate Aptitude Test in Engineering - 2013

Q. 73 During the testing of a spike-tooth type thresher for wheat crop, the throughput and the thresher capacity were found to be 750 kg h^{-1} and 445 kg h^{-1}, respectively. The grain-straw ratio (grain : straw) of the crop is 1.5 : 1. Material other than grain (MOG) collected at the main grain outlet is 0.5%. The total grain loss in percentage from all the sources will be

(A) 0.33 (B) 0.50

(C) 1.11 (D) 1.61

Graduate Aptitude Test in Engineering - 2014

Q. 74 Oil yield (Y_o) of mustard after N days of flowering is expressed as

$$Y_o = -0.0018\,N^2 + 0.1319\,N - 0.743$$

For maximum oil yield, optimum stage of harvesting, in days after flowering is,

(A) 23 (B) 29

(C) 37 (D) 39

Q. 75 While operating a pedal thresher, 14 kg of paddy seed remained unthreshed for a throughput of 112 kg. If the quantity of the threshed seed obtained was 70 kg, the threshing efficiency in percent is

(A) 62.50 (B) 80.00

(C) 83.33 (D) 87.50

Graduate Aptitude Test in Engineering – 2015

Q. 76 The grain to straw ratio of wheat crop is 1.5 : 1. The output capacity and cleaning efficiency of a thresher at an optimal operating condition are 500 km/h and 99% respectively. If the grain recovery at main grain outlet is 100%, the throughput capacity in kg/h will be

(A) 758 (B) 825

(C) 1238 (D) 1320

Graduate Aptitude Test in Engineering – 2017

Q. 77 If the real index of a grain combine is more than 1.5, it will increase

(A) Cutter bar loss (B) Shatter loss

(C) Cylinder loss (D) Straw walker loss

Q. 78 The diameter of feed rollers of a conveyor type power chaff cutter is 100 mm and they are rotating at 90 rpm for cutting dry fodder. The effective length of each feed roller is 250 mm and the average clearance between them is 15 mm. The compressed density of the material while passing through the feed rollers is 250 kg m^{-3}. The throughput capacity of the chaff cutter in ton h^{-1} is ....

Graduate Aptitude Test in Engineering – 2018

Q. 79 During field evaluation of a combine harvester, total grain collected at the grain tank was 135 kg. Amount of threshed grain found over walker and shoe were 0.65 kg and 2.4 kg, respectively. Unthreshed grain lost with straw and chaff were 4.54 kg and 0.28 kg, respectively. Threshing efficiency of the combine harvester in percent is

(A) 94.37 (B) 94.49

(C) 96.50 (D) 96.63

Q. 80 A self-propelled power reaper with a 1.2 m wide cutter bar was operated at a forward speed of 4 km h^{-1}. Turning loss during operation was observed to be 25 minutes per hectare. Neglecting any other losses, the field efficiency of the reaper will be ....%.

Graduate Aptitude Test in Engineering – 2019

Q. 81 The path traced by the material threshed between the cylinder and the concave of an axial flow thresher is

(A) Straight single pass and perpendicular to the cylinder shaft

(B) Curved and perpendicular to the cylinder shaft

(C) Helical and several times

(D) Straight and parallel to the cylinder shaft

Q. 82 The thresher 'A' has output capacity of 170 kg h^{-1} while threshing paddy crop at 14% moisture content (m.c.) with a grain to straw ratio 45:55. The thresher 'B' has output capacity of 160 kg h^{-1} while threshing paddy crop at 13% mc. with a grain to straw ratio 40:60. Both the threshers have threshing efficiency of 97%. If a farmer has to carry out threshing of paddy crop at 12% m.c. with a grain to straw ratio 40:60 in the least time, the selected thresher and its output in kg when operated for 5 hours will be

(A) A and 738 (B) A and 850

(C) B and 791 (D) B and 800

Graduate Aptitude Test in Engineering – 2020

Q. 83 A chaff cutter is operated by an electric motor running at 1440 rpm. The speed reduction from motor to the main shaft of the cutting unit is 4:1. The feed rollers of 10 cm diameter each are driven by the main shaft through a suitable gear drive with a speed reduction of 15:1. If the chaff cutter has two knives, the theoretical length of cut chaff in mm is ....

Q. 84 The grain to straw ratio of 500 kg reed material is 3:2. The blown grain loss, separation loss and cleaning efficiency of the thresher are 0.05%, 0.5% and 99.1 %, respectively. Considering 100% threshing efficiency, grain recovery at the main grain outlet in kg is ....

Graduate Aptitude Test in Engineering – 2021

Q. 85 While harvesting paddy with a self-propelled vertical conveyor reaper with a cutter bar of width 60 cm, the power required for cutting and propelling are measured to be 300 W and 350 W, respectively. If the power required for conveying the cut crop is 50% of the power required for cutting, the power required by the header unit of the vertical conveyor reaper in W will be ....

Answers Kry

1	2	3	4	5	6	7	8	9	10
C	C	A	D	D	D	A	C	C	C
11	12	13	14	15	16	17	18	19	20
B	C	B	C	A	C	D	D	D	B
21	22	23	24	25	26	27	28	29	30
A	C	C	C	BONUS	C	A	B	A	D
31	32	33	34	35	36	37	38	39	40
C	E	E	A	B	B	A	C	A	B
41	42	43	44	45	46	47	48	49	50
A	B	D	D	A	A	A	B	D	B
51	52	53	54	55	56	57	58	59	60
B	D	B	A	C	A	B	B	A	B
61	62	63	64	65	66	67	68	69	70
C	B	C	A	C	B	D	B	B	B, D
71	72	73	74	75	76	77	78	79	80
D	B	D	C	C	B	B	1.59	D	83.33
81	82	83	84	85					
C	C	10.5	298.35	450					

Explanations

Q. 11

Area to be harvested = 2.5×4000×8×0.60/10000 = 4.8 ha

Q. 50

Field capacity of mower = 4×5000×0.75/10000 = 1.5 ha/h

Hence, Time taken by mower to cover 1 hectare area = 1/1.5 = 0.67 h

or 40 minutes

Q. 50

Field capacity of mower = 2×4000×0.75/10000 = 0.6 ha/h

Q. 64

Field capacity of combine = 2×4000/10000 = 0.8 ha/h

Hence, overall efficiency = Actual area covered/ Theoretical area calculated

= 0.6/0.8 = 0.75

Q.69

Given that

Fuel consumption (FC) = 4.5 L/h or 1.25×10^{-5} m^3/s

Heating value of fuel (CV) = 40 MJ/kg = 40×10^6 J/kg

Density of the fuel (ρ) = 825 kg/m^3

∴ Power developed from burning of fuel = $1.25\times10^{-5}\times40\times10^6\times825$

= 412500 W = 412.5 kW

Brake thermal efficiency (η_{bte}) = 32%

∴ Brake horse power (BHP) = 412.5×0.32 = 13.2 kW

∵ Transmission to PTO loss is 5%.

Thus, Power consumed by thresher = 13.2×(1-0.05) = 12.54 kW

Q. 70 (i)

Given that

Bending stress (σ) = 35 N; Crop height (L) = 30 mm

Friction coefficient of knife edge

$(\mu) = 0.346;$

Angle of counter shear

$(\varnothing) = 17^{o}$

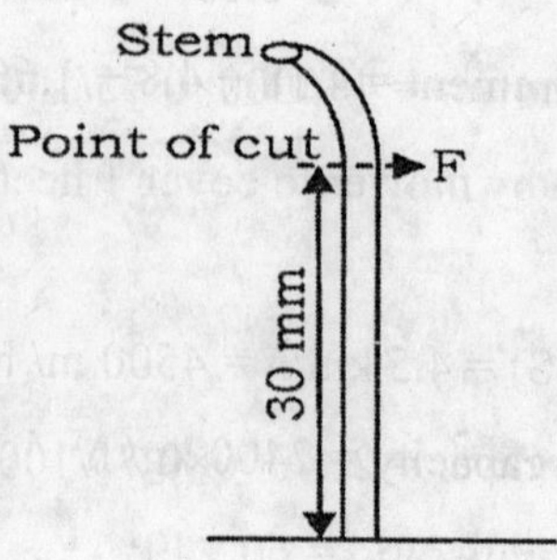

We know that

$$\text{Bending stress } (\sigma) = \pm \frac{\text{Bending moment } (M)}{I} \times y$$

Where;

$y = d/2$

And $\quad I = \text{moment of inertia} = \frac{\pi d^4}{64}$ (For circular sections)

$\therefore$ $\quad \text{Bending moment (M)} = \frac{\sigma \pi}{32}$

$$M = \frac{35 \times 3.14 \times (3)^3}{32} = 92.73 \text{ N-mm}$$

$\because$ Crop stem is acting as a cantilever beam, therefore

$$\text{Horizontal force (F)} = \frac{M}{L} = \frac{92.73}{30} = 3.09 \text{ N}$$

(ii)

$\because$ Clip angle $= \varnothing + \tan^{-1}(\mu)$

$= 17^{o} + \tan^{-1}(0.346)$

$= 36.08^{o}$

Q. 71

Cutting width of mower (w) = 6×0.4 = 2.4 m

Forward speed of mower (S) = 3 km/h = 0.83 m/s

Specific power = Specific energy × S × w

2.1×2.4×0.83 = 4.18 kW

Power loss due to air, stubble and gear train = 2×2.4 = 4.8 kW

Power loss in propelling the tractor = propelling force × forward speed of mower

$$= 2\times0.83 = 1.66 \text{ kW}$$

∴ Total power requirement = 4.18 + 4.8 + 1.66 = 10.64 kW ≈ 10.67 kW

Q. 72

Given that

Speed of harvester (S) = 4.5 km/h = 4500 m/ h; Width of cut (W) = ?

Optimal throughput capacity = 2400 kg/ h

∵ Grain to straw ratio = 60 : 40

⇒ Grain yield = 1440 kg/h

∴ Field capacity $= \frac{1440}{3000} = 0.48$ ha/ h = 4800 m²/ h

Again,

Field capacity = S.W

= 4500×W

Hence 4500×W = 4800

⇒ W = 1.07 m

Q. 73

∵ Throughput = 750 kg/ h

Therefore, gain yield at main outlet = 450 kg/ h

(∵Grain : Straw = 1.5 : 1)

But,

Thresher capacity = 445 kg/ h

∴ Grain loss = 450 – 445 = 5 kg/ h

MOG collected at main outlet $= 445\times\frac{0.5}{100} = 2.225$ kg/ h

Hence,

Total loss of grain at main outlet $= \left(\frac{5+2.225}{450}\right)\times100 = 1.61\ \%$

Q. 74

Given that,

$$Y_o = -0.0018\,N^2 + 0.1319N - 0.743$$

$$\frac{dY_o}{dN} = -0.0036N + 0.1319$$

And $$\frac{d^2Y_0}{dN^2} = -0.0036 \Rightarrow \frac{d^2Y_0}{dN^2} < 0$$

Hence, Y_0 will be maximum at the point where $\frac{dY_o}{dN} = 0$

$$\Rightarrow \quad -0.0036N + 0.1319 = 0$$

$$\Rightarrow \quad N = 36.6 \approx 37$$

Q. 75

Paddy yield = unthreshed paddy + threshed seed

= 14 + 70

= 84 kg

∵ Threshing efficiency = 100×(1 - percent unthreshed grain)

$$= 100\times(1 - \frac{14}{84})$$

= 83.33 %

Q. 76

We have

∵ Grain : Straw = 1.5 : 1 Total, grain + straw = 2.5

Grain yield at main outlet = 500 kg/ h

Therefore, Throughput capacity = $\frac{500}{1.5} \times 2.5 \times 0.99 = 825$ *kg/h*

Q. 78

Cross section area of mouth of chaff cutter mouth = 250×15 = 3750 mm^2

$= 3.75\times10^{-3}\ m^2$

Speed of feed rollers = π DN

= 3.14×0.10×(90×60) = 1695.60 m/h

Chaff carrying rate of chaff cutter is = $1695.60\times3.75\times10^{-3} = 6.36\ m^3/h$

∴ Throughput capacity = 250×6.36 = 1590 kg/h or 1.59 ton/h

Q. 79

Total Grain = 135 + 0.65 + 2.4 + 4.54 + 0.28 = 142.87 kg

Unthreshed grain lost with straw and chaff = 4.54 + 0.28 = 4.82 kg

∵ Threshing efficiency = 100×(1 − Percent unthreshed grain)

= 100×(1 – 4.82/142.87)

= 96.63 %

Q. 80

Speed = 4 km/hr = 66.67 m/s

Theoritical time per hectare = 10000/(66.67×1.2)

= 125 min

Actual time = 125 + 25 = 150 min

∵ Field Efficiency = Theoritical time per hectare × 100/ Actual time

= 125×100/150 = 83.33 %

Q. 82

Thresher A	Thresher B
Inlet capacity = 170/(0.97×0.45) = 389.46 kg/h	Inlet capacity = 160/(0.97×0.40) = 412.37 kg/h **Economically beneficial for farmer.**

Water content in grain for thresher B = 160×0.13 = 20.8 kg/h

Dry grain = 160 – 20.8 = 139.20 kg/h

Consider, Output rate for 12% moisture content = x kg

Hence,

x×(1 – 0.12) = 139.20

⇒ x = 158.09 kg/h

Total output in 5 hours = 158.09×5 = 790.45 kg

Q. 83

Consider length of cut = x cm

Hence, chaff cut in one revolution of Fly wheel = 2x cm

One revolution of feed roller = 3.14×10 = 31.4 cm

The feed roller rotation in One revolution of Fly wheel = 31.4/15

= 2.09 cm

Hence,

$2x = 2.09$

$\Rightarrow$ $x = 1.05$ cm or 10.5 mm

Q. 84

Grain = 300 kg and Straw = 200 kg

Losses = 300×0.05/100 + 300 × 0.5/100 = 1.65 kg

Grain recovery = 300 – 1.65 = 298.35 kg

Q. 85

Power required by header unit = 300 + 300 × 0.50 = 450 W

18

Farm Processing Equipments Chaff Cutter & Mowers

CGPSC State Engineering Services Exam Paper – I

Q. 1 Weight of a manually operated chaff cutter's fly wheel should not be below than

(A) 24 kg (B) 30 kg

(C) 32 kg (D) 35 kg

(E) 27 kg

UKPSC A.En. Exam 2007 Paper – II

Q. 2 In hand chaff-cutters, which of the following is used for changing the length of the cut ?

(A) Handle (B) Flywheel

(C) Worm and worm wheel (D) Spike roller

Graduate Aptitude Test in Engineering - 2007

Q. 3 A 6 bladed forage blower operates at 540 rpm. For a feed rate of 6.5×10^4 kg h^{-1}, the mass of corn silage carried on each impeller blade is

(A) 0.334 kg (B) 2.006 kg

(C) 12.037 kg (D) 20.060 kg

Graduate Aptitude Test in Engineering - 2008

Q. 4 A flail mower is operated using the PTO power of a tractor through a bevel gear drive. The tractor forward speed is 10.8 km h^{-1}. The velocity of the flail tip with respect to the ground is 18 ms^{-1}. The length of each flail is 400 mm and the diameter of the shaft carrying the flails is 100 mm. If the tractor PTO speed is 800 rpm, the required bevel gear reduction is

(A) 1.13 (B) 2.00

(C) 2.25 (D) 2.51

Graduate Aptitude Test in Engineering - 2009

Q. 5 The flywheel type chaff cutter with two cutting knives is having a feeding throat of $0.2 \times 0.1\ m^2$. The mean chaff length is found to be 20 mm for a flywheel speed of 500 rpm. If the density of the dry fodder is 120 kg m^{-3}, the capacity of the chaff cutter in kg/h at full load is

(A) 920 (B) 1440

(C) 2160 (D) 2880

Graduate Aptitude Test in Engineering - 2011

Q. 6 In a cutter bar mower, the cutter bar drive consists of an offset slider crank mechanism with a crank of radius 36 mm and a pitman of length 360 mm. If offset of the crank is 72 mm, the stroke length in mm would be

(A) 36.0 (B) 72.0

(C) 73.5 (D) 80.5

Graduate Aptitude Test in Engineering - 2013

Q. 7 The flywheel of a hand operated chaff cutter with two cutting knives is rotated at 30 rpm and is connected to a worm gear assembly for driving the feed rollers. The number of teeth of the worm gear is 24 and number of starts (threads) of the worm is 2. If diameter of each of the feed rollers is 15 cm, the chaff length in mm will be

(A) 9.8 (B) 12.8

(C) 19.6 (D) 39.2

Graduate Aptitude Test in Engineering – 2015

Q. 8 A power operated chaff cutter with two knives is rotating at 600 rpm. It has a conveyor type feeding mechanism which operates at a speed of 0.6 m/s. The theoretical chaff length in mm is

(A) 20 (B) 30

(C) 60 (D) 90

Q. 9 In a reciprocating type mower, the maximum inertia force of 2.2 kN along the pitman occurs at 35° crank angle and 25° pitman angle with the horizontal plane. The crank radius is 50 mm and the equivalent mass at the crankpin is 2.5 kg. If the crank rotates at 600 rpm, the resultant force passing through the crankpin in kN will be

(A) 2.25 (B) 2.68

(C) 2.98 (D) 3.42

Answers Key

1	2	3	4	5	6	7	8	9
A	C	A	D	D	C	C	B	B

Explanations

Q. 3

$$\text{Total feed carried} = \frac{\textit{Feed Rate}}{N}$$

$$= \frac{6.5 \times 10^4}{540 \times 60}$$

$$= 2.006 \text{ kg}$$

Therefore, silage carried on each impeller blade $= \frac{2.006}{6}$

$$= 0.334 \text{ kg}$$

Q. 4

Tractor forward speed = 10.8 km/h = 3 m/s

Velocity of flail tip with respect to the ground = 18 m/s

$\therefore$ Velocity of flail = 18 – 3 = 15 m/s

Diameter of flail tip = 400 + 100 + 400 = 900 mm = 0.9 m

$\therefore$ Speed of flail shaft (N) $= \frac{\textit{Velocity of flail shaft}}{\pi D}$ $[\because V = \pi D N]$

$$= \frac{15 \times 60}{3.14 \times 0.9} = 318.47 \text{ rpm}$$

$\therefore$ Bevel gear reduction $= \frac{\textit{Tractor PTO Speed}}{\textit{Speed of flail shaft}}$

$$= \frac{800}{318.47} = 2.51$$

Q. 5

There are two cutting knives, therefore total length of chaff passed through feeding throat is 40 mm.

Total chaff passed through the feeding throat = A.L

$= 0.2\times0.1\times0.04 = 0.0008\ m^3$

$\because$ Density of fodder = 120 kg/m^3

$\therefore$ Weight of chaff passed through the throat = 120×0.0008

$= 0.096$ kg

Capacity of chaff cutter = Weight of chaff × Rpm

$= 0.096\times500 = 48$ kg/min or 2880 kg/h

Q. 6

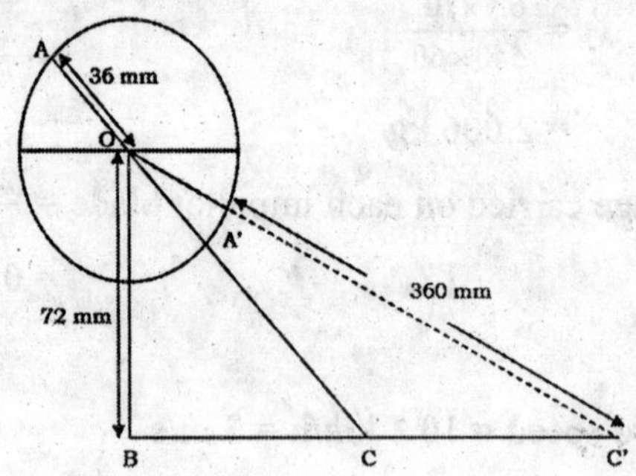

Stroke length = CC'

And CC' = BC' - BC

$$BC = \sqrt{(360-36)^2-(72)^2} = 315.9\ mm$$

$$BC' = \sqrt{(360+36)^2-(72)^2} = 389.4\ mm$$

Stroke length = 389.4 – 315.9 = 73.5 mm

Q. 7

Revolution of flywheel (N_1) = 30 rpm

Revolution of worm gear is = N_2

$\because$ $N_1/N_2 = T_2/T_1$

$30/N_2 = 24/2$

$\Rightarrow$ $N_2 = 2.5$ rpm

When there is one revolution of flywheel, revolution of worm gear (N_2) = 2.5/30 rpm

Length of chaff cut = πDN_2

= 3.14x150x2.5/30 mm = 39.2 mm

Here two knives are mounted in flywheel, therefore cutting length of chaff is

= 39.2/2

= 19.6 mm

Q. 8

Revolution of flywheel (N) = 600 rpm = 10 rps

Hence, chaff passed through chaff cutter = 0.6 m in 10 revolutions.

∴ Per revolution chaff passed = 0.6/10 = 0.06 m = 60 mm

Here, two knives are mounted in flywheel, therefore cutting length of chaff is

= 60/2

= 30 mm

Q. 9

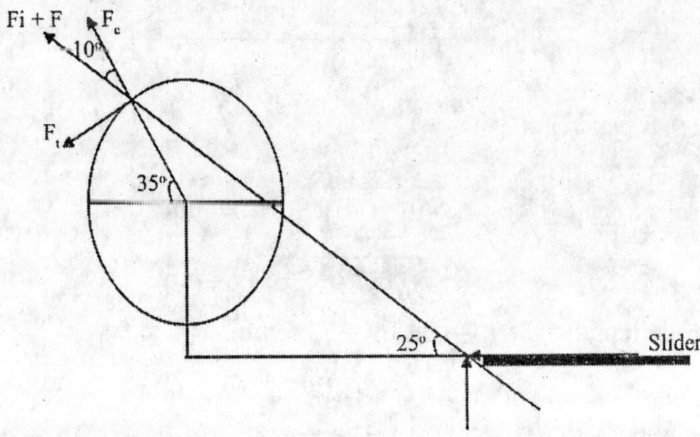

Centrifugal force acting at crankpin

$(F_c) = mw^2r$

$= 2.5\times9.81\times(2\times3.14\times600/60)^2\times0.050$

= 4831.20 N= 4.83 kN

∵ Maximum inertia force acting along the pitman (Fi) = 2.2 kN

Here, F is the resultant of other forces passing through crankpin due to centrifugal (F_c) and tangential (F_t) effect of counterweight of flywheel and power source respectively.

Referring above figure

$$F_i + F = F_c/\cos 10$$

$$\Rightarrow \quad 2.2 + F = 4.8/0.985$$

$$\Rightarrow \quad F = 2.70 \text{ kN}$$

19

Economics of Agricultural Equipments and Machinery

UKPSC Combined Junior Engineering Exam 2013 Agricultural Engineering Paper-I

Q. 1 Annual depreciation of a machine is included in

(A) Fixed cost (B) Variable cost

(C) Hiring cost (D) None of the above

Q. 2 The estimated useful life of diesel engine is

(A) 8 years (B) 10 years

(C) 12 years (D) 14 years

Q. 3 Depreciation of an IC engine is included in

(A) Fixed cost (B) Variable cost

(C) Both (A) and (B) (D) None of the above

Q. 4 Within a given HP range, the selection of tractor is based on

(A) Primary cost (B) Resale value

(C) Running cost (D) All of the above

RPSC AEn Pre Exam 2013 (Agricultural Engineering)

Q. 5 Which technical term is more appropriately used while doing cost analysis of farm equipments in agricultural farm

(A) Depreciation (B) Efficiency

(C) Both (A) and (B) (D) None of the above

UKPSC A.En. Exam 2012 Paper - II

Q. 6 Salvage or Junk value of a machine is

(A) 5% of initial cost (B) 10% of initial cost

(C) 15% of initial cost (D) None of the above

Graduate Aptitude Test in Engineering - 2009

Q. 7 The initial cost of a tractor is Rs. 4,00,000. The annual rate of depreciation is 15%. Following declining balance method, the value of the tractor at the end of 6th year is

(A) Rs. 40,000 (B) Rs. 1,51,000

(C) Rs. 1,76,000 (D) Rs. 2,01,000

Graduate Aptitude Test in Engineering - 2011

Q. 8 The cost of a tractor is Rs. 34,0000. Its service life is 10 years and the salvage value is 10% of initial cost. Its depreciation in 2nd year, following constant rate method, in rupees would be

(A) 30600 (B) 34000

(C) 50070 (D) 55550

Graduate Aptitude Test in Engineering - 2014

Q. 9 A vertical conveyor reaper costing Rs. 75000 has a useful life of 8 years. Taking salvage value of the machine as 10% of the initial cost, the depreciated value after half of its useful life following straight line method will be Rs ....

Graduate Aptitude Test in Engineering - 2017

Q. 10 The purchase price of a tractor is Rs. lakh. Considering the constant rate of depreciation as 0.2 in declining balance method, the value of the tractor at the end of 4th year in lakh will be ....

Graduate Aptitude Test in Engineering - 2019

Q. 11 The purchase price of a tractor is Rs. 5,50,000. Useful life of the tractor is 10 years and its salvage value is 10% of the purchase price. Following the sum of the years digit method, the depreciation in 3rd year in Rs. is ....

Graduate Aptitude Test in Engineering - 2021

Q. 12 Fixed cost per year and variable cost per hour of a tractor were estimated based on its annual usage of 800 h. The total cost of operation was found to be Rs. 540 per hour. It was later re-estimated and found that total cost of operation would be Rs. 510 per hour, if the annual hours of use were increased to 1000 h. Considering all the components of annual usage cost to be the same, the variable cost in Rs. per hour would be. ....

Answers Key

1	2	3	4	5	6	7	8	9	10
A	B	A	D	A	B	B	A	41250	2.05
11	12								
72000	390								

Explanations

Q. 7

Initial cost of tractor (P) = 400000 Rs.; Annual rate of depreciation (r) = 15%

∵ Value of tractor after n years $(V_n) = P(1-r)^n$

$$V_n = 400000\times(1\text{-}0.15)^6$$

$$= 150860 \approx 151000 \text{ Rs.}$$

Q. 8

Purchase price of tractor (P) = 340000 Rs.

Service life (L) = 10 years; Salvage value (S) = $340000\times\frac{10}{100} = 34000$ Rs.

∵ $$\text{Depreciation (D)} = \frac{P-S}{L}$$

$$= \frac{340000-34000}{10} = 30600 \text{ Rs.}$$

In constant rate method, Depreciation for each year remains same. Therefore, Depreciation in 2nd year is 30600 Rs.

Q. 9

Cost of Reaper (P) = 75000 Rs.; Useful life (L) = 8 years

Salvage value (S) = $75000\times\frac{10}{100} = 7500$ Rs.

$$\text{Depreciation (Rs./year)} = \frac{\textit{Purchase Price} - \textit{Salvage Value}}{\textit{Usefull life}}$$

$$= \frac{75000 - 7500}{8} = 8437.5 \text{ Rs. / year}$$

Depreciated value after 4 years = $75000 - 4 \times 8437.5 = 41250$ Rs.

Q. 10

In constant rate of depreciation

The value of tractor after n[th] year $(V_n) = P(1 - R)^n$

$$= 5 \times (1 - 0.2)^4 = 2.05 \text{ lakh.}$$

Q. 11

Depreciation after m years

$$(D_m) = \frac{L - m + 1}{SOY}$$

$$D_3 = (550000 - 55000) \frac{(10 - 3 + 1)}{(10 + 9 + 8 + \ldots\ldots + 1)} = \text{Rs. } 72000$$

Q. 12

Case I:-

Consider, Fixed Cost = x Rs/year , Operational Cost = y Rs/h or 800y Rs/year

Total Cost of operation = $x + 800y = 540 \times 800$ (i)

Case II:-

Fixed Cost remains unchanged irrespective to operation hours. Hence

Fixed cost = x Rs/year, Operational Cost = y Rs/h or 1000y Rs/year

Total Cost of operation = $x + 1000y = 510 \times 1000$ (ii)

By solving equations (i) and (ii)

Operational Cost (y) = 390 Rs/h

IV. Soil-Water Conservation and Irrigation Engineering

20

Fluid Mechanics

UKPSC Combined Assistant Engineering Exam 2013 Paper - I

Q. 1 Density of water is:-

(A) 1000 kg/m^3 (B) 500 kg/m^3

(C) 1500 kg/m^3 (D) 2000 kg/m^3

Q. 2 Specific energy diagram is plotted between specific energy and depth of flow for:-

(A) Variable discharge (B) Constant discharge

(C) Constant Froude number (D) None of the above

Q. 3 Dimension of dynamic viscosity are:-

(A) $ML^{-1}T^{1}$ (B) MLT^{-1}

(C) $ML^{-1}T$ (D) $M^{-1}LT$

Q. 4 The weight of the fluid per unit volume is also called:-

(A) Specific volume (B) Specific weight

(C) Density (D) None of the above

Q. 5 The dynamic viscosity of a fluid expressed as 'poise' equals to:-

(A) 10 Newton second per m^2 (B) 1 Newton second per m^2

(C) 1 Newton - meter per second (D) 0.1 Newton - second per m^2

Q. 6 The flow is said to be turbulent if the Reynold number in a pipe is:-

(A) 5000 (B) 1800

(C) 1500 (D) 1000

Q. 7 Bernoulli's equation is related to:-

(A) Mass conservation (B) Energy balance
(C) Heat conservation (D) Water conservation

Q. 8 Ratio of buoyant forces to the viscous forces of the fluid flow is known as:-

(A) Prandlt number (B) Grashof number
(C) Reynolds number (D) None of the above

Q. 9 Friction factor is important parameter particularly in:-

(A) Laminar flow (B) Turbulent flow
(C) Viscous flow (D) Vertical flow

Q. 10 Venturimeter is used to measure flow of fluids in pipes when pipe is :-

(A) Horizontal (B) Vertical downward
(C) Vertical upward (D) In any position

Q. 11 If a fluid which possess some viscosity is known as:-

(A) Real fluid (B) Imaginary fluid
(C) Specific fluid (D) None of the above

Q. 12 A large Reynold's number is indication of:-

(A) Highly turbulent flow (B) Laminar flow
(C) Steady flow (D) Smooth and stream line flow

Q. 13 One stoke is equal to:-

(A) 1 m^2/s (B) 0.1 m^2/s
(C) 1×10^{-2} m^2/s (D) 1×10^{-4} m^2/s

Q. 14 Absolute pressure is equal to:-

(A) Gauge pressure – Atmospheric pressure
(B) Gauge pressure + Vacuum Pressure
(C) Atmospheric pressure + Gauge pressure
(D) Atmospheric pressure - Gauge pressure

Q. 15 The dimension of specific gravity are:-

(A) MLT (B) ML^2T

(C) ML^{-1} (D) None of the above

Q. 16 The Force per unit area in fluids is exerted:-

(A) In all directions (B) Lower depths

(C) Horizontal ways (D) Upward

Q. 17 A fluid flow is said to be ideal or perfect when it has:-

(A) No frictional effect between moving fluid

(B) Viscosity

(C) Turbulence

(D) Temperature effect

Q. 18 Which one of the following is fluidised solid ?

(A) Flour (B) Fruit juice

(C) Honey (D) Carbon di oxide

Q. 19 What will be the equivalent pressure head of kerosene if its specific gravity is 0.81, and water pressure head is 100 m

(A) 123.46 m (B) 184.36 m

(C) 81.0 m (D) 162.0 m

Q. 20 Pitot tube is a simple device used to measure:-

(A) Density of flow (B) Velocity of flow

(C) Type of flow (D) None of the above

Q. 21 One atmospheric pressure is equal to:-

(A) 1.0 kg/cm^2 (B) 1.0 kg/m^2

(C) 1.0 ton/m^2 (D) 1.0 N/m^2

Q. 22 In order to avoid the tendency of separation at the throat in a venturimeter, the ratio of diameter at throat to that of the pipe should:-

(A) 1/16 to 1/8 (B) 1/8 to 1/4

(C) 1/4 to 1/3 (D) 1/3 to 3/4

Q. 23 Orifice meter is used to measure:-

(A) Roughness in pipes

(B) Roughness in open channels

(C) Flow in open channels

(D) Flow in pipes

APPSC AEES Agriculture Engineering Exam 2016

Q. 24 Darcy's Law is valid for :

(A) Laminar and turbulent flow (B) Turbulent flow

(C) Laminar flow (D) Transient flow

Q. 25 Fluid which become more fluid viscosity decreases with time as they are stirred are known as

(A) Pseudoplastic (B) Dilantant

(C) Thixotropic (D) Rheopectic

RPSC AEn Pre Exam 2013 (Agricultural Engineering)

Q. 26 Equation of continuity is based on the principle of conservation of

(A) Mass (B) Energy

(C) Momentum (D) None of the above

Q. 27 A flow in which the velocity of fluid at a particular fixed point does not change with time is known as

(A) Unsteady flow (B) Laminar flow

(C) Turbulent flow (D) Uniform flow

OCS AEn Exam 2011 (Pre) – II (Agricultural Engineering)

Q. 28 When mercury and water are separately placed in separate glass tubes and development of meniscus in both the cases means:

(A) Cohesive forces between mercury molecules are greater than the adhesive forces between mercury and glass: and adhesive forces between water and glass are greater than cohesive forces between water molecules.

(B) Adhesive forces between mercury and glass are greater than the cohesive forces between mercury molecules: and cohesive forces between water molecules are greater than adhesive forces between water and glass.

(C) Cohesive forces between mercury molecules are greater than the adhesive forces between mercury and glass: and cohesive forces between water molecules are greater than adhesive forces between water and glass.

(D) Adhesive forces between mercury and glass are greater than the cohesive forces between mercury molecules: and cohesive forces between water molecules are lesser than the adhesive forces between water and glass.

Q. 29 Unstable floating body has:

(A) Metacentre above centre of gravity

(B) Metacentre below centre of gravity

(C) Metacentre and the centre of gravity are at the same point

(D) Metacentre and centre of gravity are at different points but at same level.

Q. 30 Which of the following behaves as Newtonian fluids?

(A) Water, Air, Blood (B) Water, Oil, Blood

(C) Water, Air, Oil (D) Water, Blood, Tar

CGPSC State Engineering Services Exam (Agriculture Engineering) Part - II

Q. 31 Mean size droplets that are generated in the fog cooling system is ……….. micron

(A) Less than 15 (B) Less than 10

(C) More than 10 (D) More than 1

(E) None of these

Q. 32 In which flow meter the pressure drop remains nearly constant but the area changes

(A) Pitot tube (B) Venturi Meter

(C) Orifice meter (D) Rota meter

(E) V-notch

Q. 33 The stress-strain relation of the Newtonian fluid is

(A) Parabolic (B) Hyperbolic

(C) Linear (D) Inverse type

(E) None of the above

CGPSC State Engineering Services Exam Paper – I

Q. 34 The unit commonly used for dynamic viscosity is

(A) Pascal sec (B) Stokes

(C) cm/s^2 (D) Both (B) & (C)

(E) Both (A) & (C)

Q. 35 For pseudo plastic fluids, increases in shear rate

(A) Increases the apparent viscosity

(B) Decreases the apparent viscosity

(C) Has no effect on apparent viscosity

(D) Has unspecified effect

(E) Both (A) & (C)

Q. 36 A fluid in which the viscosity increases with increasing stirrer speed and mixing time, can be called as

(A) Newtonian fluid

(B) Pseudo plastic, thixotropic fluid

(C) Dilatant, Rheopectic fluid

(D) Dilatant, pseudo plastic fluid

(E) Non-Newtonian fluid

Q. 37 The flow with unchanging velocity distribution is called

(A) Potential flow (B) Steady flow

(C) Unsteady flow (D) Fully developed flow

(E) Redial flow

UKPSC A.En. Exam 2007 Paper – I

Q. 38 If the dynamic viscosity of a fluid is 0.5 poise and the specific gravity is 0.5, then the kinematic viscosity of the fluid in stokes is

(A) 0.25 (B) 0.50

(C) 0.75 (D) 1.00

Q. 39 In close conduits, water flows in the direction of

(A) Elevation gradient (B) Pressure gradient

(C) Seepage gradient (D) None of the above

Q. 40 What will be the dimensions of kinematic viscosity ?

(A) LT^{-1} (B) L^2T^{-2}

(C) L^2T (D) L^2T^{-1}

Q. 41 Frictional head loss in pipe is directly proportional to

(A) Length of pipe (B) Diameter of pipe

(C) Both of the above (D) None of the above

UKPSC A.En. Exam 2007 Paper – II

Q. 42 The Bernoulli's equation does not include the following

(A) Pressure head (B) Elevation head

(C) Suction head (D) Velocity head

Q. 43 Reynold's number is the ratio of inertial force to

(A) Gravitational force (B) Viscous force

(C) Drag force (D) None of the above

UKPSC A.En. Exam 2012 Paper – I

Q. 44 Pseudoplastic fluids are

(A) Shear rate thinning

(B) Shear rate thickening

(C) Time dependent

(D) Both shear rate thinning and time dependent

Q. 45 In a pipe of uniform diameter loss of head at exit is

(A) Twice that at inlet (B) Four times that at inlet

(C) One half at inlet (D) Same as that at inlet

Q. 46 An ideal fluid is

(A) Incompressible (B) Does not resist shear

(C) Has zero viscosity (D) All the above

Q. 47 If Reynold's number for a pipe flow is 1500, the flow would be

(A) Laminar flow (B) Turbulent flow

(C) Transient flow (D) None of the above

Graduate Aptitude Test in Engineering - 2007

Q. 48 A fluid in which shear stress is more than the yield value and proportional to the rate of shear strain is called

(A) Newtonian fluid (B) Non- Newtonian fluid

(C) Ideal plastic fluid (D) Real fluid

Graduate Aptitude Test in Engineering - 2008

Q. 49 Atmospheric pressure at a place is equal to 10 m of water. A liquid has a specific weight of 12 kN m^{-3}. The absolute pressure at a point 2 m below the free surface of liquid in kPa is

(A) 2.4 (B) 12.4

(C) 24.0 (D) 122.1

Q. 50 The weight of a hollow sphere is 100 N. If it floats in water just fully submerged, the external diameter of the sphere is

(A) 112 mm (B) 213 mm

(C) 269 mm (D) 315 mm

Q. 51 A spherical tank of 2 m diameter is filled with and edible oil of specific gravity 0.92. If the pressure measured at the highest point in the tank is 70 kPa, the total pressure (kPa) in the tank will be

(A) 80.5 (B) 85.3

(C) 88.1 (D) 92.2

Graduate Aptitude Test in Engineering - 2009

Q. 52 A venturimeter of 75 mm diameter is fitted to a horizontal pipe of 150 mm diameter. Gauge pressure in the venturimeter in case of no flow is 2 m of water. Taking atmospheric pressure as 10 m of water, the theoretical flow through the pipeline in litres per second, when the throat point pressure is 2.60 m of water (absolute), is

(A) 15 (B) 30

(C) 60 (D) 75

Graduate Aptitude Test in Engineering - 2010

Q. 53 A pipeline carrying a discharge of 500 litres per minute branches into two parallel pipes, X and Y, as shown in the following figure. The length and diameter of pipes X and Y are shown in the figure.

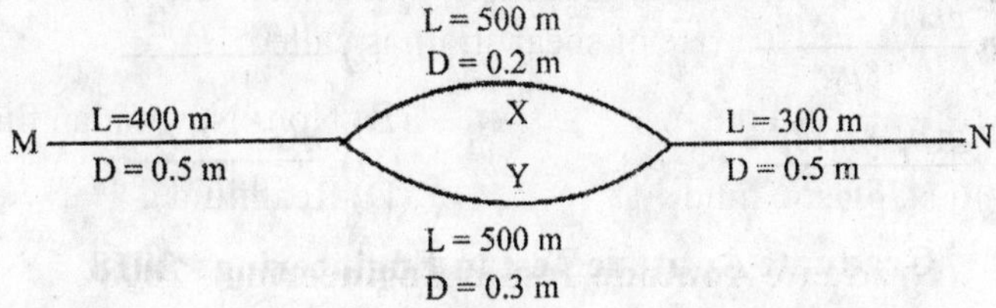

The friction factor, f for all pipes is 0.030. the ratio of flow in pipes X and Y is

(A) 0.36 (B) 0.44

(C) 0.67 (D) 1.00

Graduate Aptitude Test in Engineering - 2011

Q. 54 If the density of a fluid changes from point to point in a flow region, the flow is called

(A) Steady flow (B) Unsteady flow

(C) Non-uniform flow (D) Compressible flow

Q. 55 The viscosity of a newtonian fluid depends primarily on X and to a lesser degree on Y. X and Y are

(A) X = temperature, Y = flow velocity

(B) X = flow velocity, Y = pressure

(C) X = temperature, Y = pressure

(D) X = roughness of the surface across which the fluid flows, Y = flow velocity

Graduate Aptitude Test in Engineering - 2014

Q. 56 A 200 m long horizontal pipe carries a discharge of 50 L s^{-1}. The centre line of the pipe is 5 m above the datum. The diameter of the pipe tapers from 200 mm to 100 mm. Using g = 9.81 m s^{-2} and neglecting losses in the pipe, if the pressure at the larger end of the pipe is 100 kPa, the pressure at the other end of the pipe in kPa is.........

Graduate Aptitude Test in Engineering - 2016

Q. 57 A very small particle of diameter d_p and density ρ_p freely settles at constant velocity in a tank of depth L containing liquid of viscosity μ_l. The density of the liquid is $_l$ where $\rho_l < \rho_p$. The velocity of particle in the liquid can be expressed as

(A) $\frac{gL(\rho_p - \rho_l)d_p}{18\mu_l}$ (B) $\frac{g(\rho_p - \rho_l)d_p^{\ 3}}{18\mu_l}$

(C) $\frac{g(\rho_p - \rho_l)d_p^{\ 2}}{18\mu_l}$ (D) $\frac{g(\rho_p - \rho_l)L^2}{18\mu_l}$

Graduate Aptitude Test in Engineering - 2018

Q. 58 A cylindrical tank of 1.5 m diameter and 4 m height is filled with a liquid by a pipe of 2.5 cm internal diameter. The density and dynamic viscosity of the liquid are 1050 kg m^{-3} and 1.6×10^{-3} N s m^{-2}, respectively. If flow in the pipe is turbulent, the maximum time required to fill the tank is .... hours.

Graduate Aptitude Test in Engineering - 2020

Q. 59 Two ends of a differential mercury manometer are connected at two points on a pipe carrying oil. The manometer shows difference in mercury level of 20 cm. The specific gravity of oil and mercury are 0.8 and 13.6, respectively. The density of water is 1000 kg/m^3 at 4 °C and acceleration due to gravity (g) is 9.81 m/s^2• At the same two points in pipe, the difference of pressure in N/m^2 is

(A) 25.11 (B) 251.14

(C) 25113.60 (D) 251136.00

Graduate Aptitude Test in Engineering - 2021

Q. 60 Water is discharged from a tank through a rectangular orifice of width 1.5 m and height 1.2 m. The water level in the tank is 3.5 m above the top edge of the orifice. If the coefficient of discharge of this orifice is 0.62, the discharge through the orifice in m^3/s is. ……….. (Take acceleration due to gravity, $g = 9.81\ m/s^2$)

Answers Key

1	2	3	4	5	6	7	8	9	10
A	B	A	B	D	A	B	B	B	D
11	12	13	14	15	16	17	18	19	20
A	A	D	C	D	A	A	A	A	B
21	22	23	24	25	26	27	28	29	30
A	D	D	C	C	A	D	A	B	C
31	32	33	34	35	36	37	38	39	40
B	D	C	D	B	C	D	D	B	D
41	42	43	44	45	46	47	48	49	50
A	C	B	A	A	D	A	C	D	C
51	52	53	54	55	56	57	58	59	60
C	C	A	D	C	81.04	C	16.49	C	10.00

Explanations

Q. 6

Reynold's Number for various flow conditions

For pipe flow

Type of flow	Reynold's number
Laminar flow	< 2100
Transitional flow condition	2100 – 4000
Turbulent flow	> 4000

For open channel

Type of flow	Reynold's number
Laminar flow	< 500
Transitional flow condition	500 – 2000
Turbulent flow	> 2000

Q. 15

The specific gravity has no dimension.

Q. 19

Pressure head of kerosene = 100/0.81 = 123.46 m

Q. 25

Types of Fluids:

1. Ideal Fluid: An Ideal Fluid is a fluid that has no viscosity. It is incompressible in nature. Practically, no ideal fluid exists.
2. Real Fluid: Real fluids are compressible in nature. They have some viscosity.

   Examples: Kerosene, Petrol, Castor oil
3. Newtonian Fluid: Fluids that obey Newton's law of viscosity are known as Newtonian Fluids. For a Newtonian fluid, viscosity is entirely dependent upon the temperature and pressure of the fluid. Examples: water, air, oil, emulsions

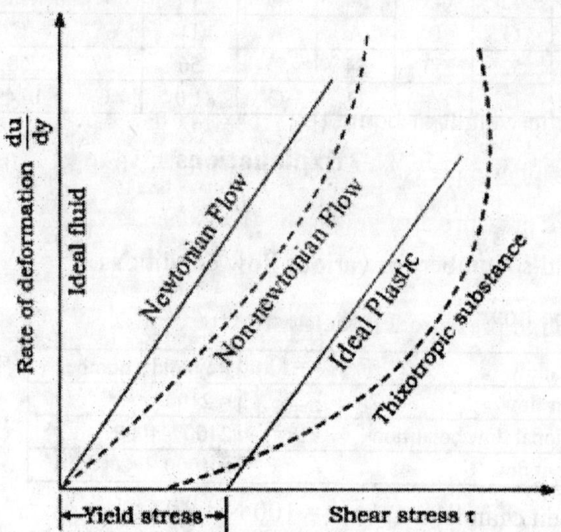

4. Non-Newtonian Fluid: Fluids that do not obey Newton's law of viscosity are non-Newtonian fluids. Examples: Flubber, Oobleck (suspension of starch in water)

- Dilatant - Viscosity of the fluid increases when shear is applied.
- For example: Quicksand, Cornflour and water, Silly putty
- Pseudoplastic - Pseudoplastic is the opposite of dilatant; the more shear applied, the less viscous it becomes. For example: Ketchup

- Rheopectic - Rheopectic is very similar to dilatant in that when shear is applied, viscosity increases. The difference here, is that viscosity increase is time-dependent. For example: Gypsum paste, Cream
- Thixotropic - Fluids with thixotropic properties decrease in viscosity when shear is applied. This is a time dependent property as well. For example: Paint, Cosmetics, Asphalt, Glue

5. Ideal Plastic Fluid: A fluid having the value of shear stress more than the yield value and shear stress is proportional to the shear strain (velocity gradient) is known as ideal plastic fluid. Example: Toothpaste

Q. 38

We know

Kinematic Viscosity = Dynamic viscosity/Specific gravity

= 0.5/0.5 = 1 strokes

Q. 49

Given

Specific weight of liquid (ρg) = 12 kN/m^3

Atmospheric pressure = 10 m of water = $10\times9.81\times1000$ = 98.1 kPa

Gauge pressure 2 m below the free surface = ρgh

= 12×2 = 24 kPa

Absolute pressure 2 m below the free surface

= Gauge pressure + Atmospheric pressure

= 24 + 98.1 = 122.1 kPa

Q. 50

Weight of hollow sphere = 100 N = 10.19 kg

Mass of water removed will be equal to 10.19 kg.

$\therefore \quad (\frac{4}{3}\pi R^3) = 10.19 \quad$ ($\therefore$ Volume×Density = Mass)

$\Rightarrow \quad R = \sqrt[3]{\frac{10.19\times3}{4\times1000\times3.14}}$

= 0.134 m = 134.5 mm ($\therefore$ Density of water = 1000 kg/m^3)

Hence, Diameter of sphere = 134.5 × 2 = 269 mm

Q. 51

Density of water = 1000 kg m^{-3}

Density of oil = 0.92 x 1000 kg m^{-3} = 920 kg m^{-3}; Z = greatest depth = 2 m

and $g = 9.81$ m s^{-2}

Now $P = Z\rho g$

$= 2 \times 920 \times 9.81$

$= 18{,}050$ Pa = 18.1 kPa.

This must be added the pressure at the surface of 70 kPa.

∴ Total pressure = 70 + 18.1 = 88.1 kPa

Q. 52

$$A_1 = \frac{3.14}{4} \times (150)^2 = 17662.5 \text{ mm}^2 = 17662.5 \times 10^{-6} \text{ m}^2$$

$$A_2 = \frac{3.14}{4} \times (75)^2 = 4415.625 \text{ mm}^2 = 4415.625 \times 10^{-6} \text{ m}^2$$

∵ Gauge pressure = 2 m; Atm. Pressure = 10 m

Throat point pressure = 2.6 m absolute

∴ Pressure difference (h) = 12 – 2.6 = 9.4 m

Therefore,

$$\text{Ideal discharge} = \frac{A_1 A_2 \sqrt{2gh}}{\sqrt{A_1^2 - A_2^2}}$$

$$= \frac{17662.5 \times 10^{-6} \times 4415.625 \times 10^{-6} \times \sqrt{2 \times 9.81 \times 9.4}}{\sqrt{(17662.5 \times 10^{-6})^2 - (4415.625 \times 10^{-6})^2}}$$

$= 0.06193$ m^3/s = 61.93 l/s

Q. 53

In this arrangement, the loss of head for each branch pipe is same.

∴ Loss of head for branch pipe X = Loss of head for branch Y

$$\frac{4 f_1 l_1 v_1^2}{2 g d_1} = \frac{4 f_2 l_2 v_2^2}{2 g d_2}$$

$$\Rightarrow \qquad \frac{V_1^2}{V_2^2} = \frac{d_1}{d_2} \qquad [\because f_1 = f_2; l_1 = l_2]$$

$$\frac{V_1}{V_2} = 0.2/0.3 = 0.8167$$

Thus, $Q_1/Q_2 = A_1V_1/A_2V_2$

$$= 0.667\times \frac{A_1}{A_2}$$

$$= 0.8167\times \frac{0.04}{0.09} = 0.36$$

Q. 56

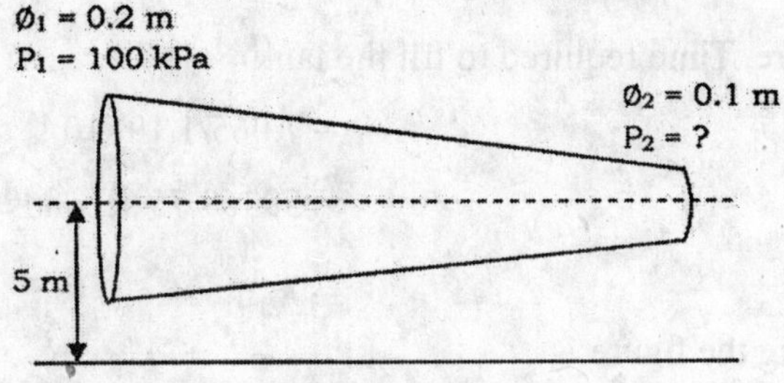

Discharge (Q) = 50 L/s = 50×10^{-3} m^3/s

At section 1

$$V_1 = \frac{50\times10^{-3}}{\frac{3.14}{4}\times(0.2)^2}$$

$$= 1.591 \text{ m/s}$$

At section 2

$$V_1 = \frac{50\times10^{-3}}{\frac{3.14}{4}\times(0.1)^2}$$

$$= 6.36 \text{ m/s}$$

Applying Bernouilli's Theorem

$$\frac{P_1}{\rho g}+\frac{V_1^2}{2g}+Z_1 = \frac{P_2}{\rho g}+\frac{V_2^2}{2g}+Z_2$$

$$\Rightarrow \qquad \frac{100\times10^3}{1000\times9.81}+\frac{(1.591)^2}{2\times9.81} = \frac{P_2}{1000\times9.81}+\frac{(6.36)^2}{2\times9.81}$$

$$\Rightarrow \qquad P_2 = 81.04 \text{ kPa}$$

Q. 58

Reynolds's number should for turbulent flow = 4000 or above 4000

Mass flow rate = Re × π × D × Viscosity/4

(∵ Reynolds's number (Re)

= Density × Fluid velocity × Diameter/ Viscosity and Mass flow rate

= Density × Fluid velocity × Cross sectional area)

$= 4000 \times 3.14 \times 0.025 \times 1.6 \times 10^{-3}/4 = 0.1256$ kg/s

Volumetric flow rate $= 0.1256/1050 = 1.19 \times 10^{-4}$ m³/s

∵ Volume of tank $= 3.14 \times 1.5^2 \times 4/4 = 7.065$ m³

Therefore, Time required to fill the tank

$= 7.065/1.19 \times 10^{-4}$

= 59369.75 seconds or 16.49 hours

Q. 59

Referring the figure

$$P_A + \rho_o g(20+h_m) = P_B + \rho_m g h_m + \rho_0 g h_1$$

$$\Rightarrow \quad P_A - P_B = 9.81 \times 0.20 \times (13600 - 800)$$

$$\Rightarrow \quad P_A - P_B = 25113.6 \text{ N/m}^2$$

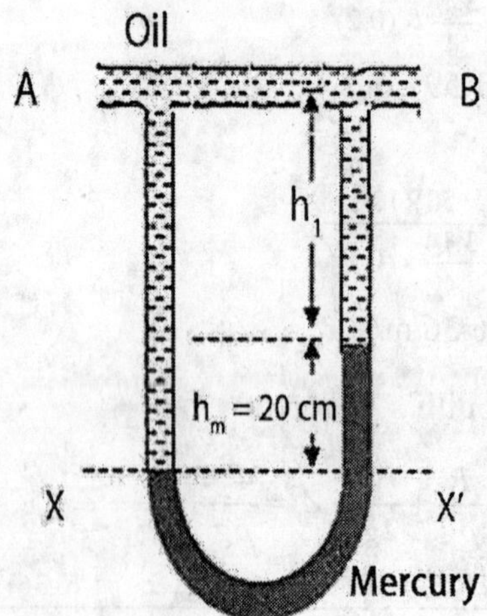

Q. 60

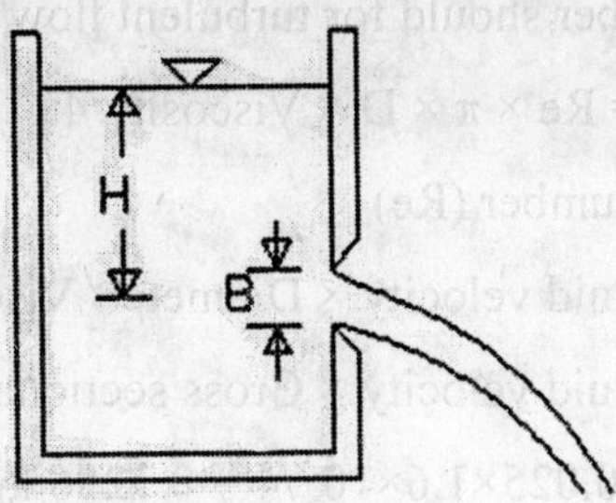

Here, H = 3.5 + 1.2/2 = 4.1 m

Discharge through the orifice $= C_d A \sqrt{2gH}$

$= 0.62 \times 1.5 \times 1.2 \times \sqrt{2 \times 9.81 \times 4.1}$

$= 10.00 \text{ m}^3/\text{s}$

21

Soil Mechanics

UKPSC Combined Assistant Engineering Exam 2013 Paper - I

Q. 1 Hygrophilic soils are those

(A) Which attract water (B) Which do not attract water

(C) Which transmit water (D) Which produce water

Q. 2 Quick sand is a state of

(A) Flow (B) Land slide

(C) Land Levelling (D) None of the above

Q. 3 Which is not the method for drawing flow nets

(A) Analytical method (B) Graphical method

(C) Electrical analogy method (D) Curve fitting method

UKPSC-AE-Agricultural-Engineering-Paper-II-2013

Q. 4 If a soil sample has porosity of 40%, its void ratio will be

(A) 0.06 (B) 0.67

(C) 0.40 (D) 0.30

Q. 5 The upper limit of a tensiometer for soil moisture tension measurement is

(A) 0.85 centibar (B) 8.5 centibar

(C) 85 centibar (D) 100 centibar

Q. 6 If a soil sample has void ratio 40 percent, its porosity will be

(A) 0.06 (B) 0.28

(C) 0.40 (D) 0.66

Q. 7 The ratio of volume of voids to the total soil volume is called

(A) Porosity (B) Void ratio

(C) Dry bulk density (D) Wet bulk density

Q. 8 The soluble soil matters are

(A) Sand and silt (B) Clay and organic matter

(C) Sand, silt, clay and organic matter (D) None of these

Q. 9 The minimum wind velocity required to initiate movement of soil particles is knows as

(A) Critical velocity (B) Dynamic velocity

(C) Intrinsic velocity (D) Threshold velocity

Q. 10 The minimum water content at which the soil just begins to crumble when rolled into threads is known as

(A) Liquid limit (B) Plastic limit

(C) Shrinkage limit (D) Permeability limit

Q. 11 In coarse sand and fine gravel layers free from silt and clay, the radius of influence of wells ranges from

(A) 300 - 600 m (B) 180 - 300 m

(C) 90 - 180 m (D) 30 - 90 m

Q. 12 The weight of 1 Kg mass on the earth will be

(A) 1 N (B) 9.8 kN

(C) 98 N (D) 9.8 N

Q. 13 Steady infiltration rate is called

(A) Basic intake rate (B) Seepage rate

(C) Percolation rate (D) Cumulative infiltration rate

Q. 14 The unit of infiltration rate is

(A) mm/hr (B) cm^3/hr

(C) m^2/hr (D) None of these

Q. 15 Which soil has a higher angle of repose?

(A) Sandy soil (B) Clay soil

(C) Sandy loam soil (D) Clay loam soil

UKPSC J.E. Exam 2013 Agricultural Engineering-II

Q. 16 If the weight of moist soil is 1.18 kg and the weight of dry soil is 1 kg, then the moisture content of soil on dry weight basis will be,

(A) 0.0018 % (B) 1.8 %

(C) 0.018 % (D) 18%

Q. 17 The amount of silt and clay in a sandy loam soil is

(A) More than 30 % (B) More than 20 %

(C) More than 50 % (D) More than 40 %

Q. 18 To determine moisture content of a soil, the soil is kept in oven for 24 hours at the temperature of

(A) 100 °C (B) 105 °C

(C) 110 °C (D) 120 °C

Q. 19 A soil solution with its pH value as 7.0 indicates that

(A) Soil is saline (B) Soil is acidic

(C) Soil is alkaline (D) Soil is neutral

Q. 20 Arrangement of soil particles in a soil mass is called

(A) Soil texture (B) Soil structure

(C) Infiltration (D) Particle density

Q. 21 Void ratio of soil is also known as

(A) Relative density (B) Relative porosity

(C) Specific gravity (D) None of the above

Q. 22 One atmospheric pressure is equal to how many cm of water column ?

(A) 1000 cm (B) 1050 cm

(C) 1023 cm (D) 1036 cm

Q. 23 Soil taxonomy deals with which of the following ?

(A) Principles of soil productivity (B) Principles of soil erosivity

(C) Principles of soil classification (D) Principles of soil formation

Q. 24 Hygrophobiic soils are those which

(A) Condense water (B) Repulse water

(C) Release water (D) Attract water

APPSC AEES Agriculture Engineering Exam 2016

Q. 25 Which of the following mineral particle size (s) is classified as silt according to ISSS:

(A) 0.05 to 0.1 mm (B) 0.002 to 0.05 mm

(C) 0.002 to 0.02 mm (D) None of the above

Q. 26 Maize is a :

(A) Soil maintaining crop (B) Soil building crop

(C) Soil depleting crop (D) None of the above

Q. 27 The ratio of the volume of voids to the total soil volume is called

(A) Void ratio (B) Porosity

(C) Dry bulk density (D) Wet bulk density

Q. 28 Most cereal foods contain mainly

(A) Protein (B) Fat

(C) Starch (D) Vitamins

Q. 29 Quicksand condition is created due to

(A) Frictionless nature of soil

(B) Low value of cohesion of soil

(C) Upward seepage force greater than submerged weight of soil

(D) Downward seepage pressure

Q. 30 At field capacity, soil moisture tension of sandy soil is approximately equal to

(A) 0.1 bar (B) 1.0 bar

(C) 10 bar (D) None of the above

Q. 31 The porosity of sandy soil usually ranges from

(A) 30 to 35 % (B) 35 to 50 %

(C) 40 to 60 % (D) 50 to 60 %

Q. 32 The indicator plant used for determining the permanent wilting percentage

(A) Wheat (B) Safflower

(C) Sunflower (D) Paddy

Q. 33 The dominant cation in most of the soils is

(A) Sodium (B) Potassium

(C) Calcium (D) Magnesium

Q. 34 The element responsible for dispersion of particles in soils is

(A) Manganese (B) Potassium

(C) Sodium (D) Calcium

Q. 35 The relative mobility of this ion is least in soil system

(A) Magnesium (B) Silicon

(C) Aluminum (D) Iron

Q. 36 This is an abundant group of rocks on the surface of the earth crust

(A) Igneous (B) Sedimentary

(C) Metamorphic (D) Organic

Q. 37 The mechanically formed sedimentary rock is

(A) Bauxite (B) Sand stone

(C) Halite (D) Graphite

Q. 38 This is an accessory mineral

(A) Microcline (B) Biotite

(C) Quartz (D) Magnetite

Q. 39 This soil structure is common in sub-surface of alkali soils

(A) Spheroidal (B) Blocky

(C) Crumby (D) Columnar

Q. 40 The net negative charge of colloidal complex is more at pH

(A) 3.0-5.0 (B) 5.0-6.5

(C) 7.0-8.0 (D) Not changed with pH

Q. 41 The chemical reaction involved in nitrogen fixation is

(A) Oxidation (B) Deamination

(C) Nitrification (D) Reduction

Q. 42 Laterite soils belong to the order

(A) Alfisols (B) Oxisols

(C) Spodosols (D) Inceptisols

Q. 43 Atterberg's limits are related to the soil property

(A) Structure (B) Plasticity

(C) Density (D) Strength

Q. 44 The dominant soil order in India

(A) Vertisols (B) Alfisols

(C) Entisols (D) Inceptisols

Q. 45 The clay enriched sub-surface horizon is called

(A) Albic (B) Argillic

(C) Cambic (D) Oxic

Q. 46 This textural class is comparatively coarser

(A) Silty clay loam (B) Silty loam

(C) Sandy clay (D) Silty clay

Q. 47 Bulk density is likely to be less in these soils

(A) Organic (B) Red

(C) Sandy (D) Black

Q. 48 The soil property for good ground water yield is :

(A) Porosity (B) Effective size > 0.1 mm

(C) Uniformity coefficient > 3 (D) Uniformity coefficient < 2

MPPSC Assistant Agricultural Engineer 2013

Q. 49 What can be said about the water content of same soil sample when subjected to 1/3 bar and 15 bar in a pressure plate apparatus ?

(A) Water content will be more at 1/3 bar pressure

(B) Water content will be more at 15 bar pressure

(C) There would not be any difference in water content

(D) Can not say from given data

RPSC AEn Pre Exam 2013 (Agricultural Engineering)

Q. 50 Soil moisture tension at field capacity varies from

(A) 0 to 0.1 atmosphere (B) 0.1 to 0. 33 atmosphere

(C) 0.33 to 1 atmosphere (D) 5 to 30 atmosphere

CGPSC State Engineering Services Exam (Agriculture Engineering) Part - II

Q. 51 The soil property for good water is

(A) Porosity (B) Effective size > 0.1 mm

(C) Uniformity coefficient > 3 (D) Uniformity coefficient < 2

(E) Uniformity coefficient > 4

Q. 52 Electrical Resistivity of the clean sand saturated with fresh water is

(A) Low (B) Medium

(C) High (D) Very High

(E) Zero

Q. 53 If porosity of aquifer is 60 % and specific retention is 20%. What is its specific yield

(A) 0.4 (B) 0.6

(C) 0.8 (D) 0.2

(E) 1.2

Q. 54 The depth of medium deep soil is

(A) < 25 cm (B) 25 – 45 cm

(C) 45 – 90 cm (D) > 90 cm

(E) > 95 cm

Q. 55 The pF of soil for 1000 cm of soil moisture tension

(A) 0 (B) 1

(C) 2 (D) 3

(E) 5

Q. 56 A soil sample has porosity of 30 % the void ratio is

(A) 0.12 (B) 0.40

(C) 0.25 (D) 0.88

(E) 0.42

UKPSC A.En. Exam 2007 Paper

Q. 57 Tension measurement by tensiometer is generally limited to matric suction values below

(A) 2 atmosphere (B) 1.50 atmosphere

(C) 1 atmosphere (D) 1.80 atmosphere

Q. 58 The porosity of a soil mass is the ratio of

(A) Volume of voids to the volume of soil mass

(B) Volume of voids to the total volume of solids

(C) Volume of solids to the total soil volume

(D) Volume of the soil to volume of voids

Q. 59 The laminar flow in soil material, the Reynold's number should be

(A) Less than 2000 (B) Greater than 2000

(C) Greater than 10 (D) Less than 1

Q. 60 Based on the sieve analysis of the aquifer material, the uniformity coefficient of a uniform material would be

(A) More than 2 (B) Less than 2

(C) More than 4 (D) None of the above

Q. 61 If "D" stands for the diameter of particles, which statement of the following is correct in sieve analysis ?

(A) $D_{60} = 3\, D_{20}$ (B) $D_{60} = D_{10}$

(C) $D_{60} < D_{20}$ (D) $D_{60} = 6\, D_{10}$

Q. 62 Plasticity index for the coarse sand is

(A) 1 (B) 0

(C) More than 1 (D) None of the above

ASCO (RPSC) Exam (Forest) 2011

Q. 63 The best soil structure for favorable physical properties is

(A) Crumby and granular (B) Platy and lamina

(C) Columnar and prismatic (D) Blocky

Q. 64 The natural aggregates of soil are known as

(A) Peds (B) Clods

(C) Fragments (D) Gravels

Q. 65 The degree to which a soil resist deformation when a force applied is termed as

(A) Field capacity (B) Capillary capacity

(C) Consistency (D) Friability

Q. 66 The following soil water is held due to adsorption forces

(A) Gravitation water

(B) Capillary water

(C) Hygroscopic water

(D) Hygroscopic and capillary water

Q. 67 The rate of decomposition of organic matter in soil reduces

(A) Acidity (B) Alkalinity

(C) Fertility (D) Salinity

UKPSC A.En. Exam 2012 Paper - II

Q. 68 The relationship between porosity (n) and void ratio (e) is given by

(A) $n = e/(1 + e)$ (B) $e = n/(1 - n)$

(C) $n = (1 + e)/e$ (D) Both (A) and (B)

Q. 69 Saline soil has a pH value

(A) > 8.5 (B) Equal to 8.5

(C) 8.5 to 10.5 (D) Less than 8.5

Q. 70 The ratio of D_{60} and D_{10} sieve sizes in sieve analysis of granular material is called

(A) Well coefficient (B) Uniformity coefficient

(C) Recharge coefficient (D) None of the above

Kerala AAO Exam 2017

Q. 71 Match the following

Particle	Diameter
P. Clay	1. 0.20 to 2 mm
Q. Finesand	2. Above 2.0 mm
R. Coarse sand	3. Less than 0.002 mm
S. Stones and gravel	4. 0.02 to 0.20 mm

(A) P-3, Q-1, R-2, S-4 (B) P-3, Q-4, R-1, S-2

(C) P-4, Q-3, R-1, S-2 (D) P-3, Q-4, R-2, S-1

Q. 72 Which type of soil exhibits self turning property ?

(A) Black soil (B) Aed soil

(C) Alluvial soil (D) Laterite soil

Q. 73 Soil colours are determined by the comparison with the

(A) Leaf colour chart (B) Fick's law

(C) Munsel's colour chart (D) Atterberg's limit

Q. 74 Percentage of sand in sandy soil

(A) > 75 % (B) > 70 %

(C) > 65 % (D) > 60%

Q. 75 The pH range of alkaline soil shall be

(A) More than 7.0 (B) Between 3.0 and 5.0

(C) Less than 7.0 (D) More than or equal to 4.0

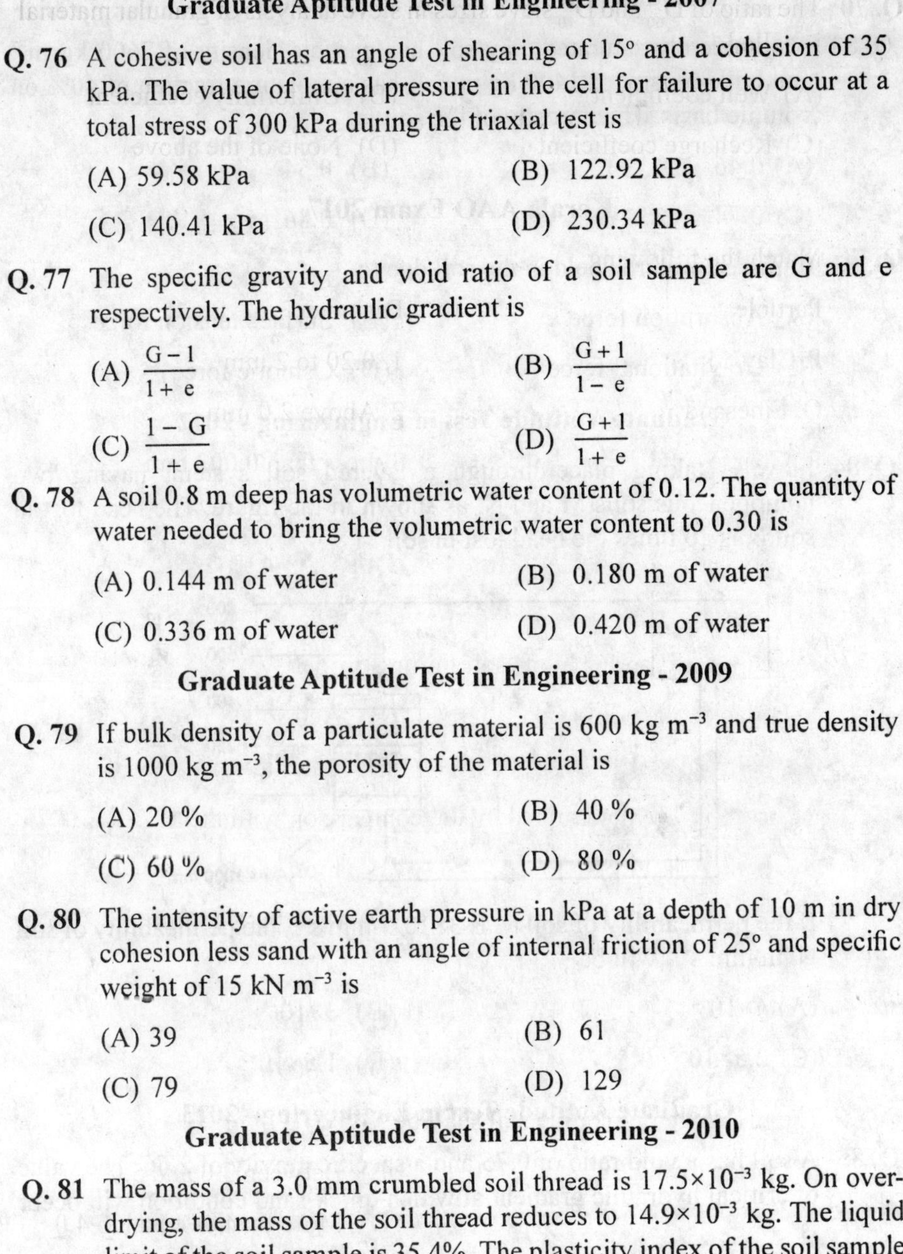

Graduate Aptitude Test in Engineering - 2007

Q. 76 A cohesive soil has an angle of shearing of 15° and a cohesion of 35 kPa. The value of lateral pressure in the cell for failure to occur at a total stress of 300 kPa during the triaxial test is

(A) 59.58 kPa (B) 122.92 kPa

(C) 140.41 kPa (D) 230.34 kPa

Q. 77 The specific gravity and void ratio of a soil sample are G and e respectively. The hydraulic gradient is

(A) $\frac{G-1}{1+e}$ (B) $\frac{G+1}{1-e}$

(C) $\frac{1-G}{1+e}$ (D) $\frac{G+1}{1+e}$

Q. 78 A soil 0.8 m deep has volumetric water content of 0.12. The quantity of water needed to bring the volumetric water content to 0.30 is

(A) 0.144 m of water (B) 0.180 m of water

(C) 0.336 m of water (D) 0.420 m of water

Graduate Aptitude Test in Engineering - 2009

Q. 79 If bulk density of a particulate material is 600 kg m^{-3} and true density is 1000 kg m^{-3}, the porosity of the material is

(A) 20 % (B) 40 %

(C) 60 % (D) 80 %

Q. 80 The intensity of active earth pressure in kPa at a depth of 10 m in dry cohesion less sand with an angle of internal friction of 25° and specific weight of 15 kN m^{-3} is

(A) 39 (B) 61

(C) 79 (D) 129

Graduate Aptitude Test in Engineering - 2010

Q. 81 The mass of a 3.0 mm crumbled soil thread is 17.5×10^{-3} kg. On over-drying, the mass of the soil thread reduces to 14.9×10^{-3} kg. The liquid limit of the soil sample is 35.4%. The plasticity index of the soil sample is

(A) 14.9 (B) 18.0

(C) 32.8 (D) 35.4

Graduate Aptitude Test in Engineering - 2011

Q. 82 A completely saturated clay soil has particle density of 2600 kg m^{-3} and bulk density of 1400 kg m^{-3}. It has a moisture fraction of 40% on volume basis. The porosity of the soil is

(A) 0.46 (B) 0.54

(C) 0.74 (D) 0.86

Q. 83 Capillary water is held in the soil due to

(A) Absorption force (B) Surface tension force

(C) Gravitational force (D) Osmotic force

Graduate Aptitude Test in Engineering - 2012

Q. 84 Flow is taking place through a layered soil system, having two homogeneous soils M and N, as shown in the figure. The head lost in soil N is 20 times the head lost in soil M.

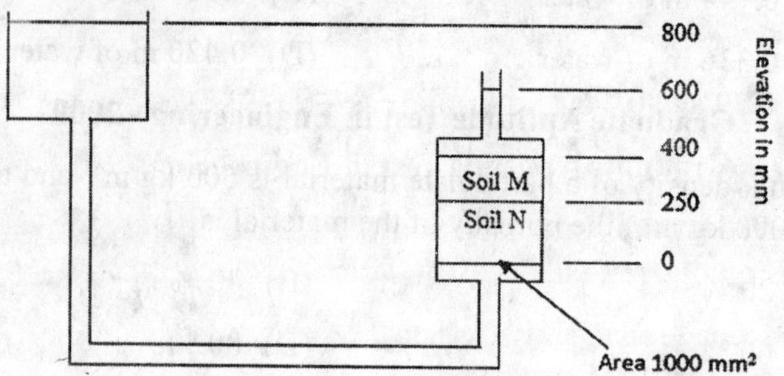

If the permeability of soil M is 3×10^{-4} mm s^{-1}, the permeability of soil N, in mm s^{-1}, will be

(A) 4×10^{-4} (B) 3×10^{-4}

(C) 2.5×10^{-5} (D) 1.5×10^{-5}.

Graduate Aptitude Test in Engineering - 2013

Q. 85 A soil has a void ratio of 0.75 and a specific gravity of 2.66. The value of critical hydraulic gradient at which quick sand condition will occur is

(A) 0.95 (B) 1.05

(C) 2.09 (D) 6.64

Q. 86 The pressure that does not have any measurable influence on the void ratio or shearing resistance of the soil mass is

(A) Pore water pressure (B) Intergranular pressure

(C) Capillary pressure (D) Surcharge pressure

Graduate Aptitude Test in Engineering - 2016

Q. 87 A 3 m high retaining wall supports sandy soil of unit weight 18.5 kN m^{-3}. The angle of shearing resistance (∅) is 30° and the surface of soil is horizontal. The magnitude of the active thrust (in kN/m) and its acting point from the top (in m) are

(A) 27.75 and 1 (B) 249.75 and 1

(C) 27.75 and 2 (D) 249.75 and 2

Q. 88 A soil has bulk density and particle density of 1.48 Mg m^{-3} and 2.64 Mg m^{-3}, respectively. The saturated volumetric moisture content of soil is 36%. The porosity and void ratio of the soil are

(A) 0.36 and 0.56 (B) 0.44 and 0.79

(C) 0.79 and 0.44 (D) 0.56 and 0.36

Graduate Aptitude Test in Engineering - 2017

Q. 89 Antecedent moisture conditions(AMC) for a soil are defined on the basis of total rainfall occurred during previous .... days.

Q. 90 The seepage analysis in earthen dams is carried out by drawing flownet which consists of equipotential and stream lines. For a homogeneous and isotropic earthen dam, these two lines are

(A) Orthogonal to each other (B) Parallel to each other

(C) Divergent to each other (D) Convergent to each other

Graduate Aptitude Test in Engineering - 2018

Q. 91 If the void ratio of a soil column is 0.43, the soil porosity is

(A) 0.30 (B) 0.40

(C) 0.70 (D) 0.75

Q. 92 The porosity and compressibility of a 7.8 m thick confined aquifer are 0.25 and 1×10^{-8} m^2 N^{-1}, respectively. The storage coefficient of the aquifer is 6.5×10^{-4}. To release 650 m^3 of water from 1 km^2 of this aquifer, the average decline in hydraulic head will be .... m.

Q. 93 The elevations of pressure gauge and porous cup of the tensiometer installed in unsaturated zone are 145.8 m and 144.2 m, respectively. Pressure measured at the gauge is - 19.62×10^3 N m^{-2}. The specific weight of water is 9810 N m^{-3}. The estimated pressure at the porous cup is .... N m^{-2}.

Graduate Aptitude Test in Engineering - 2019

Q. 94 Tensiometer installed in the soil measures

(A) Osmotic suction of soil moisture

(B) Soil permeability

(C) Soil moisture content

(D) Capillary potential of the soil

Q. 95 A soil sample has a porosity of 40%. Void ratio of the soil sample is

(A) 0.367 (B) 0.467

(C) 0.567 (D) 0.667

Graduate Aptitude Test in Engineering - 2020

Q. 96 A retaining wall of 5 m height retains cohesionless dry soil having density of 1.9 Mg/m^3 and angle of internal friction of 28°. The surface of the backfill soil is horizontal. The active and passive earth pressures per meter length of the wall in kN are and respectively. [Take g = 9.81 m/ s^2]

(A) 84.11, 645.38 (B) 645.38, 84.11

(C) 142.12, 381.63 (D) 381.63, 142.12

Graduate Aptitude Test in Engineering - 2021

Q. 97 Cohesionless soil is naturally deposited and makes a slope of infinite extent having slope angle of 25°. If the effective angle of internal friction of this soil is 30°, the factor of safety of slope is ....

Q. 98 A shear annulus with inner and outer diameters of 240 mm and 300 mm, respectively is used to measure shear strength of soil in the field. When it is inserted into the soil and rotated, the torque measured at the soil failure is 50 N.m. Shear strength of the soil in kPa is (Take π = 3.14)

(A) 14.49 (B) 18.94

(C) 21.54 (D) 28.98

Q. 99 A sample of wet sandy-clay loam soil of mass 135 kg is collected for laboratory tests. The wet density, water content (weight basis) and specific gravity of solids of this soil sample are 1.8 g/cm^3, 18%, and 2.7, respectively. The dry density (in g/cm^3) and porosity (in per cent) of the soil sample, respectively, are

(A) 1.53 and 43.50 (B) 1.53 and 77.00

(C) 1.65 and 43.50 (D) 1.65 and 77.00

Answers Key

1	2	3	4	5	6	7	8	9	10
A	A	D	B	C	B	A	B	D	B
11	12	13	14	15	16	17	18	19	20
B	D	A	A	C	D	B	B	D	B
21	22	23	24	25	26	27	28	29	30
B	D	C	D	C	C	B	C	C	A
31	32	33	34	35	36	37	38	39	40
B	C	C	C	C	B	B	D	D	C
41	42	43	44	45	46	47	48	49	50
D	B	B	D	B	B	A	B	A	B
51	52	53	54	55	56	57	58	59	60
A	C	A	C	D	E	C	A	D	B
61	62	63	64	65	66	67	68	69	70
C	B	A	A	C	C	B	D	D	B
71	72	73	74	75	76	77	78	79	80
B	A	C	B	A	B	A	A	B	B
81	82	83	84	85	86	87	88	89	90
B	A	B	C	A	A	C	B	5	A
91	92	93	94	95	96	97	98	99	
A	1	-3924	D	D	A	1.24	A	A	

Explanations

Q. 4

We know, Void ratio = Porosity/(1 - Porosity)

= 0.40(1 – 0.40) = 0.67

Q. 6

Porosity = Void ratio/(1 + Void ratio)

= 0.40/(1 + 0.40) = 0.28

Q. 10

Atterberg Limits are basic measure of the nature of a fine-grained soil. Depending on the water content in the soil, it may appear in four states: solid, semi-solid, plastic and liquid. In each state, consistency and behavior of the soil is different and so are its engineering properties. Thus, the boundary between each state can be defined based on a change in the soil's behavior.

1. **Shrinkage Limit:** Shrinkage limit (SL) is the water content where further loss of moisture will not result in any more volume reduction. (The water content that is just sufficient to fill the pores.)
2. **Liquid limit (LL):** It is defined as the percentage moisture content at which a soil changes with decreasing wetness from liquid to plastic consistency or with increasing wetness from plastic to liquid consistency. It is the water content at which a soil changes from plastic to liquid behavior. The importance of this liquid limit test is to classify soils.(The water content at which the soil has such a small shear strength that it flows to close a groove of standard width when jarred in a specified manner.)

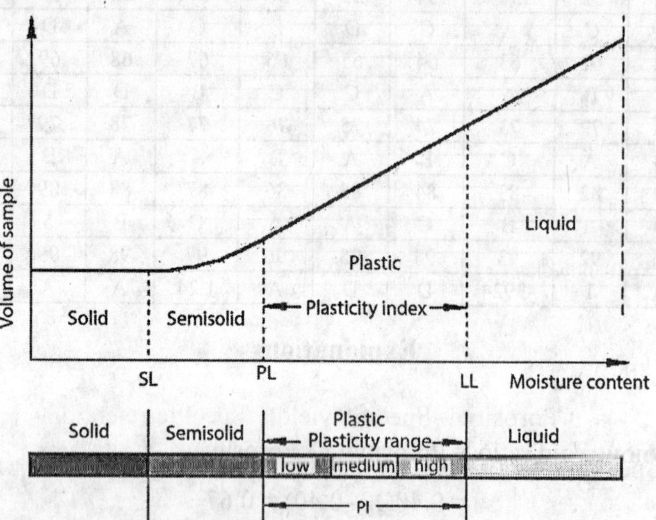

3. **Plastic limit (PL):** it is the percentage moisture content at which a soil changes with decreasing wetness from the plastic to the semi-solid consistency or with increasing wetness from the semi-solid to the plastic consistency. Plastic limit is the lower limit of the plastic state. A small increase in moisture above the plastic limit destroys

cohesion of the soil. (The water content at which the soil begins to crumble when rolled into threads of specified size.)

Q. 16

Moisture content (db) = $(1.18 - 1.00) \times 100/1.00 = 18\ \%$

Q. 25

Soil May be classified as

	USDA Classification	ISSS Classification
Soil separates	Diameter in mm	Diameter in mm
Clay	< 0.002	< 0.002
Silt	0.002 - 0.05	0.002 - 0.02
Very Fine Sand	0.05 - 0.10	-
Fine Sand	0.10 - 0.25	0.02 - 0.2
Medium Sand	0.25 - 0.50	-
Coarse Sand	0.50 - 1	0.2 - 2
Very Coarse Sand	1 - 2 mm	-

Q. 31

Typical values of soil porosity for different soils

Description	Porosity	
	Min.	Max.
Gravel	0.23	0.38
Coarse sand	0.26	0.43
Fine sand	0.29	0.46
Silty sands	0.25	0.49
Clayey sands	0.15	0.37
Silty or sandy clay	0.20	0.64

Q. 53

We know

$$\text{Porosity} = \text{Specific yield} + \text{Specific retention}$$

$\Rightarrow$ Specific retention $= 0.60 - 0.20 = 0.40$

Q. 55

PF Scale: The free energy is measured in terms of the height of a column of water required to produce necessary suction or pressure difference at a particular soil moisture level. The pF, therefore, represents the logarithm of the height of water column (cm) to give the necessary suction.

pF = log(height of water column)

A soil that is saturated with water has PF 0(Lower tension) while an oven dry soil has a PF 7(Higher tension).

Here,

$$pF = \log(1000) = 3$$

Q. 76

Given

Angle of shearing (ϕ) = 15°

$\therefore$ Angle of rupture or failure(α) = $45 + \phi/2 = 45 + 15/2 = 52.5°$

Total vertical pressure (σ_1) = 300 kPa

Cohesion stress(C) = 35 kPa

We know that,

Vertical pressure (σ_1) = $\sigma_3 \tan^2 \alpha + 2C \tan \alpha$

$$\Rightarrow \quad 300 = \sigma_3 \times \tan^2(52.5) + 2 \times 35 \times \tan(52.5)$$

$$\Rightarrow \quad \sigma_3 = 122.92 \text{ kPa}$$

Q. 78

Initial water content = 0.8×0.12 = 0.096 m

Final water content = 0.8×0.3 = 0.24 m

$\therefore$ Quantity of water needed = 0.24 – 0.096 = 0.144 m of water

Q. 79

We know

$$\text{Porosity} = 1 - \frac{\textit{Bulk density}}{\textit{Pure Density}}$$

$$= 1 - \frac{600}{1000} = 40\ \%$$

Q. 80

Given that

Depth of cohesion less sand (h) = 10 m

Angle of internal friction (∅) = 25°

Specific weight (w) = ρg = 15 kN/m^3

Coefficient of active earth pressure $(K_a) = \frac{1-sin\varnothing}{1+sin\varnothing}$

$$= \frac{1-sin25}{1+sin25} = 0.406$$

$\therefore$ Intensity of active earth pressure = $K_a \rho g h$

$$= 0.406 \times 15 \times 10 = 60.9 \text{ kPa}$$

Q. 81

Plastic limit of the soil sample is = $(17.5\times10^{-3} - 14.9\times10^{-3})/14.9\times10^{-3}$

$$= 17.4$$

Thus, Plasticity Index = Liquid limit – Plastic Limit

$$= 35.4 - 17.4 = 18$$

Q. 82

We know

Porosity $= 1 - \frac{Bulk\ density}{Pure\ Density}$

$$= 1 - \frac{1400}{2600} = 0.46$$

Q. 84

Referring figure given

	Datum head	Pressure head	Total head
Just above M	0	800	800
Just below N	400	200	600

Given

20×(Head interval in M) = Head interval in N (i)

Now,

Head interval through the samples, M and N = 200 mm

i.e. 20×(Head interval in M) + Head interval in N (ii)

Solving both the equations

Head interval in M = 9.5238 mm

Head interval in N = 190.4762 mm

We know

$$Q = K\,i\,A$$

For M

$$\text{Discharge (Q)} = 3\times10^{-4}\times \frac{9.5238}{150}\times1000$$

$$= 0.0190476 \text{ mm}^3/\text{s}$$

∵ Discharge will be same for both the soil samples.

For N

$$0.0190476 = K\times \frac{190.4762}{250}\times1000$$

$$\Rightarrow \quad K = 2.5\times10^{-5} \text{ mm/s}$$

Q. 85

$$\text{Hydraulic Gradient} = \frac{G-1}{e+1}$$

$$= \frac{2.66-1}{0.75+1}$$

$$= 0.95$$

Q. 87

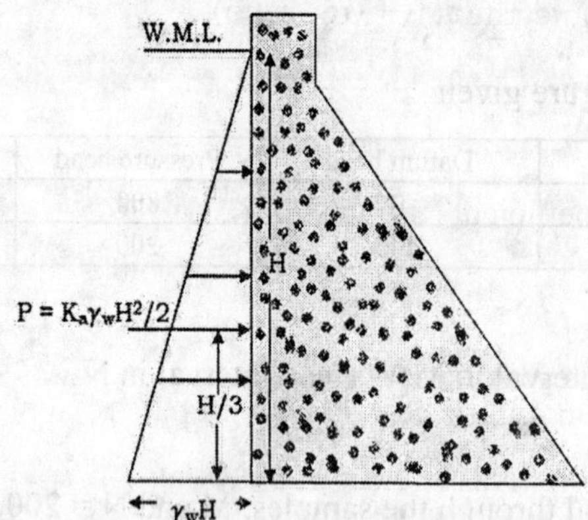

We know

Active thrust (P) = K γ_{aw} $H^2/2$

Where,

$$K_a = (1 - \sin\varnothing)/(1 + \sin\varnothing)$$

$$K_a = (1 - \sin30)/(1 + \sin30)$$

$$= 0.3333$$

Hence, $P = 0.3333 \times 18.5 \times 3^2/2$

$\Rightarrow 27.75$ kN/m

Also,

Active point = H/3 from bottom or 2H/3 from top

= 2×3/3 = 2 m from top

Q. 88

We know

$$\text{Porosity (n)} = 1 - \frac{\textit{Bulk density}}{\textit{Pure Density}}$$

$$= 1 - \frac{1.48}{2.64} = 0.44$$

And Void ratio (e) = n/(1 – n)

= 0.44/(1 – 0.44) = 0.79

Q. 91

We know

Porosity = Void ratio/(1 + Void ratio)

= 0.43/(1 + 0.43) = 0.30

Q. 92

The storage coefficient (S) = ΔV/(A×ΔH)

$\Rightarrow$ $\Delta H = 650/(6.5\times10^{-4}\times10^{6}) = 1$ m

Q. 93

Water elevation difference = 145.8 – 144.20 = 1.6 m

Pressure = ρ g h = 9810×1.6 = 15696 N/m^2

Hence,

Soil water pressure at porous cup = Pressure measured at the gauge + Pressure due to water elevation between pressure gauge and porous cup

= - 19620 + 15696 = -3924 N/m^2

Q. 94

Direct field measurements of the metric or capillary potential can be made only with tensiometers. Tensiometers can also be calibrated against soil moisture content so that readings can be converted to percentage water content.

Q. 95

We know

Void ratio = Porosity/(1 - Porosity)

= 0.40(1 – 0.40) = 0.67

Q. 96

Active earth pressure (P_a) = $K_a \rho g H^2/2$

Where,

Active earth pressure coefficient k_a = (1 – sin28)/(1+sin28)= 0.361

P_a = 1900×9.81×0.361×5²/2 = 84108.49 N or 84.11 kN

Passive earth pressure (P_p) = $K_p \rho g H^2/2$

Where

Passive earth pressure coefficient k_p = (1+sin28)/ (1 – sin28)= 2.77

P_p= 1900×9.81×2.77×5²/2 = 645375.37 N or 645.38kN

Note:- The choice also can be predicted through the below relation.

P_p/P_a = {(1+sin ∅)/(1-sin ∅)}² = (2.77)² = 7.67

Q. 97

$$\text{Factor of safety } (F_S) = \frac{tan\phi}{tan\phi'}$$

$$= \frac{tan30}{tan25} = 1.24$$

Q. 98

$$\text{Soil shear strength (S)} = \frac{3M}{2\pi(R^3 - r^3)}$$

$$= \frac{3 \times 50}{2 \times \pi \times (0.15^3 - 0.12^3)} = 14502.34 \text{ Pa or } 14.50 \text{ kPa}$$

Q. 99

Specific gravity of soil (G) = 2.7, Soil sample wet density (ρ) = 1.8 g/cm³
Water content (w) = 18%, Water density (ρ) = 1 g/cm³

We know

$$\frac{\rho}{\rho_w} = \frac{G \times (1+w)}{1+e}$$

$$\Rightarrow \quad \frac{1.8}{1} = \frac{2.7 \times (1+0.18)}{1+e}$$

$\Rightarrow$ Void ratio(e) = 0.77

Hence,

$$\text{Porosity} = \frac{e}{e+1}$$

$\Rightarrow$ n = 0.435 or 43.5 %

Again, Dry Density (ρ_d) = $\frac{G \times \rho_w}{1+e}$

$$\Rightarrow \quad = \frac{2.7 \times 1}{1+0.77} = 1.53 \text{ g/cm}^3$$

22

Hydrology

UKPSC J.E. Exam 2013 Agricultural Engineering-II

Q. 1 The Lysimeter is used to determine

(A) Sun shine hours (B) Wind velocity

(C) Flow velocity (D) Evapotranspiration

Q. 2 Discharge of water as vapour into atmosphere by leaves and stems of living plants is called

(A) Evaporation (B) Consumptive use

(C) Transpiration (D) None of the above

Q. 3 Evaporation pan is made of

(A) Brass (B) Cast iron

(C) Mild steel (D) None of the above

Q. 4 Diameter of a commonly used evaporation pan is

(A) 65 cm (B) 75 cm

(C) 95 cm (D) 120 cm

Q. 5 The rate of evaporation of water does not depend on

(A) Temperature of water

(B) Temperature of air in contact with water

(C) Humidity of air

(D) Depth of water

Q. 6 Evaporation process is a .... process.

(A) Chemical (B) Biological

(C) Physical (D) None of the above

Q. 7 Evaporation is the process of

(A) Condensation (B) Vaporization

(C) Conduction (D) Radiation

Q. 8 The depth of a commonly used evaporation pan is

(A) 20 cm (B) 50 cm

(C) 100 cm (D) 150 cm

Q. 9 A significant soil building crops group is

(A) Cereal crops (B) Leguminous crops

(C) Fodder crops (D) Fruit crops

Q. 10 Rational formula is used to estimate

(A) Peak flow rate (B) Average flow rate

(C) Uniform flow rate (D) Free flow rate

Q. 11 Small watershed are those in which

(A) Runoff is predominant

(B) Overland flow dominates the runoff

(C) Base flow predominant

(D) All of the above

Q. 12 Evapotranspiration can be measured by using which of the following instrument equipment

(A) Current meter (B) Lysimeter

(C) Coshocton wheel (D) Stage recorder

Q. 13 Water meter is used to measure

(A) Flow rate in streams (B) Flow rate in pipes

(C) Water volume storage (D) Water energy

Q. 14 One ha-cm volume of water is equal to

(A) 10 m^3 (B) 100 m^3

(C) 1000 m^3 (D) 10000 m^3

Q. 15 Pan evaporation data are recorded

(A) Twice daily (B) Thrice daily

(C) Once daily (D) Once a week

Q. 16 Tensiometers are used to measure

(A) Soil organic matter (B) Flow depth

(C) Flow velocity (D) Soil moisture

Q. 17 The part of rainfall which becomes part of the water requirement of crops is known as

(A) Gross rainfall (B) Efficient rainfall

(C) Effective rainfall (D) True rainfall

Q. 18 The method which estimates evaporation from a free water surface by examining energy balance at the water surface is

(A) Blaney-Criddle method (B) Dalton method

(C) Penman method (D) None of the above

Q. 19 The value of runoff coefficient normally is

(A) Always more than one (B) Always equal to one

(C) Always less than one (D) Any of the above

APPSC AEES Agriculture Engineering Exam 2016

Q. 20 Penman method is used to compute

(A) Consumptive use of crops

(B) Potential evapotranspiration of crops

(C) Water requirement of crops

(D) Irrigation water requirement of crops

Q. 21 Evapotranspiration in a crop field surrounded by dry fallow land will be higher than that surrounded by vegetation due to

(A) Conduction of heat (B) Oasis effect

(C) Clothesline effect (D) Convection of heat

Q. 22 When the soil moisture and rainfall are inadequate during the growing season to support healthy crop growth to maturity is called

(A) Metrological drought (B) Hydrological drought

(C) Agricultural drought (D) None of these

Q. 23 Soil strength is determined by:

(A) Penetrometer (B) Micrometer

(C) Hydrometer (D) Dynamometer

Q. 24 ……….. measures the degree of attenuation of the light beam.

(A) Infiltration (B) Evaporation

(C) Transmisometer (D) Vapour pressure

Q. 25 In the Rational formula Q = 0.0028 C.I.A., I is the intensity of rainfall in

(A) mm per hour (B) cm per hour

(C) m per hour (D) cm per minute

MPPSC Assistant Agricultural Engineer 2013

Q. 26 The imaginary line drawn through the earthen embankment to separate the zone having hydrostatic pressure and zone having no hydrostatic pressure is known as.

(A) Isocline (B) Isobar

(C) Phreatic line (D) None of these is correct

Q. 27 Albedo means

(A) The ratio of the amount of solar radiation (or visible range of radiation) reflected by a surface

(B) The amount incident on it, also expressed as a percentage

(C) Both (A) and (B)

(D) None of these

Q. 28 The ratio of sum of length of all channel segments of all orders in a watershed to the area of watershed is known as ....

(A) Constant for channel maintenance

(B) Bifurcation ratio

(C) Stream length ratio

(D) Drainage density

Q. 29 The graphical representation of stream discharge against percentage of time the flow is equaled or exceeded is called as ....

(A) Hydrograph (B) Hyetograph

(C) Flow duration curve (D) Double mass curve

Q. 30 A triangular direct runoff hydrograph due to a storm over watershed has peak flow of 20 cumec occurring at 10 hours from beginning of storm. The time base is 40 hours. If the area of watershed is 72 sq.km., what will be the effective rainfall from the watershed.

(A) 4 cm (B) 0.04 cm

(C) 2 cm (D) None of these is correct

Q. 31 Considering the standard curve number method, what should be the minimum curve number to just initiate runoff from precipitation of 40 mm.

(A) 0 (B) 86.39

(C) 65.25 (D) 100.00

Q. 32 The peak of dimensionless unit, hydrograph is given by the coordinates ....

(A) (0,0) (B) (0,1)

(C) (1,1) (D) None of these is correct

Q. 33 Ratio of area of watershed to the area of circle having perimeter equal to perimeter of watershed is known as ....

(A) Compactness coefficient (B) Elongation ratio

(C) Circularity ratio (D) Shape Index

Q. 34 Which option about the statements made below is correct ?

(1) Monitoring and evaluation of watershed is one and the same.

(2) Monitoring is done internally, whereas evaluation is done by some external agency.

(A) Statement (1) is true, but statement (2) is false

(B) Statement (2) is true, but statement (1) is false

(C) Both statements (1) and (2) are true

(D) Both statements (1) and (2) are false

Q. 35 Which option about the statements made below is correct ?

(1) People's participation in watershed development programme is extremely essential.

(2) People's participation in watershed development programme can be ensured by educating them through field trip to successful watershed development project.

(A) Statement (1) is true, but statement (2) is false

(B) Statement (2) is true, but statement (1) is false

(C) Both statements (1) and (2) are true (D) Both statements (1) and (2) are false.

Q. 36 If a forest watershed is converted to urban watershed which of the option about statements made below will be correct ?

(1) Time to peak and time base of hydrograph will be reduced.

(2) Peak flow will increase.

(A) Statement (1) is true, but statement (2) is false

(B) Statement (2) is true, but statement (1) is false

(C) Both statements (1) and (2) are true (D) Both statements (1) and (2) are false.

Q. 37 Concentric closed contours with elevations increasing inwards represent a ....

(A) Depression (B) Reservoir

(C) Hill (D) None of these is correct

Q. 38 The main application of .... is to determine the size, speed, and number of raindrops.

(A) Blaney-Criddle method (B) Disdrometer

(C) Penman - Monteith method (D) All options are correct

RPSC AEn Pre Exam 2013 (Agricultural Engineering)

Q. 39 If the radius of raindrop is doubled, then its kinetic energy will be increased by

(A) 8 times (B) 16 times

(C) 5 times (D) 4 times

Q. 40 For analysis of rainfall intensities and durations in a given watershed we need to have atleast following instrument installed therein,

(A) A simple rain gauge (B) A water stage level recorder

(C) A water current meter (D) Recording rain gauge

Q. 41 Which is the correct sequence of hydrologic processes initiated on natural catchment just after first rain storm in a monsoon season?

(A) Rainfall > Surface runoff > Stream flow > Percolation > Evaporation

(B) Rainfall > Interception > Depression storage > Infiltration > Surface runoff

(C) Ground water recharge > Percolation > Evaporation > Surface runoff

(D) Nothing can be said, sequence can be any one from above 3 options

Q. 42 Direct runoff in a watershed is generally

(A) The sum of surface runoff, inter flows and channel precipitation

(B) Surface runoff and channel flow

(C) Surface runoff and groundwater recharge

(D) None of the above

Q. 43 Watershed planning and management requires

(A) Land use data (B) Socio economic data

(C) Hydrologic data (D) All of the above

Q. 44 The most famous equation of empirical prediction model for computing peak runoff rate on small watershed is

(A) Cook's method (B) Rational method

(C) Phillip's model (D) Holton's method

Q. 45 Hydrograph is graphical representation of

(A) Runoff rate versus time (B) Infiltration rates versus time

(C) Rainfall rates versus time (D) None of the above

Q. 46 The unit hydrograph of a specified duration can be used to synthesize or evaluate the unit hydrograph of storms of

(A) Same duration only (B) Same or shorter duration only

(C) Same and longer duration (D) Any duration

OCS AEn Exam 2011 (Pre) – II (Agricultural Engineering)

Q. 47 Auger-hole method is used for the measurement of hydraulic conductivity:

(A) Below water table (B) Above water table

(C) Both (A) and (B) (D) None of the above

Q. 48 A storm hydrograph was due to 3 hours of effective rainfall. It contains 6 cm of direct run-off. The ordinates of DRH of this storm

(A) When divided by 3 give ordinates of 6 hours unit hydrograph

(B) When divided by 6 give the ordinates of 3 hours unit hydrograph

(C) When divided by 3 give the ordinates of 3 hours unit hydrograph

(D) When divided by 6 gives ordinates of 6 hours unit hydrograph

Q. 49 A plot between rainfall intensity v/s time is called as

(A) Hydrograph (B) Mass curve

(C) Hyetograph (D) Isohyet

Q. 50 In rational method, if rainfall duration exceeds the time of concentration, the resulting runoff will be

(A) Constant at peak rate (B) Less than peak runoff rate

(C) More than peak runoff rate (D) All of the above

CGPSC State Engineering Services Exam (Agriculture Engineering) Part - II

Q. 51 Watershed management involves primarily the judicious use of

(A) Human resources (B) Land and water resources

(C) Mining resources (D) Building resources

(E) All of the above

Q. 52 Watershed protection requires knowledge of

(A) Physical characteristics (B) Weather Characteristics

(C) Rainfall characteristics (D) Infiltration Characteristics

(E) Both (A) and (B)

Q. 53 To measure and map the topography of the earth above sea level is called

(A) Mass curve (B) Hydrograph

(C) Hyetograph (D) Hypsograph

(E) Both (B) and (C)

Q. 54 The area under hydrograph is equal to

(A) Depth of runoff (B) Rain depth

(C) Volume of direct runoff (D) Intensity

(E) Both (A) and (B)

Q. 55 Watersheds with good drainage network produces hydrographs with

(A) Abrupt rising limb (B) Abrupt falling limb

(C) Less area (D) More time base

(E) Both (B) and (C)

Q. 56 The return period for any hydrological event is usually expressed in

(A) Weeks (B) Months

(C) Years (D) Days

(E) Decade

Q. 57 Rainfall intensity can be obtained by data from

(A) Non-recording raingauge (B) Recording raingauge

(C) Anemometer (D) Current meter

(E) Flow meter

Q. 58 The science and practice of water flow measurement is known as

(A) Hypsometry (B) Hydro-meteorology

(C) Fluvimetry (D) Hydrometry

(E) None of the above

Q. 59 Relief of a watershed has the units

(A) m (B) m^2

(C) m^3 (D) No units

(E) m^4

Q. 60 Recommended rain gauge density for plains as per IS:4987-1968 is one station in

(A) 130 km^2 (B) 260 km^2

(C) 520 km^2 (D) 760 km^2

(E) 790 km^2

Q. 61 The most widely adopted method for determining average rain depth is

(A) Stochastically (B) Arithmetic mean

(C) Thiessen polygon (D) S-curve

(E) None of the these

Q. 62 T_p : time to peak is approximately

(A) 0.7 T_c (B) 0.6 Tc

(C) 0.4 Tc (D) 0.2 Tc

(E) 0.5 T_c

Q. 63 Unit hydrograph for different duration in a watershed can be developed by

(A) Mass curve (B) Histogram

(C) S-curve (D) Curve number

(E) All of the above

Q. 64 A part of rainfall that infiltrate and moves laterally through the upper crusts of soil is

(A) Direct runoff (B) Overland flow

(c) Sub flow (D) Inter flow

(E) None of these

Q. 65 Watershed treatment with control measures reduces the

(A) Rainfall (B) Peak runoff rate

(C) Rain intensity (D) Base period

(E) All of the above

Q. 66 Normal rainfall is the average value of rainfall for a period of

(A) 50 years (B) 40 years

(C) 30 years (D) 10 years

(E) 20 years

Q. 67 Number of inflection points in flood hydrograph

(A) 2 (B) 1

(C) 3 (D) 4

(E) 5

Q. 68 A stream does not have any base flow contribution is known as

(A) Perennial stream (B) Intermittent stream

(C) Ephemeral stream (D) Natural stream

(E) None of these

Q. 69 Which of the following flow meters is capable of giving the rate of flow as well as totalized flow

(A) Orifice meter (B) Electro magnetic flow meter

(C) Nutating disc flow meter (D) Lobbed impeller flow meter

(E) Rota meter

Q. 70 ……… is a cone made of fabric designed to indicate the direction and approximate speed of the wind.

(A) Wind sock (B) Anemometer

(C) Lidar (D) Both (B) and (C)

(E) None of the above

Q. 71 The Meteorological drought as per IMD is called as 'severe' if seasonal rainfall deficiency compared to normal value is

(A) > 25 % (B) 25 % – 50 %

(C) 75 % (D) > 50 %

(E) None of these

Q. 72 The DRH(m^3/sec) generated by flood hydrograph of 1800 m^3/sec with base flow 20 m^3/sec

(A) 1780 (B) 2000

(C) 2200 (D) 1800

(E) 1900

Q. 73 In drizzle the size of droplet is about

(A) < 0.5 mm (B) > 0.5 mm

(C) ± 0.5 mm (D) ± 1 mm

(E) > 1 mm

Q. 74 Chemical most suitable for evaporation inhabitation is

(A) Alcohol (B) Ethyl alcohol

(C) Methyl alcohol (D) Cetyl alcohol

(E) Both (A) and (C)

UKPSC A.En. Exam 2007 Paper – II

Q. 75 The line joining the equal depth of the rainfall is called

(A) Isochrone (B) Contour line

(C) Isohytes (D) None of the above

Q. 76 Synder's method is used to determine

(A) Dimensionless hydrograph (B) Distribution hydrograph

(C) Synthetic unit hydrograph (D) None of the above

Q. 77 The intensity of effective rainfall over the entire watershed for a 4 hour unit hydrograph is

(A) 1 cm per hour (B) 4 cm per hour

(C) 0.50 cm per hour (D) 0.25 cm per hour

Q. 78 The runoff coefficient is dependent on

(A) Nature of the soil surface (B) Land use pattern

(C) Rainfall intensity (D) All of the above

Q. 79 Inverse of probability of occurrence of a hydrological event is called as

(A) Frequency (B) Return period

(C) Intensity (D) Duration

Q. 80 Total run-off minus base flow gives

(A) Infiltration rate (B) Surface storage

(C) Direct runoff (D) None of the above

Q. 81 For storm run-off analysis, the base flow separation is performed on

(A) Unit hydrograph

(B) Flood hydrograph

(C) Heytograph

(D) Hydrograph of effluent streams only

Q. 82 A unit hydrograph has one unit of

(A) Peak discharge (B) Direct runoff

(C) Storm duration (D) The time base of direct run-off

Q. 83 The time required for the runoff water to flow from the most remote point of the watershed area to the outlet is called

(A) Timc of drainage (B) Time of concentration

(C) Time of precipitation (D) Time of saturation

Q. 84 Detention dams are basically constructed for

(A) Storing water foı irrigation

(B) Flood control

(C) Domestic water supply

(D) Hydroelectric power generation

Q. 85 A catchment area of 90 hectares has run-off coefficient of 0.40. A storm of duration larger than the time of concentration of the catchment area of intensity 4.50 cm/hr creates a peak discharge rate of

(A) 11.30 m^3/sec (B) 9.00 m^3/sec

(C) 45.50 m^3/sec (D) 4.50 m^3/sec

ASCO (RPSC) Exam (Engineering) 2011

Q. 86 The equation for curve number is given as

(A) CN = 2540/(25.4 + S) (B) CN = 25.40/(2540 + S)

(C) CN = 2500/(25 + S) (D) CN = (25.40 + S)/25.4

Q. 87 The maximum size of the raindrops is

(A) 6 mm (B) 10 mm

(C) 2 mm (D) 0.5 mm

Q. 88 The double mass curve technique is adopted to

(A) Check the consistency of rain gauge records

(B) To find the average rainfall over a number of years

(C) To find the number of rain gauge required

(D) To estimate the missing rainfall data

Q. 89 Which of the following statements is correct ?

(A) Evaporation from water bodies and soil masses are termed as evapotranspiration

(B) Evaporation from water bodies and soil masses together with the transpiration from vegetation is termed as evapotranspiration

(C) Evaporation, interception and depression storage combined together form evapotranspiration

(D) Evaporation, transpiration and surface runoff are main components of evapotranspiration

Q. 90 An area is classified as a drought prone area if the probability(P) of occurrence of a draught is

(A) $0.4 < P < 1.0$ (B) $0.2 < p < 0.40$

(C) $0.1 < P < 0.20$ (D) $0.0 < P < 0.20$

Q. 91 Base flow separation is performed

(A) On a unit hydrograph to get the direct runoff hydrograph

(B) On a flood hydrograph to obtain the magnitude of effective rainfall

(C) On flood hydrograph to obtain the rainfall hydrograph

(D) On hydrograph if effluent streams only

Q. 92 The flood routing by hydrologic method is based on the

(A) Energy equation

(B) Equation of motion

(C) Equation of continuity

(D) Momentum and continuity equation

Q. 93 If time of concentration is more than the rainfall duration then the rational method will

(A) Not be able to calculate run-off volume

(B) Not be able to calculate peak run-off rate

(C) Be able to calculate run-off rate

(D) Be able to calculate the peak run-off rate

Q. 94 Hydraulic routing method makes use of

(A) Energy equation only

(B) Continuity equation only

(C) Momentum equation only

(D) Both continuity and momentum equations

Q. 95 If the duration of unit hydrograph approaches zero, the resulting unit hydrograph is known as

(A) The limiting unit hydrograph

(B) The threshold unit hydrograph

(C) The instantaneous unit hydrograph

(D) The saturated unit hydrograph

Q. 96 Which statement is TRUE about the hydrologic cycle ?

(A) The hydrologic cycle occurs only once each year

(B) Evaporation is the downward movement of water through the ground

(C) All the water in our environment recycles itself over and over again

(D) The moon is the energy source that causes the water to move through the cycle

Q. 97 What type of material is most likely to be transported as suspended load ?

(A) Clay particles

(B) Sand particles

(C) Gravel particles

(D) All of these are equally likely to be transported

Q. 98 Movement of water as it passes from the ground surface in to soil is called as

(A) Infiltration (B) Percolation

(C) Evaporation (D) Interception

Q. 99 The factors which affects of Infiltration Capacity

(A) Soil density increase, Infiltration Capacity increase

(B) As degree of aggregation increases Infiltration Capacity increases

(C) Soil frost, reduces Infiltration Capacity

(D) All of the above

Q. 100 How much of the earth's water is stored in underground aquifers ?

(A) Less than 1 % (B) About 5 %

(C) About 10 % (D) About 20 %

Q. 101 Who introduced the concept of IUH (Instantaneous Unit Hydrograph) ?

(A) Clark (B) Nash

(C) Sherman (D) Bernard

Q. 102 In constructing synthetic hydrograph the information needed is/ are

(A) Time to peak (B) Peak volume

(C) Length of flow (D) All of these

Q. 103 The initial infiltration rate is at capacity rate at if the intensity of rainfall is

(A) Less than the average rate of infiltration

(B) Less than infiltration capacity of soil

(C) Equal to or more than the infiltration capacity of soil

(D) None of these

Q. 104 The formula for estimation of evapotranspiration using only temperature and day length is

(A) Thornthwaite formula (B) Penman formula

(C) Christiansen formula (D) Blaney criddle method

Q. 105 With respect to the earth's land surface, which of the following expression is correct

(A) Precipitation = evaporation – runoff

(B) Precipitation = runoff – evaporation

(C) Precipitation = evaporation + runoff

(D) Precipitation = evaporation × runoff

Q. 106 Which is the best method for calculating excess rainfall and runoff ?

(A) Hortonian (B) SCS curve number

(C) Green and Ampt (D) Snyder Unit Hydrograph

Q. 107 Rainfall intensity is measured by

(A) A hydrograph (B) Evaporimeter

(C) Simple rain gauge (D) Automatic rain guage

ASCO (RPSC) Exam (Forest) 2011

Q. 108 A convenient starting point to describe the hydrologic cycle is in the

(A) Ocean (B) Lake

(C) Earth (D) Atmosphere

Q. 109 In hydrologic analysis the volume of water is often expressed as

(A) Average depth over the catchment

(B) m^3

(C) m^3/unit area of the watershed

(D) m^3/sec

Q. 110 Recurrence interval or return period of rainfall event is expressed as

(A) $T = 1/P$ (B) $T = K.P$

(C) $t = P^{1/2}$ (D) $T = N + 1/m$

Q. 111 Tropical cyclone is called as

(A) Hurricane (B) Anticyclone

(C) Convective cyclone (D) Reverse cyclone

Q. 112 Run-off coefficient is higher of

(A) Agriculture lands (B) Barren lands

(C) Forest lands (D) Grass lands

Q. 113 K.E. > 25 index method for computing rainfall erosivity factor (R), does not consider the rainfall of intensity

(A) More than 25 mm/hr (B) 25 mm/hr

(C) Less than 25 mm/hr (D) 30 mm/hr

Q. 114 The runoff water volume of 1000 MCM (million cubic metre) equals to

(A) 100 ha. cm (B) 1 cubic km

(C) 1000 ha. cm (D) 10 cubic km

Q. 115 If the 1 mm rain water is infiltrated through ground surface of one hectare field the total volume of water infiltrated into the ground will be

(A) One m^3 (B) 10 m^3

(C) 100 m^3 (D) 1000 m^3

Q. 116 The following watershed gives highest peak discharge for the same area

(A) Fern shaped (B) Fan-shaped

(C) Square shaped (D) Rectangular shaped

Q. 117 For any watershed the values of sediment delivery ratio is always

(A) More than one (B) One

(C) Less than one (D) 2.5

Q. 118 The shape of the watershed is generally expressed by

(A) Form factor (B) Basin shape factor

(C) Drainage density (D) Elongation ratio

UKPSC A.En. Exam 2012 Paper - II

Q. 119 For a watershed having streams of order 1, 2, 3 and 4 the basin order will be

(A) 1 (B) 2

(C) 3 (D) 4

Q. 120 The basic assumptions of unit hydrograph theory are

(A) Nonlinear response and time invariance

(B) Time invariance and linear response

(C) Linear response and linear time variance

(D) Non-linear time variance and linear response

Q. 121 The time of concentration of a watershed is proportional to

(A) $L^{1.77}$ (B) $S^{-0.385}$

(C) $S^{0.385}$ (D) $L^{1.77}S^{0.385}$

Q. 122 Choose the correct statement/s is/are

(A) Cusecs: flow of cubic feet per second

(B) Cumecs: flow of cubic meter per second

(C) Both (A) and (B)

(D) None of these

Q. 123 A 4-hour unit hydrograph means

(A) 1 cm depth of rainfall over the entire watershed

(B) 1 cm of rainfall excess over the entire watershed

(C) A hydrograph resulting from an instantaneous application of 1 cm rainfall excess over the entire watershed

(D) The stream flow for four hours

Q. 124 A 4 hour rainfall in a catchment of 250 sq. km, produces rainfall depths of 6.2 cm and 5.0 cm in successive 2-hour unit period. Assuming the phi-index of the soil to be 1.2 cm per hour, the runoff volume in ha.m will be

(A) 16 (B) 22

(C) 1600 (D) 2200

Q. 125 Flood routing techniques are generally used for

(A) Flood forecasting

(B) Design of spillways and reservoir

(C) Flood protection works

(D) None of the above

Graduate Aptitude Test in Engineering - 2007

Q. 126 A 4-h unit hydrograph (UH) is used to derive S-hydrograph. The ordinates of 4-h UH are given below:

Time (h)	0	4	8	12	16	20	24	28	32	36	40	44
Ordinates(m^3/s)	0	20	80	130	150	130	90	52	27	15	5	0

(i) Equilibrium discharge and its time of occurrence for the derived S-hydrograph are

(A) 150 m^3 s^{-1} and 16 h (B) 380 m^3 s^{-1} and 16 h

(C) 699 m^3 s^{-1} and 40 h (D) 699 m^3 s^{-1} and 44 h

(ii) Area of watershed is

(A) 215.98 km^2 (B) 251.61 km^2

(C) 547.15 km^2 (D) 1006.47 km^2

Q. 127 The normal annual rainfall at stations I, II, III and IV in a basin are 155, 150, 120 and 105 cm respectively. In the year 2000, stations I, II and III received annual rainfalls of 156, 140 and 104 cm respectively. Estimated value of rainfall at station IV during the year 2000 is

(A) 98.2 cm (B) 105.0 cm

(C) 133.3 cm (D) 141.7 cm

Q. 128 The maximum rainfall with a return period of 25 years is given below for a watershed having a time of concentration of 47.65 minutes:

Time (min)	10	20	30	40	60
Rainfall depth(mm)	52.50	55.00	57.50	60.00	65.00

In this watershed, 2.0 km^2 area has cultivated sandy soil (C = 0.2) and the remaining 3.0 km^2 has cultivated clay soil (C = 0.7). The peak rate of runoff from the watershed is

(A) 4.29 $m^3\ s^{-1}$ (B) 5.41 $m^3\ s^{-1}$

(C) 42.99 $m^3\ s^{-1}$ (D) 54.13 $m^3\ s^{-1}$

Q. 129 A drop spillway is subjected to horizontal and vertical forces of 40.8 kN and 36.5 kN respectively. The area of plane of sliding is 10 m^2. Angle of internal friction and cohesive resistance of foundation material are 25 and 4.9 kPa respectively. The factor of safety against sliding is

(A) 0.53 (B) 0.61

(C) 1.62 (D) 1.86

Graduate Aptitude Test in Engineering - 2008

Q. 130 The analysis of maximum one-day rainfall in a city indicated that a depth of 280 mm has a return period of 50 years. The probability of a one–day rainfall depth equal to or greater than 280 mm in the city occurring two times in 15 successive years is

(A) 0.032 (B) 0.323

(C) 0.042 (D) 0.272

Q. 131 A catchment with an area of 756 km^2 has a 6 h unit hydrograph which is triangular with a base of 70 h. The peak discharge of direct runoff hydrograph due to 5 cm of rainfall excess in 6 h from the catchment is

(A) 60 m^3/s (B) 535 m^3/s

(C) 300 m^3/s (D) 756 m^3/s

Graduate Aptitude Test in Engineering - 2009

Q. 132 An earthen embankment with a pipe spillway is constructed to create temporary storage in gully. The pipe of 10 m length has to carry a peak discharge of 1 $m^3\ s^{-1}$ at an available head of 4 m. Entrance loss coefficient for the square entrance is 0.50 and the friction loss coefficient is 0.15. Required diameter of the pipe in m is

(A) 0.50 (B) 0.60

(C) 0.75 (D) 1.00

Q. 133 A 10 ha watershed received 100 mm uniformly distributed rainfall. Land use pattern consists of 25% residential area with soil group C and curve number 82, good meadow condition in 50% of the area with soil group D and curve number 78. There is also good open space condition in 25% of the area with soil group D and the curve number 80. Assuming AMC-II condition, the volume of runoff in m^3 from the watershed will be

(A)4955 (B) 5705

(C)5755 (D) 6555

Q. 134 If the probability of occurrence of rainfall on any day during June to September is 0.15, the probability that 4 out of 20 days in the month of August to remain dry will be

(A)0.162 (B) 0.182

(C) 0.192 (D) 0.228

Graduate Aptitude Test in Engineering - 2010

Q. 135 An effective rainfall of 20 mm h^{-1} occurs for 2 hours in a catchment. The time of concentration of the catchment is 1.5 hour. the peak of the resulting direct runoff hydrograph, in mm h^{-1}, is

(A)10 (B) 20

(C)30 (D) 40

Q. 136 The peak of a flood hydrograph due to a 1-h duration isolated storm in a catchment of area 13.5 km^2 is 135 $m^3 s^{-1}$. The total depth of rainfall is 54 mm. Assume a constant base flow of 10 $m^3 s^{-1}$ and phi-index equal to 4 mm h^{-1}.

(i) The peak of 1-h unit hydrograph for the catchment in $m^3 s^{-1}$ is

(A)15 (B) 20

(C)25 (D) 30

(ii) Assuming the above 1-h unit hydrograph to be triangular in shape with the time to peak as 1 hour, the peak of the 2-h unit hydrograph for the catchment in $m^3 s^{-1}$ is

(A)13.25 (B) 18.75

(C)21.25 (D) 26.75

Graduate Aptitude Test in Engineering - 2011

Q. 137 For a catchment with an area of 400 km², the equivalent discharge of the S-curve obtained by summation of 4-h unit hydrograph in $m^3\ s^{-1}$ is

(A) 100 (B) 139

(C) 200 (D) 278

Q. 138 The following figure presents the hyetograph for a 3-hour storm. The surface runoff for the event is estimated to be 38 mm. The phi - index for the event in mm h^{-1} is

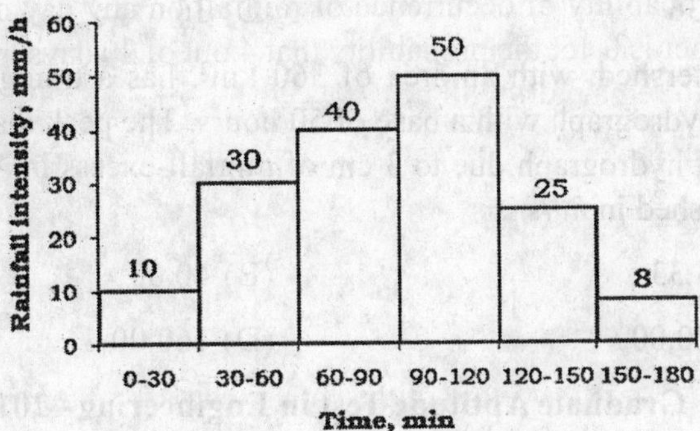

(A) 11.50 (B) 14.25

(C) 15.80 (D) 17.25

Graduate Aptitude Test in Engineering - 2012

Q. 139 The hourly discharge observations at the mouth of a watershed due to 2 cm excess rainfall during 0 to 1 h and 3 cm excess rainfall during 1 to 2 h are given in the table below. Assume a constant base flow of 1 $m^3\ s^{-1}$.

Time (h)	0	1	2	3	4	5	6
Discharge(m^3s^{-1})	1	7	26	37	27	13	1

(i) The area of the watershed, in km² is

(A) 7.56 (B) 8.24

(C) 8.35 (D) 8.86

(ii) The peak of 1 h unit hydrograph in m^3s^{-1} for the watershed and its time of occurrence in h, respectively are

(A) 6, 1 (B) 7, 2

(C) 8, 2 (D) 9, 1

Graduate Aptitude Test in Engineering - 2013

Q. 140 The Rational method is used to estimate

(A) Runoff volume (B) Peak runoff rate

(C) Runoff depth (D) Direct surface runoff

Q. 141 A 100 ha watershed received rainfall at a rate of 5 cm/h for 2 hours. If the runoff generated by the storm was at the rate of 1 m^3/s for 10 hours, the runoff coefficient for the watershed would be

(A) 3.6×10^{-3} (B) 6.0×10^{-2}

(C) 0.36 (D) 36

Q. 142 A watershed, with an area of 360 km^2, has a triangular shaped 4-h unit hydrograph with a base of 50 hours. The peak discharge of direct runoff hydrograph due to 3 cm of rainfall-excess in 4 hours from the watershed in m^3/s is

(A) 13.33 (B) 40.00

(C) 120.00 (D) 160.00

Graduate Aptitude Test in Engineering - 2014

Q. 143 A triangular 5-hour unit hydrograph (UH1) of a catchment area of 400 km^2 has a peak discharge of 60 $m^3 s^{-1}$. Another triangular 5-hour unit hydrograph (UH2) having the same base width as UH1 has a peak flow of 90 $m^3 s^{-1}$. The catchment area corresponding to UH2 in km^2 is

(A) 267 (B) 600

(C) 750 (D) 867

Graduate Aptitude Test in Engineering - 2015

Q. 144 The incorrect statement from the following is

(A) The peak runoff from an agricultural watershed is generally less than that from an urban watershed of the same area.

(B) The horizontal hydraulic conductivity of soil is less than its vertical hydraulic conductivity.

(C) The magnitude of a 75-year flood is less than that of a 100-year flood.

(D) The rating curve due to an unsteady flood event forms a loop.

Q. 145 If x be the highest mean monthly precipitation and y be the mean annual precipitation, then the rainfall aggressiveness to soil erosion is

(A) x^2/y (B) x/y^2

(C) x^2/y^2 (D) y^2/x

Q. 146 A catchment of 720 ha area has 25- year mean rainfall intensity of 100 mm/h occurring for a duration equal to its time of concentration. During a storm event, the catchment received a total of 7.5 cm design rainfall for 6 hours. Assuming ∅-index of 0.25 cm/h and runoff coefficient of 0.6, the peak ordinate of the 6-h unit hydrograph in m^3/s will be ....

Q. 147 Match the following items between **Column-I** and **Column-II** with the most appropriate combinations.

Column - I	Column - II
(i) Direct Runoff	(1) T – year rainfall depth
(ii) Peak runoff	(2) Curve number
(iii) Tensiometer	(3) Aquifer
(iv) Isoerodent map	(4) Rational formula
(v) Isopluvial map	(5) 30-minute rainfall intensity
(vi) Zone of saturation	(6) Field capacity

(A) i - 4, ii-2, iii - 6, iv-5, v-1, vi-3

(B) i - 2, ii-4, iii - 6, iv-5, v-1, vi-3

(C) i - 4, ii-2, iii - 6, iv-1, v-5, vi-3

(D) i - 2, ii-4, iii - 6, iv-1, v-5, vi-3

Graduate Aptitude Test in Engineering - 2016

Q. 148 A watershed area of 1851 hectare has maximum distance of 7.12 km from the outlet to the farthest point on the divide line. The form factor of the watershed is

(A) 2.60 (B) 2.73

(C) 0.365 (D) 0.385

Q. 149 The normal annual rainfall for 5 rain guage stations A, B, C, D and E in a watershed were 112.7, 120.4, 118.3, 125.2 and 110.6 cm, respectively. In a particular year, the rain gauge installed at station C failed to record rainfall. In the same year the rain gauges at stations A, B, D and E recorded annual rainfall of 114.9, 118.3, 122.6 and 114.5 cm, respectively. The estimated rainfall at station C in that particular year in cm was ....

Q. 150 Annual average soil loss from a watershed has been measured as 20 Mg $ha^{-1}year^{-1}$. Watershed has 8% land slope and 84 m maximum slope length. Assume all other factors same and dimensionless exponent for slope factor is 0.5. To reduce soil loss from the watershed to 10 Mg ha $ha^{-1}year^{-1}$, the maximum slope length in m should be ....

Q. 151 The maximum one day rainfall depth at 20 year return period of a city is 150 mm. The probability of one day rainfall equal to or greater than 150 mm in the same city occurring twice in 20 successive years is ....

Q. 152 The observed rainfall of a 12 h duration event is given in the table below. If the phi (∅) index of the storm is 0.46, the total direct runoff of the event in mm will be ....

Time (h)	0	2	4	6	8	10	12
Cumulative rainfall (cm)	0	0.64	2.64	5.64	6.00	6.80	10.80

Graduate Aptitude Test in Engineering - 2017

Q. 153 Remedial measure generally adopted for controlling stream-bank erosion is

(A) Shelter belts
(B) Spurs
(C) Brushwood dams
(D) Drop spillways

Q. 154 A watershed of 100 km^2 is underlain by an unconfined aquifer having hydraulic conductivity of 15 m day^{-1} and specific yield of 0.20. If 30 million m^3 of water is pumped through uniformly distributed wells, the average drop of water table over the watershed in meter will be

(A) 1.50
(B) 0.75
(C) 0.06
(D) 6.00

Q. 155 Match the following items between Column-I and Column-II with the most appropriate combinations:

Column I	Column - II
1) Neutron probe	P) Open channel flow
2) Pressure Plate Apparatus	Q) Deep percolation
3) Tipping Bucket	R) Aquifer parameters
4) Current Meter	S) Soil Moisture
5) Pumping Test	T) Rainfall intensity
6) Lysimeter	U) Soil-moisture characteristic curve

(A) 1-Q, 2-P, 3-T, 4-S, 5-R, 6-U (B) 1-S, 2-U, 3-T, 4-P, 5-R, 6-Q

(C) 1-S, 2-R, 3-T, 4-P, 5-Q, 6-U (D) 1-P, 2-U, 3-S, 4-T, 5-R, 6-Q

Q. 156 A watershed of 4.8 km^2 generates 4.3 cm runoff from a rain storm of 4-hour duration. The measured rainfall intensities for this storm in successive 30-minute durations are given below:

Time interval (min)	0-30	30 -60	60 -90	90 -120	120 -150	150-180	180 -210	210 -240
Rainfall intensity ($cm\ h^{-1}$)	1.6	4.8	3.2	3.4	2.2	5.0	4.2	1.2

The value of ∅-index for the watershed in $cm\ h^{-1}$ will be

(A) 6.5 (B) 5.3

(C) 2.4 (D) 2.1

Graduate Aptitude Test in Engineering - 2018

Q. 157 The 6-hour unit hydrograph of a watershed is represented by an isosceles triangle with the peak of 180 $m^3\ s^{-1}$ and time to peak of 18 hours. The phi-index (ø) of this watershed is 3.0 $mm\ h^{-1}$ and the constant baseflow is 20 $m^3\ s^{-1}$. The accumulated rainfall received in this watershed at 6 h and 12 h from the start of the storm are 38 mm and 106 mm, respectively. The resulting peak of the flood hydrograph due to this storm event is .... $m^3\ s^{-1}$.

Graduate Aptitude Test in Engineering - 2019

Q. 158 A watershed of area 80 ha has a runoff coefficient of 0.3. A storm of intensity 5 cm/ h occurs for a duration more than the time of concentration of the watershed. The peak discharge in m^3/s is ....

Q. 159 A catchment has eight rain gauge stations. In a year, the annual rainfall recorded by the gauges (in cm) are 93.8, 106.5, 170.6. 138.7, 87.8, 156.2, 180.9 and 110.3. For a 10% error in the estimation of the mean rainfall, the optimum number of stations in the catchment is

(A) 4 (B) 6

(C) 8 (D) 10

Q. 160 The peak of a flood hydrograph due to a 5-hour storm is 670 m^3/s. The total depth of rainfall is 9 cm. Assuming an average infiltration loss of 0.2 cm/h and a constant base flow of 30 m^3/s, the peak discharge of the 5-hour unit hydrograph for this catchment in m^3/s is ....

Graduate Aptitude Test in Engineering - 2020

Q. 161 In a basin, rainfall is recorded by five automatic weather stations A, B, C, D and E with respective average annual rainfall of 1020, 810, 675, 940 and 780 mm. In a particular year, the station A was non-operational and the remaining stations B, C, D and E recorded annual rainfall of 890, 725, 980 and 850 mm, respectively. The estimated rainfall at the station A in that particular year in mm is

(A) 758 (B) 878

(C) 1038 (D) 1098

·Graduate Aptitude Test in Engineering - 2021

Q. 162 The shape of the Instantaneous Unit Hydrograph (IUH) of a catchment is an isosceles triangle with a peak of 60 m^3/s and time to peak of 3 h. If the constant baseflow is 7.5 m^3/s, the peak of the 3 h Unit Hydrograph (UH) in m^3/s is

(A) 43.33 (B) 50.83

(C) 52.50 (D) 60.00

Q. 163 A small watershed receives rainfall of 90 mm in a day. For this watershed, irrespective of the land use, the amount of initial abstraction can be considered as 25% of the potential maximum retention (S) of soil. Initially, the entire watershed was under forest with S = 136 mm, which was converted into cultivated land with S = 64 mm. The change in the daily runoff volume due to this land use alteration for this specific rainfall event in percent is ....

Answers Key

1	2	3	4	5	6	7	8	9	10
D	C	C	D	D	C	B	A	B	A
11	12	13	14	15	16	17	18	19	20
B	B	B	B	C	D	C	C	C	B
21	22	23	24	25	26	27	28	29	30
B	B	A	C	A	C	C	D	C	C
31	32	33	34	35	36	37	38	39	40
B	C	C	B	C	C	C	B	A	D
41	42	43	44	45	46	47	48	49	50
B	A	D	B	A	D	A	B	C	A
51	52	53	54	55	56	57	58	59	60
B	E	D	C	B	C	B	C	A	C
61	62	63	64	65	66	67	68	69	70
C	A	C	D	B	C	A	C	D	A
71	72	73	74	75	76	77	78	79	80
D	A	A	D	C	C	D	D	B	C
81	82	83	84	85	86	87	88	89	90
B	B	B	B	D	A	A	A	B	B
91	92	93	94	95	96	97	98	99	100
B	C	B	D	C	C	A	A	D	A
101	102	103	104	105	106	107	108	109	110
A	D	C	D	C	B	D	A	A	A
111	112	113	114	115	116	117	118	119	120
A	B	C	B	B	B	C	A	D	B
121	122	123	124	125	126	127	128	129	130
B	C	B	C	A	C, D	A	D	C	A
131	132	133	134	135	136	137	138	139	140
C	A	A	BONUS	B	C, B	D	D	A, C	B
141	142	143	144	145	146	147	148	149	150
C	C	B	B	A	20	B	C	118.79	21
151	152	153	154	155	156	157	158	159	160
0.1887	62	B	A	B	C	BONUS	3.30	C	80
161	162	163							
D	A	142.99							

Explanations

Q. 28

Bifurcation ratio denotins the ratio between the number of streams of one order and those of the next-higher order in a drainage network. It may be a useful measure of proneness to flooding: the higher the bifurcation ratio, the greater the probability of flooding.

Drainage density is the total length of all the streams and rivers in a drainage basin divided by the total area of the drainage basin. It is a measure of how well or how poorly a watershed is drained by stream channels.

Q. 30

We know, For triangular shaped direct runoff hydrograph

$$\frac{1}{2} \times \text{Peak discharge of UH} \times \text{Base period of UH}$$

$$= \text{Watershed area} \times \text{Effective rainfall}$$

$$\Rightarrow \text{Effective rainfall} = (20\times3600\times40)/(2\times72000000) = 0.02 \text{ m or } 2 \text{ cm}$$

Q. 31

$$\text{Depth of runoff (S)} = \frac{25400}{CN} - 254$$

$$CN = 25400/(40 + 254) = 86.39$$

Q. 39

The kinetic energy of a raindrop is easily expressed in terms of his properties as:

$$K.E. = mV^2/2 = \pi\rho D^3V^2/12$$

where ρ is the water density, D is the diameter of the equivalent spherical drop and V is its terminal velocity. So, the KE of a rainfall event per unit of surface can be directly computed knowing the D and the V of each raindrop falling on a surface unity.

Q. 60

The Indian Standards IS: 4987-1968 has recommended the following densities

One gauge per 520 km^2 for plain area.

One gauge per 260 to 390 km^2 for region of average elevation.

One gauge per 130 km^2 hilly area with heavy rainfall.

Q. 71

The meteorological Droughts are classified into (a) moderate and (b) severe based on rainfall deficiency, i.e. 26 to 50% and more than 50% respectively.

Q. 72

Direct runoff = Flood runoff – Base flow

$= 1800 - 20 = 1780\ m^3/sec$

Q. 73

Forms of Precipitation is classified as-

Rain- Precipitation in the form of water drops of size larger than 0.5 mm. The maximum raindrop size is about 6 mm. Drops of larger size break up into smaller drops as it falls down. It is the major form of precipitation in India.

Snow- Consists of ice crystals in flaky form (average density ~ 0.1g/cc). It is also an important form of precipitation. Occurs in the Himalayan region in India.

Drizzle- A fine sprinkle of tiny water droplets of size < 0.5 mm and intensity < 1 mm/h. The tiny drops forming a drizzle appear to float in air.

Glaze (Freezing Rain)– Formed when rain or drizzle comes in contact with cold ground at around 0 °C. The water drops freeze to form an ice coating.

Sleet – Frozen rain drops formed when rain falls through air at subfreezing temperature.

Q. 85

Peak discharge rate = CIA

$= 0.40 \times (0.045/3600) \times 90 \times 10000 = 4.50\ m^3/sec$

Q. 124

We know

∅ index = (Total precipitation – Effective Rainfall

Total precipitation = (6.2 + 5.0)/4 = 2.8 cm/h

⇒ 1.20 = 2.8 – Effective Rainfall

⇒ Effective Rainfall (ER) = 1.6 cm

Hence, Runoff volume = $1.6 \times 250 \times 1000000 \times 4/100$

$= 16000000\ m^3$ or 1600 ha.m

Q. 126 (i)

Ordinates of 4-h unit hydrograph & S – hydrograph (in m^3/s) is as shown in table

Time(h)	0	4	8	12	16	20	24	28	32	36	40	44
4-h UH	0	20	80	130	150	130	90	52	27	15	5	0
S-hydrograph	0	20	100	230	380	510	600	652	679	694	699	699

According to above table

Equilibrium discharge for S-hydrograph is = 699 m^3/s and its time of occurrence = 40 h

(ii)

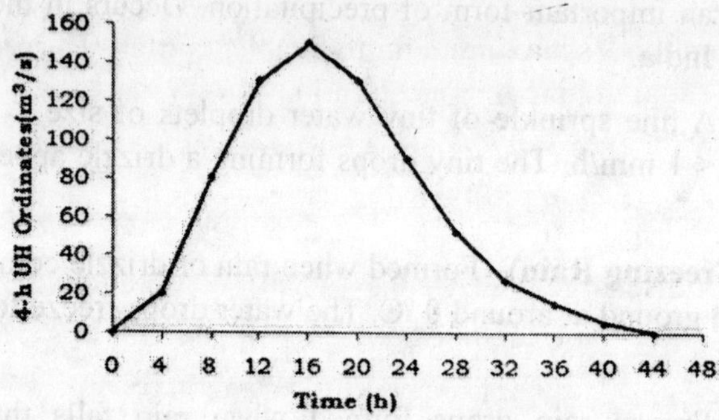

Total area below curve gives total volume of runoff using trapezoidal formula

$$V = \frac{h}{2}\left[Q_1+Q_{12}+2(Q_2+Q_3+Q_4+Q_5+Q_6+Q_7+Q_8+Q_9+Q_{10}+Q_{11})\right]$$

$$V = \frac{4\times3600}{2}\left[0+0+2\,(20+80+130+150+130+90+52+27+15+5)\right]$$

$$V = 10065600 \text{ m}^3$$

We know that

Total volume of runoff – effective rainfall × Watershed area

$$\Rightarrow \qquad 10065600 = \frac{1}{100} \times \text{Watershed area}$$

$$\Rightarrow \qquad \text{Watershed area} = 1006.5 \text{ km}^2$$

Q. 127

We know

$$\frac{P_x}{N_x} = \frac{1}{n-1}\left(\frac{P_1}{N_1} + \frac{P_2}{N_2} + \frac{P_3}{N_3}\right)$$

$$\Rightarrow \quad P_x = \frac{105}{4-1}\left(\frac{156}{155} + \frac{140}{150} + \frac{104}{120}\right)$$

$$\Rightarrow \quad P_x = 98.2 \text{ cm}$$

Q. 128

Runoff coefficient (C) of watershed is $= \dfrac{A_1C_1 + A_2C_2}{A_1 + A_2}$

$$= \frac{2\times0.2+3\times0.7}{2+3} = 0.5$$

From Duration Vs maximum rainfall table, rainfall (P) corresponding to duration equal to time of concentration 47.65 min is given by,

$$P = 60 + (7.65\times5/20) = 61.9125 \text{ mm}$$

$$\therefore \text{ Rainfall intensity} = \frac{61.9125}{47.65} = 1.30 \text{ mm/min} = 2.13\times10^{-5} \text{ m/s}$$

$$\therefore \quad \text{Peak discharge} = CIA = 0.5\times2.13\times10^{-5}\times5000000$$

$$= 53.25 \text{ m}^3/\text{s} \approx 54.13 \text{ m}^3/\text{s}$$

Q. 129

Given that

Horizontal force (F_h) = 40.8 kN

Vertical force (F_v) = 36.5

Angle of internal friction (ϕ) = 25°

⇒ Friction coefficient (μ) = tan25 = 0.47

Cohesive resistance of foundation(C) = 4.9 kPa

Area of sliding (A) = 10 m²

Factor of safety

$$\frac{\textit{Actual horizontal force}}{\textit{Horizontal force}} = \frac{\textit{Frictional force} + \textit{Cohesive force}}{\textit{Horizontal force}} = \frac{\mu F_v + CA}{F_h}$$

$$= \frac{0.47\times36.5+4.9\times10}{40.8} = 1.62$$

Q. 130

Rainfall return period = 50 years

Probability of rainfall occurrence (p) = 1/50 = 0.02

Probability of rainfall non-occurrence (q) = 1 − 0.02 = 0.98

∴ Probability of rainfall occurring two times in 15 successive years

$$= {}^{15}C_2''(q)^{13}(p)^2$$

$$= {}^{15}C_2(0.98)^{13}(0.02)^2 = 0.032$$

Q. 131

We know

$$\frac{1}{2}\times\text{Peak discharge of UH}\times\text{Base period of UH} = \text{Watershed area}\times\frac{1}{100}$$

$$\Rightarrow \quad \text{Peak discharge of unit hydrograph} = \frac{756000000\times 2}{100\times 252000} = 60\ m^3/s$$

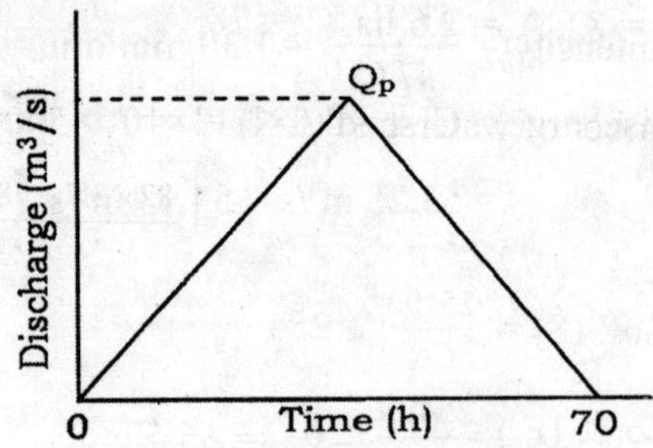

∵ Peak discharge of direct runoff hydrograph

= Peak discharge of UH × Excess rainfall in cm

∴ Peak discharge of direct runoff hydrograph = 60×5 = 300 m^3/s

Q. 132

Given that

Length of pipe (L) = 10 m

Peak discharge (Q) = 1 m^3/s

Head (h) = 4 m

Entrance loss coefficient (K_e) = 0.50

Friction loss coefficient (K_f) = 0.15

We know that velocity of flow in pipe spillway

$$V = \sqrt{\frac{2gh}{1 + K_e + LK_f}}$$

$$V = \sqrt{\frac{2 \times 9.81 \times 4}{1 + 0.50 + 10 \times 0.15}}$$

$$V = 5.11 \text{ m/s}$$

Applying equation of continuity

$$Q = AV$$

$$\Rightarrow \quad 1 = \frac{\pi d^2}{4} \times 5.11$$

$$\Rightarrow \quad d = 0.50 \text{ m}$$

Q. 133

Given that

Total watershed area = 10 ha

$C_1 = 82$, $A_1 = 2.5$ ha, $C_2 = 78$, $A_2 = 5$ ha, $C_3 = 80$, $A_3 = 2.5$ ha

$$\therefore \text{ Curve number of watershed (CN)} = \frac{C_1A_1 + C_2A_2 + C_3A_3}{A_1 + A_2 + A_3}$$

$$= \frac{82 \times 2.5 + 78 \times 5 + 80 \times 2.5}{10} = 79.5$$

$$\text{Depth of runoff (S)} = \frac{25400}{CN} - 254$$

$$= \frac{25400}{79.5} - 254 = 65.5 \text{ mm}$$

$$\text{Hence,} \quad \text{Runoff volume (Q)} = \frac{(P - 0.2S)^2}{P + 0.8S} = 49.55 \text{ mm}$$

In terms of m^3;

$$Q = \frac{49.55}{1000} \times 10 \times 10000 = 4955 \text{ m}^3$$

Q. 134

Probability of rainfall occurrence on any day during June to September

(p) = 0.15

Probability of rainfall not occurrence on any day during June to September

(q) = 1 − 0.15 = 0.85

Probability of rainfall occurrence on 4 out of 20 days = $^{20}C_4p^4q^{16}$

= $^{20}C_4(0.15)^4(0.85)^{16}$ = 0.182

Probability of remain dry on 4 out of 20 days = 1 – 0.182 = 0.818

Q. 136 (i)

Depth of rainfall = 54 mm

Phi index = 4 mm/h

Effective Rainfall = 54 – 4 = 50 mm = 5 cm

Peak of direct rainfall hydrograph (DRH) = (135 – 10) m³/sec = 125 m³/sec

Peak of unit hydrograph = Peak of DRH/Effective Rainfall

= 125/5 = 25 m³/s

(ii)

Area of the catchment × Rainfall excess under UH

= (Peak of unit hydrograph × Time)/2

$\Rightarrow$ 13.5×10⁶× (1/100) = (25×t)/2

$\Rightarrow$ t = 3 hours

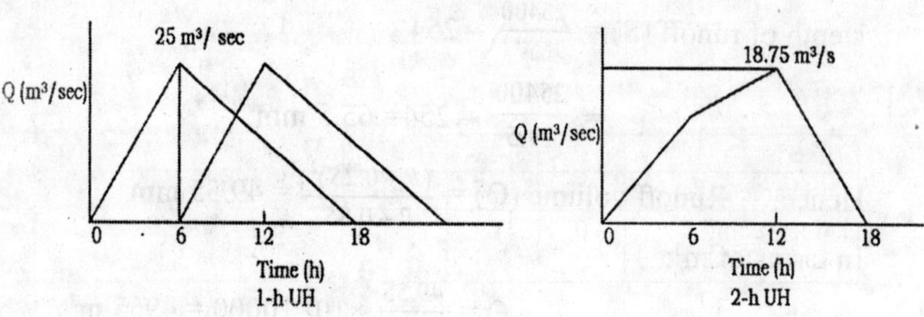

Q = (25 + 25/2)/2 = 18.75 m³/sec

Q. 137

We know

Equivalent discharge = $2.778\frac{A}{D}$

$= 2.778\times\frac{400}{4} = 278$ m³/s

Q. 138

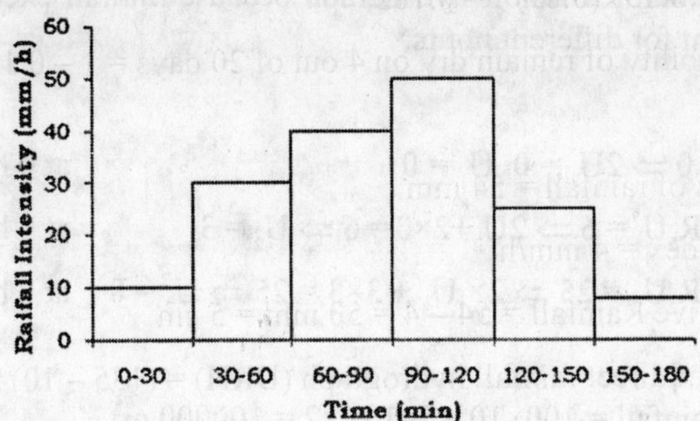

Total Rainfall (P) = 10×0.5 + 30×0.5 + 40×0.5 +50×0.5 +25×0.5 +8×0.5

= 81.5 mm

Surface runoff (R) = 38 mm

Thus, infiltration rate = $\frac{P-R}{T}$

$= \frac{81.5-38}{3} = 14.5$ mm/h

Thus, rainfall intensity below 14.5 mm/h will be not counted in runoff.

Thus, Effective Rainfall (P) = 81.5-10 × 0.5-8 ×0. 5 = 72.5mm

and ∅ index = (72.5 – 38)/2 = 17.25 mm/h

Q. 139 (i)

Time(h)	0	1	2	3	4	5	6
Discharge(m^3/s)	1	7	26	37	27	13	1
Direct runoff discharge(m^3/s)	0	6	25	36	26	12	0

Total discharge = 0 + 6 + 25 + 36 + 26 + 12 + 0 = 105×3600 = 378000 m^3

Excess rainfall = 2 + 3 = 5 cm = 0.05 m

We know that

Watershed area × Excess rainfall = Total runoff discharge

⇒ Watershed area = 378000/0.05

= 7560000 m^2 = 7.56 km^2

(ii)

It is a complex runoff hydrograph because rainfall excess is different-different for different hours.

Thus

$R_1U_1 = 0 \Rightarrow 2U_1 = 0 \;\; U_1 = 0$ at 0 hrs

$R_1U_2 + R_2U_1 = 6 \Rightarrow 2U_2 + 2\times0 = 6 \Rightarrow U_2 = 3$ at 1 hrs

$R_1U_3 + R_2U_2 = 25 \Rightarrow 2\times U_3 + 3\times3 = 25 \Rightarrow U_3 = 8$ at 2 hrs

Q. 141

Total rainfall $= 100\times10^4\times5\times10^{-2}\times2 = 100000\ m^3$

Total Runoff $= 1\times3600\times10 = 36000\ m^3$

Runoff Coefficient = Runoff/Rainfall

$= 36000/100000 = 0.36$

Q. 142

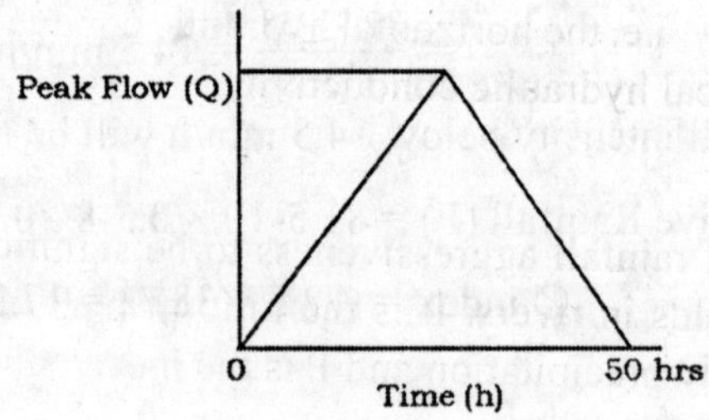

Runoff Volume = Area of triangle

$= Q\times50\times3600/2$

Rainfall volume $= 3\times10^{-2}\times360\times10^6$

Thus

$Q\times50\times3600/2 = 3\times10^{-2}\times360\times10^4$

$\Rightarrow \quad Q = 120\ m^3/sec$

Q. 143

For Unit Hydrograph(UH_1)

Total Discharge of water in 5 hours = Area under triangle

$$A \times 1\ cm = \frac{1}{2} \times \text{base width} \times \text{peak discharge}$$

Therefore $A_1 = \frac{1}{2} \times b \times Q_1$

Similarly, $A_2 = \frac{1}{2} \times b \times Q_2$

Now $\frac{A_1}{A_2} = \frac{}{}$

$\Rightarrow$ $A_2 = \frac{90}{60} \times 400 = 600\ km^2$

Q. 144

We know

Horizontal Hydraulic conductivity

$k_H = (K_1H_1 + K_2H_2 + K_1H_1 + \ldots\ldots\ldots\ldots)/(H_1 + H_2 + H_3 + \ldots\ldots\ldots\ldots)$

Vertical Hydraulic conductivity

$k_V = (H_1 + H_2 + H_3 + \ldots\ldots\ldots\ldots)/((H_1/K_1 + H_2/K_2 + H_1/K_1 + \ldots\ldots\ldots)$

Here, $k_V < k_h$ i.e. the horizontal hydraulic conductivity of soil is greater than its vertical hydraulic conductivity.

Q. 145

The index of rainfall aggressiveness to be significantly correlated with sediment yields in rivers. It is the ratio of p^2/P, where p is the highest mean monthly precipitation and P is the mean annual precipitation.

Hence, Rainfall aggressiveness = x^2/y

Q. 146

Peak of direct runoff hydrograph (Q_{DRH})

$= CIA$

$$Q_{DRH} = 0.6 \times \left(\frac{100}{1000 \times 3600}\right) \times 720 \times 10000$$

$= 120\ m^3/s$

We know

$\varnothing$ index = (Total precipitation – Effective Rainfall)/Time

$\Rightarrow$ 0.25 = (7.5 – Effective Rainfall)/6

$\Rightarrow$ Effective Rainfall (ER) = 6 cm

We also know

Peak ordinate of unit hydrograph = Q_{DRH} (in m^3/s)/E.R.(in cm)

$= 120/6 = 20\ m^3/s$

Q. 147

Iso-erodent map – The lines on the map join points with the same erosion-index value (which shows equally erosive average annual rainfall) and are called iso-erodent lines. The iso-erodent map presents average annual Energy Intensity (EI) values.

Iso-pluvial map - – The lines on the map join points with the same pluvial-index for a given length of storm and a given period of time, are called iso-pluvial lines.

Q. 148

Form factor = Average width of the basin/Axial length of the basin

$= B/l = (A/l)/l = A/l^2$

$= 1851 \times 10000/(7120)^2 = 0.365$

Q. 149

We know

$$\frac{P_C}{N_C} = \frac{1}{n-1}\left(\frac{P_A}{N_A} + \frac{P_B}{N_B} + \frac{P_D}{N_D} + \frac{P_E}{N_E}\right)$$

$$\Rightarrow \quad P_x = \frac{118.3}{4}\left(\frac{114.9}{112.7} + \frac{118.3}{120.4} + \frac{122.6}{125.2} + \frac{114.5}{110.6}\right)$$

$$\Rightarrow \quad P_x = 118.79 \text{ cm}$$

Q. 150

Given, $m = 0.5$

And $A_1 = 20$ Mg/ha-year, $A_2 = 10$ Mg/ha-year

We know,

A = RKLSCP (USLE)

Where, $L = (L_{p1}/22.13)^m$

$\Rightarrow$ $A_1/A_2 = (L_{p1}/L_{p2})^{0.5}$ (Assuming all other factors same.)

or $20/10 = (84/L_{p2})^{0.5}$

$\Rightarrow$ $L_{p2} = 21$ m

Q. 151

Probability of occurrence of 100 mm rainfall in 20 years (p)

$$= \frac{1}{20} = 0.05$$

Probability of non-occurrence in 20 years (q) = 1 – 0.05 = 0.95

Probability of rainfall occurrence two times in 20 successive years

$$= {}^{20}C_2(q)^{18}(p)^2$$

$$= {}^{20}C_2(0.95)^{18}(0.05)^2 = 0.1887$$

Q. 152

We know

$\varnothing$ index = (Precipitation – Direct Runoff)/Rainfall Duration

0.46 = (10.80 – Direct Runoff)/10

$\Rightarrow$ Direct Runoff = 62 mm

Q. 154

Specific yield (S_y) = Volume of water released from storage/Surface area×water table drop

$\Rightarrow$ Water table drop = $0.20 \times 100 \times 10^6/(30 \times 10^6)$ = 1.50 m

Q. 156

Total rainfall (P) = 1.6×0.5 + 4.8×0.5 + 3.2×0.5 + 3.4×0.5 + 2.2×0.5 + 5×0.5 + 4.2×0.5 + 1.2×0.5 = 12.80 cm

Thus, Infiltration rate = (Rainfall – Runoff)/Time

= (12.80 – 4.3)/8 = 2.125 cm/h

Hence, Rainfall intensity below 2.125 cm/h will not be counted in runoff.

Effective rainfall

= 4.8×0.5 + 3.2×0.5 + 3.4×0.5 + 2.2×0.5 + 5×0.5 + 4.2×0.5

= 11.4 cm

Thus, ø-index = (11.4 – 4.3)/3 = 2.40 cm/h

Q. 157

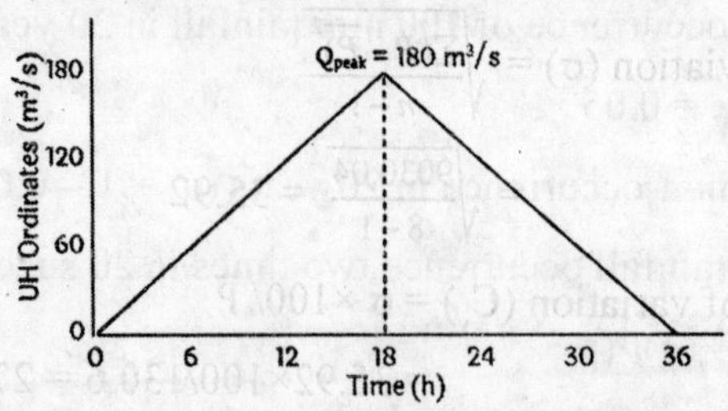

Excess rainfall for first 6 hours = 38 - 6×3 = 20 mm or 2 cm

Excess rainfall for second 6 hours = 106 - 38 - 6×3 = 50 mm or 5 cm

Time (h) A)	UH Ordinates (Q/m^3) (B)	Leg of UH ordinates for 2 cm of rainfall excess (Q/m^3) (C) = 2×(B)	Leg of UH ordinates for 5 cm of rainfall excess (Q/m^3) (D) = 5×(B)	Base Flow (Q/m^3) (E)	Flood hydrograph ordinates (Q/m^3) (C)+(D)+(E)
0	0	0	0	20	20
6	60	120	0	20	20
12	120	240	300	20	560
18	180	360	600	20	980
24	120	240	900	20	1160
30	60	120	600	20	740
36	0	0	300	20	320
			0	20	20

Hence, Peak of flood hydrograph = 1160 m3/s

Q. 158

Peak Discharge = CIA

$$= 0.3\times(0.05/3600)\times80\times10000 = 3.30\ m^3/s$$

Q. 159

Rainfall P (cm)	$P - \bar{P}$	$(P - \bar{P})^2$
93.8	-36.8	1354.24
106.5	-24.1	580.81
170.6	40	1600
138.7	8.1	65.61
87.8	-42.8	1831.84
156.2	25.6	655.36
180.9	50.3	2530.09
110.3	-20.3	412.09
$\sum P = 1044.8$		$\sum(P - \bar{P})^2 = 9030.04$

Average Rainfall($\overline{P}$) = 1044.8/8 = 130.6 cm

Standard deviation (σ) = $\sqrt{\frac{\Sigma(P-\overline{P})^2}{n-1}}$

$= \sqrt{\frac{9030.04}{8-1}} = 35.92$

Coefficient of variation (C_v) = $\sigma \times 100/\overline{P}$

$= 35.92 \times 100/130.6 = 27.50$

Optimum number of stations = $(C_v/\text{error})^2$

$= (27.5/10)^2 = 7.56$ or 8 Raingauge stations

Q. 160

Discharge after base flow seperation = 670 – 30 = 640 m³/s

Effective rainfall over 5 hours storm = 9 – 0.2×5 = 8 cm

Hence,

Peak of 5 hours unit hydrograph = 640/8 = 80 m³/s

Q. 161

We know

$$\frac{P}{N} = \frac{}{n}\left(\frac{P_1}{N_1}+\frac{P_2}{N_2}+\frac{P_3}{N_3}\right)$$

$$\Rightarrow \quad P_x = \frac{1020}{5-1}\left(\frac{890}{810}+\frac{725}{675}+\frac{980}{940}+\frac{850}{780}\right)$$

$$\Rightarrow \quad P_x = 1098 \text{ mm}$$

Q. 162

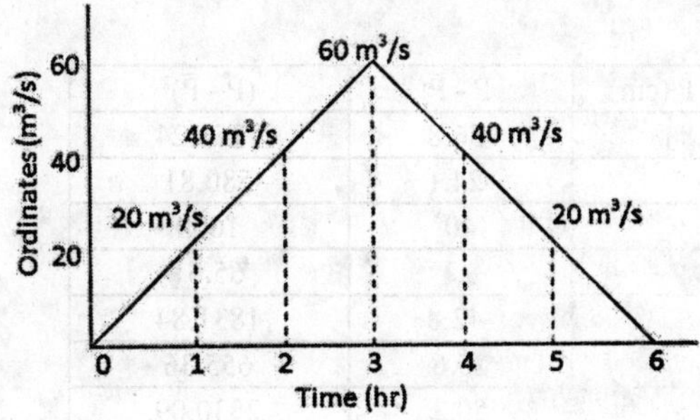

Time	IUH Ordinates	IUH lagged by 1 hr	1 hr Unit hydrograph	S_1	S_2 S_1 lagged by 3-hr	3 hr DRH	3 UH Ordinates
1	2	3	4 =(3+2)/2	5	6	7 = 6–5	8/3
0	0		0	0		0	0.00
1	20	0	10	10		10	3.33
2	40	20	30	40		40	13.33
3	60	40	50	90	0	90	30.00
4	40	60	50	140	10	130	43.33
5	20	40	30	170	40	130	43.33
6	0	20	10	180	90	90	30.00
7		0	0	180	140	40	13.33
8				180	170	10	3.33
9				180	180	0	0.00

Peak of 3 UH = 43.33 m^3/s

Q. 163

We know,

Volume of runoff (Q) $= \dfrac{(P-0.25S)^2}{P+0.75S}$

Hence, $Q_1 = \dfrac{(90-0.25\times 136)^2}{90+0.75\times 136} = 16.33\ m^3$

Again, $Q_2 = \dfrac{(90-0.25\times 64)^2}{90+0.75\times 64} = 39.68\ m^3$

$\Rightarrow$ Precentage change in runoff volume $= \dfrac{39.68-16.33}{16.33}\times 100 = 142.99\ \%$

23

Groundwater Hydrology and Wells

UKPSC-AE-Agricultural-Engineering-Paper-II-2013

Q. 1 The development of a tube well should be started from :-

(A) Middle of screen (B) Top of screen

(C) Bottom of screen (D) Bottom of casing pipe

Q. 2 Unconfined aquifer is also called as :-

(A) Semi-confined aquifer (B) Water table aquifer

(C) Aquifuge (D) All of above

Q. 3 The difference between a shallow tube-well and a deep tube well is made on the basis of -

(A) Depth of aquifer (B) Depth of tube-well

(C) Position of water table (D) Type of aquifer

Q. 4 The wells in which the water levels remain at the water table level are :-

(A) Non-artesian wells (B) Artesian wells

(C) Confined wells (D) Tube wells

Q. 5 A geological formation, which has more water holding capacity but impermeable for flow of water, is known as:-

(A) Aquifer (B) Aquifuge

(C) Aquitard (D) Aquiclude

Q. 6 If all other parameters are fixed, then doubling the diameter of a tubewell increases the discharge by:-

(A) 75 % (B) 50 %

(C) 20 % (D) 10 %

Q. 7 The hydraulic diffusivity of an aquifer 'D' is expressed in terms of transmissibility 'T' and storage coefficient 'S' as:-

(A) D = T×S (B) D = T/S

(C) D = T+S (D) D = S/T

Q. 8 In terms of discharge 'D' the well loss will be proportional to:-

(A) D (B) $D^{1/2}$

(C) D^2 (D) D^{-2}

Q. 9 The product of hydraulic conductivity and aquifer thickness is called:-

(A) Storage coefficient (B) Specific yield

(C) Transmissibility (D) Drainable porosity

Q. 10 Line joining equal depths of water table are called:-

(A) Isobath lines (B) Isohyete lines

(C) Isobar lines (D) None of these

Q. 11 Cavity well is most suitable under the following condition:-

(A) Aquifer with fine sand and small thickness

(B) Aquifer with coarse sand and small thickness

(C) Aquifer with coarse sand and large thickness

(D) Aquifer with coarse sand and hard covering layer

Q. 12 For a well, yield per unit of draw down is called:-

(A) Specific yield (B) Specific capacity

(C) Well yield (D) Safe yield

Q. 13 The distance between the centre of a well and the outer point of the cone of depression is called :-

(A) Well radius (B) Aquifer radius

(C) Radius of draw down (D) Radius of influence

UKPSC J.E. Exam 2013 Agricultural Engineering-II

Q. 14 The common size of screen opening for a tube well is

(A) Less than 1.5 mm (B) 1.5 mm to 5.0 mm

(C) 5.0 mm to 7.5 mm (D) 10.0 mm to 15.0 mm

Q. 15 One cubic meter volume is equal to

(A) 10 litre (B) 1000 litre

(C) 100 litre (D) 10000 litre

Q. 16 The water table in an unconfined aquifer depends on

(A) Permeability of soil only (B) Pumpage from well only

(C) Area of recharge only (D) All of the above

Q. 17 Depth of deep tube wells is kept as, ...

(A) 20 m (B) 40 m

(C) 60 m (D) 100 m or more

Q. 18 The Planimeter measures

(A) Area (B) Volume

(C) Density (D) Land roughness

Q. 19 A well screen is used in

(A) Open well (B) Cavity well

(C) Tube well (D) None of the above

Q. 20 The recorded description of materials encountered in sequence throughout drilling with reference to the distance from the ground surface is known as,

(A) Well log (B) Well casing

(C) Well depth (D) Well description

MPPSC Assistant Agricultural Engineer 2013

Q. 21 q = kiA is mathematical representation of ....

(A) Theis law (B) Dupuit's law

(C) Darcy's law (D) Kirchoff's law

Q. 22 The dimensions of transmissibility of an aquifer are ....

(A) LT^{-1} (B) L^2T^{-1}

(C) L^3T^{-1} (D) None of these is correct

Q. 23 Gravel packing in a tube well is a thin layer of ....

(A) Coarse material (B) Coarse material with concrete

(C) Fine material (D) None of these is correct

RPSC AEn Pre Exam 2013 (Agricultural Engineering)

Q. 24 A tubewell with an average discharge of 250 litres per minute is used to irrigate one hectare wheat crop in 40 hours. What is the average depth of irrigation

(A) 40 mm (B) 25 mm

(C) 50 mm (D) 60 mm

Q. 25 When a channel receive supply from groundwater it is known as

(A) Ground water recharging (B) Effluent system

(C) Effective stream (D) None of the above is correct

Q. 26 An artesian aquifer is the one where

(A) Water surface under the ground is at atmospheric pressure

(B) Water is under pressure between two impervious layers

(C) Water table serves as upper surface of zone of saturation

(D) None of the above

Q. 27 Maximum practical drawdown in a well is the distance between

(A) Static water level and middle of well screen

(B) Static water level and top of well screen

(C) Static water level and pumping water level

(D) Static water level and end of well screen

OCS AEn Exam 2011 (Pre) – II (Agricultural Engineering)

Q. 28 A semi-confined aquifer is also called .... aquifer.

(A) Semi pervious (B) Artesian

(C) Leaky (D) Phreatic

Q. 29 Electrical resistivity method is an example of .... method of groundwater investigation.

(A) Seismographic (B) Geophysical

(C) Electrical (D) Direct

Q. 30 The coefficient of storage in ……… aquifer is due to compression of aquifer and expansion of contained water for reduced pressure while pumping

(A) Unconfined (B) Confined

(C) Semi-confined (D) Leaky

Q. 31 Geological formation having high porosity but capability to transmit water is called as

(A) Aquiclude (B) Aquifuse

(C) Aquitard (D) False aquifer

Q. 32 A well development technique through pumping that can also be used for cleaning a clogged strainer is known as

(A) Surging (B) Bailing

(C) Flushing (D) Backwashing

Q. 33 Down the hole hammer drill is a tool used in ……… drilling of well.

(A) Air percussion rotary (B) Reverse rotary

(C) Core drilling (D) Hydraulic drilling

Q. 34 Hydraulic resistance of semi-pervious layer of a semi-confined aquifer is reciprocal to ………

(A) Specific yield (B) Specific retention

(C) Leakage factor (D) Porosity

CGPSC State Engineering Services Exam (Agriculture Engineering) Part - II

Q. 35 A geological formation which is essentially impermeable for flow of water even though it may contain water in it's pores is called

(A) Aquifer (B) Acquifuge

(C) Aquitard (D) Aquiclude

(E) Both (A) and (B)

Q. 36 The single drawn or float glass has a uniform thickness of

(A) 3 – 4 mm (B) 3 – 5 mm

(C) 3 – 6 mm (D) 3 – 8 mm

(E) 3 – 7 mm

Q. 37 Specific yield is a property of

(A) Unconfined aquifer (B) Confined aquifer

(C) Semi Confined aquifer (D) Perched aquifer

(E) None of these

Q. 38 Water enters through bottom only in case of

(A) Filter points (B) Tube wells

(C) Cavity wells (D) Driven wells

(E) Dugged wells

Q. 39 Perched aquifer is a specific case of

(A) Unconfined aquifer (B) Confined aquifer

(C) Semi-confined aquifer (D) Leaky aquifer

(E) None of these

Q. 40 Diameter of observation well generally range from

(A) 15 – 45 cm (B) 10 – 15 cm

(C) 5 – 10 cm (D) 2.5 – 5 cm

(E) 20 – 25 cm

Q. 41 Piezometre is used to measure

(A) Hydrostatic pressure (B) Water table

(C) Hydraulic conductivity (D) Infiltration

(E) None of these

Q. 42 A part of the precipitation moves laterally through upper crust of the soil and returns to the surface at some location away from the point of entry into the soil

(A) Interflow (B) Seepage

(C) Percolation (D) Depression storage

(E) Infiltration

Q. 43 The dimension of intrinsic permeability is

(A) $M^oL^2T^{-1}$ (B) $M^oL^2T^o$

(C) $M^oL^2T^{-3}$ (D) $M^oL^2T^{-2}$

(E) $M^oL^2T^{-4}$

Q. 44 Pressure plate apparatus is used for measurement of soil moisture tension upto

(A) 15 bars (B) 10 bars

(C) 50 bars (D) 75 bars

(E) 100 bars

Q. 45 Curve number for impervious surface is

(A) 15 (B) 100

(C) 20 - 40 (D) 70

(E) 10

Q. 46 Time Domain Reflectometer (TDR) is used for measuring

(A) Soil moisture (B) Vapour pressure

(C) Soil salinity (D) Solar radiation

(E) Relative humidity

Q. 47 While constructing wells generally the bottom one third portion of aquifer is provided with well screen in

(A) Tube well constructed in gravity aquifer

(B) Dug well constructed in confined aquifer

(C) Tube well constructed in homogeneous unconfined aquifer

(D) Tube well constructed in a multi aquifer

(E) Open well

Q. 48 Darcy's law is applicable when Reynolds number is

(A) > 1 (B) < 1

(C) < 2 (D) > 2

(E) > 3

Q. 49 A Completely saturated aquifer that is bounded by a semi pervious layer and below by a layer that is either impervious or semi-pervious is

(A) Semi-confined aquifer (B) Confined aquifer

(C) Unconfined aquifer (D) Perched aquifer

(E) None of these

UKPSC A.En. Exam 2007 Paper – I

Q. 50 Hydraulic grade line is obtained by joining the points of

(A) Piezometric head (B) Equal elevation

(C) Equal velocity (D) Equal discharge

Q. 51 The non permeable aquifers which neither can keep water nor discharge it are called

(A) Aquiclude (B) Aquifuge

(C) Unconfined aquifer (D) Confined aquifer

UKPSC A.En. Exam 2007 Paper – II

Q. 52 The soil having more infiltration capacity is

(A) Sandy soil (B) Silty-loam soil

(C) Clayey soil (D) Silty-clay soil

Q. 53 Spacing between shallow tubewells should not be less than

(A) 50 m (B) 16 m

(C) 26 m (D) 36 m

Q. 54 Hydraulic resistance for the groundwater flow has the dimension of

(A) Ohm (B) Ohms/m

(C) Days (D) Metre

Q. 55 A cavity well has got

(A) No strainer (B) Brass strainer

(C) Bamboo strainer (D) Iron strainer

Q. 56 Which one of the following has got upper layer as impermeable ?

(A) Confined aquifer (B) Unconfined aquifer

(C) Both of the above (D) None of the above

Q. 57 The transmissibility(T) of an unconfined aquifer of thickness(B) and hydraulic conductivity(K) are related as

(A) T = B/K (B) T = K/B

(C) T = K×B (D) T = K + B

Q. 58 If a soil has the coefficient of permeability in two mutually perpendicular directions as K_1 and K_2, the effective permeability would be

(A) $K_1 + K_2$ (B) K_1 / K_2

(C) $K_1 \times K_2$ (D) $(K_1 \times K_2)^{1/2}$

Q. 59 Well development is the process of

(A) Lowering casing pipe

(B) Drilling of tube well

(C) Gravel packing

(D) Reversal of flow through the casing pipe

Q. 60 Gravel packing in the well is done to

(A) Decrease the effective diameter of the well

(B) Increase the effective diameter of the well

(C) Increase the screen length

(D) None of the above

Q. 61 The hydraulic conductivity of soils varies directly with

(A) Capillary porosity (B) Non-capillary porosity

(C) Total-porosity (D) Water holding capacity

Q. 62 Surging is used in

(A) Rainfall analysis (B) Terrace construction

(C) Well development (D) None of the above

Q. 63 For a water bearing-strata, the sum of specific yield and specific retention is equal to

(A) Void ratio (B) Bulk density

(C) Porosity (D) Specific gravity

Q. 64 Difference between the static water level and the pumping water level is called as

(A) Drawdown (B) Aquifer

(C) Well log (D) None of the above

Q. 65 Water held tightly to the surface of soil particles by absorption forces is called

(A) Capillary water (B) Hygroscopic water

(C) Gravitational water (D) None of the above

Q. 66 A well is made by driving or jetting a pipe and well point through the soil into an aquifer is known as

(A) Dug well (B) Driven well

(C) Drilled well (D) None of the above

Q. 67 In homogeneous artesian aquifer, the length of screen for the tube-well is kept

(A) 20 to 30 % of the thickness of aquifer

(B) 35 to 45 % of the thickness of aquifer

(C) 50 to 60 % of the thickness of aquifer

(D) 70 to 80 % of the thickness of aquifer

Q. 68 Theis method is applicable for well hydraulics under the condition

(A) Unsteady state flow in a confined aquifer

(B) Unsteady state flow in an unconfined aquifer

(C) Steady state flow in confined aquifer

(D) Steady state flow in unconfined aquifer

Q. 69 The piezometric head is higher than the ground surface, water flows out from the well which is called

(A) Deep well (B) Dug well

(C) Artesian well (D) Gravity well

ASCO (RPSC) Exam (Engineering) 2011

Q. 70 The ability of rock or soil to allow water to flow through it is called as

(A) Non point pollution (B) Irrigation

(C) Permeability (D) Water conservation

Q. 71 Most groundwater withdrawn in the world is used for

(A) Industry (B) Irrigation

(C) Drinking water (D) Swimming pools

Q. 72 The boundary between the saturated zone and the unsaturated zone is called the

(A) Water table (B) Aquifer

(C) Aquiline (D) Porosity

Q. 73 Excessive pumping in relation to groundwater well can cause

(A) The water table to decline (B) A cone of depression

(C) The well to go dry (D) All of these

Q. 74 Which of the following is the term given to water moving horizontally through soil eventually reaching river ?

(A) Percolation (B) Ground flow

(C) Interception (D) Through flow

Q. 75 What is difference between the saturated and the unsaturated zones of the groundwater ?

(A) The saturated zone has a higher porosity than the unsaturated zone

(B) The saturated zone has a lower porosity than the unsaturated zone

(C) The pore spaces in the saturated zone are completely full of water, the pore spaces in the unsaturated zone are not completely full of water

(D) The pore spaces in the saturated zone are not completely full of water, the pore spaces in the unsaturated zone are completely full of water

UKPSC A.En. Exam 2012 Paper - II

Q. 76 Boulton's well function is associated with semi-confined aquifer having

(A) Excessive yield (B) Normal yield

(C) Delayed yield (D) None of the above

Q. 77 Leakage factor and hydraulic resistance are important properties of the

(A) Unconfined aquifer (B) Semi-confined aquifer

(C) Confined aquifer (D) All of the above

Q. 78 Which of the following is a method of tube well drilling ?

(A) Percussion drilling (B) Rotary drilling

(C) Core drilling (D) All of the above

Q. 79 In alluvial formations the economical method of drilling is

(A) Direct rotary (B) Reverse rotary

(C) Cable tool (D) DTH

Q. 80 Which of the following aquifer is related to the term specific yield ?

(A) Unconfined aquifer

(B) Confined and unconfined aquifers

(C) Confined aquifer

(D) None of the above

Q. 81 Which of the following cases is associated with surging ?

(A) Gravel packing (B) Well drilling

(C) Well development (D) Sealing of saline water

Q. 82 Specific yield of clay formation varies from

(A) 1 – 10 % (B) 5 – 15 %

(C) 10 – 15 % (D) 15 – 20 %

Q. 83 Based on the sieve analysis of the aquifer material, the uniformity coefficient of a uniform material is

(A) More than 2 (B) Less than 2

(C) More than 4 (D) More than 8

Q. 84 Diameter of a dug well varies from

(A) 2 to 10 m (B) 5 to 10 m

(C) 10 to 20 m (D) 10 to 15 m

Q. 85 For development of well using compressed air, the capacity of the compressor is usually

(A) 1.5 kg/cm^2 (B) 20 kg/cm^2

(C) 25 kg/cm^2 (D) 30 kg/cm^2

Graduate Aptitude Test in Engineering - 2007

Q. 86 A recharge well of 300 mm diameter is constructed in a confined aquifer of 1000 m^2 d^{-2} transmissibility. From the top of impermeable bed, the water level in the well is 50 m and the height of constant water level is 40 m. The constant water level occurs at a distance of 150 m from the center of the well. The possible maximum recharge rate is

(A) 3.16 m^3 min^{-1} (B) 6.32 m^3 min^{-1}

(C) 9.48 m^3 min^{-1} (D) 12.64 m^3 min^{-1}

Q. 87 An unconfined aquifer is pumped at a constant rate of 10 L s^{-1}. Steady state draw downs measured at radial distances of 30 m and 60 m are 0.80 m and 0.70 m, respectively. Original Thickness of aquifer is 30 m. Transmissibility of the aquifer is

(A) 19 m^2 d^{-1} (B) 760 m^2 d^{-1}

(C) 952 m^2 d^{-1} (D) 982 m^2 d^{-1}

Graduate Aptitude Test in Engineering - 2008

Q. 88 The following data were collected from two piezometers P and Q located adjacent to each other in a groundwater basin.

Description	Piezometers	
	P	Q
R.L. of ground surface, m	220	220
Depth of piezometer, m	120	90
Depth to groundwater level from ground surface, m	60	50

(i) Hydraulic heads in m at P and Q respectively will be

(A) 100, 130 (B) 160, 170

(C) 60,40 (D) 170, 160

(ii) Hydraulic gradient between the piezometers is

(A) 0.33 (B) 3.00

(C) 0.94 (D) 1.06

Graduate Aptitude Test in Engineering - 2009

Q. 89 Water is recharged to an aquifer through a well of 200 mm diameter penetrating up to the base of the aquifer. The hydraulic conductivity of the aquifer is 19 m d^{-1}. During recharge, the water level in the well is 32 m from the base of the aquifer. A constant height of 27 m above the same base is obtained at a distance of 200 m from the well.

(i) If the aquifer is confined with a thickness of 20 m, the theoretical recharge rate in litre per second is

(A) 18.2 (B) 25.2

(C) 28.8 (D) 30.5

(ii) If the aquifer is unconfined, the theoretical recharge rate in litre per second is

(A) 10.5 (B) 17.7

(C) 26.8 (D) 30.5

Q. 90 For construction of a tubewell, the following formations were obtained from drilling: an unconfined aquifer between 12 m and 16 m and a confined aquifer between 30 m and 40 m below ground level. A horizontal centrifugal pump installed on the ground level can pump water from the constructed tubewell. Probable static water level from the ground surface in m is

(A) 5 (B) 12

(C) 16 (D) 30

Graduate Aptitude Test in Engineering - 2010

Q. 91 In an irrigation command area, the irrigation interval, gross application in an irrigation and the application efficiency are 20 days, 75 mm and 60% respectively. the soil in homogeneous with K = 0.9 m day^{-1}. The impermeable layer is at a depth of 7 m from the ground surface. The area is to be tile drained with tiles at a depth of 2 m below the ground surface. the maximum permissible steady state water table height mid-way between the drains, from the plane of the drain, is 1.2 m. Using the steady state approach of Hooghoudt, assuming an equivalent depth of 4.12 m, the drain spacing in m will be

(A) 115.25 (B) 131.75

(C) 146.25 (D) 186.35

Q. 92 A tube well in a confined aquifer has a diameter of 0.30 m for a certain yield, the radius of influence is 400 m. all conditions remaining the same. If the diameter of the well is doubled, then the percentage increase in the yield is

(A) 9.28 (B) 9.63

(C) 10.00 (D) 10.23

Graduate Aptitude Test in Engineering – 2011

Q. 93 In a falling head permeameter test, the initial head is 0.30 m. The head drops to 0.10 m in 40 min. The permeability of a soil sample, 0.06 m high and 50×10^{-4} m² in cross-sectional area, is found to be 1.0×10^{-6} m s^{-1}. The size of the stand pipe in m² is

(A) 1.82 (B) 1.82×10^{-1}

(C) 1.82×10^{-2} (D) 1.82×10^{-4}

Q. 94 A fully penetrating tubewell of 0.2 m diameter is operational in a confined aquifer of 20 m thickness. The hydraulic conductivity of the formation (K) is 20 m per day and the radius of influence is 1000 m. Water table is 50 m above the bottom of screen and the maximum drawdown is 10 m. Subsequently two more tubewells of the same size are installed in such a way that the three tubewells form an equilateral triangle of side 300 m. All three tubewells have equal discharge when operated independently. If all tubewells are operated simultaneously, the percent loss in the discharge of the first well is

(A) 20.73 (B) 35.75

(C) 48.85 (D) 61.90

Graduate Aptitude Test In Engineering - 2012

Q. 95 A 200 mm well fully penetrates a confined aquifer. After a long period of pumping at a rate of 1400 litres per minute, the drawdowns in the observation wells located at 25 m and 40 m from the pumping well are found to be 2.6 m and 1.9 m, respectively. The transmissivity of the aquifer in m² day^{-1} is

(A) 190 (B) 198

(C) 206 (D) 215

Graduate Aptitude Test in Engineering - 2014

Q. 96 A 150 mm diameter well is driven fully in a confined aquifer of thickness 12 m. When pumping is done at a constant rate of 150 L min^{-1} for 10 h, the steady-state draw-downs are found to be 2.20 m and 0.02 m at distances of 14 m and 50 m, respectively from the centre of the well. The hydraulic conductivity of the aquifer in m day^{-1} is ....

Q. 97 An imaginary surface obtained by joining the water levels in several observation wells driven in a confined aquifer is known as

(A) Phreatic surface (B) Piezometric surface

(C) Capillary fringe (D) Water table

Graduate Aptitude Test in Engineering - 2015

Q. 98 Dupit-Forchheimer assumptions are used for analyzing groundwater flow in

(A) Confined aquifers

(B) Leaky confined acquifers

(C) Unconfined aquifers

(D) Both confined and unconfined acquifers

Q. 99 An unsteady time-drawdown pumping test was conducted in a confined acquifer and the drawdown was measured with time in an observation well located at a certain distance away from the pumping well. Using these time-drawdown data, we can determine

(A) Transmissivity only

(B) Storage coefficient

(C) Specific storage only

(D) Both transmissivity & storage coefficient

Q. 100 An unconfined acquifer covering an area of 50 ha has a hydraulic conductivity of 20 m/day and specific yield of 12%. After a significant rainfall event, the water table rises from 17 m to 14.5 m below the ground level. Assuming no abstraction and outflow of groundwater during the recharge period, the amount of groundwater recharge contributed by the rainfall in m^3 is ....

Graduate Aptitude Test in Engineering - 2016

Q. 101 A fully penetrating tube well in a 30 m deep confined aquifer with hydraulic conductivity of 4×10^{-4} ms^{-1} has 50 Ls^{-1} discharge. The drawdown and radius of influence are 5 m and 250 m, respectively. Diameter of the tube well in mm is ……….

Graduate Aptitude Test in Engineering - 2017

Q. 102 Thickness of capillary zone above the water table varies

(A) Linearly with the pore size of soil

(B) Inversely with the height of water table

(C) Linearly with the height of water table

(D) Inversely with the pore size of soil

Q. 103 A cavity well is a tubewell which has

(A) Gravel pack and a strainer (B) A PVC strainer

(C) No strainer (D) A bamboo strainer

Q. 104 The sum of 'specific yield' and 'specific retention' for an unconsolidated geologic formation is equals to its

(A) Effective porosity (B) Total porosity

(C) Micro-porosity (D) Macro-porosity

Graduate Aptitude Test in Engineering - 2018

Q. 105 The drawdown at the well face of a 30 cm diameter fully penetrating well in a confined aquifer is 3 m. The radius of influence of this well is 3 km. For the same discharge and aquifer condition, the drawdown at 300 m away from the centre of the well is ………. m.

Q. 106 Three geometrically identical Lysimeters (A, B and C) installed in a paddy field have uniform initial ponding depths of 12 cm. After a week, the recorded water depths in A, B and C are 10.4 cm, 8.6 cm, and 7.0 cm, respectively. Rainfall received during this week is 15 mm, and the crop coefficient of paddy is 0.94. Lysimeter 'A' has a closed bottom with no plant. Lysimeter 'B' has an open bottom with no plant. Lysimeter 'C' has an open bottom with paddy plants. Neglecting the boundary effects and groundwater contribution in the Lysimeters, the weekly potential evapotranspiration will be ………. cm.

Graduate Aptitude Test in Engineering - 2019

Q. 107 A 20 cm diameter well fully penetrates a confined aquifer of thickness 20 m. After a long period of pumping at the rate of 1500 L/min, the steady drawdowns in the wells at 30 m and 50 m from the pumping well are found to be 2.0 m and 1.5 m, respectively. Assuming $\pi = 3.14$, the transmissivity of the aquifer in $m^2 s^{-1}$ is

(A) 0.244 (B) 14.676

(C) 352.224 (D) 880.560

Graduate Aptitude Test in Engineering - 2020

Q. 108 A tube well has a discharge of 40 m^3/h and operates daily for 20 h during irrigation season. The irrigation interval is 20 days and depth of irrigation is 8 cm. The command area of tube well in ha is ....

Q. 109 A 40 cm diameter tubewell is constructed in a 10 m thick confined aquifer having hydraulic conductivity of 25 m/day. The peizometric surface is observed to be 40 m high from the impervious stratum at the radius of influence of 500 m. The drawdown in the tubewell is 30 m. If the thickness of aquifer is doubled and diameter of tubewell is reduced to half, keeping all other parameters and conditions same, the change in discharge from the well is [Take $\pi = 3.14$]

(A) Increased by 83.72% (B) Decreased by 83.72%

(C) Increased by 82.28% (D) Decreased by 82.28%

Q. 110 Match the following items between Column I and Column II with the most appropriate combinations:

Column I	Column II
P. Properties of confined and semi- confined aquifers	1. Perched aquifer
Q. Property of an unconfined aquifer	2. Coefficient of storage
R. Property of semi-confined aquifer	3. Specific yield
S. A special case of an unconfined aquifer	4. Leakage factor

(A) P-3, Q-2, R-1, S-4 (B) P-2, Q-3, R-1, S-4

(C) P-2, Q-3, R-4, S-1 (D) P-3, Q-2, R-4, S-1

Q. 111 Match the following items between Column I and Column II with the most appropriate combinations:

Column I	Column II
P. Hydraulic Conductivity	I. Upper limit of moisture available to plant
Q. Permeability	2. All soil pores are filled with water
R. Viscosity	3. Soil capillarity
S. Surface Tension	4. Properties of fluid as well as soil
T. Saturation Capacity	5. Property of the medium
U. Field Capacity	6. Internal friction that brings about resistance to flow

(A) P-4, Q-5, R-6, S-3, T-1, U-2 (B) P-5, Q-4, R-6, S-3, T-2, U-1

(C) P-4, Q-5, R-3, S-6, T-2, U-1 (D) P-4, Q-5, R-6, S-3, T-2, U-1

Graduate Aptitude Test in Engineering - 2021

Q. 112 Pumping test is carried out at a constant discharge of 5400 L/min for 24 h in a main well of 30 cm diameter penetrated 25 m below the static water table. The water level in observation wells located at 30 m and 90 m away from the main well are lowered by 1.11 m and 0.53 m, respectively. Considering steady state flow condition, drawdown estimated in the main well in m is .... (Take $\pi = 3.14$)

Q. 113 Two fully penetrating wells are dug 1.4 km apart in a homogenous confined aquifer. The difference in their piezometric levels is 4.0 m. The groundwater flow is steady and unidirectional. If the aquifer has a hydraulic conductivity of 3.5 m/day and effective porosity of 40%, the time taken for water to move from one well to the other in days is..........

Answers Key

1	2	3	4	5	6	7	8	9	10
C	B	A	A	D	D	B	C	C	A
11	12	13	14	15	16	17	18	19	20
D	A	D	B	B	D	D	A	C	A
21	22	23	24	25	26	27	28	29	30
C	B	A	D	B	B	B	C	B	B
31	32	33	34	35	36	37	38	39	40
A	D	A	C	D	A	A	C	A	D
41	42	43	44	45	46	47	48	49	50
A	A	B	A	B	A	C	B	A	A
51	52	53	54	55	56	57	58	59	60
B	A	B	C	A	A	C	D	D	B
61	62	63	64	65	66	67	68	69	70
B	C	C	A	B	B	D	A	C	C
71	72	73	74	75	76	77	78	79	80
B	A	D	D	C	C	B	D	B	A
81	82	83	84	85	86	87	88	89	90
A	A	B	A	A	B	D	B, A	A, C	Bonus
91	92	93	94	95	96	97	98	99	100
B	B	D	A	D	1.67	B	C	D	150000
101	102	103	104	105	106	107	108	109	110
266.80	D	C	B	0.70	5	Bonus	20	Bonus	C
111	112	113							
D	4.12	56000							

Explanations

Q. 2

Water-bearing layers are classified according to their water-transmitting properties into aquifers, aquitards or aquicludes.

An **aquifer** is a water-bearing layer in which the vertical flow component is so small with respect to the horizontal flow component that it can be neglected. The groundwater flow in an aquifer is assumed to be predominantly horizontal.

The four types of aquifer are such as

1. **Confined aquifer** is a completely saturated aquifer bounded above and below by aquicludes. The pressure of the water in confined aquifers is usually higher than atmospheric pressure, which is why when a well is bored into the aquifer the water rises up the well tube, to a level higher than the aquifer.

2. **Unconfined aquifer** is a partly saturated aquifer bounded below by an aquiclude and above by the free water table or phreatic surface. At the free water table, the groundwater is at atmospheric pressure. In general, the water level in a well penetrating an unconfined aquifer does not rise above the water table, except when there is vertical flow.

3. **Leaky aquifer**, also known as a semi-confined aquifer, is a completely saturated aquifer that is bounded below by an aquiclude and above by an aquitard. If the overlying aquitard extends to the land surface, it may be partly saturated, but if it is overlain by an unconfined aquifer that is bounded above by the water table, it will be fully saturated. If there is hydrological equilibrium, the piezometric level in a well tapping a leaky aquifer may coincide with the water table.

4. **Perched Aquifer** is a special case of an unconfined aquifer. This type of aquifer occurs when an impervious or relatively impervious layer of limited area in the form of a lens is located in the water bearing unconfined aquifer. The water storage created above the lens is perched aquifer and its top layer is called perched water table.

Aquitard is a water-bearing layer in which the horizontal flow component is so small with respect to the vertical flow component that it can be neglected. The groundwater flow in an aquitard is assumed to be predominantly vertical. Examples of common aquitards are clays, shales, loam, and silt.

Aquiclude is a water-bearing layer in which both the horizontal and vertical flow components are so small that they can be neglected. The groundwater flow in an aquiclude is assumed to be zero. Common aquifers are geological formations of unconsolidated sand and gravel, sandstone, limestone, and severely fractured volcanic and crystalline rocks.

Q. 6

Refer explanation for Q. 92.

Q.24

Water to be applied in 40 hours = $250\times10^{-3}\times40\times60 = 600\ m^3$

Hence, average depth of irrigation = $600/(1\times10000) = 0.06$ m or 60 mm

Q. 86

Given

Well radius(r) = 0.15 m; Transmissibility (T) = 1000 m^2/d

H_2 = 50 m and H_1 = 40 m

R = 150 m

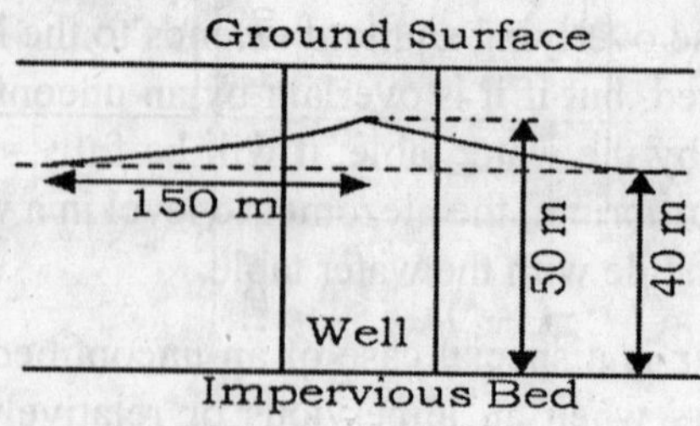

Applying formula

$$Q = \frac{2\pi T(H_2 - H_1)}{ln\frac{R}{r}}$$

$$= \frac{2 \times 3.14 \times 1000 \times (50-40)}{ln(\frac{150}{0.15})} = 9091.23 \text{ m}^3/\text{d} = 6.32 \text{ m}^3/\text{min}$$

Q. 87

R_1 = 30 m; S_1 = 0.80 m ⇒ h_1 = H-S_1 = 30 - 0.80 = 29.2m

R_2 = 60 m; S_2 = 0.70 m ⇒ h_2 = H-S_2 = 30 - 0.70 + 29.30m

Discharge (Q) = 10 l/ min = 864 m^3/ day

We know

$$Q = \frac{\pi K(h_2^2 - h_1^2)}{ln\left(\frac{R_2}{R_1}\right)}$$

$$\Rightarrow \quad K = \frac{864 \times ln\left(\frac{60}{30}\right)}{3.14 \times (29.3^2 - 29.2^2)}$$

$$\Rightarrow \quad K = 32.60 \text{ m/ day} \Rightarrow T = KB \Rightarrow 32.60 \times 30 = 978 \text{m}^2/\text{day}$$

$$\approx 982 \text{m}^2/\text{day}$$

Q. 88 (i)

Hydraulic head in piezometers P and Q are 160 m (220 – 60) and 170 m (220 – 50) respectively.

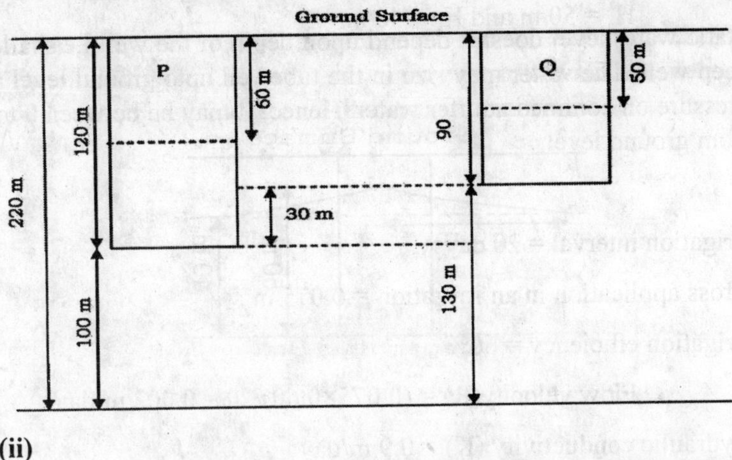

(ii)

$$\text{Hydraulic gradient } (i_h) = \frac{\textit{Difference between hydraulic heads in Piezometers}}{\textit{Difference between piezometer depth}}$$

$$= \frac{170-160}{120-90} = 0.33$$

Q. 89 (i)

K = 19 m/day = 2.2×10^{-4} m/s; R = 200 m; B = 20 m

H = 32 m; h – 27 m; r – 100/1000 = 0.1 m

According to Dupit theory

$$Q = \frac{2\pi KB(H-h)}{ln\left(\frac{R}{r}\right)}$$

$$= \frac{2\times3{,}14\times2.2\times10^{-4}\times(20-5)}{ln\left(\frac{200}{0.1}\right)}$$

= 18.2 l/s

(ii)

For unconfined aquifer

$$Q = \frac{\pi K(H^2-h^2)}{ln\left(\frac{R}{r}\right)}$$

$$= \frac{3,14 \times 2.2 \times 10^{-4} \times (32^2 - 27^2)}{ln\left(\frac{200}{0.1}\right)}$$

$$= 26.8 \text{ l/s}$$

Q. 90

Static water level doesn't depend upon depth of the well i.e.shallow or deep well. The water may rize in the tubewell upto ground level as per pressure on confined aquifer water. Hence, It may be between 0 to 12m from ground level.

Q. 91

Irrigation interval = 20 days

Gross application in an irrigation = 0.075 m

Irrigation efficiency = 60%

∴ Flow velocity(R) = (0.075×0.60)/20 = 0.002 m/day

Hydraulic conductivity (K) = 0.9 m/day

Static water table height (H) = 1.2 m

Equivalent depth (d) = 4.12 m

Applying Hooghoudt's Formula;

$$S^2 = \frac{8KHd}{R}$$

$$= \frac{8 \times 0.9 \times 4.12 \times 1.2}{0.002}$$

$$S = \sqrt{17798.4} = 133.41 \text{ m } 131.75 \text{ m}$$

Q. 92

Case 1

$$\text{Discharge from well } (Q_1) = \frac{2\pi kbS_w}{ln(R / r_w)}$$

Case 2

$$\text{Discharge from well } (Q_2) = \frac{2\pi kbS_w}{ln(R / 2r_w)}$$

$$\therefore \quad \text{Increase in discharge} = \frac{\frac{2\pi kbS_w}{ln(R/2r_w)} - \frac{2\pi kbS_w}{ln(R/r_w)}}{\frac{2\pi kbS_w}{ln(R/r_w)}} \times 100$$

$$= \frac{\dfrac{1}{ln\left(\dfrac{R}{2r_w}\right)} - \dfrac{1}{ln\left(\dfrac{R}{r_w}\right)}}{\dfrac{1}{ln\left(\dfrac{R}{r_w}\right)}} \times 100$$

$$= \frac{\dfrac{1}{ln\left(\dfrac{400}{2\times 0.15}\right)} - \dfrac{1}{ln\left(\dfrac{400}{0.15}\right)}}{\dfrac{1}{ln\left(\dfrac{400}{0.15}\right)}} \times 100 = 9.63\ \%$$

Q. 93

$h_0 = 0.30$ m; $h_1 = 0.10$ m; $A = 50\times10^{-4}$ m²; $t = 40$ min $= 2400$ seconds

$H = 0.06$ m; $K = 1\times10^{-6}$ m/ s

We know

$$K = \frac{aH}{At}\, ln\left(\frac{h_0}{h_1}\right)$$

$$\Rightarrow \quad 1\times10^{-6} = \frac{a\times 0.06}{50\times10^{-4}\times 2400}\, ln\left(\frac{0.30}{0.10}\right)$$

$$\Rightarrow \quad a = 1.82\times10^{-4}\ m^2$$

Q. 94

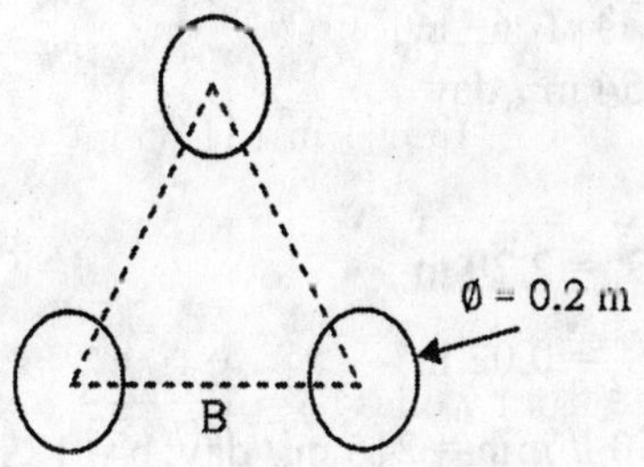

Discharge without interference (Q) = $\dfrac{2\pi KH(D-h_w)}{ln\left(\dfrac{R}{r_w}\right)}$

$$= \frac{2\times 3.14\times 20\times(20-10)}{ln\left(\dfrac{1000}{0.1}\right)}$$

$= 2727.37 \text{ m}^3/\text{ day}$

Discharge with interference (Q_i) $= \dfrac{2\pi KH(D-h_w)}{ln\left(\dfrac{R^3}{r_w B}\right)}$

$= \dfrac{2\times 3.14\times 20(20-10)}{ln\left(\dfrac{1000^3}{0.1\times 300^2}\right)}$

$= 2161.11 \text{ m}^3/\text{ day}$

$\therefore$ Percent decrease $= \dfrac{Q-Q_i}{Q}\times 100$

$= \dfrac{2728.75-2163.20}{2728.75}\times 100 = 20.76\ \%$

Q. 95

$R_1 = 25$ m; $S_1 = 2.6$ m

$R_2 = 40$ m; $S_2 = 1.9$ m

Discharge (Q) = 1400 l/ min = 2016 m³/ day

We know

$$Q = \frac{2\pi T(S_1-S_2)}{ln\left(\dfrac{R_2}{R_1}\right)}$$

$$\Rightarrow \quad T = \frac{2016\times ln\left(\dfrac{40}{25}\right)}{2\times 3.14\times(2.6-1.9)}$$

$$\Rightarrow \quad T = 215.54 \text{ m}^2/\text{ day}$$

Q. 96

$R_1 = 14$ m; $S_1 = 2.20$ m

$R_2 = 50$ m; $S_2 = 0.02$ m

Discharge (Q) = 150 l/ min = 216 m³/ day; b = 12 m

We know

$$Q = \frac{2\pi T(S_1-S_2)}{ln\left(\dfrac{R_2}{R_1}\right)}$$

$$\Rightarrow \quad T = \frac{216\times ln\left(\dfrac{50}{14}\right)}{2\times 3.14\times(2.20-0.02)}$$

$\Rightarrow$ $T = 20.08\ m^2/\ day$

Therefore $K = T/b = 20.08/12 = 1.67$ m/ day

Q. 100

We know

Groundwater storage = area of basin × water table rise× specific yield

$$= 50 \times 10^4 \times 2.5 \times 0.12$$

$$= 150000\ m^3$$

Q. 101

Discharge from well $(Q_1) = \dfrac{2\pi kbS_w}{ln(R/r_w)}$

$\Rightarrow$ $50 \times 10^{-3} = 2 \times 3.14 \times 4 \times 10^{-4} \times 30 \times 5/\ln(250/r_w)$

$\Rightarrow$ $r_w = 0.1334$ m

Therefore, Diameter of the well = 2×0.1334

$= 0.2668$ m or 266.80 mm

Q. 105

Discharge from well $(Q_1) = \dfrac{2\pi kbSw}{\ln(R/r_w)}$

For both the conditions

$$Q_1/Q_2 = 1 = \dfrac{\dfrac{2\pi kbSw_1}{\ln(R/r_w)}}{\dfrac{2\pi kbSw_2}{\ln(R/r)}}$$

$(\because Q_1 = Q_2 = 1)$

$$\Rightarrow \quad \dfrac{2\pi kbSw_1}{\ln(R/r_w)} = \dfrac{2\pi kbSw_2}{\ln(R/r)}$$

$$\dfrac{2\pi kb \times 3}{\ln(3000/0.15)} = \dfrac{2\pi kbSw_2}{\ln(3000/300)}$$

$\Rightarrow$ $Sw_2 = 0.70$ m

Q. 106

Evaporation = Initial ponding + Rainfall – Recorded water depth in Lysimeter A

$= 12 + 1.5 - 10.4 = 3.1$ cm

Transpiration = Recorded water depth in Lysimeter B – Recorded water depth in Lysimeter C
= 8.6 – 7.0 = 1.6 cm

Hence, Evapotranspiration measured from lysimeter (ET)= 3.1 + 1.6 = 4.7 cm

We know Crop Coefficient(kC) = ET/ETo

Reference crop Evapotranspiration (ETo) = 4.7/0.94 = 5 cm

Q. 107

According to Dupit theory

$$Q = \frac{2\pi T(H-h)}{ln\left(\frac{R}{r}\right)}$$

$$\Rightarrow \quad 1500/(1000\times60) = \frac{2\times3,14\times T(2-1.5)}{ln(50/30)}$$

$$\Rightarrow \quad T = 4.067\times10^{-3}\ m^2/s$$

Q. 108

Irrigation required = 8/20 = 0.4 cm/day or 0.004 m/day

Tube well discharge = 40×20 = 800 m^3/day

Thus, Command area = 800/0.004 = 200000 m^2 or 20 ha

Q. 109

$$\text{Discharge from well } (Q_1) = \frac{2\pi kbS_w}{ln(R/r_w)}$$

$$= \frac{2\pi\times25\times10\times30}{ln(500/0.20)} = 6020\ m^3/day$$

In second condition, The well drawdown is below the upper surface of aquifer. Hence, it is the condition of fully penetrating well.

$$\text{Discharge from well } (Q_2) = \frac{\pi k(2bH-b^2-h^2)}{ln(R/r_w)}$$

$$= \frac{\pi\times25(2\times20\times40-20^2-10^2)}{ln(500/0.10)} = 10138\ m^3/day$$

$$\text{Change in Discharge} = \frac{10138-6020}{6020}\times100 = 68.40\ \%\ \text{increased}$$

Note:- Answer given as per answer key is wrong. The Question Paper maker didn't keep in mind that the second condition is of fully penetrating well. Hence, He/she taken it as below.

$$\text{Discharge from well } (Q_2) = \frac{2\pi kbS_w}{ln(R/r_w)}$$

$$\text{Discharge after given assumptions } (Q_2) = \frac{2\pi \times 25 \times 20 \times 30}{ln(500/0.10)}$$

$$= 11060 \text{ m}^3/\text{day}$$

$$\text{Change in Discharge} = \frac{11060-6020}{6020} \times 100 = 83.72\ \% \text{ increased}$$

Q. 112

We know

$$Q = \frac{2\pi T(S_w - S_1)}{ln\left(\frac{R_1}{R_w}\right)}$$

For both the conditions

$$\Rightarrow \quad \frac{2\pi T(S_w - S_1)}{ln\left(\frac{R_1}{R_w}\right)} = \frac{2\pi T(S_w - S_2)}{ln\left(\frac{R_2}{R_w}\right)}$$

$$\Rightarrow \quad \frac{(S_w - S_1)}{ln\left(\frac{R_1}{R_w}\right)} = \frac{(S_w - S_1)}{ln\left(\frac{R_1}{R_w}\right)}$$

$$\Rightarrow \quad \frac{(S_w - 1.11)}{ln\left(\frac{30}{0.15}\right)} = \frac{(S_w - 0.53)}{ln\left(\frac{90}{0.15}\right)}$$

$$\Rightarrow \quad S_w = 4.12 \text{ m}$$

Q. 113

We know

Actual Velocity (v) = k×i/n

& Hydraulic gradient (i) = Δh/L

& Time taken to move the water from one well to another (t)

= L/v

= n×L^2/(k Δh)

Here,

k= 3.5 m/day; h = 4 m; n = 0.40; L = 1400 m;

Hence, t = 0.40×1400^2 /(3.5×4) = 56000 days

24

Water Lifting Devices, Pumps

24

Water Lifting Devices: Pumps

UKPSC Combined Assistant Engineering Exam 2013 Paper - I

Q. 1 A submersible pump may be

(A) Single stage pump (B) Multi stage pump

(C) Centrifugal pump (D) Both (A) and (B)

Q. 2 Closed impeller is suitable to handle the liquids of type

(A) Viscous (B) Non-viscous

(C) Dense (D) Thin

Q. 3 Static head of a pump is equal to

(A) Sum of suction head and delivery head

(B) Suction head

(C) Delivery head

(D) Difference of delivery head and suction head

Q. 4 Semi open impeller is used for

(A) Clean water (B) Hot water

(C) Sewage water (D) None of these

Q. 5 For which type of pump verticality of tube well is critical

(A) Jet pump

(B) Submersible pump

(C) Directly connected centrifugal pump

(D) Reciprocating pump

Q. 6 The parameter which is least important for a selection of pump is

(A) Discharge (B) Horse power

(C) Head (D) Quality of water

Q. 7 The relationship to determine break horse power in a centrifugal pump is

(A) Water horse power x pump efficiency

(B) Water horse power / pump efficiency

(C) Water horse power + pump efficiency

(D) Pump efficiency / water horse power

Q. 8 Water can be lifted from a depth beyond 45 meter by

(A) Centrifugal pump (B) Jet pump

(C) Submersible pump (D) None of the above

Q. 9 The hand pump used in houses is a type of pump

(A) Submersible pump (B) Reciprocating pump

(C) Centrifugal pump (D) Jet pump

Q. 10 Practically the suction head of a centrifugal pump should not exceed

(A) 8 meters (B) 12.5 meters

(C) 32.5 meters (D) 13.20 meters

Q. 11 Nozzle and venturi are main components of

(A) Centrifugal pump (B) Jet pump

(C) Reciprocating pump (D) Submersible pump

Q. 12 Reciprocating pumps are found to be best suited for

(A) Low discharge and high heads (B) Low discharge and low heads

(C) High discharge and high heads (D) High discharge and low heads

Q. 13 High lift centrifugal pump are generally

(A) Multi stage pumps (B) Used in deep wells

(C) Both (A) and (B) (D) Single stage pumps

Q. 14 In the centrifugal pump, the liquid enters through

(A) Eye of the impeller (B) Delivery pipe

(C) By-pass line (D) None of the above

Q. 15 Axial flow pump delivers large quantity of water at

(A) High heads (B) Medium heads

(C) Low heads (D) Zero heads

Q. 16 Which of the following type of impeller is used for centrifugal pump dealing with mud

(A) One side shrouded (B) Two side shrouded

(C) Double action (D) Open from both side

Q. 17 Filling water before its operation in a centrifugal pump is called

(A) Testing of pump (B) Energizing of pump

(C) Priming of pump (D) All of the above

Q. 18 Propeller pumps are used for

(A) Low head and low discharge (B) Low head and high discharge

(C) High head and low discharge (D) High head and high discharge

Q. 19 If a pump vibrates, the possible reasons may be

(A) Misalignment (B) Poor suction conditions

(C) Bearing failure (D) All of the above

Q. 20 In centrifugal pumps, the number of vanes made in impeller varies from

(A) 3 to 12 (B) 5 to 10

(C) 10 to 12 (D) 7 to 12

Q. 21 The vertical height to which the liquid can be raised by the pump above center line of pump is called

(A) Suction head (B) Total head

(C) Discharge head (D) Total differential head

Q. 22 Which of the following is not a type of impeller of a centrifugal pump

(A) Semi-open (B) Closed

(C) Open (D) High speed

Q. 23 Discharge of reciprocating pump is constant regardless of

(A) Temperature (B) Speed

(C) Delivery head (D) Suction head

Q. 24 A pump may fail to prime due to

(A) Air leakage in suction line or pump

(B) Foot value may be leaking

(C) Leakage in rotary shaft seals (D) All of the above

Q. 25 An "Archimedean screw", is device used for

(A) To tight a screw (B) To measure electric current

(C) To lift water (D) To lift earth material

Q. 26 Elbow joints are generally used to

(A) Increase the pressure (B) Reduce flow

(C) Increase flow (D) Reduce pressure

Q. 27 Which of the following is not the characteristic of centrifugal pump

(A) Capacity is inversely proportional to speed

(B) Horse power is the function of capacity, head and efficiency of pump

(C) Smooth and even flow

(D) Adaptive to high speed operation

Q. 28 Pump is operating but not delivering water, possible causes are

(A) Pump head is too high (B) Rotation of driver is incorrect

(C) Pump speed is too low (D) All of the above

Q. 29 The depth from which the liquid has to be lifted by pumps is known as

(A) Suction head (B) Delivery head

(C) Velocity head (D) Static head

Q. 30 Relation between specific speed (Ns), discharge (Q), Head (H) and velocity (N) for a pump is given by

(A) $Ns = \frac{N\sqrt[3]{Q}}{H^{4/3}}$ (B) $Ns = \frac{2N\sqrt{Q}}{H^{3/4}}$

(C) $Ns = \frac{N\sqrt[3]{Q}}{H^{3/4}}$ (D) $Ns = \frac{N\sqrt{Q}}{H^{3/4}}$

UKPSC-AE-Agricultural-Engineering-Paper-II-2013

Q. 31 Hydraulic ram is

(A) A centrifugal pump (B) An impulse pump

(C) A volute pump (D) A rotary pump

Q. 32 As compared to an electrical motor driven pump, the cost of lifting water by an engine driven pump is higher by

(A) 5 to 10% (B) 10 to 15%

(C) 15 to 25% (D) 25 to 40%

Q. 33 The S.I. unit of torque is

(A) Jule (B) Pascal

(C) Newton-Meter (D) Watt

Q. 34 The S.I. unit of pressure is

(A) Kg / cm^2 (B) mm of Hg column

(C) Pascal (D) Bar

Q. 35 The brake horse power of a centrifugal pump has a direct relationship with

(A) The speed of the impeller

(B) The square of the speed of impeller

(C) The cube of the speed of impeller

(D) None of these

Q. 36 The sum of static suction head and static discharge head is called

(A) Pressure head (B) Velocity head

(C) Total static head (D) Total discharge head

Q. 37 The capacity of a centrifugal pump varies with its speed 'N' as

(A) N^2 (B) $N^{1/2}$

(C) N (D) N^{-1}

UKPSC J.E. Exam 2013 Agricultural Engineering-II

Q. 38 Which of the following is not a water lift ?

(A) Persian wheel (B) Swing basket

(C) Counterpoise (D) Turbine

Q. 39 One horse power is equal to

(A) 724 watts (B) 746 watts

(C) 714 watts (D) 764 watts

APPSC AEES Agriculture Engineering Exam 2016

Q. 40 Self priming centrifugal pump has impellor of following type

(A) Backward curved vane impeller

(B) Radial vane impeller

(C) Forward curved vane impeller

(D) Combination of radial and forward curved vane impeller.

MPPSC Assistant Agricultural Engineer 2013

Q. 41 An indigenous water lifting device has the discharge of 4 lps. During a day of 8 hours 0.2 ha area is irrigated by this device. What is the average depth of water applied ?

(A) 57.6 cm (B) 5.76 cm

(C) 0.576 cm (D) None of these is correct

Q. 42 Which of the option is correct about the statements made below ?

(1) Submersible pump is placed under water.

(2) Submersible pump can give high discharge at high head.

(A) Statement (1) is true, but statement (2) is false.

(B) Statement (2) is true, but statement (1) is false.

(C) Both statements (1) and (2) are true.

(D) Both Statements (1) and (2) are false.

Q. 43 Specific speed of a pump expresses relationship between ....

(A) Pump speed and total head

(B) Pump speed and pump discharge

(C) Pump speed, pump discharge and total head

(D) Pump discharge, total head and efficiency

RPSC AEn Pre Exam 2013 (Agricultural Engineering)

Q. 44 Specific speed of pump is

(A) Directly proportional to the discharge

(B) Proportional to the square of the discharge

(C) Proportional to cube root of the discharge

(D) Proportional to square root of the discharge

Q. 45 The major component used in a centrifugal pump is

(A) Turbine (B) Impeller

(C) Valve inlet (D) All above

Q. 46 The pumps which are installed completely under water, including motor is called

(A) Displacement pump (B) Solar pump

(C) Submersible pump (D) None of the above

Q. 47 Propeller pump is used for

(A) Low head and low discharge (B) High head and low discharge

(C) Low head and high discharge (D) High head and high discharge

OCS AEn Exam 2011 (Pre) – II (Agricultural Engineering)

Q. 48 Which type of pump is most suitable to lift water from canals and rivers to open channels ?

(A) Reciprocating pump (B) Centrifugal Pump

(C) Propeller Pump (D) Mixed Flow Pump

CGPSC State Engineering Services Exam (Agriculture Engineering) Part - II

Q. 49 Which impeller has a shroud, or wall on one side only

(A) Closed impeller
(B) Semi-open impeller
(C) Open impeller
(D) Non-clog impeller
(E) Fixed Impeller

Q. 50 To get high efficiency, the lower end of the Archimedean screw water lift is to be submerged in water at an angle less than

(A) 10°
(B) 30°
(C) 45°
(D) 60°
(E) 90°

Q. 51 Maximum number of vanes on the impeller used

(A) 12
(B) 15
(C) 18
(D) 8
(E) 6

Q. 52 A centrifugal pump delivers 10 lps against total head of 7.5 m. The water power is

(A) 0.746 W
(B) 0.746 kW
(C) 7.46 W
(D) 74.6 W
(E) 735 W

Q. 53 If the speed of centrifugal pump is doubled the power required will increased by

(A) 8 times
(B) 6 times
(C) 4 times
(D) 2 times
(E) 12 times

Q. 54 D'Aubussion efficiency ratio is related to

(A) Centrifugal pump
(B) Reciprocating pump
(C) Propeller pump
(D) Hydraulic ram
(E) None of these

Q. 55 The vertical distance between pump and water surface should not exceed

(A) 5 m
(B) 7.5 m
(C) 10 m
(D) 12.5 m
(E) 14.5 m

UKPSC A.En. Exam 2007 Paper – I

Q. 56 In a pumping system, a foot valve is placed at the

(A) End of delivery pipe
(B) End of suction pipe
(C) Between suction pipe
(D) Inside Impeller of the pump

Q. 57 Insufficient discharge by the centrifugal pump is due to

(A) Speed too high
(B) Suction lift too low
(C) Impeller passages too clean
(D) None of the above

Q. 58 When two similar centrifugal pumps are operated in series, the discharge

(A) Remains constant
(B) Increases
(C) Decreases
(D) None of the above

Q. 59 For satisfactory operation of a centrifugal pump, the suction head should not exceed

(A) 1 – 2 m
(B) 10 – 12 m
(C) 4.5 – 6.0 m
(D) 3.0 – 4.0 m

Q. 60 Mechanical efficiency of centrifugal pump is about

(A) 40 %
(B) 60 %
(C) 90 %
(d) None of the above

Q. 61 The ratio between dynamic head and actual head developed by a pump is called

(A) Hydraulic efficiency
(B) Manometric efficiency
(C) Overall efficiency
(D) None of the above

Q. 62 Vertical turbine pump is adopted for

(A) High lift with high efficiency
(B) High lift with low efficiency
(C) Low lift with high efficiency
(D) Low lift with high discharge

Q. 63 Pump characteristic curve is the inter-relationship between

(A) Capacity – head

(B) Capacity – power

(C) Capacity – head, power and efficiency

(D) Capacity – head and efficiency

Q. 64 A centrifugal pump running at 1450 rpm discharges 20 lps at 30 m total head. The specific speed of the pump will be

(A) 12 (B) 16

(C) 20 (D) 24

Q. 65 If there is more difference between WHP and BHP the pump is considered as

(A) 100 % efficient (B) Less efficient

(C) More efficient (D) None of the above

Q. 66 The casing around an impeller wheel in a centrifugal pump is called

(A) Housing (B) Cylinder

(C) Volute (D) None of these

Q. 67 The head developed by a centrifugal pump increases with its speed 'N' as

(A) N (B) N^2

(C) $N^{1/2}$ (D) $N^{3/4}$

Q. 68 Gear pump is a

(A) Non-positive displacement pump

(B) Positive displacement pump

(C) Centrifugal pump

(D) None of the above

Q. 69 A pump requires 100 HP to pump water (with specific gravity of 1.0) at a certain capacity to a given elevation. What HP is required if the capacity and elevation remain the same but the fluid pumped has a specific gravity of 0.8 ?

(A) 60 HP (B) 80 HP

(C) 100 HP (D) 120 HP

Q. 70 A one cubic metre per second discharge is equal to

(A) 100 litre per second (B) 10 litre per second

(C) 1000 litre per second (D) 50 litre per second

Q. 71 One horsepower is equal to

(A) 76 kg-m/sec (B) 750 kg-m/sec

(C) 45 kg-m/sec (D) 450 kg-m/sec

Q. 72 At constant speed of positive displacement pumps how the discharge rate changes with increase in head ?

(A) Increases (B) Decreases

(C) Remains constant (D) None of the above

Q. 73 A single acting reciprocating pump has cross section area of piston as 60 cm^2, stroke length as 10 cm and crank speed as 20 rpm. The discharge will be

(A) 100 cm^3/sec (B) 100 cm^3/min

(C) 200 cm^3/sec (D) 100 cm^3/hour

Q. 74 Hydraulic ram is a device to pump

(A) Small quantity of water to higher elevation

(B) Large quantity of water to higher elevation

(C) Water from higher elevation to lower elevation

(D) None of the above

Q. 75 If there is a leakage in the pumping system suction line, what will happen to pump ?

(A) Discharge will increase (B) Consume more power

(C) Will not lift water (D) None of the above

Q. 76 If the discharge pressure is increased, then Net Positive Suction Head (NPSH) will

(A) Decrease (B) Increase

(C) Remain constant (D) None of the above

Q. 77 If the speed and diameter of the impeller are doubled, the capacity of pump will be increased

(A) 4 times (B) 8 times

(C) 16 times (D) 0 times

Q. 78 If 'S' is speed of impeller, the Brake Horse Power (BHP) of a centrifugal pump varies as

(A) BHP is proportional to S (B) BHP is proportional to S^2

(C) BHP is proportional to S^3 (D) BHP is proportional to S^4

Q. 79 Doubling the diameter of holes of screen for a tubewell in an unconfined aquifer, the increase in discharge is only by about

(A) 20 % (B) 10 %

(C) 50 % (D) 40 %

UKPSC A.En. Exam 2007 Paper – II

Q. 80 Practically, the suction lift of a centrifugal pump is

(A) Less than 2 atmosphere (B) Equal to 2 atmosphere

(C) More than 2 atmosphere (D) Less than 1 atmosphere

Q. 81 If the speed of a centrifugal pump is doubled, the discharge will be

(A) The same (B) Doubled

(C) Half (D) None of the above

Q. 82 When two centrifugal pumps are operated in series, the discharge of the pump will be

(A) Increased (B) Decreased

(C) Remain constant (D) None of the above

UKPSC A.En. Exam 2012 Paper – I

Q. 83 At a constant speed of positive displacement pumps, the discharge rate .... with increase in heat

(A) Increases (B) Decreases

(C) Remains constant (D) None of the above

Q. 84 Pumps based on method of energy imparted to the liquid are

(A) Positive displacement pump (B) Variable displacement pump

(C) Both (A) and (B) (D) None of the above

Q. 85 Where is the submersible pump located ?

(A) Over water in the well

(B) Submerged in water but above draw down level

(C) Submerged in water below draw down level

(D) Kept over ground

Q. 86 A suitable pump for a situation where flooding is a hazard is

(A) Turbine pump (B) Centrifugal pump

(C) Submersible pump (D) None of the above

Q. 87 If the speed of a pump is doubled then the head developed by it will be

(A) Double (B) Three times

(C) Four times (D) Half

Q. 88 A pump will not lift water, when

(A) Direction of impeller rotation is reverse

(B) Pump is not primed

(C) Speed of impeller is not high

(D) Both (A) and (B)

Q. 89 Pump efficiency is higher when

(A) Impeller blades are bent backward

(B) Impeller are with straight blades

(C) Impeller blades are ending radically

(D) All of the above

Q. 90 In a multistage centrifugal pump

(A) Two or more impellers are keyed to a single shaft

(B) Impellers are surrounded by guide vanes

(C) Pressure is built in steps

(D) All of the above

Q. 91 Elbows of large radius should be used to reduce friction and to have a flow of type

(A) Turbulent (B) Uniform

(C) Non-uniform (D) Gravity flow

Q. 92 In vertical turbine pump the shaft used made of stainless steel in the length of only

(A) 1 m (B) 2 m

(C) 2.5 m (D) 3 m

Q. 93 In a pumping system, a float valve is placed at the

(A) End of delivery pipe (B) End of suction pipe

(C) Between suction pipe (D) Inside impeller of the pump

Q. 94 The specific energy diagram is a plot between specific energy and flow depth for a constant

(A) Turbulence (B) Discharge

(C) Specific force (D) Specific yield

Q. 95 Gauge pressure at a point is equal to

(A) Absolute pressure plus atmospheric pressure

(B) Absolute pressure minus atmospheric pressure

(C) Vacuum pressure plus absolute pressure

(D) None of the above

Q. 96 The discharge capacity and efficiency of multistage pump is almost

(A) Same as for a single stage of the pump

(B) Less than as single stage pump

(C) More than single stage pump

(D) None of the above

Q. 97 In gear pump number of gears fitted in a housing are

(A) One (B) Two

(C) Three (D) Four

Q. 98 The impeller used for pumping sewerage is

(A) Open (B) Semi-open

(C) Non-clog (D) Closed

Q. 99 Which of these pumps are also called the axial flow pump ?

(A) Centrifugal pump (B) Gear pump

(C) Propeller pump (D) None of the above

Q. 100 The impeller of a centrifugal pump is generally made of

(A) Steel (B) Aluminum

(C) Cast Iron (D) Copper

Q. 101 In pipe flows, the frictional head loss is inversely proportional to

(A) Length of pipe (B) Square of length of pipe

(C) Diameter of pipe (D) Square of diameter of pipe

Q. 102 Which pump operates using the compressed air injected into water

(A) Vertical turbine pump (B) Air lift pump

(C) Submersible pump (D) Gear pump

Q. 103 The speed of a centrifugal pump N, and its discharge Q are related as

(A) $Q \propto N^2$ (B) $Q \propto N^3$

(C) $Q \propto N^{1/2}$ (D) $Q \propto N$

Q. 104 Power requirement of a centrifugal pump will become maximum, when increasing

(A) Diameter only (B) Speed of pump only

(C) Diameter and speed both (D) None of the above

Q. 105 Head developed by a centrifugal pump is proportional to

(A) Square of speed of pump (B) Speed of pump

(C) Cube of speed of pump (D) Square root of speed of pump

Q. 106 Open impeller is used in pumps to draw ....

(A) Clear water

(B) Water mixed with sand and clay

(C) Viscous liquid

(D) None of the above

Q. 107 If pump is not lifting required amount of water then the pump is suffering with the problem of

(A) Speed may be very high (B) Suction head may be very low

(C) Suction pipe may be chocked (D) None of the above

Q. 108 Water can be lifted from depth beyond 45 meters by

(A) Centrifugal pump (B) Submersible pump

(C) Propeller pump (D) Jet pump

Q. 109 Net Positive Suction Head (NPSH) .... with increase in discharge pressure

(A) Decreases (B) Increases

(C) Remains constant (D) None of the above

Q. 110 In a submersible pump, the motor and pump are such that

(A) Only pump is submerged

(B) Only motor is submerged

(C) Both motor and pump are submerged

(D) None of the above

UKPSC A.En. Exam 2012 Paper - II

Q. 111 A centrifugal pump has a discharge of 5400 litres per minute with total head of 20 metres.

If the overall efficiency of the pumpset is 64 percent the size of the required electric motor will be

(A) 24 HP (B) 37 HP

(C) 28 HP (D) 20 HP

Graduate Aptitude Test in Engineering - 2007

Q. 112 A centrifugal pump delivers 30 L s^{-1} of water against static suction and delivery heads o 6 m and 10 m respectively. The length and diameter of delivery pipe are 100 m and 100 mm respectively. The outlet of delivery pipe is submerged. Friction factor for the pipe is 0.03. If the minor losses in the delivery pipe amount to 1.0 m, pressure at delivery end of the pump is

(A) 327 kPa (B) 385 kPa

(C) 680 kPa (D) 984 kPa

Q. 113 When the water level in a well is at a depth of 7 m from the surface, the most suitable pump to lift water for irrigation is

(A) Submersible pump (B) Horizontal centrifugal pump

(C) Axial flow pump (D) Reciprocating pump

Graduate Aptitude Test in Engineering - 2008

Q. 114 A centrifugal pump delivers 0.03 $m^3 s^{-1}$ of water through a 100 mm diameter pipe to a vertical height of 14 m from the centerline of the pump. The pump is installed 6 m above the water level in the sump and the head loss in the pipeline is found to be 5 m of water. If the overall efficiency is 72%, the power required to run the pump will be

(A) 7.36 kW (B) 10.22 kW

(C) 8.18 kW (D) 5.89 kW

Q. 115 A double acting single cylinder reciprocating pump has a cylinder diameter of 150 mm and stroke 300 mm. Suction and delivery heads for the pump are 3 and 30 m respectively. If the pump delivers 0.01033 $m^3 s^{-1}$ of water at 60 rpm, the percentage slip is

(A) 97.43 (B) 1.57

(C) 2.57 (D) 0.0257

Graduate Aptitude Test in Engineering - 2009

Q. 116 A three stage centrifugal pump discharges water at a rate of 2400 litre per minute at a total head of 36 m. If the pump is directly connected to an electric motor operating at 1440 revolutions per minute, its specific speed will be

(A) 41.92 (B) 44.67

(C) 51.16 (D) 54.65

Q. 117 The inside diameter and stroke length of a single acting reciprocating pump are 120 mm and 400 mm respectively. The speed of the piston is 50 strokes per minute. the suction and delivery heads are 5 m and 10 m respectively. If the efficiency of both suction and delivery strokes is 60%, the actual power required in kW by the pump is

(A) 0.55 (B) 0.65

(C) 0.94 (D) 1.12

Graduate Aptitude Test in Engineering - 2010

Q. 118 A pump installed in an existing irrigation system deliver 3200 litres per minute flow at a total head of 60.0 m. The impeller diameter is 0.26 m and it is rotated at 1800 rpm. A motor with an output shaft power of 54 kW is required to drive the pump. The existing irrigation system, however, is modified in such a way that the discharge pressure requirement is reduced to 52.0 m while keeping the flow rate unchanged. If the existing pump is to be utilized, then to meet the new system requirement, the impeller diameter in m will be

(A) 0.24 (B) 0.25

(C) 0.26 (D) 0.27

Graduate Aptitude Test in Engineering - 2011

Q. 119 A 5 ha field under wheat crop is irrigated at 40% depletion of available moisture content. The field capacity, wilting point and bulk density of the soil in the field are 32% (on weight basis), 20% (on weight basis) and 1400 kg m^{-3} respectively. To irrigate the field in a day of 10 hours, a pump is used which lifts 270 m^3 of water per hour against a total head of 20 m.

(i) If the root zone depth is 0.8 m, the volume of water required to irrigate the field in m^3 will be

(A) 2700 (B) 3150

(C) 3600 (D) 4050

(ii) The pump is driven by an electric motor. If the pump, drive and motor efficiencies are 80%, 82% and 90% respectively, the required size of electric motor in horse-power will be

(A) 20.0 (B) 24.5

(C) 30.0 (D) 33.5

Graduate Aptitude Test in Engineering - 2012

Q. 120 A pumping device that combines the advantages of both centrifugal and reciprocating pumps is known as

(A) Air lift pump (B) Hydraulic ram

(C) Jet pump (D) Rotary pump

Graduate Aptitude Test in Engineering - 2014

Q. 121 The discharge of a single suction centrifugal pump operating against a total head of 12 m is 50 L s^{-1}. If the pump is directly connected to a motor operating at 1440 rpm, the specific speed of the pump will be.........

Q. 122 The brake power of a centrifugal pump having an impeller diameter of 200 mm is 1.86 kW. If the impeller is replaced with another impeller of 180 mm diameter, the brake power of the pump in kW will be ....

Graduate Aptitude Test in Engineering - 2016

Q. 123 Match the following

(P) Waste valve	(1) Jet pump
(Q) Plunger	(2) Centrifugal pump
(R) Foot valve	(3) Reciprocating pump
(S) Nozzle and venturi	(4) Hydraulic ram
(A) P-3, Q-2, R-4, S-1	(B) P-4,Q-2, R-3, S-1
(C) P-1, Q-3, R-4, S-2	(D) P-4, Q-3, R-2, S-1

Q. 124 A gear pump discharges 100 L min^{-1} against a system pressure of 15 MPa. The overall efficiency of the pump is 0.75. Input power to run the pump in kW is ....

Q. 125 The discharge of a centrifugal pump is 25 L s^{-1} against the delivery head of 10 m. The outlet of the delivery pipe is submerged. A 200 m long 100 mm diameter pipe is connected with the delivery end of the pump. The friction factor for the pipe is 0.03. The minor losses in the delivery pipe are 1 m. The pressure at the delivery end of the pump in kPa is ....

Q. 126 A water pumping system is being driven by a propeller type wind turbine having the power coefficient of 0.4. The total pumping head and rate of discharge are 20 m and 7.0 L s^{-1}, respectively. Mean wind velocity is 18 km h^{-1} and the density of air is 1.2 kg m^{-3}. The required diameter of the propeller in m is.........

Graduate Aptitude Test in Engineering – 2017

Q. 127 The pump used in high pressure orchard sprayer is

(A) Centrifugal pump

(B) Rotary pump

(C) Plunger type positive displacement pump

(D) Turbine pump

Q. 128 A six-stage centrifugal pump delivers 120L s^{-1} against a total of 510 m. If the design speed of this pump is 1450 rpm, the specific speed of the pump will be ....

Q. 129 A hydraulic system comprising of a pump and a single acting cylinder lifts 11 kN load. The pump flow rate is 25 L min^{-1} and its overall efficiency is 80%. The cylinder diameter is 80 mm and its efficiency is 90%. If the pressure drop in the hydraulic circuit is 500 kPa, the power required to drive the pump in kW will be ....

Graduate Aptitude Test in Engineering – 2018

Q. 130 Which one of the following is a **WRONG** statement?

(A) The head generated by a centrifugal pump at zero discharge is the 'shutoff head'.

(B) To avoid cavitation in centrifugal pumps, the net positive suction head should be more than the theoretical atmospheric pressure.

(C) According to the affinity laws of centrifugal pumps, the head varies with the square of the impeller speed.

(D) Most of the turbine pumps have operational characteristics closer to those of the centrifugal pumps than the propeller pumps.

Graduate Aptitude Test in Engineering – 2019

Q. 131 Water is flowing at a velocity of 1.6 m/s in a pipe of diameter 8 cm and length 100 m. Assuming the value of coefficient of friction for pipe, f = 0.005 and acceleration due to gravity, g = 9.81 m/s^2, the head loss (in meter) due to friction in the pipe is

(A) 1.28 (B) 2.28

(C) 2.78 (D) 3.26

Graduate Aptitude Test in Engineering – 2020

Q. 132 A water pumping system is driven by a horizontal axis multi-bladed wind turbine at a power coefficient (Cp) of 0.4. The total pumping head and discharge are 20 m and 15 L/s, respectively. The mean wind velocity is 8 m/s and the pump efficiency is 70%. The density of air and water are 1.2 and 1000 kg/m^3, respectively. If the transmission efficiency from wind turbine to the pump is 90%, the required diameter of the wind turbine in m is ....

Q. 133 At a speed of 1800 rpm, a centrifugal pump discharges 50 L/s at its best point of efficiency for a total head of 25 m. The specific speed of the pump in rpm is ....

Graduate Aptitude Test in Engineering – 2021

Q. 134 A gear pump has a displacement of 120 cm^3/rev and it runs at 1500 rpm against a system pressure of 18 MPa. If the torque efficiency of the pump is 90%, actual torque required to run the pump in N.m is .... (Take $\pi = 3.14$)

Q. 135 A pump, discharging water at a rate of 80 L/s, is used to irrigate 2 ha of land in 10 h. On irrigation, moisture content of the soil (on weight basis) in the root zone depth of 50 cm is increased from 18% to 30%. If bulk density of the soil is 1500 kg/m^3, water application efficiency in percent is ....

Answers Key

1	2	3	4	5	6	7	8	9	10
D	B	A	C	B	D	B	C	B	A
11	12	13	14	15	16	17	18	19	20
B	A	C	A	C	D	C	B	D	A
21	22	23	24	25	26	27	28	29	30
C	D	A	D	C	D	A	D	A	D
31	32	33	34	35	36	37	38	39	40
B	C	C	C	C	C	C	D	B	A
41	42	43	44	45	46	47	48	49	50
B	C	C	D	B	C	C	C	B	B
51	52	53	54	55	56	57	58	59	60
A	E	A	D	B	B	D	A	C	C
61	62	63	64	65	66	67	68	69	70
B	A	C	B	B	C	B	B	B	C
71	72	73	74	75	76	77	78	79	80
A	C	C	A	C	B	A	C	B	D

81	82	83	84	85	86	87	88	89	90
B	C	C	C	C	C	C	D	A	D
91	92	93	94	95	96	97	98	99	100
B	D	B	B	B	A	B	C	C	C
101	102	103	104	105	106	107	108	109	110
C	B	D	C	A	B	C	B	B	C
111	112	113	114	115	116	117	118	119	120
B	B	D	B	C	B	C	A	A, D	D
121	122	123	124	125	126	127	128	129	130
50	1.356	D	34	411.08	7.64	C	Bonus	1.529	B
131	132	133	134	135					
D	6.96	36	382.17	62.50					

Explanations

Q. 41

Water to be applied in 40 hours = $4\times10^{-3}\times8\times3600 = 115.2\ m^3$

Hence, average depth of irrigation = $115.2/(0.2\times10000) = 0.0576$ m or 5.76 cm

Q. 52

Water power = PQ = ρgHQ

$= 1000\times9.81\times7.5\times10\times10^{-3} = 735$ W

Q. 53

As per affinity law

$$(P_1/P_2) = (N_1/N_2)^3$$

Hence, doubling the speed of centrifugal pump, power will increase by 8 times.

Q. 60

Many medium and larger centrifugals offer efficiencies of 75 – 90% and even the smaller ones usually fall into the 50 – 70% range.

Q. 64

Given that

Discharge (Q) = 20 L/sec = $20\times10^{-3}\ m^3/s$

N = 1450 rev/min

Head (H) = 30 m

We know

$$\text{Specific speed} = \frac{N\sqrt{Q}}{H^{3/4}}$$

$$= \frac{1450\times\sqrt{20\times10^{-3}}}{(30)^{3/4}} = 16$$

Q. 69

Case- I

Water power = P×Q

$\Rightarrow$ 1000×gHQ = 100 $\Rightarrow$ gHQ = 1/10

Case- II

Water power = P×Q

= 800×gHQ

= 800×(1/10) = 80 hp

Q. 73

Discharge = 60×10×20/60 = 200 cm^3/sec

Q. 77

According to affinity law

Q α N D

Hence, if both D and N are doubled, Q will increase 4 times.

Q. 81

According to affinity law

Q α N

Hence, if N is doubled, Q will increase 2 times.

Q. 87

According to affinity law

H α N^2

Hence, if N is doubled, Q will increase 4 times.

Q. 111

Water power = PQ = ρgHQ

$= 1000 \times 9.81 \times 20 \times (5400 \times 10^{-3}/60) = 17658$ W or 23.67 hp

Hence, Required size of electric motor = 23.67/0.64 = 36.98 hp

Q. 112

Suction head (H_s) = 6 m; Delivery head (H_d) = 10 m;

Minor loss of head (H_l) = 1 m

We know

$$Q = AV$$

$$\Rightarrow \quad 30 \times 10^{-3} = \frac{\pi}{4} \times (0.1)^2 \times V$$

$$\Rightarrow \quad V = 3.822 \text{ m/s}$$

Frictional head loss through pipe (H_f) = $\dfrac{f \cdot LV^2}{2gD}$

$$= \frac{0.03 \times 100 \times (3.822)^2}{2 \times 9.81 \times 0.1}$$

$$= 22.34 \text{ m}$$

Therefore, total head (H) = $H_s + H_d + H_l + H_f$

$= 6 + 10 + 22.34 + 1 = 39.34$ m

∴ Pressure at delivery head = g H

$= 1000 \times 9.81 \times 39.34 = 385.92$ kPa

Q. 114

Suction head (H_s) = 6 m; Delivery head (H_d) = 14 m;

Loss of head in pipeline (H_l) = 5 m

Therefore, Total head (H) = $H_s + H_d + H_l + H_f$

$= 6 + 14 + 5 = 25$ m

∴ Pressure at delivery head (P) = ρ.g.H

$= 1000 \times 9.81 \times 25 = 245.25$ kPa

∴ Hydraulic Power = P × Q

$= 245.25 \times 0.03 = 7.36$ kW

∵ Overall efficiency (η) = 72 %

∴ Power required to run the pump = 7.36/0.72 = 10.22 kW

Q. 115

Theoretical discharge (Q_t) = 2ALN

$$= 2 \times \frac{\pi}{4} \times (0.150)^2 \times 0.3 \times \left(\frac{60}{60}\right)$$

$= 0.0106$ m³/ s

Actual discharge (Q_a) = 0.01033 m³/ s

$$\therefore \quad \text{Slip percentage (S)} = \frac{Q_t - Q_a}{Q_t} \times 100$$

$$= \frac{0.0106 - 0.01033}{0.0106} \times 100 = 2.55\ \%$$

Q. 116

Given that

Discharge (Q) = 2400 L/min = 40×10^{-3} m³/s

N = 1440 rev/min

Head (H) = 36/3 = 12 m (Because it is a three stage centrifugal pump)

We know

$$\text{Specific speed} = \frac{N\sqrt{Q}}{H^{3/4}}$$

$$= \frac{1440 \times \sqrt{40 \times 10^{-3}}}{(12)^{3/4}} = 44.67 \text{ rpm}$$

Q. 117

Given that

Inside diameter (D) = 120 mm = 0.12 m

Stroke length (L) = 400 mm = 0.4 m

Speed (N) = 50 strokes/min = 0.83 strokes/s

$$\text{Discharge (Q)} = \frac{3.14}{4} \times (0.120)^2 \times 0.4 \times 0.83$$

$= 0.00375$ m³/ sec

Suction head (H_s) = 5 m

Delivery head (H_d) = 10 m

∴ Suction pressure (P_s) =1000×9.81×5 = 49050 Pa

Delivery pressure (P_d) = 1000×9.81×10 = 98100 Pa

Efficiency of both suction and delivery stroke (η_s,η_d) = 60 %

Thus total power required (P) = Q(P_s + P_d)

= 0.00375×(49050 + 98100) = 0.551 kW

Here, efficiency is 60 %, thus

∴ Actual power required = 0.551/0.60 = 0.92 kW

Q. 118

In first case

Diameter of impeller (D_0) = 0.26 m; Discharge head (H_0) = 60 m

In second case

Diameter of impeller (D) = ? m; Discharge head(H_0) = 52 m

We know that

$$N\left(\frac{D}{D_0}\right)^2 = \frac{H}{H_0}$$

$$\Rightarrow \quad \left(\frac{D}{0.26}\right)^2 = \frac{52}{60}$$

$$\Rightarrow \quad D = 0.24 \text{ m}$$

Q. 119 (i)

Total volume of soil = 0.8×5×10⁴ = 4×10⁴ m³

Weight of the soil = 1400×4×10⁴ = 56×10⁶ kg

∴ Field Capacity of soil = 56×10⁴×0.32

= 17.92×10⁶ kg

Wilting Point = 56×0.20 = 11.2×10⁶ kg

Therefore, weight of available moisture content = 17.92×10⁶ – 11.2×10⁶

= 6.72×10⁶ kg

∴ Volume of available moisture content = (6.72)/1000 = 6720 m^3

Hence,

Volume of water required to irrigate the field = 6720×0.40

= 2688 m^3 ≈ 2700 m^3

(ii)

Power used to lift the water = Q×P = QρgH

= $(270\times10^3\times1\times9.81\times20)/(3600\times746)$

= 19.72 Hp

∵ Pump, drive and motor efficiency are 80%, 82% and 90% respectively thus size of electric motor required will be

= 19.72/(0.80×0.82×0.90) = 33.5 Hp

Q. 121

Specific speed of the pump (S_s)= $\frac{SQ^{1/2}}{H^{3/4}}$

$= \frac{1440\times(50\times10^{-3})^{1/2}}{(12)^{3/4}} = 49.9 \text{ rpm} \approx 50 \text{ rpm}$

Q. 122

According to Affinity's Law

$$\frac{P_1}{P_2} = \left(\frac{D_1}{D_2}\right)^3$$

$$\Rightarrow \quad P_2 = P_1 \times \left(\frac{D_2}{D_1}\right)^3$$

$$P_2 = P_1 \times \left(\frac{180}{200}\right)^3 = 1.356 \text{ kW}$$

Q. 124

Discharge (Q) = 100 litre/min = 0.0017 m^3/s

Hydraulic Power = Q×P

= 0.0017 ×15×1000000

= 25500 W = 25.5 kW

∵ Overall efficiency (η) = 75 %

∴ Power required to run the pump = 25.5/0.75

= 34 kW

Q. 125

We know

$$\text{Disharge} = AV$$

$$25\times10^{-3} = 3.14\times(0.1)^2\times V/4$$

$$V = 3.18 \text{ m/s}$$

Head loss due to friction $= fLV^2/(2gD)$

$= 0.03\times200\times(3.18)^2/(2\times9.81\times0.1)$

$= 31.02$ m'

Hence, Total head loss $= 10 + 31.02 + 1$

$= 42.02$ m

$\therefore$ Pressure $= \rho gh$

$= 1000\times9.81\times42.02 = 411.08$ kPa

Q. 126

Discharge (Q) = 7 L/s = 7×10^{-3} m^3/s

Density of the air (ρ) = 1.2 kg/m^3

Velocity of wind (v) = 18 km/h = 5 m/s

Cross sectional area (A) = $\pi D^2/4 = 0.785D^2$

Wind power = $K(\rho\, Av^3/2)$

$= 0.4\times(1.2\times0.785\, D^2\times5^3/2) = 23.55\, D^2$ W

Hydraulic power for water lifting $= QP = Q\,\rho_w gh$

$= 7\times10^{-3}\times1000\times9.81\times20 = 1374.4$ W

Here, power generated by wind turbine is being utilized in lifting of water.

Hence, $23.55\, D^2 = 1374.4$

$\Rightarrow$ $D = 7.64$ m

Q. 128

Given that

Discharge (Q) = 120 L/min = 120×10^{-3} m^3/s

N = 1450 rev/min

Head (H) = 510/6 = 85 m (Because it is a six stage centrifugal pump)

We know

$$\text{Specific speed} = \frac{N\sqrt{Q}}{H^{3/4}}$$

$$= \frac{1450 \times \sqrt{120 \times 10^{-3}}}{(85)^{3/4}} = 17.94$$

Q. 129

Load on cylinder piston = 11 kN

Cross section area of cylinder = $\pi D^2/4 = 3.14 \times (0.08)^2/4 = 5.024 \times 10^{-3}\ m^2$

Pressure of oil due to load = $11/(0.90 \times 5.024 \times 10^{-3})$ = 2432.77 kPa

($\because$ Cylinder efficiency = 0.90)

Total pressure drop into hydraulic system = 2432.77 + 500

= 2932.77 kPa

Volumetric Flow rate = 25 L/min or $4.17 \times 10^{-4}\ m^3/s$

We know

Power required to drive the pump

= pressure drop into hydraulic system × Flow rate

= $2932.77 \times 4.17 \times 10^{-4}/0.80$ = 1.529 kW

Q. 131

Head loss = $4fLv^2/(2gD)$

= $4 \times 0.005 \times 100 \times 1.6^2/(2 \times 9.81 \times 0.08)$ = 3.26 m

Q. 132

$$\text{Power generated by windmill (P)} = \frac{1}{2} \times C_p\, \rho A V^3$$

$$= \frac{1}{2} \times 0.4 \times 1.2 \times A \times 8^3$$

= 122.88A W

Power used to lift the water = 122.88A×0.90×0.70 = 77.41A W

Water Horse Power = 1000×9.81×20×15/1000= 2943 W

Hence,

$$77.41A = 2943$$

$$\Rightarrow \quad A = 38.02 \text{ m}^2$$

$$\Rightarrow \quad d = (38.02\times4/3.14)^{1/2} = 6.96 \text{ m}$$

Q. 133

Given that

Discharge (Q) = 50 L/sec = 50×10^{-3} m^3/s

N = 1800 rev/min

Head (H) = 25 m

We know

$$\text{Specific speed} = \frac{N\sqrt{Q}}{H^{3/4}}$$

$$= \frac{1800\times\sqrt{50\times10^{-3}}}{(25)^{3/4}} = 36 \text{ rpm}$$

Q . 134

Discharge (Q) = 120 cm^3 rev^{-1} or $120\text{x}10^{-6}\text{x}1500$ m^3 / min

Pressure(P) =18 MPa=$18\text{x}10^6$ Pa

Speed (N) = 1500 rpm

We know

$$\text{Work done} = 2\pi NT = P\times Q$$

$$\Rightarrow \quad 2\times\pi\times1500\times T = 18\text{x}10^6\text{x}120\text{x}10^{-6}\times1500$$

$$\text{Torque(T)} = 343.95 \text{ N.m}$$

$\because$ Torque efficiency of the pump = 90%

$\therefore$ Actual torque to operate the pump = 343.95/0.90 =382.17 N.m

Actual torque required to run the pump = 343.95/0.90 = 382.17 N-m

Q. 135

Total volume of soil = $0.50\times2\times10^4 = 1\times10^4$ m^3

Weight of the soil = $1500\times1\times10^4 = 15\times10^6$ kg

Hence, weight of available moisture content = $15\times10^6(0.30 - 0.18)$

$= 1.8\times10^6$ kg

∴ Volume of available moisture content = $(1.8\times10^6)/1000$

$= 1.8\times10^3\ m^3$

Hence,

Volume of water applied to irrigate the field = $80\times10\times3600/1000$

$= 2.8\times10^3\ m^3$

⇒ Water application efficiency = $1.8\times10^3 \times100 / 2.8\times10^3 = 62.50\ \%$

25

Irrigation Engineering

UKPSC Combined Assistant Engineering Exam 2013 Paper - I

Q. 1 When flow parameters change with time but remain constant with space the flow is called as:-

(A) Unsteady uniform flow (B) Steady uniform flow

(C) Steady non-uniform flow (D) Unsteady non-uniform flow

Q. 2 Which of the following is not a type of fluid flow?

(A) Steady (B) Direct

(C) Laminar (D) Compressible

Q. 3 Froude number for a triangular channel is given as:-

(A) $F = V\sqrt{2}/\sqrt{gY}$ (B) $F = V/\sqrt{gY}$

(C) $F = V/\sqrt{g/Y}$ (D) $F = V^2/\sqrt{gY}$

Q. 4 A laminar flow is associated to.-

(A) Low flow velocity

(B) Moving of fluid without lateral mixing

(C) Absence of eddies\

(D) All of the above

Q. 5 A uniform flow generally occurs in:-

(A) An inclined channel of constant bed slope

(B) A level channel

(C) A channel with negative slope

(D) All of the above

Q. 6 The Manning's roughness (n) decreases with:-

(A) Decrease in flow depth (B) Increase in flow depth

(C) Decrease in turbulence (D) Both (B) and (C)

Q. 7 At the critical flow condition the velocity head is equal to:-

(A) Half of the flow depth

(B) Half of the hydraulic depth

(C) Two times of hydraulic radius

(D) None of the above

Q. 8 Bernoulli's equation is applicable for:-

(A) Frictionless steady flow (B) Incompressible flow

(C) Ideal flow (D) Both (A) and (B)

Q. 9 A trapezoidal weir is said to be cipolettie when its side slope is:-

(A) $\tan(\theta/2) = 1/4$ (B) $\theta° = 30°$

(C) $\sin\theta = 1/4$ (D) None of the above

Q. 10 Frictional head loss in a pipe is directly proportional to:-

(A) Length of pipe (B) Diameter of pipe

(C) Both of the above (D) None of the above

Q. 11 Current meters are used to determine:-

(A) Flow type (B) Flow turbulence

(C) Flow direction (D) Flow velocity

Q. 12 Seepage force always acts in:-

(A) Upward direction (B) Downward direction

(C) In the direction of flow (D) In the opposite direction of flow

Q. 13 Under open flow conditions, the hydraulic jump is formed when the Froude number is:-

(A) Equal to one (B) More than one

(C) Less than one (D) None of the above

Q. 14 At critical depth of flow :-

(A) Specific energy is minimum (B) Specific energy is maximum

(C) Both (A) and (B) (D) None of the above

Q. 15 The flow in the open channel may be characterized as laminar flow when Reynold's number is:-

(A) Re < 500 (B) Re > 2000

(C) Re > 1000 (D) Re > 800

Q. 16 The velocity of water through a circular channel will be maximum when the depth of water is:-

(A) 0.95 times the diameter of channel

(B) 0.81 times the diameter of channel

(C) 0.67 times the diameter of channel

(D) 0.50 times the diameter of channel

Q. 17 The hydraulic radius for a fully flowing pipe of 1 meter diameter will be:-

(A) 1 meter (B) 0.5 meter

(C) 0.25 meter (D) 2.0 meter

Q. 18 Manning's formula for mean flow velocity is:-

(A) V = C (B) V = $\sqrt{8gRS}$ /f

(C) V = 1/n ($R^{3/2} S^{1/2}$) (D) V = 1/n ($R^{2/3} S^{1/2}$)

where V = velocity of flow, m/s

R = hydraulic radius, m

S = slope

n, f, g, c = constants

Q. 19 Manning's formula is used to determine flow velocity in which type of flow:-

(A) Rapidly varied flow (B) Non-uniform flow

(C) Uniform flow (D) Critical flow

Q. 20 At critical state of flow the kinetic energy of flow is equal to:-

(A) One fourth of its potential energy

(B) One half of its potential energy

(C) One third of its potential energy

(D) One fifth of its potential energy

Q. 21 The hydraulic radius (R) of a channel in terms of its wetted perimeter (P) and cross sectional area (A) is given on:-

(A) R = A x P (B) R = P / A

(C) R = A / P (D) $R = A^2 / P$

Q. 22 Hydraulic jump formula is based on:-

(A) Froude number of flow (B) Reynold number of flow

(C) Critical number of flow (D) Specific number of flow

Q. 23 The section factor (Ƶ) for uniform flow condition is given as:-

(A) $Ƶ = AR^{1/3}$ (B) $Ƶ = AR^{5/3}$

(C) $Ƶ = AR^{3/5}$ (D) $Ƶ = AR^{2/3}$

A = Cross sectional area

R = Hydraulic radius

Q. 24 All natural channels are generally:-

(A) Non-prismatic (B) Prismatic

(C) Ideal (D) None of the above

Q. 25 At critical flow condition, the section factor (zc) can be expressed as:-

(A) $zc = Q^2 / g$ (B) zc = Q/ g

(C) zc = Q / g (D) $zc = Q / g^{1.5}$

Where Q = discharge

g = Acceleration due to gravity

Q. 26 Most efficient channel section is:-

(A) Triangular (B) Trapezoidal

(C) Rectangular (D) Semi circular

Q. 27 For maximum discharge in a circular channel the depth of flow should be equal to:-

(A) 0.70 times diameter (B) 0.91 times diameter

(C) 0.82 times diameter (D) 1.15 times diameter

Q. 28 Hydraulic jump is based on the principle of:-

(A) Energy conservation (B) Velocity conservation

(C) Momentum conservation (D) Turbulence conservation

Q. 29 For a given cross sectional area the hydraulic radius will increase with:-

(A) Increase in wetted perimeter

(B) Decrease in wetted perimeter

(C) Increase in turbulence of flow

(D) Decrease in turbulence of flow

Q. 30 If Froude number of flow is 'F' inflow and outflow depths are y_1 & y_2, then the jump formula will be:-

(A) $\frac{y_2}{y_1} = \frac{1}{2}\left(\sqrt{1+8F^2} - 1\right)$ (B) $\frac{y_2}{y_1} = \frac{1}{2}\left(\sqrt{1+8F^2} - 1\right)$

(C) $\frac{y_2}{y_1} = \frac{1}{2}\left(\sqrt{1+8F^2} - 4\right)$ (D) $\frac{y_2}{y_1} = \frac{1}{2}\left(\sqrt{1+8F^2} - 4\right)$

UKPSC-AE-Agricultural-Engineering-Paper-II-2013

Q. 31 Which of the following is the most correct match?

(A) Manning's formula – Estimation of seepage loss in a canal

(B) Kennedy theory = Open channel flow

(C) Darcy-weisbach formula = Friction factor

(D) Seepage meter = Design of open channel

Q. 32 If the hydraulic radius is more than 5.0 m, then for designing an open ditch, the value of Manning's 'n' should be taken as :-

(A) 0.002 – 0.030 (B) 0.04 – 0.045

(C) 0.035 – 0.040 (D) 0.040 – 0.045

Q. 33 In light sandy soil, with slope 1 to 5 %, the maximum water application rate of a sprinkler irrigation system is :-

(A) 2.0 cm/hr (B) 2.5 cm/hr

(C) 3.0 cm/hr (D) 3.5 cm/hr

Q. 34 The water present in the soil between field capacity and permanent wilting point is known as:-

(A) Ground water (B) Gravitational water

(C) Hygroscopic water (D) Available water

Q. 35 Concentration of nitrate in unpolluted surface water is generally less than :-

(A) 1 ppm (B) 10 ppm

(C) 15 ppm (D) 20 ppm

Q. 36 Water loss through seepage in earthen channels amounts to about :-

(A) 15 % (B) 20 %

(C) 25 % (D) 30 %

Q. 37 Evaporation pan coefficient of class 'A' pan is generally taken as :-

(A) 7.0 (B) 0.7

(C) 0.007 (D) 5.5

Q. 38 The ratio of total volume of water delivered to a crop, to area on which it has been spread is called :-

(A) Duty (B) Delta

(C) Critical depth (D) Specific volume

Q. 39 The unit of delta is :-

(A) cm (B) $cm^{1/2}$

(C) cm^2 (D) cm^3

Q. 40 As per BIS recommendation, the shape of lined channels should be :-

(A) Semi-circular (B) Trapezoidal

(C) Parabolic (D) Rectangular

Q. 41 The maximum rate of emission by drippers under normal working condition is :-

(A) 15 lit /hr (B) 10 lit /hr

(C) 5 lit /hr (D) 20 lit /hr

Q. 42 The effect of gravity force upon the state of flow is represented by :-

(A) Reynold's number (B) Darcy's number

(C) Kennedy's number (D) Froude's number

Q. 43 Sub irrigation is useful in a situation where :-

(A) Saline water is used for irrigation

(B) Soil is heavy clay to permit capillary rise

(C) Capillary movement is rapid around roots

(D) No hard pan is present

Q. 44 Chute spillway is superior to drop inlet spillway when :-

(A) Large discharge is required

(B) No opportunity to provide temporary storage

(C) Channel is to be brought down a steep slope

(D) All of these

Q. 45 Sheet of water which overflows a weir is called :-

(A) Head (B) Static head

(C) Nappe (D) None of these

Q. 46 Most common method for irrigating the potato crop is :-

(A) Sprinkler method (B) Check-basin method

(C) Furrow method (D) Border method

Q. 47 Infiltration rate during irrigation :-

(A) Increases (B) Decreases

(C) Both (A) and (B) (D) None of these

Q. 48 Darcy's law is valid under the conditon of :-

(A) Laminar flow with Reynold's number > 10.0

(B) Reynold's number < 1.0

(C) Newtonian flow

(D) Steady uniform flow

Q. 49 The sprinkler system operates at following pressure range :-

(A) 1 - 4 kg /cm^2 (B) 4 - 8 kg /cm^2

(C) 4 - 12 kg /cm^2 (D) 8 - 12 kg /cm^2

Q. 50 The average water extraction of plants from the first 25% of soil depth is:-

(A) 50 % (B) 45 %

(C) 40 % (D) 35 %

Q. 51 At field capacity, soil moisture tension is :-

(A) 10 - 30 centi bar (B) 10 - 30 bar

(C) 7 - 32 centi bar (D) 7 - 32 bar

Q. 52 Froude's number is the ratio of the :-

(A) Inertia force to the shearforce

(B) Inertia force to the viscous force

(C) Inertia force to the gravity force

(D) Viscous force to the gravity force

Q. 53 The drip irrigation is very effective for :-

(A) Water conservation

(B) Management of large cultivable land for high production

(C) Energy conservation

(D) All of the above

Q. 54 Irrigation interval of drip irrigation varies from :-

(A) 3 - 4 days (B) 1 - 3 days

(C) 7 days (D) 15 days

Q. 55 The device used to apply liquids in a crop is called:-

(A) Duster (B) Ridger

(C) Sprayer (D) Pump

Q. 56 Emitters are used in:-

(A) Sprinkler irrigation (B) Surface irrigation

(C) Drip irrigation (D) Sub-surface irrigation

Q. 57 The form of water which is not available to plants, is:-

(A) Unavailable water (B) Hygroscopic water

(C) Capillary water (D) Free water

Q. 58 In irrigation, tensiometer is used for:-

(A) The measurement of soil moisture content

(B) The measurement of soil moisture tension

(C) Irrigation scheduling

(D) Both (B) and (C) are correct

Q. 59 At optimum moisture content, the dry density of soil will be:-

(A) Minimum (B) Maximum

(C) Average (D) Zero

Q. 60 Movement of irrigation water from the soil surface into the soil is called:-

(A) Percolation (B) Interflow

(C) Infiltration (D) Water intake

Q. 61 Most efficient rectangular cross section of a channel is defined as follows when hydraulic radius is 'R' and flow depth is 'd':-

(A) R = d (B) R = d/2

(C) R = 2d (D) R = 4d

Q. 62 The empirical method for computing consumptive use of a crop using the mean monthly temperature and day light hours is:-

(A) Thornth waite (B) Blanney criddle

(C) Hargreaves (D) Lowry Johnson

Q. 63 If the Froude's number during a hydraulic jump is 1-1.7, then it is called:-

(A) Oscillating jump (B) Weak jump

(C) Steady jump (D) Undular jump

Q. 64 Usually, the discharge capacity of minor distributary in India is :-

(A) > 1000 lit/sec (B) < 750 lit/sec

(C) 0.75 – 5.5 cumec (D) 4 – 8.5 cumec

Q. 65 Horizontal to vertical side slope in a cipolette weir is:-

(A) 1 : 2 (B) 1 : 4

(C) 2 : 1 (D) 4 : 1

Q. 66 If 'H' is the head over the crest of a triangular weir, the discharge will be proportional to :

(A) $H^{1/2}$ (B) $H^{3/2}$

(C) $H^{5/2}$ (D) H^2

UKPSC J.E. Exam 2013 Agricultural Engineering-II

Q. 67 Which method should be used for irrigation in undulating fields ?

(A) Flooding method of irrigation

(B) Furrow method of irrigation

(C) Check basin method of irrigation

(D) Sprinkler irrigation method

Q. 68 Which of the following is not a method of surface irrigation ?

(A) Free flooding irrigation (B) Border irrigation

(C) Critical irrigation (D) Check basin irrigation

Q. 69 Which of the following is not a method of irrigation ?

(A) Drip irrigation (B) Accelerated irrigation

(C) Sub-surface irrigation (D) Pitcher irrigation

Q. 70 Due to evaporation, the liquid water is transferred to

(A) Atmosphere (B) River

(C) Pond (D) Well

Q. 71 Over irrigation results in

(A) Increasing crop yield

(B) Maintaining soil fertility

(C) Economic utilization of water

(D) None of the above

Q. 72 Efficient irrigation to the crops will provide

(A) Increased crop yield (B) Increased water logging

(C) Increased salt accumulation (D) None of the above

Q. 73 Water conveyance efficiency is related to which of the following ?

(A) Water stored in the root zone

(B) Water needed in the root zone

(C) Uniformity of the water applied

(D) None of the above

Q. 74 Water distribution efficiency indicates

(A) Water supplied at source

(B) Water delivered at field

(C) Average depth of water stored during irrigation

(D) None of the above

Q. 75 Which of the following is not related to the 'duty' of water

(A) Quantity of water used (B) Area to the irrigated

(C) Height of crop (D) None of the above

Q. 76 Which of the following is not a component of a drip irrigation system ?

(A) Head unit

(B) Perforated pipe

(C) Water conveyance and distribution pipes

(D) Drippers

Q. 77 In which country the drip irrigation system was originally developed ?

(A) India (B) Pakistan

(C) Israel (D) China

Q. 78 Which of the following is not a component of a sprinkler irrigation system ?

(A) Pump (B) Sprinkler nozzle

(C) Main and lateral pipe lines (D) Drip nozzles

Q. 79 The water requirement of a crop is mainly estimated for

(A) Crop planning on the farm (B) Determining transpiration

(C) Determining water loss (D) None of the above

Q. 80 The permissible mean flow velocity in an earthen channel in sandy loam soil is

(A) 40 cm/sec (B) 60 cm/sec

(C) 80 cm/sec (D) 100 cm/sec

Q. 81 Border slope for clay to clay loam soils is kept generally

(A) 0.05 to 0.20 % (B) 0.20 to 0.40 %

(C) 0.25 to 0.65 % (D) none of the above

Q. 82 Chezy's formula is used to determine

(A) Flow direction (B) Flow velocity

(C) Flow duration (D) None of the above

Q. 83 The satisfactory operating pressure for a sprinkler system ranges between

(A) 1.0 – 1.5 kg/ cm^2 (B) 1.5 – 2.5 kg/ cm^2

(C) 2.5 – 3.5 kg/ cm^2 (D) 3.5 – 4.0 kg/cm^2

Q. 84 Weirs are classified based on ....

(A) Shape of the notch only (B) Type of the crest only

(C) Both (A) and (B) (D) None of the above

Q. 85 Minor irrigation schemes include

(A) Land reclamation schemes

(B) Surface water irrigation scheme

(C) Ground-water irrigation scheme

(D) Both (B) and (C)

Q. 86 The flow rate from a canal outlet depends on

(A) Size of outlet opening only (B) Head of water above outlet only

(C) Both the above (D) None of the above

Q. 87 Ponds are small water storage structures used for

(A) Water supply only (B) Ground-water recharge only

(C) Irrigation only (D) All of the above

Q. 88 Water required by the border strip is 42.86 cubic meter, if the discharge available is 36 m^3/hr, then the time required to irrigate the border will be

(A) 4.19 hour (B) 3.19 hour

(C) 2.19 hour (D) 1.19 hour

Q. 89 Emitter pipes are classified based on pressure rating into

(A) Two classes (B) Three classes

(C) Four classes (D) Five classes

Q. 90 The material of emitting pipes should be resistant to

(A) Fertilizers (B) Chemicals

(C) Both (A) and (B) (D) None of the above

Q. 91 Water application rate from sprinkler system depends on

(A) Discharge of sprinkler only (B) Sprinkler spacing only

(C) Lateral movement speed (D) All of the above

Q. 92 Sprinklers are lubricated by

(A) Oil (B) Grease

(C) Oil and grease (D) Water

Q. 93 Culturable command area for a minor irrigation project is

(A) More than 10000 hectare (B) More than 5000 hectare

(C) Upto 5000 hectare (D) Upto 2000 hectare

Q. 94 Sprinkler irrigation method can be preferably used

(A) In places with high wind velocity.

(B) For clay soil with very low infiltration rate.

(C) For soils with heavy texture and low infiltration rate.

(D) For soils with light texture and high infiltration rate.

Q. 95 If the net amount of irrigation is 10 cm and the field irrigation efficiency is 80%, the gross amount of water applied is

(A) 10.66 cm (B) 12.50 cm

(C) 8.0 cm (D) 800 cm

Q. 96 The depth of a most efficient rectangular channel is how many times of its width ?

(A) 2.0 times (B) 0.75 times

(C) 0.5 times (D) 1.5 times

Q. 97 The relation between the area irrigated and the quantity of water used to irrigate it for the purpose of maturing its crop is known as,

(A) Irrigation frequency (B) Irrigation period

(C) Irrigation requirement (D) Duty of water

Q. 98 The pressure head in a sprinkler system is converted into velocity head at the

(A) Sprinkler head (B) Sprinkler jet

(C) Nozzle (D) Riser head

Q. 99 The period for which water is supplied from canal to a crop is called

(A) Base (B) Duty

(C) Delta (D) Irrigation period

Q. 100 The factor which affects irrigation efficiency is

(A) Design of irrigation system (B) Degree of land preparation

(C) Skill and ease of irrigation (D) All of the above

Q. 101 To carry irrigation channels across the road, the appropriate structure is

(A) Culvert (B) Siphon

(C) Spillway (D) All of the above

Q. 102 In an open channel water flows under the influence of

(A) Gravity force (B) Inertia force

(C) Suction force (D) None of the above

Q. 103 For a given cross sectional area of channel, the hydraulic radius decreases with

(A) Increase in wetted perimeter (B) Decrease in wetted perimeter

(C) Decrease in flow velocity (D) Increase in flow velocity

Q. 104 The common water distribution pattern used in canal command areas of northern India is

(A) Shejpali system (B) Tail to tail system

(C) Warabandi system (D) All of the above

Q. 105 Rate of increase in crop yield per unit increase in water application is known as,

(A) Water use efficiency (B) Irrigation response

(C) Project efficiency (D) Gross water requirement

Q. 106 Strip cropping systems has alternate rows of

(A) Erosion creating and erosion permitting crops

(B) Erosion permitting and erosion tolerating crops

(C) Erosion permitting and erosion resisting crops

(D) None of the above

Q. 107 According to U.S. Dept. of Agriculture, the size of particles of coarse sand varies between

(A) 2.0 to 1.0 mm (B) 1.0 to 0.5 mm

(C) 0.5 to 0.25 mm (D) None of the above

Q. 108 The drip irrigation system is very effective in case of

(A) Paddy fields (B) Sugarcane fields

(C) Orchards (D) All of the above

Q. 109 If the depth of flow in a river is 'd', then the average flow velocity occurs at a depth equal to

(A) 0.4 d from free surface (B) 0.6 d from free surface

(C) 0.2 d from free surface (D) At the free surface

Q. 110 Advantages of minor irrigation projects are

(A) Low cost (B) Easily manageable

(C) Higher irrigation efficiency (D) All of the above

Q. 111 In drip irrigation system, the drippers are equipped on

(A) Main line (B) Sub-main line

(C) Lateral line (D) All of the above

Q. 112 Drip irrigation is also known as

(A) Sub-surface irrigation (B) Complete irrigation

(C) Trickle irrigation (D) Surge irrigation

Q. 113 The term extensively used to express the performance of a complete irrigation system or components of the system is called

(A) Irrigation effect (B) Irrigation efficiency

(C) Irrigation facility (D) Irrigation availability

Q. 114 Uniformity of water application in sprinkler irrigation is affected by

(A) Water temperature (B) Wind velocity

(C) Soil characteristics (D) Time of irrigation

Q. 115 Which of the following is a measure for reclamation of alkali soil ?

(A) Green manuring with dhaincha in summer

(B) Use of gypsum

(C) Provision of drainage system

(D) All of the above

Q. 116 Under irregular topographic field conditions which of the following strip cropping system is preferred ?

(A) Contour strip cropping system

(B) Field strip cropping system

(C) Buffer strip cropping system

(D) Either of the above

Q. 117 In open channels, the flow is due to

(A) Pressure difference (B) Elevation difference

(C) Velocity difference (D) All of the above

Q. 118 Estimation of design discharge is done in

(A) Hydraulic design (B) Structural design

(C) Hydrologic design (D) Final design

Q. 119 Chezy's formula is represented as,

(A) $V = R\sqrt{CS}$ (B) $V = C\sqrt{RS}$

(C) $V = \sqrt{CRS}$ (D) $V = S\sqrt{RC}$

Where : V = Flow velocity

C = Coefficient

R = Hydraulic radius

S = Slope of land

Q. 120 The critical state of flow occurs when the value of Froude number is

(A) More than one (B) Less than one

(C) Equal to one (D) All of the above

APPSC AEES Agriculture Engineering Exam 2016

Q. 121 The ratio of water stored in the root zone of the plants to water delivered to the field is termed as

(A) Water application efficiency (B) Water storage efficiency

(C) Water distribution efficiency (D) Water use efficiency

Q. 122 Sprinkler irrigation is recommended in:

(A) Highly undulating area

(B) Soils having high infiltration rate

(C) Irrigation source is a tube well

(D) In all the above situations

MPPSC Assistant Agricultural Engineer 2013

Q. 123 When flow changes from .... hydraulic jumps occurs.

(A) Critical to Super critical (B) Sub critical to Super critical

(C) Super critical to Sub critical(D) None of these is correct

Q. 124 Safety against overturning of water retaining structure can be ensured by ....

(A) Increasing uplift pressure

(B) Reducing restoring moment

(C) Reducing self weight of structure

(D) None of these is correct

Q. 125 .... Water measuring device do not need free flow conditions at outlet.

(A) H-flume (B) Rectangular weir

(C) Parshall flume (D) All options are correct

Q. 126 The planning of conjunctive use of surface and ground water is done for ....

(A) Facilitating irrigation with poor quality ground water

(B) Augmenting canal supplies

(C) Combating water logging in the area\

(D) All options are correct

Q. 127 The discharge in distributaries is usually ....

(A) Less than 2.5 cumec (B) Less than 30 cumec

(C) More than 30 cumec (D) None of these is correct

Q. 128 Sprinkler irrigation system's performance is considered satisfactory, if the uniformity coefficient is equal to or more than ....

(A) 55 % (B) 65 %

(C) 75 % (D) 85 %

Q. 129 In order to determine moisture content of soil at permanent wilting point in the field which of the following crop is sown as indicator crop ?

(A) Safflower (B) Sunflower

(C) Alfalfa (D) None of these is correct

Q. 130 .... is the component of drip irrigation system, which requires major cost.

(A) Drippers (B) Pumping system

(C) Filtering system (D) Laterals

Q. 131 The drip irrigation system is designed by considering .... per cent variation in pressure in lateral.

(A) 1 (B) 5

(C) 10 (D) 20

Q. 132 The emitter discharge in drip irrigation system is decided based on ....

(A) Water holding capacity of soil

(B) Evapotranspiration demand of crop

(C) Discharge capacity of pumping system

(D) Infiltration capacity of soil

Q. 133 In India .... sprinkler system is most commonly used.

(A) Central Pivot (B) Portable

(C) Permanent (D) Raingun

Q. 134 Drip irrigation system gives better yields of crops because ....

(A) It is costly.

(B) It operates at low pressure.

(C) It provides nearby equal water to entire filed at shorter interval.

(D) It was invented in Israel.

RPSC AEn Pre Exam 2013 (Agricultural Engineering)

Q. 135 Drip fertigation depicts the process of

(A) Applying fresh irrigation water

(B) Applying soluble fertilizer through drip pipes while irrigation fields

(C) Applying combination of drip and sprinkler method

(D) All above

Q. 136 The ratio of total cropped land and the actual net cultivated and in a given catchment is known as

(A) Cropped area (B) Cropping intensity

(C) Sown area (D) All of the above

Q. 137 The capacity of a farm pond is computed by adopting which formula

(A) Trapezoidal rule (B) Clark's formula

(C) Rational formula (D) All of the above

Q. 138 Water use efficienc, is simply defined as

(A) Crop yield per unit volume of irrigation water utilized

(B) Percent of water actually consumed by the plants

(C) Ratio of water delivered at field and the water remained unutilized at the end

(D) None of the above

Q. 139 Depth of irrigation water required to bring the soil moisture content of a given soil upto its field capacity is called

(A) Hygroscopic water (B) Equivalent moisture

(C) Soil moisture deficiency (D) Peculiar water

Q. 140 The velocity of flow in an open channel can be measured with the help of

(A) V-notch (B) Infiltrometer

(C) Water current meter (D) None of the above

Q. 141 The difference between advance and recession curve in border strip denotes

(A) Overlapping time (B) Opportunity time

(C) Time of concentration (D) Depth of ponding

Q. 142 Depth of water required for a crop excluding rainfall is called

(A) Base (B) Duty

(C) Delta (D) None of these

Q. 143 Which pairs of components are most vital while designing a grassed waterway ?

(A) Permissible flow velocity and type of grass

(B) Type of soil and volume of water

(C) Permissible velocity of flow and most economical channel cross section

(D) All above are equally important

Q. 144 Command area can be described as extent of area, where

(A) Encompassing all network of small and big natural channels \

(B) Total area whose runoff water passes through a single outlet point

(C) Both (A) and (B)

(D) None of the above

Q. 145 The ratio of quantity of water delivered to the field and water diverted from the source is known as

(A) Water conveyance efficiency

(B) Water application efficiency

(C) Water use efficiency

(D) None of the above

Q. 146 The conveyance efficiency of canal usually depends upon

(A) Length of canal, soil types, and the permeability

(B) Only canal flow rates

(C) Gates of canal

(D) None of the above

OCS AEn Exam 2011 (Pre) – II (Agricultural Engineering)

Q. 147 Chezy's roughness coefficient C and Manning's roughness coefficient n are related as:

(A) $C = R^{1/6}/n$ (B) $C = 1/n$

(C) $C = R^{2/3}/n$ (D) $C = n/R^{2/3}$

Q. 148 The most efficient hydraulic cross-section of trapezoidal channel with side slope of 1 horizontal to vertical is (b : bed width ; d = depth of water flow)

(A) $b = 2d/\sqrt{3}$ (B) $b = d/\sqrt{3}$

(C) $b = 2d$ (D) $b = 2\sqrt{3}\,d$

Q. 149 Uniformity coefficient C_u of 100 percent for sprinkler irrigation indicates :

(A) 100 % application efficiency

(B) 100 % storage efficiency

(C) Uniform depth of application

(D) Both (A) and (B)

Q. 150 In an open channel flow, the critical depth at which maximum discharge will occur for a specific energy of 2 m is .... m.

(A) 3 (B) 1.33

(C) 2 (D) 4

CGPSC State Engineering Services Exam (Agriculture Engineering) Part - II

Q. 151 The gravity dam of base width 'b' the maximum value of eccentricity 'e' to be safe against tension is

(A) b/8 (B) b/4

(C) b/3 (D) b/6

(E) b/7

Q. 152 Soil moisture extraction pattern from root

(A) 40 : 20 : 30 : 10 (B) 10 : 20 : 30 : 40

(C) 40 : 10 : 20 : 30 (D) 40 : 20 : 10 : 30

(E) 40 : 30 : 20 : 10

Q. 153 Riser pipes and collection trays are component of

(A) Perimeter watering (B) Drip Irrigation

(C) Over head sprinkler (D) Boom watering

(E) Both (B) and (D)

Q. 154 Maize is a

(A) Soil depleting crop (B) Soil building crop

(C) Soil maintaining crop (D) Soil erosion crop

(E) Soil transport crop

Q. 155 A water sample has Na, Ca, Mg with quantity 4, 5 and 7 meq/lit respectively. Then what will be the SAR of the water

(A) 1.6 (B) 0.6

(C) 2.6 (D) 6.3

(E) 4.6

Q. 156 Only water available to plants is

(A) Available water (B) Hygroscopic water

(C) Capillary water (D) Gravitation water

(E) None of these

Q. 157 Ratio of cultivated land for a particular group to the total cultivable command area

(A) Time factor (B) Cultivable command area

(C) Intensity of irrigation (D) Capacity factor

(E) None of these

Q. 158 The seepage line may be defined as

(A) Line above which there is no hydrostatic pressure

(B) Line below which there is no hydrostatic pressure

(C) Line above which there is no hydrostatic pressure and below which there is hydrostatic pressure

(D) Both (A) and (B)

(E) Both (B) and (C)

Q. 159 Penman method is used to compute

(A) Consumptive use of crop

(B) Evaporation from open water surface

(C) Water requirement of crops

(D) Irrigation water requirement of crops

(E) None of these

Q. 160 Sub-irrigation is a situation where

(A) Saline water is used for irrigation

(B) Soil is heavy clay to permit high capacity rise

(C) Capillary movement in the root zone

(D) No hard pan is present below the root zone

(E) None of these

Q. 161 The conjunctive use of water in a basin means

(A) Combined use of water for irrigation and hydro power generation

(B) Use of water by co-operative farmers

(C) Use of water for irrigating both rabi and kharif crops

(D) Combined use of surface and ground water resources

(E) None of these

Q. 162 Sprinkler irrigation system performance is considered satisfactory when the minimum value of the uniformity coefficient is

(A) 75 % (B) 80 %

(C) 90 % (D) 60 %

(E) 50 %

Q. 163 For satisfactory performance of sprinkler system except gravity fed, it's operating pressure should not be

(A) Less than 1 kg/ sq. cm (B) More than 5 kg/ sq. cm

(C) Between 1 -2 kg/ sq. cm (D) Between 2.5 – 3.5 kg/ sq. cm

(E) Between 5 – 10 kg/ sq. cm

Q. 164 Which of the following mineral particle size is classified as silt according to ISSS

(A) 0.05 – 0.1 mm (B) 0.1 – 0.2 mm

(C) 0.002 – 0.05 mm (D) 0.06 – 0.1 mm

(E) 1.0 – 1.5 mm

Q. 165 Tillage practices for moisture conservation includes

(A) Stubble tillage (B) Deep tillage

(C) Surface Tillage (D) Zero Tillage

(E) Puddling

Q. 166 A commonly used indigenous water lift in the rice growing belts of eastern and southern India is

(A) Swing basket (B) Archmedean screw

(C) Water wheel (D) Persian wheel

(E) Chain pump

Q. 167 The discharge over a Rectangular-notch is proportional to

(A) $H^{5/2}$ (B) $H^{-3/2}$

(C) $H^{3/2}$ (D) $H^{-5/2}$

(E) $H^{1/3}$

UKPSC A.En. Exam 2007 Paper – I

Q. 168 The critical depth (d_c) of flow in an open rectangular channel with width(b) the discharge (Q) is expressed as

(A) $d_c - \{Q^2/(b^2g)\}^{1/3}$ (B) $d_c = \{Q^2/(b^2g)\}^{1/2}$

(C) $d_c = \{b^2g/Q^2\}^{1/2}$ (D) $d_c = \{Q^2/(bg^2)\}^{1/3}$

Where 'g' is the acceleration due to gravity

Q. 169 Hydraulic radius of rectangular open channel section of a width 'b' and depth of water 'y' will be

(A) by/(b + 2y) (B) by/(b + y)

(C) by/(2b + y) (D) $by^2/(b + 2y)$

Q. 170 In case of rectangular channel if 'D' is the depth of flow and 'V' is the flow velocity, then the specific energy of flow is written as

(A) $D + V^2/2g$ (B) $D^2 + V/2g$

(C) $D^2 + V^2/2g$ (D) $D - V^2/2g$

Q. 171 When upper surface of flow is exposed to atmosphere, it is called as

(A) Pipe flow (B) Open channel flow

(C) Submerged flow (D) None of the above

Q. 172 The upper most line of seepage in an Earth dam is also called as

(A) Water pressure line (B) Hydraulic gradient line

(C) Phreatic line (D) Parabolic line

Q. 173 Manning's formula for mean flow velocity is

(A) $V = C(RS)^{1/2}$ (B) $V = (8gRS/f)^{1/2}$

(C) $V = R^{2/3}S^{1/2}/n$ (D) $V = R^{3/2}S^{1/2}/n$

Where, V – velocity; R - hydraulic radius; S – slope; n,f,g - constants

Q. 174 For a uniform flow, the velocity of flow

(A) Changes with time (B) Changes with space

(C) Does not change with time (D) Does not change with space

Q. 175 In a hydraulic efficient circular channel flow, the ratio of the hydraulic radius to the diameter of the pipe is

(A) 1.00 (B) 0.50

(C) 2.00 (D) 0.25

Q. 176 If the value of Froude number is less than one, the flow will be

(A) Critical flow (B) Supercritical flow

(C) Subcritical flow (D) Laminar flow

Q. 177 The specific energy diagram is a plot between specific energy and flow depth for a constant

(A) Turbulence (B) Discharge

(C) Specific force (D) Specific yield

Q. 178 A separate arrangement for aeration of the nappe is necessary in a

(A) Contracted rectangular weir

(B) Suppressed rectangular weir

(C) Sub-merged rectangular weir

(D) Triangular weir

Q. 179 At critical depth of fluid discharge is

(A) Minimum for a given specific energy

(B) Maximum for a given specific force

(C) Maximum for a given specific energy

(D) Minimum for a given specific force

Q. 180 A Hydraulic jump occurs when there is a break in grade from

(A) Mild slope to steep slope

(B) Steep slope to mild slope

(C) Steep slope to steeper slope

(D) Mild slope to very steep slope

Q. 181 If the Froude number of a hydraulic jump is 5.5, it would be called as

(A) An oscillating jump (B) A weak jump

(C) A strong jump (D) A steady jump

Q. 182 As per Francis formula, the value of Coefficient of discharge in rectangular weir will be about

(A) 0.554 (B) 0.482

(C) 0.623 (D) 0.854

Q. 183 In an open channel flow, the hydrostatic pressure at the top of water surface is

(A) Zero (B) Negative

(C) Infinite (D) None of the above

Q. 184 If the Reynold's number for water flowing in a closed pipe is 1800, the flow will be

(A) Turbulent flow (B) Laminar flow

(C) Uniform flow (D) None of the above

Q. 185 For accurate measurement of small discharge in water channels, which weir is better suitable ?

(A) 90 degree V-notch weir (B) Rectangular weir

(C) Cipoletti weir (D) Trapezoidal weir

UKPSC A.En. Exam 2007 Paper – II

Q. 186 Available water to the crops in the field lies between

(A) Field capacity and saturation point

(B) Field capacity and permanent wilting point

(C) Field capacity and gravitation water

(D) None of the above

Q. 187 The most efficient method of irrigation is

(A) Furrow irrigation (B) Sprinkler irrigation

(C) Drip irrigation (D) Basin irrigation

Q. 188 A crop requires 4 cm of irrigation water. If 5 cm of water was applied to this crop, the application efficiency will be

(A) 60 % (B) 70 %

(C) 80 % (D) 90 %

Q. 189 The Laccy's theory for irrigation channel design considers the following as a variable

(A) Depth of flow (B) Hydraulic mean radius

(C) Both of the above (D) None of the above

Q. 190 A cut-throat flume is a modified form of

(A) Aipolette weir (B) Parshall flume

(C) H-flume (D) None of the above

Q. 191 The friction coefficient in a drip irrigation line can be estimated by the use of equation given by

(A) Williams and Hazen (B) Howell and Hilel

(C) Wit and Gitlin (D) Blassius

Q. 192 The net amount of irrigation is 12 cm and the field efficiency is 80 %, the gross amount to be applied to the field will be

(A) 10 cm (B) 12 cm

(C) 15 cm (D) 18 cm

Q. 193 The actual area irrigated from an outlet in an year is known as

(A) Gross command area (B) Culturable command area

(C) Intensity of irrigation (D) None of the above

Q. 194 Emitters are used in

(A) Sprinkler irrigation (B) Water pump

(C) Surface irrigation (D) Drip irrigation

Q. 195 If a saline water is used to irrigate the crop which is sensitive to salinity, the leaching requirement should be

(A) Lesser (B) Higher

(C) Moderate (D) Lesser than moderate

Q. 196 The sprinkler irrigation can be used for almost all crops, except

(A) Wheat and gram crops (B) Oilseed crops

(C) Rice and jute crops (C) None of the above

Q. 197 A stream of 20 litres per second is available for applying 4 cm of irrigation water in one hectare field. The time taken for irrigation will be

(A) 5.55 hours (B) 6.55 hours

(C) 4.55 hours (D) 3.55 hours

Q. 198 In a particular location, the evapotranspiration of paddy crop is 99.80 cm and effective rainfall is 75.50 cm. If the irrigation efficiency is 60 %, the irrigation requirement will be

(A) 80.83 cm (B) 40.50 cm

(C) 60.83 cm (D) 90.83 cm

Q. 199 The capacity of main Canals in India varies, usually from

(A) 110 to 230 cumec (B) 110 to 230 cusec

(C) 280 to 425 cumec (D) 280 to 425 cusec

Q. 200 For a rectangular channel the minimum specific energy is

(A) 2/3 times of the critical depth

(B) 3/2 times of the critical depth

(C) 5/4 times of the critical depth

(D) None of the above

Q. 201 Hydraulic jump causes

(A) Increase in flow velocity (B) Decrease in flow velocity

(C) Energy dissipation (D) Increase in discharge

Q. 202 Chute spillways are used for a drop of

(A) Less than 1 m (B) More than 3 m but less than 6 m

(C) More than 6 m (D) Less than 2 m

Q. 203 For a channel flow, the Froude's number is 0.70, this flow is

(A) Critical flow (B) Subcritical flow

(C) Supercritical flow (D) None of the above

Q. 204 Actually evaporation of a crop is generally

(A) Equal to the potential evapotranspiration

(B) Less than the potential evapotranspiration

(C) More then the potential evapotranspiration

(D) None of the above

UKPSC A.En. Exam 2012 Paper – I

Q. 205 A pipe of circular cross section of diameter 'd' is flowing half-full. It's hydraulic radius will be

(A) d (B) d/2

(C) d/3 (D) d/4

Q. 206 The hydraulic depth for a full-flowing semi-circular channel of radius 'r' will be

(A) r (B) r/2

(C) 3.14×r/4 (D) 3.14×r/2

Q. 207 If the Froude number for hydraulic jump is 4.0, it is called

(A) Oscillating jump (B) Weak jump

(C) Steady jump (D) Strong jump

Q. 208 The velocity with which water approaches a weir is called

(A) Velocity of flow (B) Velocity of approach

(C) Velocity of Nappe (D) None of the above

Q. 209 A one cubic meter per second discharge is equal to .... litre/second

(A) 100 (B) 10

(C) 1000 (D) 50

Q. 210 If the slopes of channel bottom, hydraulic line and energy line are equal the flow would be

(A) Uniform (B) Non-uniform

(C) Steady (D) None of the above

Q. 211 In a rectangular open channel, the critical flow occurs at 2 m depth of flow, the specific energy at this depth will be .... m

(A) 1 (B) 2

(C) 3 (D) 4

Q. 212 In an open channel flow, the specific force for critical flow condition will be

(A) Zero (B) Average

(C) Maximum (D) Minimum

Q. 213 The channels of first kind have a unique depth in uniform flow for a given discharge. This unique depth is called as

(A) Special depth (B) Critical depth

(C) Uniform depth (D) Alternate depth

Q. 214 The sum of pressure head and elevation head is known as

(A) Kinetic head (B) Critical head

(C) Piezometric head (D) Acceleration head

UKPSC A.En. Exam 2012 Paper - II

Q. 215 At permanent wilting point soil moisture tension is

(A) 5 to 50 bars (B) 50 bars

(C) 5 to 10 bars (D) 7 to 32 atmosphere

Q. 216 If the net amount of irrigation is 8 cm and field efficiency is 75%, the gross amount of water to be applied to the field will be

(A) 10.66 cm (B) 8.75 cm

(C) 10.66 mm (D) 8.75 mm

Q. 217 The velocity of water through pipes and open channels may be estimated by

(A) Chezy's formula (B) Manning's formula

(C) Both (A) and(B) (D) None of the above

Q. 218 If gross irrigation is 15 cm and irrigation efficiency is 80%, then net irrigation will be

(A) 10 cm (B) 12 cm

(C) 15 cm (D) 19 cm

Q. 219 Evapotranspiration is directly measured by

(A) Penman – Monteith method

(B) Blaney-Criddle method

(C) Penman method

(D) Lysimeter method

Q. 220 The gauge used to measure the depth of flow over a weir is located upstream of a weir at a distance of about

(A) Twice the approximate head

(B) Thrice the approximate head

(C) Four times the approximate head

(D) Equal to approximate head

Q. 221 For medium loam soils, the length of the border should be

(A) 60 – 120 m (B) 100 – 180 m

(C) 150 – 300 m (D) 300 – 500 m

Q. 222 For better uniformity of water application in sprinkler irrigation percent overlap should

(A) Decrease with the increase in wind velocity

(B) Increase with the increase in wind velocity

(C) Be 50 %

(D) Be 70 %

Q. 223 In sprinkler irrigation normally water application is considered to be satisfactory when

(A) Uniformity coefficient (Cu) is 100 percent

(B) Uniformity coefficient is 80% or above

(C) Water is uniformly distributed

(D) All the above

Q. 224 The depth of flow over a sharp crested rectangular weir should not be more than

(A) Width of weir

(B) Half of the crest width

(C) Two-third of the crest width

(D) Three-fourth of the crest width

Q. 225 Minimum slope of an earthen channel for polyethylene lining is

(A) 1.5 : 1 (B) 2 : 1

(C) 2.5 : 1 (D) 3 : 1

Q. 226 The principle of working of a meter gate is

(A) Free flow circular orifice

(B) Free flow rectangular orifice

(C) Submerged flow circular orifice

(D) Submerged flow rectangular orifice

Q. 227 The capacity of major distributory in India usually varies from

(A) 0.75 to 5.5 cusec (B) 4.0 to 8.5 cumec

(C) 4.0 to 8.5 cusec (D) 0.75 to 5.5 cumec

Q. 228 Laminar flow condition in open channel exists if the Reynold's number is less than

(A) 500 (B) 2000

(C) 1500 (D) 1000

Q. 229 How the emitter flow varies with kinematic viscosity in non-orifice type emitters ?

(A) Directly proportional

(B) Inversely proportional

(C) Square root of kinematic viscosity

(D) None of the above

Q. 230 A parshall flume can be used conveniently on

(A) Flatter grades (B) Mild slopes

(C) Both (A) and (B) (D) None of the above

Q. 231 Wheat crop requires 45 cm of water during 120 days of its irrigation period. The total area of land irrigated with a flow of 20 lps for 22 hours a day shall be

(A) 102.24 ha (B) 42.24 ha

(C) 204.24 ha (D) None of the above

Q. 232 Sides of an earthen channel will be stable if its slope

(A) Does not exceed the angle of repose

(B) Exceeds the angle of repose

(C) Is kept same as angle of repose

(D) None of the above

Q. 233 The best hydraulic cross-section of a trapezoidal channel under favourable structural condition is

(A) $b = 2d$ (B) $b = d/2$

(C) $b = 2d \tan \theta/2$ (D) $b = 4d$

Q. 234 In check basin design the useful thumb rule is that the ratio of water spread time required to cover the entire basin to the time required to infiltrate the net depth of irrigation is

(A) 0.25 (B) 0.75

(C) 2.00 (D) None of the above

Q. 235 The thickness of a spillway apron depends on

(A) Hydraulic jump (B) Water velocity

(C) Uplift pressure (D) All of the above

Q. 236 Darcy's law is valid for

(A) Turbulent flow (B) Laminar flow

(C) Both (A) and (B) (D) None of the above

Q. 237 The maximum non-erosive velocity of a water stream which is used for designing furrows laid at 0.2 percent longitudinal slope should be

(A) 1 litre/sec (B) 3 litres/sec

(C) 4 litres/sec (D) 5 litres/sec

Q. 238 Which of the following may be used for measurement of water discharge ?

(A) By rectangular weir (B) By 90° V notch

(C) By orifice (D) All of the above

Q. 239 The phenomenon of removal of material from the foundation of structure through seepage water is called as

(A) Piping (B) Sliding

(C) Overturning (D) None of the above

Q. 240 In a drop structure safety against tension can be achieved by assuming that the resultant of the forces on the head wall pass through

(A) The middle one-third of the base

(B) The left one-third of the base

(C) The right one-third of the base

(D) Any point on the base

Kerala AAO Exam 2017

Q. 241 The working pressure required for rain gun

(A) 2 to 10 kg/cm^2 (B) 0.1 to 0.2 kg/cm^2

(C) 8 to 10 kg/cm^2 (D) 2 to 5 kg/cm^2

Q. 242 Match the following match the following

List-I	**List-II**
P. Check irrigation	1. Maize, sorghum
Q. Furrow irrigation	2. Wheat
R. Surge irrigation	3. Finger millet, pulses
S. Corrugation	4. Cotton, vegetables

(A) P-4, Q-3, R-2, S-1 (B) P-3, Q-4, R-1, S-2

(C) P-2, Q-1, R-3, S-4 (D) P-3, Q-2, R-1, S-4

Q. 243 1.50 gram of sodium bi-metasulphate dissolved in 1 litre of water will give strength of

(A) 15 ppm (B) 150 ppm

(C) 1500 ppm (D) 15000 ppm

Graduate Aptitude Test in Engineering - 2007

Q. 244 A 50 km long canal with an average width of 25 m is used for irrigation. Mean daily evaporation as measured from a Class A evaporation pan is 5 mm d^{-1}. Considering the pan coefficient as 0.80, the mean daily evaporation loss from this canal is

(A) 5.00×10^3 m^3 d^{-1} (B) 6.25×10^3 m^3 d^{-1}

(C) 5.00×10^4 m^3 d^{-1} (D) 6.25×10^4 m^3 d^{-1}

Q. 245 To deliver 1.3 litre min^{-1} discharge, the operating pressure of a 3 m long, 3 mm diameter bubbler tube is

(A) 1.64 kPa (B) 16.46 kPa

(C) 164.61 kPa (D) 1646.20 kPa

Q. 246 A field is irrigated by constructing 100 m long furrows spaced at 0.75 m apart. The advance time to the end of furrow was 30 min with an inflow rate of 2 litre s^{-1}. After that the inflow rate was cutback to 0.5 litre s^{-1} and continued for one hour. The average depth of irrigation is

(A) 2.4 cm (B) 7.2 cm

(C) 9.0 cm (D) 18.0 cm

Q. 247 A sprinkler system consists of two 192 m long laterals. On each lateral, sixteen sprinklers are located at an interval of 12 m. The spacing between the laterals is 10 m. The required capacity (in litre s^{-1}) of sprinkler system for application rate of 1.0 cm h^{-1} is

(A) 5.33 (B) 10.66

(C) 14.22 (D) 17.06

Q. 248 The discharge through a 90 V- notch for a head of 0.5 m and coefficient of discharge of 0.6 is

(A) 0.25 $m^3\ s^{-1}$ (B) 0.50 $m^3\ s^{-1}$

(C) 0.65 $m^3\ s^{-1}$ (D) 0.75 $m^3\ s^{-1}$

Q. 249 A hydraulically efficient trapezoidal drainage channel has to be designed for draining 400 ha of land with a drainage coefficient of 20 mm. If the recommended side slope and depth are 2:1 and 1.06 m respectively, the bottom width is

(A) 0.25 m (B) 0.50 m

(C) 0.75 m (D) 1.00 m

Graduate Aptitude Test in Engineering - 2008

Q. 250 A clayey soil has a field capacity of 0.38m^3m^{-3} and wilting point of 0.24 m^3m^{-3}. If the specific weight of the soil is 12.75 kN and the effective root-zone depth is 0.8 m, the available moisture holding capacity is

(A) 15.6 cm (B) 11.2 cm

(C) 1.12 cm (D) 20.8 cm

Q. 251 In the Moody diagram, the third parameter is ϵ/D, Here, is

(A) The equivalent uniform sand grain roughness

(B) An arbitrarily chosen roughness magnitude

(C) Median size in a non-uniform sand grain roughness

(D) Mean height of the actual roughness of commercial pipes

Q. 252 A flow of 150 Ls^{-1} was supplied for 8 hours from a tank to irrigate 2 ha of land. It was found that the actual delivery rate at the farm was less than 150 Ls^{-1}. If the conveyance loss was 864 m^3 and percolation and runoff losses in the field were 240 and 760 m^3 respectively, the water application efficiency of this system is

(A) 80 % (B) 61 %

(C) 77 % (D) 71 %

Q. 253 A 50g L^{-1} solution of tracer was discharged into a stream at a constant rate of 20 mLs^{-1}. At a downstream section, the tracer was completely mixed and attained an equilibrium concentration of 10 parts per billion. Assuming the background concentration as zero, the stream discharge m^3s^{-1} is

(A) 100 m^3s^{-1} (B) 200 m^3s^{-1}

(C) 800 m^3s^{-1} (D) 1000 m^3s^{-1}

Q. 254 The velocity of flow of water through a drop inlet pipe spillway is 4 m and the friction loss coefficient is 0.12. Maximum slope that can be provided to the pipe to maintain pipe flow condition is

(A) 8.9 % (B) 9.8 %

(C) 10.3 % (D) 10.8 %

Q. 255 A sandy loam soil has a water holding capacity of 140 mm/m depth between field capacity and wilting point. The area to be irrigated is 60 ha and the depth of effective root zone is 0.30 m. The management allowed soil moisture depletion is 40% and the consumptive use is 6 mm per day. The conveyance and application efficiencies are estimated to be 80% and 50%. There are no leaching requirements as well as rainfall and groundwater contribution to the crop water requirement.

(i) The frequency of irrigation will be

(A) 1 day (B) 3 days

(C) 7 days (D) 5 days

(ii) The field irrigation requirement will be

(A) 21600 m^3 (B) 10800 m^3

(C) 2.16 m^3 (D) 27000 m^3

Graduate Aptitude Test in Engineering - 2009

Q. 256 A check basin of size 15 m×12 m is to be irrigated using a stream of 26 litre per second. The depth of crop root zone is 1.3 m and the apparent specific gravity of the root zone soil is 1.6. The water holding capacity of the soil is 16%. Irrigation is to be applied when the soil moisture content in the crop root zone attains 12%. Deep percolation loss is neglected.

(i) The net irrigation requirement in mm is

(A) 43.2 (B) 63.2

(C) 73.2 (D) 83.2

(ii) The duration of irrigation in minutes to replenish up to field capacity is

(A) 4.8 (B) 9.6

(C) 16.6 (D) 24.6

Q. 257 A hydraulically efficient trapezoidal drainage channel with a side slope of 2:1 has been designed in a sandy loam soil for a catchment 600 ha. Taking a drainage coefficient of 16 mm, the flow velocity in mm s^{-1} in the drainage channel with a flow depth of 1 m is

(A) 427 (B) 450

(C) 497 (D) 527

Q. 258 The empiricla method for computing the consumptive use of a crop using the mean monthly temperature and day light hours is

(A) Thornthwaite (B) Blaney Criddle

(C) Hargreaves (D) Lowry Johnson

Graduate Aptitude Test in Engineering - 2010

Q. 259 In order to evaluate irrigation distribution, an irrigator estimates the depth of infiltration, in mm, around a field as given below

42	36	32	38
40	32	35	34
25	28	29	31
36	30	32	28
40	38	34	44

The distribution uniformity for the irrigation is

(A) 80.4 (B) 81.9

(C) 87.9 (D) 88.1

Q. 260 A parabolic shaped grassed waterway has a top width of 4 m, a maximum depth of 0.40 m, and a slope of 2.5%. The Manning's 'n' value is 0.035 and there is no provision of freeboard. The discharge carrying capacity of the waterway in m^3s^{-1} is

(A) 1.38 (B) 1.52

(C) 1.76 (D) 1.96

Graduate Aptitude Test in Engineering - 2011

Q. 261 The wedge storage in a river reach during the passage of a flood wave is

(A) Positive during rising phase

(B) Negative during rising phase

(C) Positive during falling phase

(D) Constant

Q. 262 A concrete-lined non-circular tunnel has a semi-circle at the top and a rectangular section at the bottom. The semi-circle has a diameter of 6 m whereas the rectangular section is 6 m wide and 3 m high. The tunnel carries a discharge of 128 $m^3\ s^{-1}$. If the friction factor f = 0.017, then the head loss in 1 km length of the tunnel in m will be

(A) 0.57 (B) 1.15

(C) 2.29 (D) 4.58

Q. 263 A 100 ha reservoir receives 2500 mm of rainfall during a period of 2 years. During this period the mean inflow to the stream is 1.0 $m^3\ s^{-1}$, the mean outflow from the stream is 0.8 $m^3\ s^{-1}$, and the increase in the storage is 500 ha-m. Assuming that there is no seepage loss, the total evaporation during the period in m is

(A) 1.011 (B) 10.11

(C) 101.1 (D) 1011

Graduate Aptitude Test in Engineering - 2012

Q. 264 In a semi-modular outlet, the discharge

(A) Is independent of water levels in the distributary and the water course

(B) Depends upon the water levels of both distributary and water course

(C) Depends upon the water level in the distributary

(D) Depends upon the water level in the water course

Q. 265 A trapezoidal grassed waterway is constructed along a longitudinal gradient of 4%. If the crosssectional area of flow is 1.52 m^2, wetted perimeter is 12.5 m and Manning's n for the waterway is 0.04 $m^{-1/3}$ s, the flow through the waterway in m^3 s^{-1} is

(A) 1.9 (B) 2.1

(C) 2.3 (D) 2.5

Q. 266 A border strip of 8×250 m is being irrigated by a border stream of 50 lps. The infiltration capacity of the soil is 25 mm h^{-1} (assumed to be constant throughout the period of irrigation). The average depth of the advancing sheet of water over the land is 70 mm. The time required to irrigate the border strip, in minutes, will be

(A) 16.7 (B) 25.7

(C) 54.7 (D) 67.7

Q. 267 A regime channel carrying a discharge of 25 m^3 s^{-1} is designed using Lacey's regime theory. The side slope of the channel is 1/2 H : 1 V, and Laccy's silt factor is unity. The bottom width and depth of flow in the channel, in m, respectively are

(A) 20.26, 1.38 (B) 20.26, 1.56

(C) 23.75, 1.56 (D) 32.78, 1.56

Q. 268 A trapezoidal canal, having a bottom width of 5.0 m and a side slope of 1 H : 1 V, is carrying a discharge of 20 m^3 s^{-1}. The critical depth, in m, is

(A) 1.09 (B) 1.18

(C) 2.12 (D) 2.62

Graduate Aptitude Test in Engineering - 2013

Q. 269 During land leveling of agricultural land for irrigation and drainage purposes, the acceptable deviation in elevation from the design value in metre is

(A) 0.015 (B) 0.025

(C) 0.055 (D) 0.150

Q. 270 Pan evaporation data recorded at a certain location over a period of one week are 4.0, 4.3, 4.6, 4.9, 5.12, 5.18, and 6.21 mm. If irrigation scheduling based on ratio of irrigation water (IW) to cumulative pan evaporation (CPE) is practiced, the depth of irrigation at an interval of a week for IW/CPE = 0.9 is

(A) 3.60 (B) 4.41

(C) 5.59 (D) 30.88

Q. 271 In an experimental setup, the discharge through a triangular notch is 0.0074 $m^3\ s^{-1}$ at an operating head of 0.1 m. The coefficient of discharge for the notch is 0.7. If the required discharge is 0.1 m^3/s,

(i) the corresponding head in m is

(A) 0.00015 (B) 0.035

(C) 0.283 (D) 67.13

(ii) the corresponding width of water surface in m is

(A) 0.80 (B) 0.099

(C) 0.00043 (D) 0.00086

Graduate Aptitude Test in Engineering - 2014

Q. 272 A crop has effective root zone depth of 1200 mm and monthly (30 days) crop evapotranspiration of 260 mm. The effective rainfall during 30 days period is 20 mm. The field capacity and permissible soil moisture depletion (volume basis) are 16% and 8%, respectively. The irrigation interval in days for the crop will be

(A) 30 (B) 18

(C) 12 (D) 8

Q. 273 In a citrus orchard, planting is done at a spacing of 5 m×5 m. The daily pan evaporation of the orchard is 6 mm. The pan coefficient, wetting factor (crop canopy factor) and crop coefficient are 0.8, 0.6 and 0.6, respectively. Four drippers each of 4 L h^{-1} discharge are used to irrigate each plant. The time of operation of drip irrigation system in hours will be

(A) 2.7 (B) 4.5

(C) 16.2 (D) 19.8

Graduate Aptitude Test in Engineering - 2015

Q. 274 Discharge through an irrigation outlet is independent of the water levels in the distributary and water courses in case of

(A) Non-modular outlet (B) Semi-modular outlet

(C) Kennedy's gauge outlet (D) Gubb's module outlet

Q. 275 The USDA classification of irrigation water with regard to alkali and salinity hazards is based on

(A) Exchangeable sodium percentage and pH

(B) Electrrical conductivity and Sodium adsorption ratio

(C) Electrical conductivity and pH

(D) Sodium percentage and pH

Q. 276 A steady discharge of 60 m^3/s is passing through a trapezoidal irrigation channel with a bottom width of 5 m, bed slope of 0.01% and Manning's roughness coefficient of 0.025. The conveyance of the channel in m^3/s is ....

Q. 277 Two irrigation channels are designed in the same type of alluvial soil having the bed slopes of 1.52×10^{-4} and 1.6×10^{-4}, respectively. If the design discharge of the first channel, using Lacey's regime theory, is 30 m^3/s, then the discharge of the second channel in m^3/s will be

(A) 22.05 (B) 29.74

(C) 30.26 (D) 40.81

Q. 278 A Cipolletti weir has the crest length of 1.2 m and a crest water level of 0.5 m. The average approach velocity of water in m/s on the crest is

(A) 0.49 (B) 1.18

(C) 1.19 (D) 1.32

Q. 279 If an irrigation water source has the concentrations of Na^+, Ca^{++} and Mg^{++} as 28. 10 and 5 milli equivalents per litre, repectively, then the Sodium adsorption ratio of this water is ....

Q.280 In a canal command, maize crop is grown in an area of 30 ha. The crop evapotranspiration (ET_c) of maize is 840 mm per season and the effective rainfall during growing season is 20 mm. It is irrigated with water having salinity of 1.1 dS/m by a surface irrigation method. If the leaching efficiency of the field soil is 0.8 and the average soil salinity tolerated by the maize crop for 100% Yield is 1.7 dS/m, the depth of irrigation water in mm per season required to meet the seasonal ET_c and leaching requirement will be ....

Q.281 The depth of the impounded water in a 72 m long earthen dam is 6.2 m, while the tail water is 2.2 m deep. The hydraulic conductivity of the isotropic and homogeneous soil-fill of the dam is 0.53 m/day. Flow net method is used to estimate seepage wherein the number of the flow channels is 6 and the number of potential drops is 21. The seepage rate through the dam in m^3/ day is ....

Graduate Aptitude Test in Engineering - 2016

Q. 282 In a cropped field, the following data are observed.

Moisture content at Field capacity (weight basis) = 36%

Current moisture content (weight basis) = 24%

Bulk density of soil = 1.5 Mg m^{-3}

Effective root zone depth = 0.8 m

Conveyance efficiency = 80%

Application efficiency = 90%

To bring soil moisture content to field capacity, the depth of irrigation in mm will be ....

Q. 283 The correct conditions for which the hydraulically efficient rectangular channel will deliver maximum discharge are

P – depth of water is equal to half the breadth of channel

Q – depth of water is equal to breadth of channel

R – depth of water is equal to twice the breadth of channel

S – hydraulic radius is equal to half the depth of water

(A) P and R (B) P and Q

(C) P and S (D) R and S

Q. 284 A solid set permanent micro-irrigation system is installed in a vegetable field of 1 ha area. The spacing between the micro sprinklers is 2.5 m and spacing between laterals is 5 m. The peak evapotranspiration rate is 10 mm day^{-1}. The application efficiency is 80%. Irrigation system operates 5 hours in a day. The total operating head of the pump is 30 m. At 65% pump efficiency, the horse power of the pump is ....

Graduate Aptitude Test in Engineering - 2017

Q. 285 Graded furrow of 80 m length and 0.75 m spacing are used for irrigating a field with an initial furrow stream of 100 L min^{-1}. The initial furrow stream flow reaches the lower end of the field in 40 min. Thereafter, the furrow stream flow is reduced to 30 L min^{-1} and the cutback stream flow is continued for 1 hour. The average depth of irrigation over the field in cm will be ....

Q. 286 The soil of a cropped field has field capacity of 25% and wilting point of 13% on weight basis. The effective root zone depth of the crop is 0.70 m and consumptive use of water by the crop is 5 mm day^{-1}. Apparent specific gravity of the soil is 1.50 m. if the allowable soil moisture depletion is 40%, the permissible moisture depletion between irrigations and the frequency of irrigation are

(A) 5 cm; 10 days (B) 10 cm; 7 days

(C) 8 cm; 12 days (D) 4 cm; 8 days

Q. 287 A rectangular channel having bed slope of 0.05% and Manning's roughness coefficient of 0.01 carries a discharge of 5 m^3 s^{-1}. if the channel is designed as the most economical section, the width of the channel in meter will be ....

Q. 288 A trapezoidal notch, placed over an emergency spillway, has the following details:

Top width = 2 m

Bottom width = 1 m

Height = 0.5 m

Co-efficient of discharge for the triangular portion = 0.65

Co-efficient of discharge for the rectangular portion = 0.68

For a flow head of 0.4 m over the notch, the discharge in Ls^{-1} will be about

(A) 155 (B) 353

(C) 508 (D) 663

Graduate Aptitude Test in Engineering - 2018

Q. 289 A permanent matured orchard has a tree spacing of 4 m×5 m. Each tree has a shading area of 40% to be irrigated with the 'pig tail' pattern multi-exit drip emitters. The effective wetting geometry of each emitter is 2 m×2 m. The emitters have discharge constant and exponent of 0.3 and 0.6, respectively. The coefficient of variation of emitter discharge is 0.06. The average and minimum operating pressures are 120 kPa and 100 kPa, respectively. The emission uniformity of the emitters is .... %.

Q. 290 Furrows of 120 m length with 0.5% slope are made at 90 cm spacing. The maximum non erosive stream flow rate is applied in a furrow that takes 1.0 hour to reach the lower end. Then this flow rate is reduced to half of its size and, subsequently, continued for another 1.0 hour. The average depth of applied water is .... cm.

Graduate Aptitude Test in Engineering - 2019

Q. 291 A lateral has 12 sprinklers spaced 14 m apart in a sprinkler irrigation system. The laterals are spaced 20 m apart on the main line. If the recommended fertilizer dose is 80 kg/ha, the amount of fertilizer to be applied at each setting in kg is ....

Q. 292 In the Muskingum method of channel routing, the routing equation is written as $Q_2 = C_0 I_2 + C_1 I_1 + C_2 Q_1$. If the storage-time constant K = 12 h, weighting factor x = 0.15 and the time step for routing Δt = 4 h, the coefficient C_o is

(A) 0.016 (B) 0.048

(C) 0.328 (D) 0.656

Q. 293 An irrigation stream of 27 L/s is diverted to a check basin of size 12 m × 12 m. The water holding capacity of the soil is 15% and the average soil moisture content in the crop root zone prior to applying water is 7.5%. The depth of crop root zone is 1.2 m and apparent specific gravity of the soil is 1.5. Assuming no loss due to deep percolation, irrigation time (in minute) required to replenish the root zone moisture to its field capacity is ....

Q. 294 Undisturbed soil sample is collected from a field when the soil moisture is at field capacity. The inside diameter of the core sampler is 7.5 cm with a height of 15 cm. Weight of the core sampling cylinder with moist soil is 2.81 kg and that with oven dry soil is 2.61 kg. The weight of the core sampling cylinder is 1.56 kg. Assuming $\pi = 3.14$, the water depth in centimeter per meter depth of soil is ....

Graduate Aptitude Test in Engineering - 2020

Q. 295 From the performance evaluation of drippers, the discharge exponent value and the coefficient of variation were obtained as 0.5 and 0.04, respectively. The drippers are categorized as

(A) Pressure compensating drippers of excellent quality

(B) Turbulent flow-tortuous path orifice type drippers of good quality

(C) Turbulent flow-tortuous path orifice type drippers of marginal quality

(D) Laminar flow drippers of excellent quality

Q. 296 A field crop is irrigated when the available soil water reduces to 60%. The moisture content at field capacity and wilting point are 32% and 12%, respectively. The bulk density of the soil is 1.5 g/cm³. The field water application efficiency is 75% and the crop root zone depth is 50 cm. The gross depth of irrigation required to bring soil moisture content to field capacity in cm

(A) 6 (B) 8

(C) 9 (D) 12

Q. 297 In an irrigation channel of uniform section, water passes through a 90° triangular weir measuring 36 cm head over the crest. After traveling certain distance in the same channel, water passes through I.0 m long rectangular weir. There is no loss of water in between two weirs. Using Francis' formula, the head over the crest of rectangular weir in cm is

(A) 22.1 (B) 18.4

(C) 15.0 (D) 11.8

Graduate Aptitude Test in Engineering - 2021

Q. 298 In a field test of drip irrigation system having an application efficiency of 90%, the minimum, maximum and average flow rates are found to be 45 L/h, 65 L/h and 50 L/h, respectively. The manufacturer's coefficient of variation of the emitter is 0.07. If there is one emitter per plant, the drip irrigation efficiency in percent is ……..

Q. 299 The observed concentrations of magnesium (Mg^{2+}), sodium (Na^+), and bicarbonate (HCO_3^-) in saturated extract of a soil sample taken from the root zone are 5.68 meq/L, 9.90 meq/L, and 11.20 meq/L, respectively. If the concentration ratio of HCO_3^-/Ca^{2+} is 2.8, the sodium adsorption ratio is ………

Q. 300 The most economical trapezoidal channel section with 1:1 (horizontal:vertical) side slope is designed to carry a maximum of 40 cm depth of water at its full capacity. If the bed slope of the channel is 1:2500 and the Manning's roughness coefficient of channel section is 0.01, the estimated discharge capacity of the channel in m^3/s is……….

Answers Key

1	2	3	4	5	6	7	8	9	10
A	B	A	D	A	B	A	A	A	A
11	12	13	14	15	16	17	18	19	20
D	C	B	A	A	B	C	D	C	B
21	22	23	24	25	26	27	28	29	30
C	A	D	A	B	D	B	C	B	B
31	32	33	34	35	36	37	38	39	40
C	A	B	D	A	C	B	B	A	B
41	42	43	44	45	46	47	48	49	50
A	D	C	D	C	C	B	B	A	C
51	52	53	54	55	56	57	58	59	60
A	D	D	B	C	C	B	D	B	D
61	62	63	64	65	66	67	68	69	70
B	B	D	B	B	C	D	C	B	A
71	72	73	74	75	76	77	78	79	80
D	A	D	C	A	B	C	D	A	B
81	82	83	84	85	86	87	88	89	90
A	B	C	C	D	C	D	D	C	C
91	92	93	94	95	96	97	98	99	100
D	D	D	D	B	C	D	C	A	D
101	102	103	104	105	106	107	108	109	110
B	A	A	C	B	C	B	C	B	D
111	112	113	114	115	116	117	118	119	120
C	C	B	B	D	B	B	C	B	C
121	122	123	124	125	126	127	128	129	130

A	D	C	D	C	D	B	D	B	D
131	132	133	134	135	136	137	138	139	140
C	B	B	C	B	B	A	A	C	C
141	142	143	144	145	146	147	148	149	150
B	C	C	D	A	A	A	A	C	B
151	152	153	154	155	156	157	158	159	160
D	E	C	A	A	C	C	C	B	C
161	162	163	164	165	166	167	168	169	170
D	B	A	C	D	A	C	A	A	A
171	172	173	174	175	176	177	178	179	180
B	C	C	D	D	C	B	B	C	B
181	182	183	184	185	186	187	188	189	190
D	C	A	B	A	B	C	C	B	B
191	192	193	194	195	196	197	198	199	200
D	C	C	D	B	C	A	B	C	B
201	202	203	204	205	206	207	208	209	210
C	B	B	B	D	C	A	B	C	A
211	212	213	214	215	216	217	218	219	220
C	D	C	C	D	A	C	B	D	C
221	222	223	224	225	226	227	228	229	230
B	B	B	C	B	C	D	A	B	A
231	232	233	234	235	236	237	238	239	240
B	A	C	A	C	B	B	D	A	A
241	242	243	244	245	246	247	248	249	250
C	B	C	A	C	B	B	A	B	B
251	252	253	254	255	256	257	258	259	260
D	D	A	B	C,A	D, B	B	B	B	D
261	262	263	264	265	266	267	268	269	270
A	C	B	C	A	C	B	A	A	D
271	272	273	274	275	276	277	278	279	280
C, A	C	A	D	B	6000	A	1.1914	10.23	1009.23
281	282	283	284	285	286	287	288	289	290
43.61	200	C	4.22	9.70	A	2.488	D	84.79	6
291	292	293	294	295	296	297	298	299	300
26.88	A	12	30.2	B	B	C	73.8	4.5	0.20

Explanations

Q. 3

We know

$$F = V/\sqrt{gD}$$

Where:

V = Water velocity

g = Acceleration due to gravity

D = Hydraulic depth (cross sectional area of flow / top width)

For 90° V-notch, if Depth = Y, Top width = Y + Y = 2Y

Hence, $D = (2Y \times Y/2)/2Y = Y/2$

Therefore, $F = V\sqrt{2}/\sqrt{gD}$

When:

Fr = 1, Critical flow,

Fr > 1, Supercritical flow (fast rapid flow),

Fr < 1, Subcritical flow (slow / tranquil flow)

Q. 63

Hydraulic Jump Characteristics

Froude Number	Ratio of height after to height before jump	Descriptive characteristics of jump
≤ 1.0	1.0	No jump; flow must be supercritical for jump to occur
1.0–1.7	1.0–2.0	Standing or undulating wave
1.7–2.5	2.0–3.1	Weak jump (series of small rollers)
2.5–4.5	3.1–5.9	Oscillating jump
4.5–9.0	5.9–12.0	Stable clearly defined well-balanced jump
> 9.0	> 12.0	Clearly defined, turbulent, strong jump

Q. 71

Excessive and irregular irrigation has the following effects:

1. Increase in Saline and Alkaline Elements in Soil or Increase in Salinity
2. Problem of Water logging
3. Reduction in Temperature of Soil
4. More Nitrate Formation
5. Shortage of Soil Nutritive Elements and Decrease in Productivity

Q. 88

Time required = 42.86/36 = 1.19 hours

Q. 95

Amount of water to be applied = 10/0.80 = 12.50 cm

Q. 144

The area which can be irrigated from a scheme and is fit for cultivation.

Q. 148

Fro most efficient trapezoidal section

$b = 2d \tan(\theta/2)$ Where $\theta = 60°$

Where $\tan(\theta/2) = 1/\sqrt{3}$

Hence, $b = 2 \times d/\sqrt{3}$

Q 152

The moisture extraction pattern shows the relative amounts of moisture extracted from different depths within the crop root zone.

It is seen that about 40 percent of the total moisture used is extracted from first quarter of the root zone, 30 percent from the second, 20 percent from third and only 10 percent from last quarter.

Q. 155

Refer explanation for Q.279.

Q. 188

Application efficiency = 4/5 = 0.80

Q. 192

Water to be applied = 12/0.80 = 15 cm

Q. 197

Time taken = $0.04 \times 10000/(20 \times 10^{-3} \times 3600) = 5.55$ hours

Q. 198

Irrigation requirement = (99.80 – 75.50)/0.60 = 40.50 cm

Q. 243

Concentration of sodium bi-metasulphate = 1.50 g/L = 1500 mg/L or 1500 ppm

Q. 244

Canal length (L) = 50 km = 50000 m

Canal width (W) = 25 m

$\therefore$ Water evaporating surface area (A) = 25×50000

$= 1250000 \text{ m}^2$

$\because$ Pan coefficient(C) = 0.80 and pan evaporation = 5 mm/d = 0.005 m/d

We know that

Evaporation from pond or canal = pan coefficient × pan evaporation

$\therefore$ Evaporation loss from canal = $0.80 \times 0.005 \times 1250000 = 5 \times 10^3 \text{ m}^3/\text{d}$

Q. 245

According to Empirical Formula

$$\text{Bubbler tube discharge (Q)} = 5.52 \times D^{2.71} \times \left(\frac{Pressure\ (kPa)}{Length\ (cm)}\right)^{0.57}$$

$$\Rightarrow \quad 1.3 \times 60 = 5.52 \times (3)^{2.71} \times \left(\frac{Pressure\ (kPa)}{300}\right)^{0.57}$$

$$\Rightarrow \quad \text{Pressure} = 168.47 \text{ kPa} \approx 164.61 \text{ kPa}$$

Q. 246

Area between furrows (A) = $100 \times 0.75 = 75 \text{ m}^2$

Total inflow after cutback in one hour (Q_2) = $0.5 \times 60 \times 60 = 1800$ L

Total Inflow before cutback in (Q_1) = $2 \times 30 \times 60 = 3600$ L

$$\therefore \quad \text{Depth of irrigation} = \frac{Q_1 + Q_2}{A}$$

$$= \frac{1800 + 3600}{1000 \times 75} = 0.072 \text{ m} = 7.2 \text{ cm}$$

Q. 247

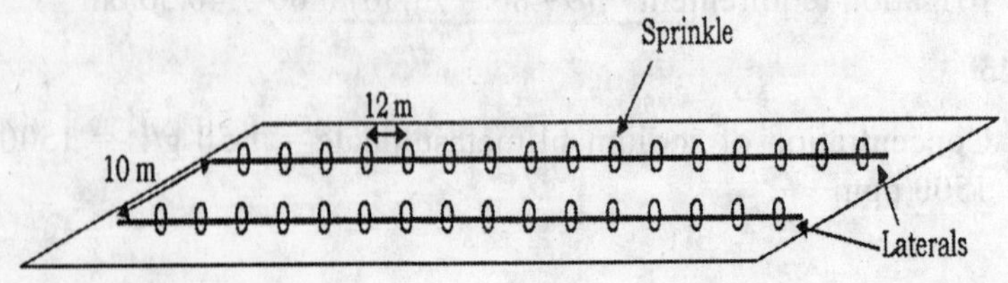

Water application rate (I) = 1 cm/h = 1/360000 m/s

Lateral spacing (S_l) = 10 m

Sprinkler interval/spacing (S_S) = 12 m

Number of sprinklers (n) = 16×2 = 32

∴ Capacity of sprinkler system (Q)

$$= n \times S_l \times S_S \times I$$

$$= 32 \times 12 \times 10 \times \frac{1}{360000} = 0.01066 \text{ m}^3/\text{s}$$

$$= 10.66 \text{ L/s}$$

Q. 248

Head (H) = 0.5 m

Coefficient of discharge (C_d) = 0.6

And θ = 90°

$$\therefore \quad \text{Discharge (Q)} = \frac{8}{15} C^d H^{5/2} \sqrt{2g} \tan(\theta/2)$$

$$= \frac{8}{15} \times 0.6 \times (0.5)^{5/2} \sqrt{2 \times 9.81} \tan(45)$$

$$= 0.25 \text{ m}^3/\text{s}$$

Alternate

Discharge through rectangular channel(Q) = $1.38(H)^{5/2}$

$$= 1.38 \times (0.5)^{5/2} = 0.246 \text{ m}^3/\text{s}$$

Q. 249

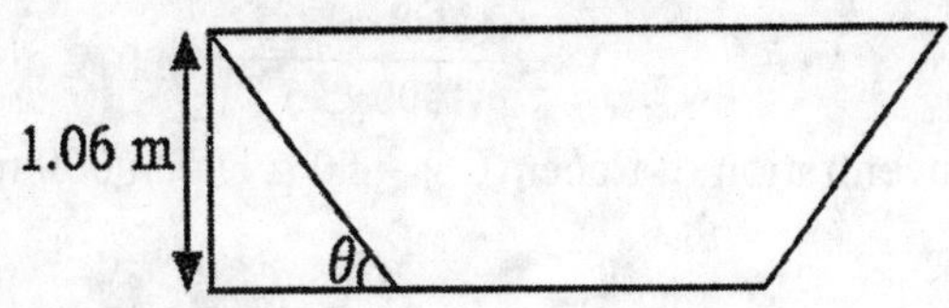

For most efficient channel

Bottom width (b) = 2d tan($\frac{\theta}{2}$)

Here, tan($\frac{\theta}{2}$) = $\frac{b}{2\times1.06}$ = 0.236 ($\because$ tanθ = $\frac{1}{2}$ ⇒ θ = 26.57°)

⇒ b = 0.50 m

Q. 250

Field capacity moisture content (M_{FC}) = 0.38 m³/ m³ of soil

Wilting point moisture content (M_{WP}) = 0.24 m³/ m³ of soil

Specific weight (w) = 12.75 kN/ m³

Effective root zone depth (d_r) = 0.8 m

∴ Available moisture holding capacity (AMHC)

= (M_{FC} - M_{WP})×d_r

= (0.38 – 0.24)×0.8 = 11.2 cm

Q. 252

Total water discharged from the tank in 8 hrs = $150\times10^{-3}\times8\times3600$ = 4320 m³

$\because$ Conveyance loss = 864 m³

Therefore,

Water delivered at the farm = 4320 – 864 = 3456 m³

Water stored in root zone = 3456 – 240 – 760 = 2456 m³

Hence,

Water application efficiency = $\frac{2456}{3456}\times100$ = 71 %

Q. 253

Initial concentration of tracer (C_0) = 50 g/L = 50000 mg/L = 50000 ppm = 5×10^{-1}

Tracer discharge rate (Q_0) = 20 mL/s = 20×10^{-3} L/s = 20×10^{-6} m³/s

Concentration after mixing with stream (C_2) = 10 ppb = 10×10^{-9}

Consider stream discharge = Q L/s

We know that

$$C_0Q_0 + QC_1 = (Q_0 + Q)C_2 \qquad (\because C_1 = 0 \text{ gm/ l})$$

$$\Rightarrow \quad 50\times10^{-1}\times20\times10^{-6} = (20\times10^{-6} + Q)\times10\times10^{-9}$$

$$\Rightarrow \quad Q = 100 \text{ m}^3/\text{ s}$$

Q. 254

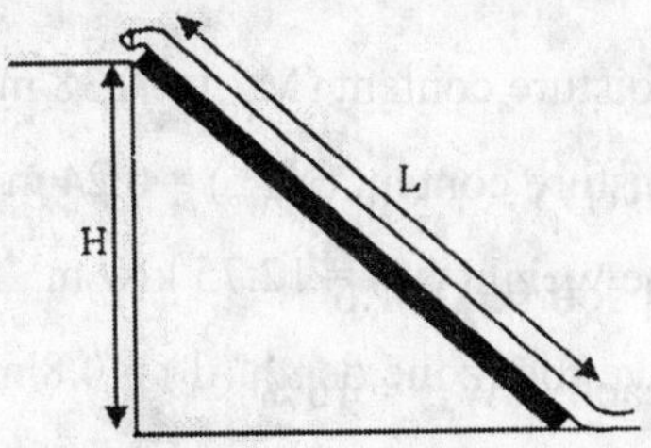

Velocity of flow (v) = 4 m/s

Friction loss coefficient (K_f) = 0.12

We know that

Head loss due to friction (H_f) = $L\times K_f\times(V^2/2g)$

$\therefore$ Neutral slope (H_f/L) = $K_f\times(V^2/2g)$

$$= 0.12\times(4^2/2\times9.81) = 9.8\ \%$$

Q. 255 (i)

Water holding capacity = 140×0.3 = 42 mm

$\because$ Moisture depletion = 40 %

Thus, Water removed = 42×0.40 = 16.8 mm

$\because$ Consumptive use = 6 mm/ day

Therefore, Frequency of irrigation = $\dfrac{16.8}{6\times0.80\times0.50} = 7$ days

($\because$ Conveyance and application efficiencies are 80 % and 50 % respectively.)

(ii)

Net irrigation requirement (NIR) = 6×3 = 18 mm

($\because$ Irrigation period = $\dfrac{16.8}{6} = 2.8 \approx 3$)

$\therefore$ Field irrigation requirement (FIR) $= \frac{NIR}{Application efficiency}$

$= \frac{18}{0.50} = 36$ mm

In terms of m³;

$$FIR = \frac{36}{1000} \times 60 \times 10000 = 21600 \text{ m}^3$$

Q. 256(i)

Given that:

Depth of root zone (D_r) = 1.3 m

Specific gravity of soil (G) = 1.6

Water holding capacity (W_C) = 16%

Soil moisture content retention (W_r) = 12%

$$\text{Depth of irrigation } (D_i) = \frac{(W_c - W_r) \times G \times D_r}{100}$$

$$= \frac{(16-12) \times 1.6 \times 1.3}{100} = 0.0832\text{m}$$

$= 83.2$ mm

(ii)

Size of check basin (A) = 15×12 = 180 m²

Stream discharge (Q_s) = 26 L/s = 26×10^{-3} m³/s

Total water that has to be applied (Q) = A×D_i

$= 180\times0.0832 = 14.976$ m³

Duration of irrigation to replenish up to field capacity (t)

$$= \frac{\textit{Total water that has to be applied}}{\textit{Stream discharge}}$$

$$= \frac{Q}{Q_s}$$

$$= \frac{14.976}{26\times10^{-3}} = 576 \text{ s} = 9.6 \text{ min}$$

Q. 257

Catchment area = 600 ha = 6×10^6 m²

$$\text{Drainage coefficient} = 16 \text{ mm/day} = \frac{16\times6\times10^6}{24\times3600\times1000} = 1.11 \text{ m}^3/\text{sec}$$

$\therefore$ Volume of drainable water (Q) = 1.11 m³/sec

For most efficient channel

Bottom width (b) = 2d $\tan(\frac{\theta}{2})$

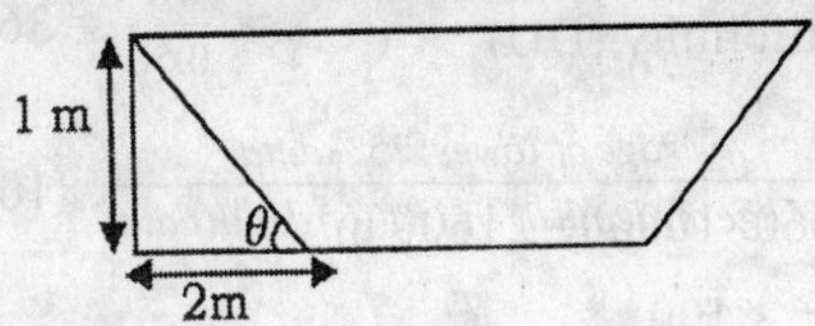

Hence, $\tan(\frac{\theta}{2}) = \frac{b}{2\times 1} = 0.236$ ($\because$ $\tan\theta = \frac{1}{2} \Rightarrow \theta = 26.57^o$)

$\Rightarrow$ b = 0.472 m

Applying equation of continuity

$$V = Q/A = \frac{Q}{\frac{1}{2}d(2b+4d)}$$

$$= \frac{1.11}{\frac{1}{2}\times 1\times(2\times 0.472+4\times 1)}$$

= 450 mm/ s

Q. 259

Depth of irrigation	Lowest 25%	Difference
42		7.8
40		5.8
25	25	9.2
36		1.8
40		5.8
36		1.8
32		2.2
28	28	6.2
30	30	4.2
38		3.8
32		2.2
35		0.8
29	29	5.2
32		2.2
34		0.2
38		3.8
34		0.2
31		8.2
28	28	6.2
44		9.8
Total - 684	Total - 140	Total - 82.7

Average of depth of irrigation = 684/20 = 34.2

Average of lowest 25% = 140/5 = 28

Distribution Uniformity (DU)

$$= \frac{\textit{Average of lowest 25\% data}}{\textit{Average of depth of irrigation depth data}} \times 100$$

$$= \frac{28}{34.2} \times 100$$

$$= 81.9\ \%$$

Note: - Coefficient of uniformity (CU)

$$= (1 - \frac{\textit{total difference from average value}}{\textit{total of all readings}}) \times 100$$

$$= (1 - \frac{82.7}{648}) \times 100 = 87.9\ \%$$

Q. 260

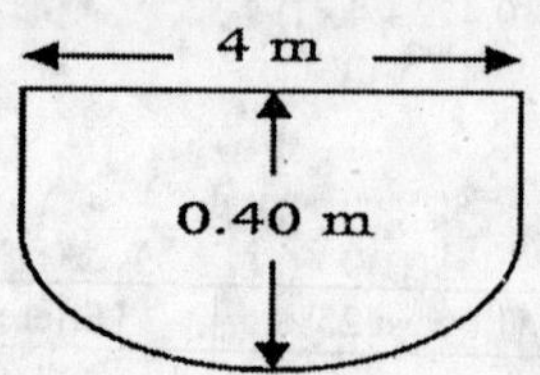

Slope (S) = 0.025

Manning's Value (n) = 0.035

Top width (b) = 4m

Maximum Depth (d) = 0.4 m

Wetted Perimeter (P) = $b + \frac{8d^2}{3b}$

$$= 4 + \frac{8(0.4)^2}{3 \times 4} = 4.107 \text{ m}$$

Cross Sectional Area (A) $= \frac{2}{3}$ b d

$$= \frac{2}{3} \times 0.4 \times 4 = 1.067 \text{ m}^2$$

Thus,

Hydraulic Radius (R) = A/P

R = 1.067/4.107 = 0.26 m

Velocity of discharge (V) = $\frac{R^{2/3}S^{1/2}}{n}$

$$= \frac{(0.26)^{2/3}(0.025)^{1/2}}{0.035} = 1.84 \text{ m/s}$$

Thus

Discharge (Q) = A×V

= 1.65×1.067 = 1.96 m³/s

Q. 262

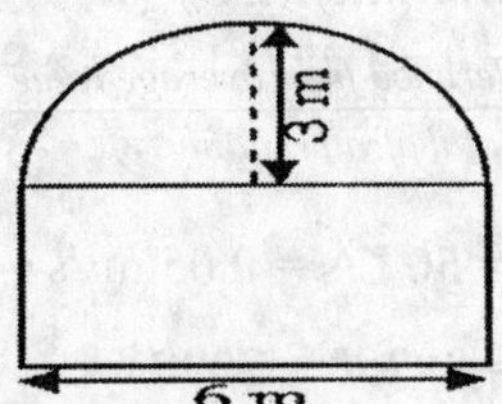

Applying Darcy Weisbach's equation

Head loss (h_f) = $\frac{fLV^2}{2gD} = \frac{\left(\frac{f'}{4}\right)LV^2}{2gD}$

Referring above figure

D = hydraulic mean depth = $\frac{wetted\ area}{wetted\ perimeter} = \frac{32.137}{21.4277}$ m

Now,

$$h_f = \frac{\left(\frac{0.017}{4}\right)\times 1000\times(128)^2}{2\times 9.81\times\frac{32.137}{21.4277}}$$

= 2.29 m

Q. 263

Inflow to reservoir = 2.5×100×10000 = 2500000 m³

Net inflow = 2500000 + (1 − 0.8)×2×365×24×3600

= 15114400 m³

∵ Increase in storage = 500×10000 = 5000000 m³

Therefore,

Evaporation = $\frac{15114400-5000000}{100\times 10000}$ = 10.11 m

Q. 265

We know

Flow velocity (V) = $\frac{1}{n}$ $(A)^{2/3}(S)^{1/2}$

$$V = \frac{1}{0.04} \times (1.52/12.5)^{2/3} \times (0.04)^{1/2}$$

$$= 1.227 \text{ m/ s}$$

Therefore,

Discharge (Q) = AV

$$= 1.227 \times 1.52 = 1.9 \text{ m}^3\text{/ s}$$

Q. 266

Stream discharge (Q) = 50 L/s = 0.05 m^3/s

Border strip area (A) = 8×250 = 2000 m^2

Infiltration rate (f) = 25 mm/h = 6.9×10^{-6} m/s

Depth of water over land (y) = 70 mm = 0.07m

∴ Time required to irrigate the border strip (t)

$$= \frac{y}{f} \ln\left(\frac{Q}{Q-fA}\right)$$

$$= \frac{0.07}{6.9\times10^{-6}} \times \ln\left(\frac{0.05}{0.05-6.9\times10^{-6}\times2000}\right)$$

$$= 3276.45 \text{ s} = 54.7 \text{ min}$$

Q. 267

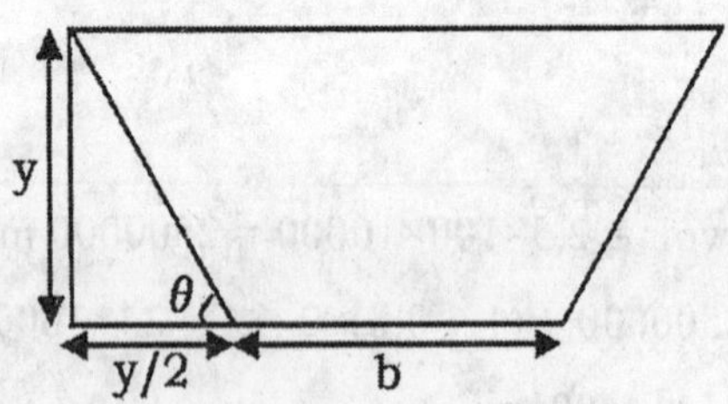

Discharge (Q) = 25 m^3/ s; Silt factor (F) = 1

Consider, Bottom width = b; Depth of flow = y

We know

$$\text{Velocity (V)} = \left(\frac{QF^2}{140}\right)^{1/6}$$

$$= \left(\frac{25 \times (1)^2}{140}\right)^{1/6} = 0.750 \text{ m/ s}$$

Again

$$\text{Wetted perimeter (P)} = 4.75\sqrt{Q}$$

$$= 4.75 \times \sqrt{25} = 23.75 \text{ m}$$

Also

$$\text{Cross sectional area (A)} = \frac{Q}{V}$$

$$= \frac{25}{0.75} = 33.33 \text{ m}^2$$

Referring above figure

$$A = (2b + y) \times \frac{y}{2}$$

$$\Rightarrow \quad 33.3 = (2b + y) \times \frac{y}{2} \qquad \text{(i)}$$

Also

$$P = b + \sqrt{y^2 + (y/2)^2}$$

$$\Rightarrow \quad 23.75 = b + \sqrt{y^2 + (y/2)^2} \qquad \text{(ii)}$$

Solving both the equations

$$b = 20.26 \text{ m (Approx.) and } d = 1.56 \text{ m (Approx.)}$$

Q. 268

Consider, critical depth (D_c) = d; g = 9.81 m/s²

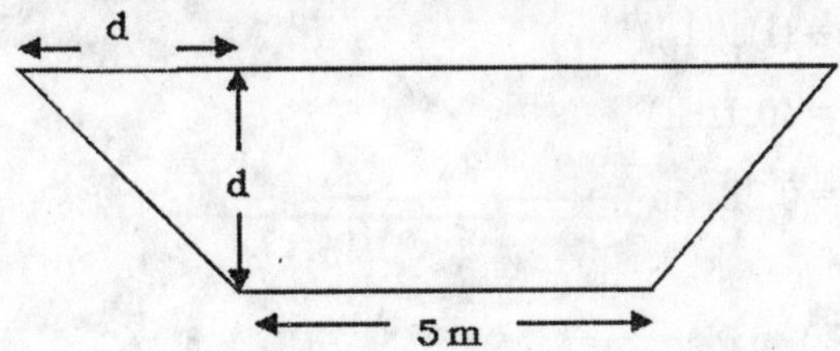

Referring above figure

Discharge (Q) = 20 m^3/ s; Top width (T) = 5 + 2d

Cross sectional area (A) = 5d + d^2 m^2

We know

$$\frac{A^3}{T} = \frac{Q^2}{g}$$

$$\Rightarrow \quad \frac{(5d + d^2)^3}{5 + 2d} = \frac{(20)^2}{9.81}$$

$$\Rightarrow \quad d = 1.09 \text{ m}$$

Q. 270

Depth Of Irrigation = 0.9×(4+4.3+4.6+4.9+5.12+5.18+6.21)

= 30.88 mm

Q. 271 (i)

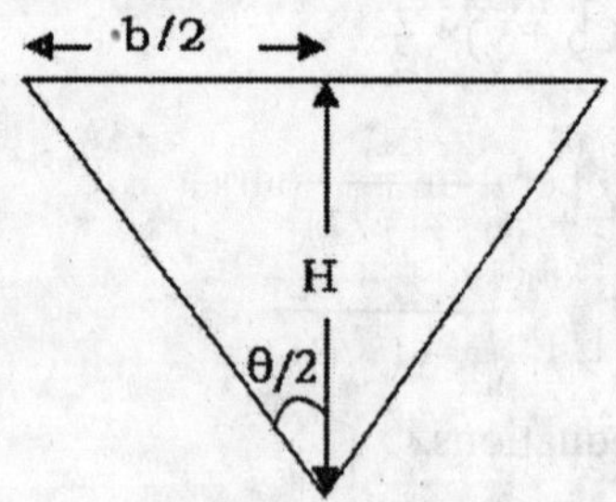

We know

Discharge through V notch (Q) = 0.0138×$H^{5/2}$

Therefore

$Q_1/Q_2 = (H_1/H_2)^{5/2}$

$0.0074/0.1 = (0.1/H_2)^{5/2}$

$\Rightarrow \quad H_2 = 0.283$ m

(ii)

We know

$$Q = \frac{8}{15} C^d H^{5/2} \sqrt{2g} \tan(\theta/2)$$

$$0.1 = \frac{8}{15} \times 0.7 \times (0.283)^{5/2} \sqrt{2 \times 9.81} \tan(\theta/2))$$

$\Rightarrow \quad \tan(\theta/2) = 1.419$

Now, $\quad b/2 = H\times \tan(\theta/2)$

$\Rightarrow \quad b = 2\times0.283\times1.419 = 0.80$ m

Q. 272

Effective root zone depth (d) = 1200 mm

Evapotranspiration (ET) = 260 mm; Effective rainfall (P) = 20 mm

We know

$$\text{Water depletion depth} = \frac{1200\times8}{100} = 96 \text{ mm}$$

$$\text{Consumptive use per day} = \frac{ET-P}{30}$$

$$= \frac{260-20}{30} = 8 \text{ mm}$$

Therefore,

$$\text{Irrigation interval} = \frac{96}{8} = 12 \text{ days}$$

Q. 273

Consider time of operation = t hours

Therefore, total discharge = $4\times4\times10^{-3}\times t$ m^3

$\because 4\times4\times10^{-3}\times t = \text{Water applied into the orchard} = 5\times5\times6\times10^{-3}\times0.8\times0.6\times0.6$

$\Rightarrow \quad t = 2.7$ hrs

Q. 274

(A) **Non-modular outlet** – This is an outlet in which the discharge depends upon the difference in level between the water levels in the distributing channel and water course (field channel).

Types of non-modular outlet –

(i) Submerged pipe outlet

(ii) Masonry sluice and orifices

(iii) Wooden shoots

(B) **Semi-modular or flexible type** – The discharge through semi-modular outlet is only affected by the change in the water level of the distributing canal.

Types of semi-modular outlet –

(i) Kennedy's gauge outlet

(ii) Crump's open flume outlet

(iii) Pipe-cum open flume outlet

(C) **Kennedy's gauge outlet** – It is a semi-modular type outlet.

(D) **Rigid modular outlet** – This type of outlet delivers a constant discharge within working limits, irrespective of the water level fluctuations in the distributing channel and/or field channel.

Types of modular outlet

(i) Gibb's rigid modular outlet

(ii) Khanna's rigid modular outlet

Q. 275

The USDA classification of the irrigation water comes between medium and high salinity hazards and the values for Sodium adsorption ratio (SAR), indicate the danger of sodification. Therefore, this water shouldn't be used for irrigation purpose without special management practices or treatments.

Q. 276

Applying Manning's Formula

$$\text{Discharge (Q)} = \frac{A R^{2/3} S^{1/2}}{n}$$

Again, $Q = K\,S^{1/2}$ $\quad (\because \text{Conveyance } K = \frac{A R^{2/3}}{n})$

$\Rightarrow$ $60 = K\,(0.01/100)^{1/2}$

$\Rightarrow$ $K = 6000$

Q. 277

According to Lacey's regime theory

Bed slope (S) = $(f^{5/3})/(3340\,Q^{1/6})$

Where, f = Lacey's number (constant)

Hence, $S_1 \times Q_1^{\,1/6} = S_2 \times Q_2^{\,1/6}$

$\Rightarrow$ $1.52\times10^{-4}\times(30)^{1/6} = 1.6\times10^{-4}\times(Q_2)^{1/6}$

$$Q_2 = 22.05\ \text{m}^3/\text{s}$$

Q. 278

For Cipolletti weir, side slope (H : V) = 1 : 4 = tan(θ/2)

Depth of water (H) = 0.5 m; Crest length (B) = 1.2 m

Coefficient of discharge (C_d) = 0.63 (Normally taken)

Applying Cipolletti's Formula

$$\text{Flow rate (Q)} = \frac{2}{3}\,\text{Cd B } H^{3/2}\,(2g)^{1/2}$$

$$= \frac{2}{3}\times 0.63\times 1.2\times(0.5)^{3/2}\times(2\times 9.81)^{1/2}$$

$$= 0.7893\ \text{m}^3/\text{s}$$

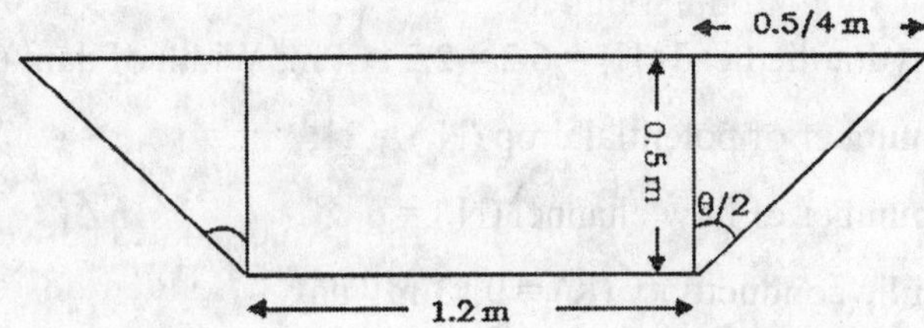

Referring above figure

Cross sectional area of weir (A) = 0.5×0.5/4 + 0.5×1.2

= 0.6625 m²

We know

Flow rate (Q) = AV

$\Rightarrow$ 0.7893 = 0.6625×V

$\rightarrow$ Velocity of water (V) = 0.7893/0.6625 = 1.1914 m/s

Q. 279

$$\text{Sodium Adsorption Ratio} = \frac{Na^{+}}{\sqrt{\dfrac{Mg^{+2}+Ca^{+2}}{2}}}$$

$$= \frac{28}{\sqrt{\dfrac{10+5}{2}}} = 10.23$$

Q. 280

We know

Irrigation requirement (IR) = Evapotranspiration requirement – Effective rainfall

IR = 840 – 20 = 820 mm

∵ Leaching efficiency = 0.8

Therefore,

Water percolation/drainage for leaching (D_d) = 820×0.8 = 656 mm

Again,

Leaching requirement = $EC_i/EC_d = D_d/D_i$

⇒ $1.1/1.7 = 656/D_i$

⇒ $0.65 = 656/D_i$

⇒ Depth of irrigation water (D_i)= 1009.23 mm

Q. 281

Total hydraulic head (H) = 6.2 – 2.2 = 4 m; Width of dam (b) = 72 m

Total number of potential drop (N_d) = 21;

Total number of flow channel (N_f) = 6

Hydraulic conductivity (K) = 0.53 m³/ day

According to flow net theory

Seepage rate through dam (Q) $= KH \frac{N_f}{N_d} b$

$= 0.53\times4\times(6/21)\times72$

$= 43.61$ m³/ day

Q. 282

Given that:

Depth of root zone (D_r) = 0.8 m

Bulk density of soil = 1.5 Mg m^{-3} = 1500 kg/m³

Hence, Specific gravity of soil (G) = 1.5 kg/m³

Field capacity (W_C) = 36%

Soil moisture content (W_r) = 24%

Depth of irrigation (D_i) $= \frac{(W_c - W_r)\times G\times D_r}{100\times\eta_a\times\eta_c}$

$= \frac{(36-24)\times1.5\times0.80}{100\times0.90\times0.80}$

$= 0.2$ m = 200 mm

Q. 283

Considering a rectangular section of width B and depth of flow h, one can determine the dimensions of the most efficient rectangular section as follows:

$$A = Bh$$

$$P = B + 2h$$

$$= A/h + 2h$$

For P to be minimum, $dp/dh = -A/h^2 + 2 = 0$

i.e., $A = 2h^2$

or $B = 2h$

$\therefore$ $P = 4h$

Thus, Hydraulic radius (R) = A/P

$$R = 2h^2/4h$$

$$R = h/2$$

Q. 284

The peak evapotranspiration rate is 10 mm day^{-1} = 10/5 = 2 mm/h

Evapotranspiration form 1 ha = $2\times10000/(1000\times3600)$

$= 6.94\times10^{-3}\ m^3/s$

The water horse power = ρghQ

$= 1000\times9.81\times30\times6.94\times10^{-3}$

$= 2.042$ kW

$\because$ Pump efficiency = 65%

$\therefore$ Horse power of the pump = 2.042/0.65

= 3.142 kW

or = 3.142/0.746 = 4.22 hp

Q. 285

Avg. depth (d)

$= d_1 + d_2$

$d = Q_1t_1/A_1 + Q_2t_2/A_2$

$= (Q_1t_1 + Q_2t_2)/A \qquad (\because A_1 = A_2 = 1)$

$= (100\times10^{-3}\times40 + 30\times10^{-3}\times60)/(80\times0.75) = 0.097$ m or 9.70 cm

Q. 286

Depth of root zone (D_r) = 0.70 m

Depth of irrigation = $(m_{fc} - m_{wp}) \times G \times D_r$

$= (0.25 - 0.13) \times 1.50 \times 0.70 = 0.126$ m

∵ Moisture depletion is 40% allowable.

Hence,

Permissible moisture depletion = 0.126×0.40 = 0.0504 m = 50.04 mm

Thus

Frequency of irrigation = Permissible moisture depletion/ Crop Consumptive Use

= 50.04/5 = 10.08 days

Q. 287

According to Manning's Equation

$$\text{Discharge}(Q) = (1/n)AR_h^{2/3}S^{1/2}$$

For most efficient rectangular channel

Width of channel = 2×Depth of the channel (d)

= 2d

Hydraulic radius (R_h) = Cross section area/wetted perimeter = $d^2/4d$ = d/2

Hence, $5 = (1/0.01)(2d^2)(d/2)^{2/3}(0.0005)^{1/2}$

⇒ d = 1.244 m

Hence, width of the channel = 2d = 2×1.244 = 2.488 m

Q. 288

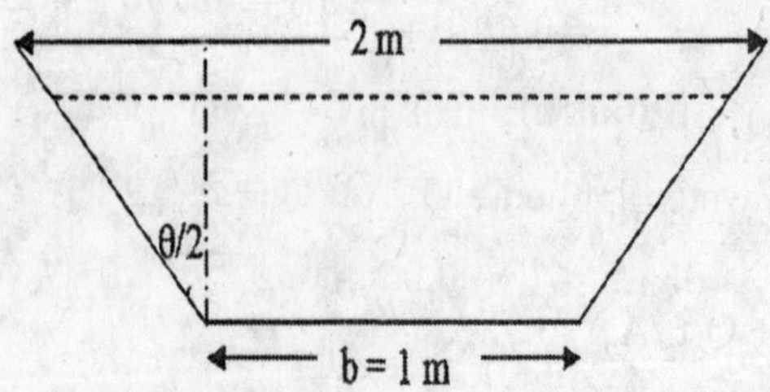

Referring figure

$\tan(\theta/2) = \frac{2-1}{2\times 0.5} = 1$

and Flow head (H) = 0.4 m

We know

Discharge over trapezoidal notch = Discharge over rectangular notch + Discharge over triangular notch

Hence,

$$\text{Discharge} = \frac{2}{3}C_d b\sqrt{2g}\,H^{3/2} + \frac{8}{15}C_d b\sqrt{2g}\,\tan(\theta/2)\,H^{5/2}$$

$$= \frac{2}{3}\times 0.68\times 1\times\sqrt{2\times 9.81}\times(0.4)^{3/2} + \frac{8}{15}\times 0.65\times 1\times\sqrt{2\times 9.81}\times 1\times(0.4)^{5/2}$$

$= 0.663\ m^3/s$ or 663 L/s

Q. 289

Permanent matured orchard area = 4×5 = 20 m^2

∵ 40 % of area is irrigated with drip emitters that is 0.40×20 = 8 m^2

Number of emitters per plant(N) = 8/(2×2) = 2

Average discharge (Q_{av}) = 0.3×(120)0.6 = 5.30 l/h (∵ Discharge = Discharge constant ×(Pressure)^{Exponent})

And Minimum discharge (Q_{min}) = 0.3×(100)$^{0.6}$ = 4.75 l/h

$$\text{Emission uniformity (EU)} = \left[1-\frac{1.27Cv}{\sqrt{N}}\right]\frac{Q_{min}}{Q_{avg}}\times 100$$

$$= \left[1-\frac{1.27\times 0.06}{\sqrt{2}}\right]\frac{4.75}{5.30}\times 100 = 84.79\ \%$$

Q. 290

For Furrow stream Maximum non erosive flow rate (Q_{max})

= 0.6/S

= 0.6/0.5 = 1.2 l/s or 4.32 m^3/h

∵ Furrow area(A) = 120×0.90 = 108 m^2

Average depth of water applied = Q_1/A + Q_2/A = 4.32/108 + 2.16/108

= 0.06 m or 6 cm

Q. 291

Area covered under each lateral = 12×14×20 = 3360 m²

Therefore,

Amount of fertilizer to be applied = 80×3360/10000 = 26.88 kg

Q. 292

In Muskingum Method

$$C_0 = -\frac{kx - 0.5\Delta t}{k(1-x) + 0.5\Delta t}$$

$$= -\frac{12 \times 3600 \times 0.15 - 0.5 \times 4 \times 3600}{12 \times 3600 \times (1-0.15) + 0.5 \times 4 \times 3600} = 0.016$$

Q. 293

Water to be applied = 1.2×(0.15 - 0.075)×1.5

= 0.135 m

Quantity of water to be applied = 12×12×0.135 = 19.44 m³

Irrigation time = 19.44/(27×10⁻³×60) = 12 minutes

Q. 294

Volume of water = 0.20/1000 m³

$$\text{Volume of soil} = \frac{\pi D^2}{4} L$$

= 3.14×0.075²×*0.15/4 = 6.62/10000 m³*

Hence,

Percentage of water drained or depth of water = (0.20/1000)/(6.62/10000)

= 0.302 meter of water/ meter depth of soil

= 30.2 cm of water/ meter depth of soil

Q. 295

According to ASAE(American Society of Agricultural Engineers) standards, drippers are categorized as:-

Emitter type	C_v	Interpretation
Point Source	< 0.05	Excellent
	0.05 - 0.07	Average
	0.07 - 0.11	Marginal
	0.11- 0.15	Poor
	> 0.15	Unacceptable

Line Source	< 0.10 0.10 - 0.20 > 0.20	Good Average Marginal to Unacceptable

Table 1: ASAE Interpretation on Manufacturing Coefficient of Variation

Flow Regime	x - Value	Emitter Type
Variable flow path	0.0 0.1 0.2 0.3	Pressure compensating
Vortex flow	0.4	Vortex
Fully turbulent flow	0.5	Orifice tortuous
Mostly turbulent flow	0.6 0.7 0.8	Long or spiral path
Mostly laminar flow	0.9	Micro tube
Fully laminar flow	1.0	Capillary

Q. 296

Given that:

Depth of root zone (D_r) = 50 cm

Specific gravity of soil (G) = 1.5

Field capacity (W_C) = 32 %

Wilting point (W_r) = 12%

Depth of irrigation (D_i) = 50×(0.32-0.12)×0.60/0.75 = 8 cm

Q. 297

Discharge of 90° V shape weir = Discharge of Rectangular shape weir

$$\Rightarrow \quad 0.138\, H^{5/2} = 1.84\, LH^{3/2}$$

$$0.138 \times 36^{5/2} = 1.84 \times 1 \times H^{3/2}$$

$$\Rightarrow \quad H = 15 \text{ cm}$$

Q. 298

$$\text{Emission uniformity (EU)} = [\,1 - \frac{1{\cdot}27\, C_v}{N^{0.5}}] \times \frac{Q_{min}}{Q_{avg}}$$

$$= 1 - \frac{1{\cdot}27 \times 0.07}{1^{0.5}}] \times \frac{45}{50} = 0.82$$

Drip irrigation efficiency = 0.90×0.82 = 0.738 or 73.8 %

Q. 299

$$\text{SAR} = \frac{Na^{+2}}{\sqrt{\frac{Ca^{+2}+Mg^{+2}}{2}}}$$

Concentration of HCO^{-3} = 11.20 meq/L

Concentration of Ca^{+2} = 11.20/2.8 = 4 meq/L

We know

$$\text{SAR} = \frac{Na^{+2}}{\sqrt{\frac{Ca^{+2}+Mg^{+2}}{2}}}$$

$$= \frac{9.90}{\sqrt{\frac{4+5.68}{2}}} = 4.5$$

Q. 300

Slope (S) = 0.0004

Manning's Value (n) = 0.01

Maximum Depth (d) = 0.4 m

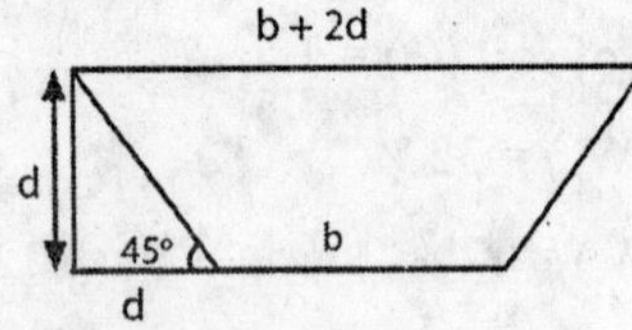

For most efficient channel

Bottom width (b) = 2d $\tan(\frac{\theta}{2})$

∵ $b = 2\times0.4\times\tan(\frac{45}{2})$ (∵ θ = 45°)

∵ b = 0.33 m

Top width (b) = 1.13 m

Wetted Perimeter (P) = $2\times(\sqrt{2}\times0.4) + 0.33 = 1.46$ m

Cross Sectional Area (A) = $\frac{1}{2}\times0.4\times(0.33 + 1.13) = 0.292\ m^2$

Thus,

Hydraulic Radius (R) = A/P

R = 0.292/1.46 = 0.2 m

Velocity of discharge (V) = $\frac{R^{2/3}S^{1/2}}{n}$

$$= \frac{(0.2)^{2/3}(0.0004)^{1/2}}{0.01} = 0.68 \text{ m/s}$$

Thus,

Discharge (Q) = A×V

$$= 0.68 \times 0.292 = 0.20 \text{ m}^3/\text{s}$$

26

Drainage System

UKPSC-AE-Agricultural-Engineering-Paper-II-2013

Q. 1 Which of the following is a correct match?

(A) Mole drain = pump drainage

(B) Vertical drainage = well drain age

(C) Pipe drain = well drainage

(D) Horizontal drain = bio-drainage

Q. 2 Dimension of leakage factor is equivalent to

(A) Time (B) Velocity

(C) Resistance (D) Length

Q. 3 Subsurface drains removes

(A) Gravitational water (B) Capillary water

(C) Hygroscopic water (D) All above

Q. 4 Spacing of drains in flat land under parallel field drain system varies from

(A) 150 - 300 m (B) 100 - 500 m

(C) 100 - 200 m (D) 300 - 450 m

Q. 5 Spacing of mole drains varies from

(A) 2 to 5 m (B) 5 to 7.5 m

(C) 7 to 15 m (D) 2 to 10 m

Q. 6 Dimension of leakage factor is

(A) L (B) L/T

(C) T^{-2} (D) LT^{-2}

Q. 7 The uppermost line of seepage in an earth embankment is called

(A) Phreatic line (B) Lower most flow line

(C) Runoff line (D) Water pressure line

Q. 8 If drainage canal discharge is 0.2 m^3/sec and the drainage area is 250 hectares, then the drainage coefficient will be

(A) 70 cm (B) 70 mm

(C) 0.5 cm (D) 7 mm

Q. 9 A soil has hydraulic conductivity in horizontal and vertical directions as K_1 and K_2, then its effective hydraulic conductivity would be

(A) $K_1 + K_2$ (B) K_1 / K_2

(C) $K_1 \times K_2$ (D) $K_1 \times K_2/2$

Q. 10 20–40 rule is used to compute

(A) Head loss in a drainage pipe

(B) Slope of drainage ditch

(C) Design discharge below the junction point of ditches

(D) Hydraulic radius of drainage ditch

Q. 11 To ensure surface drainage, a minimum furrow grade needed is

(A) 0.5 % (B) 0.05 %

(C) 5 % (D) 10 %

Q. 12 The nature of Hougoudt's equation for drain spacing is

(A) Parabolic (B) Hyperbolic

(C) Elliptical (D) Circular

Q. 13 Depth of open ditch for surface drainage varies from

(A) 60 to 100 cm (B) 100 to 150 cm

(C) 150 to 250 cm (D) 30 to 60 cm

Q. 14 The dimensions of transmissibility of an aquifer will be

(A) L/T (B) L^2/T

(C) L/T^2 (D) None of these

UKPSC AE Agricultural Engineering Paper-II 2013

Q. 15 Unit of Drainage coefficient is

(A) kg (B) ton

(C) cm (D) percent

Q. 16 Spacing of tile drains is generally kept as

(A) 1 – 2 m (B) 4 – 5 m

(C) 9 – 60 m (D) 70 – 95 m

Q. 17 The depth of tile drains in uniformly permeable soil is kept as

(A) 0.25 to 0.5 m (B) 0.75 to 1.5 m

(C) 2.0 to 3.5 m (D) 4.0 to 5.5 m

Q. 18 Which of the following is not a drainage system ?

(A) Drip system (B) Surface system

(C) Sub-surface system (D) All of the above

Q. 19 Drainage coefficient is defined as the depth of water in centimeter drained from any area in

(A) 12 hours (B) 18 hours

(C) 06 hours (D) 24 hours

Q. 20 Most important factor influencing the value of drainage coefficient is

(A) Crop (B) Field size

(C) Rainfall (D) None of the above

Q. 21 Collector drains must have their diameter

(A) Larger than lateral drains (B) Less than lateral drains

(C) Equal to lateral drains (D) None of the above

Q. 22 Sub-surface drainage system includes

(A) Drainage outlet and collector drains only

(B) Drainage outlet and laterals only

(C) Drainage outlet, collectors and laterals

(D) None of the above

Q. 23 In irrigated regions, water logging of low lands is due to

(A) Seepage from canals

(B) Seepage from unlined irrigation channels

(C) Both of the above

(D) None of the above

Q. 24 Open drains have disadvantage because

(A) Divide farm area into fragments.

(B) Interfere with the movement of machinery.

(C) Waste considerable area of the farm.

(D) All of the above

Q. 25 Mole drains are underground unlined circular or oval earthen channels formed by a mole plough in

(A) Sandy-loam soils (B) Non-cohesive soils

(C) Sandy soils (D) Cohesive soils

Q. 26 A drainage canal draining 250 hectare area has a discharge of 0.3 m^3/s, the drainage coefficient of the area will be

(A) 0.7 cm (B) 1.036 cm

(C) 1.412 cm (D) 2.178 cm

Q. 27 The water which can be removed by drainage is

(A) Hygroscopic water (B) Gravitational water

(C) Capillary water (D) Perched water

Q. 28 Which of the following trees can be used for bio-drainage work ?

(A) Mango (B) Eucalyptus

(C) Guava (D) Sisal

Q. 29 In tile drain installation, the envelope material is required for

(A) Increasing the flow rate

(B) Increasing the life of tile drains

(C) Increasing the strength of tile drains

(D) Decreasing the length of tile drains

Q. 30 Which of the following is the surface drain ?

(A) Mole drain (B) Tile drain

(C) Vertical drain (D) Open ditch drain

Q. 31 For drainage of land with concave surface or a valley at the centre, the most suitable drainage system is

(A) Herringbone system (B) Bedding system

(C) Random system (D) Gridiron system

Q. 32 In water logged lands, the soil pores are saturated

(A) Within a depth of 30 cm (B) Within a depth of 40 cm

(C) Within a depth of 60 cm (D) Beyond root zone of a crop

Q. 33 The removal of excess water from the ground surface is called

(A) Ground water flow (B) Surface drainage

(C) Sub-surface drainage (D) Water table control

Q. 34 Which major nutrient is very soluble and can be expected in drainage effluent ?

(A) Nitrogen (B) Phosphorous

(C) Potash (D) None of the above

Q. 35 In water-logged areas, when organic matter accumulates in the soil ?

(A) Before water logging (B) After water logging

(C) During water logging (D) None of the above

Q. 36 In which type of soil, the drainage will be poor ?

(A) Sandy soils (B) Clayey soils

(C) Silty soils (D) Loamy soils

Q. 37 Which of the following is not a salt tolerant crop ?

(A) Barley (B) Dhaincha

(C) Gram (D) Spinach

Q. 38 Internal drainage is the common problem in these soils

(A) Deltaic alluvium (B) Coastal sands

(C) Black cotton soils (D) Red loam soils

APPSC AEES Agriculture Engineering Exam 2016

Q. 39 Mole drains are suitable for

(A) Very coarse soil (B) Medium coarse soil

(C) Sandy loam soil (D) Fine texture soil

MPPSC Assistant Agricultural Engineer 2013

Q. 40 The structure used for passing the canal over the natural drain is called as ....

(A) Aqueduct (B) Syphon

(C) Cross regulator (D) None of these is correct

Q. 41 A field of 120 ha is drained by drainage channel. The discharge of drainage channel is 0.1 m^3/s. What is the drainage coefficient?

(A) 0.0072 cm (B) 0.072 cm

(C) 0.72 cm (D) 7.2 cm

Q. 42 Alkaline soils are better reclaimed by ....

(A) Providing sub surface drainage

(B) Addition of sodium salts to soil

(C) Addition of gypsum to soil and leaching

(D) Addition of gypsum to soil

Q. 43 The drain spacing for steady state drainage is decided by using .... equation.

(A) Ernst's (B) Hooghoudt's

(C) Schilfgoarde's (D) None of these is correct

Q. 44 In which of the following types of soils, mole drainage is best suited ?

(A) Sandy soil (B) Silty soil

(C) Clayey soil (D) Loamy soil

Q. 45 A soil having EC more than 4 dS/m and ESP more than 15 is termed as ....

(A) Sodic soil (B) Saline soil

(C) Saline sodic soil (D) Normal soil

RPSC AEn Pre Exam 2013 (Agricultural Engineering)

Q. 46 Drainage Coefficient in a given area can be judged as

(A) Ratio of total channels length to the total drainage area

(B) Ratio of total channel length to number of channels in the area

(C) Both (A) and (B)

(D) None of the above

OCS AEn Exam 2011 (Pre) – II (Agricultural Engineering)

Q. 47 Spacing between laterals of tile drainage system when installed in heavy-textured soil in comparison to when installed in light-textured soil is kept:

(A) More (B) Less

(C) Equal (D) Not governed by soil type

Q. 48 Vertical drainage system is more suitable for installation when :

(A) Aquifer has good water quality and hydraulic conductivity of the aquifer is high

(B) Aquifer has poor water quality and hydraulic conductivity of the aquifer is low

(C) For immediate removal of excess water from root zone

(D) For immediate removal of excess water from land surface

Q. 49 Drainage system is installed for:

(A) To control only soil moisture and soil temperature in the root zone

(B) To control only soil salinity and soil temperature in the root zone

(C) To control only soil salinity and soil moisture content in the root zone

(D) To control soil salinity, soil temperature and soil moisture content in the root zone

CGPSC State Engineering Services Exam (Agriculture Engineering)
Part - II

Q. 50 In tile drainage system which layout type is used to drain seepage water across hillside'

(A) Grid-ron (B) Herring bone

(C) Interceptor drain (D) Random

(E) Parallel field

Q. 51 Glover-Dumm equation used in which state for drain spacing ?

(A) Unsteady state (B) Steady state

(C) Uniform (D) Non uniform

(E) None of these

Q. 52 Drainage coefficient for a drainage having 1 m^3/sec for 100 ha field is

(A) 0.864 cm (B) 8.64 cm

(C) 864 cm (D) 8640 cm

(E) 8641 cm

Q. 53 The flow when water starts disappear at head end until it eventually recedes from field after cut off flow

(A) Depletion flow (B) Cut back stream

(C) Advance flow (D) Recession flow

(E) Subsurface flow

Q. 54 Type of flow into a tile drain

(A) Horizontal (B) Vertically downward

(C) Vertically upward (D) Radial flow

(E) Both (A) and (B)

Q. 55 Sub surface drains remove

(A) Excess surface water

(B) Capillary sub surface water

(C) Sub surface gravitational water

(D) Excess runoff water from rainfall

(E) Rainfall water

UKPSC A.En. Exam 2007 Paper – II

Q. 56 The critical depth in a mole drain usually occurs at a depth corresponding to an aspect ratio of about

(A) 5 : 7 (B) 2 : 3

(C) 1 : 3 (D) 2 : 5

Q. 57 A mole plough is used for

(A) Ploughing the land (B) Harrowing the land

(C) Draining the land (D) Ridging the land

Q. 58 Herringbone pattern is seen in

(A) Surface drainage system (B) Surface irrigation system

(C) Tile drainage system (D) None of the above

Q. 59 Hooghoudt's equation for tile drainage is based on

(A) Manning's equation (B) Darcy's equation

(C) Chezy's equation (D) None of the above

Q. 60 Slow lateral movement of gravitational water through the soil is known as

(A) Percolation (B) Infiltration

(C) Seepage (D) None of the above

Q. 61 Water in a subsurface drainage system is generally carried through

(A) G.I. pipes (B) Open drains

(C) Tile drains (D) None of the above

UKPSC A.En. Exam 2012 Paper - II

Q. 62 A tile drainage system consists of

(A) Tile drain (B) Submain

(C) Drainage outlet (D) All of the above

Q. 63 The drainable porosity of clay soil varies from

(A) 10 – 15 % by volume (B) 18 – 35 % by volume

(C) 3 – 11 % by volume (D) 15 – 20 % by volume

Q. 64 In drainage projects the design return period usually ranges from

(A) 5 – 25 years (B) 25 – 50 years

(C) 50 – 75 years (D) 75 – 100 years

Q. 65 Steady state drainage design equation assumes that the drain discharge

(A) Exceeds the recharge to the ground water

(B) Equals the recharge to the ground water

(C) Both (A) and (B)

(D) None of the above

Q. 66 Blinding of tiles refers to

(A) Trenching of the drains

(B) Laying the tiles

(C) Joining the tiles

(D) Covering the tiles with loose earth

Graduate Aptitude Test in Engineering - 2007

Q. 67 In a sub-surface drainage system, tile drains are laid with a slope of 0.28% to carry a peak discharge of 3 liter s^{-1} per drain. If the Manning's n is 0.011, the practical diameter of tile required is

(A) 50 mm (B) 75 mm

(C) 100 mm (D) 150 mm

Graduate Aptitude Test in Engineering - 2008

Q. 68 The following data were obtained from an agricultural land requiring a pipe drainage system for groundwater control:

Hydraulic conductivity = 8.3 cm/h , drainable porosity = 5%, reaction factor = 0.31 per day and equivalent depth to the impermeable layer = 2.8 m.

The drain spacing computed by the Glover-Dumm formula will be

(A) 60 m (B) 190 m

(C) 50 m (D) 6 m

Q. 69 A tile drainage system draining 12 ha flows at the design capacity for two days in response to a storm. If the system is designed using a drainage coefficient of 1.25 cm, the amount of water removed from the drainage area during two days is

(A) 150 (B) 1500

(C) 30 (D) 3000

Graduate Aptitude Test in Engineering - 2009

Q. 70 The nature of Hooghoudt's equation for drain spacing is

(A) Parabolic (B) Hyperbolic

(C) Elliptic (D) Circular

Graduate Aptitude Test in Engineering - 2011

Q. 71 A crop grown over an area of 50 ha can tolerate 5 decisiemens m^{-1} of electrical conductivity in the drainage water. The consumptive use of the crop is 1.0 m, 30% of which is obtained from the rainfall and the remainder is met from the irrigation water having electrical conductivity of 2 decisiemens m^{-1}. For efficient crop production, the leaching requirement and the quantity of water that must be drained from the area are

(A) 40 % and 0.23 Mm^3 (B) 60 % and 0.23 Mm^3

(C) 40 % and 0.21 Mm^3 (D) 60 % and 0.21 Mm^3

Graduate Aptitude Test in Engineering - 2013

Q. 72 The gridiron pipe drainage system is more economical than the herringbone pipe drainage system because

(A) It is adopted in the fields which do not require complete drainage

(B) The number of main or sub-main lines is reduced

(C) The number of junctions and the double-drained area are reduced

(D) It has only main or sub-main lines

Q. 73 If the drainable porosity of a command area is 5% and the design rate of drop of the water table is 0.25 m day^{-1}, the drainage coefficient of the command area in mm day^{-1} will be

(A) 250 (B) 12.5

(C) 1.25 (D) 0.0125

Graduate Aptitude Test in Engineering - 2014

Q. 74 A tile drainage system having 200 mm diameter lateral of 400 m length is used to drain an area with a drainage coefficient of 40 mm. Manning's roughness coefficient for the drain pipe is 0.01. Drain pipes are laid at 0.3% slope. The spacing of the tile drain in m is.........

Graduate Aptitude Test in Engineering – 2015

Q. 75 Mole drain is the most suitable drainage system for

(A) Heavy clay soil (B) Loamy soil

(C) Sandy soil (D) Silty soil

Graduate Aptitude Test in Engineering – 2016

Q. 76 In a subsurface drainage system, the peak discharge through tile drain under full flow condition is given by

$Q = 5.715\times10^{-4}S^{0.5}n^{-1}$

Where, Q = discharge, m^3s^{-1}, S = drain bed slope and n = Manning's roughness coefficient.

Size of the drain in mm is ....

Graduate Aptitude Test in Engineering – 2017

Q. 77 Modified Hooghoudt's equation for the computation of drain spacing is applicable to

(A) Homogeneous soils (B) Anisotropic soils

(C) Heavy clay soils only (D) Layered soils

Graduate Aptitude Test in Engineering – 2018

Q. 78 One of the key assumptions made by Dupuit-Forchheimer about the flow behavior in the porous medium towards subsurface drains is

(A) Homogeneous medium (B) Isotropic medium

(C) Horizontal streamlines (D) Uniform recharge

Q. 79 The depths from the soil surface to subsurface tile drains, impermeable soil layer and the highest water table are measured as 2.8 m, 5.0 m and 0.8 m, respectively. The effective hydraulic head for drainage in meter is

(A) 0.8 (B) 2.0

(C) 2.2 (D) 4.2

Graduate Aptitude Test in Engineering – 2019

Q. 80 In a drainage area of 15 ha, the slope and drainage coefficient are 0.4% and 11 mm/ day. respectively. The value of Manning's roughness coefficient is 0.016. The inside diameter (in mm) of the corrugated plastic tubing used for drainage is

(A) 200.51 (B) 205.52

(C) 209.51 (D) 215.23

Graduate Aptitude Test in Engineering – 2020

Q. 81 The drainage coefficient or a watershed of 720 ha area is 1.2 cm. The design discharge of the drain in m^3/s is ………

Graduate Aptitude Test in Engineering – 2021

Q. 82 The approximate relationship between Sediment Delivery Ratio (SDR) and drainage area (A) shows that SDR varies

(A) Directly with $A^{0.2}$ (B) Inversely with $A^{0.2}$

(C) Directly with A (D) Inversely with A

Answers Key

1	2	3	4	5	6	7	8	9	10
B	D	A	C	D	A	A	D	D	C
11	12	13	14	15	16	17	18	19	20
B	C	A	B	C	C	B	A	D	C
21	22	23	24	25	26	27	28	29	30
A	C	C	D	D	B	B	B	A	D
31	32	33	34	35	36	37	38	39	40
A	D	B	A	B	B	C	A	D	A
41	42	43	44	45	46	47	48	49	50
C	C	B	C	C	D	A	A	D	C
51	52	53	54	55	56	57	58	59	60
A	B	D	D	C	A	C	C	B	C
61	62	63	64	65	66	67	68	69	70
C	D	C	A	B	D	C	A	D	C
71	72	73	74	75	76	77	78	79	80
A	C	B	126	A	100	D	C	B	C
81	82								
1	B								

Explanations

Q. 8

Drainage Coefficient = 0.2×3600×24/(250×10000) = 0.0070 m/day or 7 mm

Q. 26

Refer explanation for Q. 8.

Q. 41

Refer explanation for Q. 8 and Q. 26.

Q. 46

Refer Q. 19 for definition.

Q. 52

Refer explanation for Q. 8 and Q. 26 and Q. 41.

Q. 67

Consider diameter of tile drain = D m

$\therefore$ Hydraulic radius(R) = D/4

Cross sectional area of tile drain (A) = $\frac{\pi}{4}D^2$

Applying Manning's Formula

$$\text{Discharge}(Q) = \frac{1}{n} \times A \times (R)^{2/3}(S)^{1/2}$$

or
$$3\times10^{-3} = \frac{3.14\times D^2\times(D/4)^{2/3}(0.0028)^{1/2}}{4\times0.011}$$

$\Rightarrow$
$$D = 0.097 \text{ m} = 97 \text{ mm} \approx 100 \text{ mm}$$

(Practically, no tile drain of 97 mm diameter is available.)

Q. 68

Hydraulic conductivity (K) = 8.3 cm/h = 0.083 m/h = 1.992 m/d

Drainable porosity (μ) = 5 %

Reaction factor (α) = 0.31 /day =

Equivalent depth of impermeable layer (d) = 2.8 m

According to Glover-Dumm

$$\text{Reaction factor}(\alpha) = \frac{10Kd}{\mu L^2} \Rightarrow L = \sqrt{\frac{10Kd}{\mu\alpha}}$$

$$\therefore \quad L = \sqrt{\frac{10 \times 1.992 \times 2.8}{0.31 \times 0.05}}$$

$$= 59.99 \text{ m} \approx 60 \text{ m}$$

Q. 69

Drain area (A) = 12 ha = 120000 m^2

Drainage coefficient = 1.25 cm

Drainage coefficient is the depth of water drained in 24 hours or per day.

$\therefore$ Drainage rate (v) = 0.0125 m/day

Amount of water removed in two days = 120000×2×0.0125 = 3000 m^3

Q. 71

Leaching Requirement (L.R.)

$$= \frac{\textit{Electrical conductivity of irrigation water}}{\textit{Electrical conductivity of water to be drained}} = \frac{D_d}{D_i}$$

$$= \frac{2}{5} = 40\ \%$$

Depth of irrigation water = d (consider)

Now,

Depth of water to be drained(D_d) = 0.40×d ($\because$ L.R. = $\frac{D_d}{D_i}$)

Therefore,

d – 0.40×d = 1×0.30

$\Rightarrow$ d = 1.167 m

Again,

Depth of water to be drained (D_d) = 0.40×1.167 = 0.466 m

Quantity of water to be drained = 0.466×50×10000 = 0.233 M m^3

Q. 73

$$\text{Drainage coefficient} = 0.25 \times 1000 \times \frac{5}{100}$$

$$= 12.5 \text{ mm day}^{-1}$$

Q. 74

Drainage coefficient = water to be drained per day

$= 40$ mm/day $= 5.55\times10^{-7}$ m/s

Consider, Drain spacing = S_d m

Again, Area of field to be drained = L× S_d

$= 400\ S_d\ m^2$

Discharge (Q) = Drainage coefficient ×Area

$= 5.55\times10^{-7}\times400\ S_d\ m^3/s$

Now, for tile drains

$$Q = AV$$

$$\Rightarrow \quad 5.55\times10^{-7}\times400\ S_d = \frac{1}{n}\left(\frac{\pi d^2}{4}\right)\times(D/4)^{2/3}(S)^{1/2}$$

$$\Rightarrow \quad 5.55\times10^{-7}\times400\ S_d = \frac{1}{0.01}\times\frac{\pi(0.2)^2}{4}\ (0.2/4)^{2/3}(S)^{1/2}$$

$$\Rightarrow \quad S_d = 126\ m$$

Q. 76

Given that

$$Q = 6.715\times10^{-4}\left(\frac{1}{n}S^{1/2}\right) \quad \text{(i)}$$

According to Maning's formula

$$Q = AV = A\left(\frac{1}{n}\ R^{2/3}\ S^{1/2}\right) \quad \text{(ii)}$$

Comparing both the equation

$$AR^{2/3} = 6.715\times10^{-4}$$

$$\Rightarrow \quad (\pi D^2/4)(D/4)^{2/3} = 6.715\times10^{-4}$$

$$(\because R = D/4 \text{ and } A = \pi D^2/4)$$

$$\Rightarrow \quad D = 0.1\ m = 100\ mm$$

Q. 79

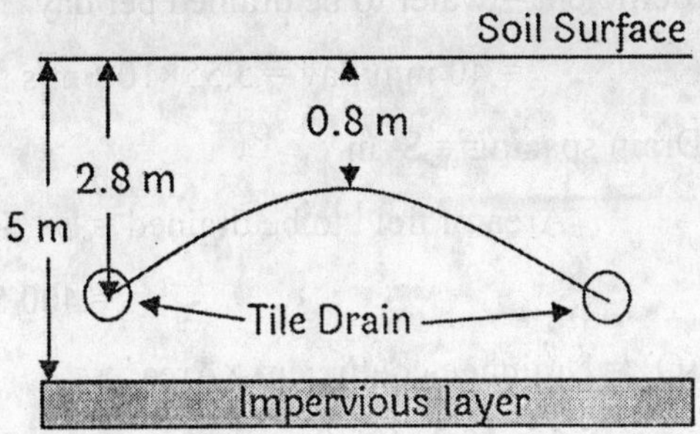

Effective Hydraulic Head = 2.8 – 0.8 = 2 m

Q. 80

Discharge from the drainage area = 15×10000×11/(1000×24×3600)

= 0.0191 m^3/s

Use Manning's Formula

$$\text{Discharge (Q)} = \frac{A R^{2/3} S^{1/2}}{n}$$

$$\Rightarrow \quad 0.0191 = \frac{\pi D^2}{4}\left[\frac{(D/4)^{2/3}(0.004)^{1/2}}{0.016}\right]$$

$$\Rightarrow \quad D = 0.20651 \text{ m or } 209.51 \text{ mm}$$

Q. 81

Drainage coefficient = 1.2 cm in 24 hrs

= 1.39×10^{-7} m/s

Design discharge = $1.39\times10^{-7}\times720\times10000$ = 1 m^3/s

27

Soil Conservation Measures

UKPSC-AE-Agricultural-Engineering-Paper-II-2013

Q. 1 Which type of gully is formed on flat land?

(A) U-shaped (B) V-shaped

(C) L-shaped (D) T-shaped

Q. 2 Check dams are provided with the main objective of

(A) Ponding of water

(B) Causing upstream sedimentation

(C) Reducing channel (gully) slope

(D) All above

Q. 3 Contour farming is recommended for lands having slope

(A) Less than 1 % (B) 2 – 7 %

(C) 7 – 12 % (D) More than 12 %

Q. 4 If rainfall occurs on a perfectly impermeable surface (such as glass), then the Curve Number (CN) for computing runoff will be assumed to be

(A) 50 (B) 75

(C) 100 (D) 25

Q. 5 The average soil loss in India from agricultural lands is about

(A) 6-8 t/ha/year (B) 10-12 t/ha/year

(C) 20-30 t/ha/year (D) 80-100 t/ha/year

Q. 6 Contour bunds are adopted

(A) In low rainfall regions

(B) In highly pervious soils

(C) In high rainfall regions

(D) Where conservation of moisture is needed

Q. 7 If the velocity of overland flow is doubled, the erosive capacity is increased as

(A) 2 times (B) 4 times

(C) 6 times (D) 32 times

Q. 8 The energy of a raindrop causing erosion is expressed in

(A) m-mg / ha.mm (B) m.mg / ha

(C) m.t / ha (D) t / ha-m

Q. 9 The kinetic energy of raindrops compared to that of runoff is

(A) 100 times more (B) 100 times less

(C) 250 times more (D) 30 times more

Q. 10 Dicken's formula is used to determine

(A) Peak discharge (B) Monthly runoff

(C) Periodic runoff (D) Fortnightly runoff

Q. 11 Tension on a hydraulic structure develops because of

(A) Load of water (B) Action of uneven forces

(C) Structural weight (D) None of these

Q. 12 Erodibility is basically associated to

(A) Soil properties (B) Rainfall properties

(C) Crops (D) Cultural practices

Q. 13 The relationship between bench width (W) and depth of cut (D) in bench terraces is given by

(A) W = 100D/S (B) W = 100S/D

(C) W = 200S/D (D) None of above

Q. 14 Velocity Measurement of stream flow is carried out by the device called

(A) Current meter (B) Planimeter

(C) Tachometer (D) Hydrometer

Q. 15 In large watersheds, the peak discharge (Q) is computed by

(A) $Q \alpha A^{-n}$ (B) $Q \alpha \sqrt{S^{-1}}$

(C) $Q \alpha \log A$ (D) $Q \alpha A^{n}$

Q. 16 Inward slope of a bench terrace is in the range of

(A) 1 - 5 Percent (B) 5 - 10 Percent

(C) 2 - 10 Percent (D) 8 - 20 Percent

Q. 17 Soil erosion will be more in

(A) Clay soil (B) Loam soil

(C) Sandy soil (D) Silty clay soil

Q. 18 If a fourth order watershed has bifurcation ratio as 4, the number of second order streams would be

(A) 4 (B) 8

(C) 16 (D) 32

Q. 19 In treatment of catchment area of gullies, contour bunds are recommended for lands with slopes upto

(A) 4 % (B) 6 %

(C) 8 % (D) 10 %

Q. 20 Soil formation that have the maximum specific yield is

(A) Clay (B) Sand

(C) Gravel (D) Sand stone

Q. 21 Runoff coefficient will be higher in

(A) Sandy soil (B) Sandy loam soil

(C) Clay and silt loam (D) Stiff clay

Q. 22 The ratio of the basin area to the area of circle having the same perimeter as the basin is known as

(A) Circulatory ratio (B) Elongation ratio

(C) Compactness coefficient (D) Farm factor

Q. 23 A practical method of reducing sheet erosion from slopy land is

(A) Keep the land fallow (B) Farming on contour strips

(C) Construction of small reservoir (D) Use plastic sheet

Q. 24 The percentage of soil particles transported by saltation process in case of wind erosion varies from

(A) 3 to 40 % (B) 50 to 75 %

(C) 80 to 90 % (D) 5 to 25 %

Q. 25 The boundary line along a topographic ridge separating two adjacent drainage basins is called as

(A) Contour line (B) Drainage divide

(C) Iso bath (D) Iso chrome

Q. 26 For a given condition, a high intensity rain produces

(A) Less splash (B) More infiltration

(C) More splash (D) Crusting of soil

Q. 27 Soil degradation takes place due to

(A) Geological erosion (B) Accelerated erosion

(C) Glacial erosion (D) Tidal erosion

Q. 28 In a watershed, when the streams of second and third order meet, the resulting stream is of order

(A) Second (B) Third

(C) Fourth (D) Fifth

Q. 29 In general, the contour bunding practice is not recommended for the following type of soil

(A) Loamy soil (B) Deep black soil

(C) Sandy soil (D) Highly permeable soil

Q. 30 Permissible flow velocity for design of grassed waterway to be under a good grass cover is taken as

(A) 1 – 1.5 m/s (B) 1.5 – 1.8 m/s

(C) 2 – 2.5 m/s (D) 2.5 – 3.0 m/s

Q. 31 The number of stream segments per unit area of watershed is known as

(A) Watershed density (B) Stream density

(C) Stream order (D) Basin relief

Q. 32 The value of recession constant (kr) is

(A) 1 (B) < 1

(C) >1 (D) – 1

Q. 33 The value of circulatory ratio of watershed varies from

(A) 0.2 to 0.8 (B) 0.4 to 1.0

(C) 1.5 to 2.0 (D) 4 to 4.5

UKPSC J.E. Exam 2013 Agricultural Engineering-II

Q. 34 Which of the following is not used for gully erosion control ?

(A) Check dams (B) Brush dams

(C) Loose rock dams (D) High dams

Q. 35 Which of the following is not a mechanical method of erosion control ?

(A) Contour farming (B) Conservation tillage

(C) Free cropping (D) Small dams

Q. 36 Which of the following is not a stage of gully development ?

(A) Zero stage (B) Formation stage

(C) Development stage (D) Healing stage

Q. 37 Which of the following is not a permanent gully control structure ?

(A) Contour bund (B) Drop spillway

(C) Permanent check dams (D) Drop (inlet) spillway

Q. 38 Which of the following is not a type of soil erosion due to water ?

(A) Raindrop erosion (B) Gully erosion

(C) Final erosion (D) Sheet erosion

Q. 39 Which of the following is not a process of soil erosion ?

(A) Detachment of soil particles (B) Transportation of soil particles

(C) Both of the above (D) Shining of soil particles

Q. 40 Which of the following is not a type of soil erosion ?

(A) Accelerated soil erosion (B) Geological soil erosion

(C) Water erosion (D) Single erosion

Q. 41 Which of the following is not related to wind erosion ?

(A) Speed of wind (B) Age of the soil

(C) Moisture content of soil (D) Texture of soil

Q. 42 The first step in planning a terrace system is the selection of

(A) Inlet location (B) Terrace slope

(C) Watershed area (D) Outlet location

Q. 43 Temporary gully control structures are designed for a recurrence interval of

(A) Upto 10 years (B) Upto 8 years

(C) Upto 15 years (D) Upto 20 years

Q. 44 Primary cause of failure of permanent gully control structures is

(A) Insufficient hydraulic capacity (B) Insufficient flow velocity

(C) Both of the above (D) None of the above

Q. 45 In heavy rainfall areas which type of bench terraces should be preferred ?

(A) Level benches (B) Slopping outward benches

(C) Sloping inward benches (D) None of the above

Q. 46 Universal Soil Loss Equation (USLE) is used to calculate

(A) Average annual soil loss (B) Average weekly soil loss

(C) Average daily soil loss (D) Average monthly soil loss

Q. 47 The vertical interval of a terrace system does not depend on

(A) Land slope (B) Crop height

(C) Surface topography (D) Soil type

Q. 48 The longitudinal slope in a terrace is kept as

(A) 0.75 % (B) 1.50 %

(C) 2.50 % (D) 10.0 %

Q. 49 The drainage area above the top most terrace is generally kept

(A) 4.0 hectare (B) 10.0 hectare

(C) 2.0 hectare (D) 1.0 hectare

Q. 50 Earth work for bench terrace is calculated by

(A) 100 WS/8 (B) 100 WS/15

(C) (100 + WS)/8 (D) None of the above

Where, W = width of terrace, m

S = land slope, %

Q. 51 The gully development is accomplished under how many stages ?

(A) Only two stages (B) Only three stages

(C) Only four stages (D) Only five stages

Q. 52 Which of the following is due to wind erosion ?

(A) Delta (B) Sea coast

(C) Sedimentation (D) Sand dune

Q. 53 Gully erosion is an advanced stage of

(A) Sheet erosion (B) Channel erosion

(C) Splash erosion (D) Rill erosion

Q. 54 Contour trenching should be done in

(A) Winter season (B) Summer season

(C) Throughout the year (D) None of the above

Q. 55 Wind erosion is more in

(A) Cohesive soils (B) Non-cohesive soils

(C) Rocky soils (D) Wet soil

Q. 56 In splash erosion, the raindrop energy acts in the form of

(A) Kinetic energy (B) Potential energy

(C) Chemical energy (D) None of the above

Q. 57 The property of rainfall which mainly affects the soil erosion is

(A) Direction of rainfall (B) Time of rainfall

(C) Form of rainfall (D) Intensity of rainfall

Q. 58 Which of the following soil property is mainly affected by wind erosion ?

(A) Soil texture (B) Soil structure

(C) Soil temperature (D) Soil moisture

Q. 59 If 'E' is the soil loss due to water erosion and '*l*' is the slope length, then which of the following statement is true ?

(A) $E \alpha \ell^2$ (B) $E \alpha \sqrt{\ell}$

(C) $E \alpha \ell^3$ (D) $E \alpha \ell$

Q. 60 Which of the following phase is more significant to cause soil erosion ?

(A) Transportation of particles (B) Detachment of particles

(C) Deposition of particles (D) Abrasion of particles

Q. 61 The 'gabion' is constructed by using which material ?

(A) Vegetative materials (B) R.C.C.

(C) Brick mortar (D) Wire-net and stones

Q. 62 Check dams help in ....

(A) Decrease in flow velocity (B) Minimize channel erosion

(C) Increase infiltration of water (D) All of the above

Q. 63 Removal of thin and fairly uniform layer of the soil from the land surface by runoff water is called as,

(A) Torrent erosion (B) Sheet erosion

(C) Glacial erosion (D) Uniform erosion

Q. 64 A high intensity rainfall causes severe erosion because of

(A) Greater detachment of soil particles only

(B) More runoff generation only

(C) Better transportation of eroded soil only

(D) All of the above

Q. 65 Which of the following aspects are kept in consideration while designing a drop structure

(A) Hydrologic (B) Hydraulic

(C) Structural (D) All above

Q. 66 Runoff coefficient depends on

(A) Nature of soil surface (B) Type of land use

(C) Rainfall intensity (D) All the above

Q. 67 Free board is provided to prevent the embankment against

(A) Overtopping (B) Sliding

(C) Overturning (D) All of the above

Q. 68 The agronomic measures increase infiltration rate and thereby there is

(A) Reduction in runoff (B) Increase in soil erosion

(C) Formation of gullies (D) None of the above

Q. 69 Graded bunds are useful in areas where

(A) Rainfall is very low (B) Rainfall is low

(C) Rainfall is high (D) None of the above

Q. 70 Due to shrinkage of soil during land grading, the cut-fill ratio should be

(A) Greater than one (B) Less than one

(C) Either of (A) and (B) (D) None of (A) and (B)

Q. 71 The desired cut or fill at any point can be obtained by comparing

(A) Original and proposed elevations

(B) Original and surrounding elevations

(C) Proposed and surrounding elevations

(D) All of the above

Q. 72 Control of soil erosion is vital for

(A) Sustenance of agriculture (B) Maintaining crop productivity

(C) Control of sedimentation (D) All of the above

Q. 73 Mulching helps in

(A) Controlling soil erosion (B) Maintaining soil moisture

(C) Controlling weeds (D) All of the above

Q. 74 The drop in case of a straight drop spillway should not exceed to

(A) 5 m (B) 3 m

(C) 10 m (D) 8 m

Q. 75 The primary purpose of contour farming in low rainfall areas is

(A) To increase runoff (B) To increase soil loss

(C) To conserve flow energy (D) To conserve rain water

Q. 76 In a field contour lines are obtained by joining the points of

(A) Same rainfall (B) Same elevation

(C) Same erosion (D) Same infiltration

Q. 77 Gully plugs help in

(A) Reducing infiltration (B) Reducing flow velocity

(C) Reducing sediment deposition (D) Reducing percolation

Q. 78 In the design of a drop structure which of the following force is considered ?

(A) Gravitational forces (B) Viscous forces

(C) Frictional forces (D) Drag forces

Q. 79 Straight and parallel contour lines represent

(A) Uniformly sloping undulating surface

(B) Uniformly sloping plane surface

(C) Undulating surface

(D) None of the above

Q. 80 In contour cultivation tillage operations are performed

(A) Normal to the contours of the field

(B) Along the contours of the field

(C) Across the contours of the field

(D) In any direction

Q. 81 Up and down the slope farming results in

(A) Minimizing soil erosion

(B) Increasing soil erosion

(C) Improving moisture conservation

(D) None of the above

Q. 82 The term "EROSIVITY" refers to

(A) Erosive time of rain (B) Erosive potential of rain

(C) Erosive resistance of rain (D) None of the above

APPSC AEES Agriculture Engineering Exam 2016

Q. 83 Class IV type lands are :

(A) Suitable for cultivation (B) Not suitable for cultivation

(C) Partly suitable for cultivation (D) None of the above

Q. 84 Vertical interval (VI) of bench terrace for a vertical cut is given as :

(A) (100S)/W (B) (100W)/S

(C) (100/(WS) (D) (WS/100)

Q. 85 Sediment yield is the

(A) Total soil loss from an area

(B) Product of gross erosion and delivery ratio

(C) Both (A) and (B)

(D) None of the above

Q. 86 Chute spillway is used in control a drop of

(A) 0-3 m (B) 1-4 m

(C) 2-4 m (D) 3-6 m

Q. 87 Soil loss in geologic erosion is

(A) Les than accelerated erosion

(B) Greater than accelerated erosion

(C) Medium

(D) None of the above

MPPSC Assistant Agricultural Engineer 2013

Q. 88 Contour bunds are recommended for the areas of ....

(A) High rainfall and low infiltration capacity

(B) Low rainfall and high infiltration capacity

(C) Medium rainfall and low infiltration capacity

(D) None of these is correct

Q. 89 Drop structures are designed in respect of ....

(A) Hydrologic (B) Hydraulic

(C) Structural (D) All options are correct

Q. 90 Detachment of soil particle from parent material by the beating action of rain drop is ....

(A) Rill erosion (B) Sheet erosion

(C) Splash erosion (D) Gully erosion

Q. 91 In clay soils the particle detachment is less as compared to sandy soils because

(A) Sandy soils have more cohesive force than clay soils

(B) Sandy soil particles are larger in size than clay soil particles

(C) Clay soils have more cohesive force than sandy soils

(D) None of these is correct

Q. 92 Soil erosion will be more from the land-whose surface is ....

(A) Covered by forest

(B) Covered by thick bushy growth

(C) Covered by grass

(D) Covered by cultivated crop

RPSC AEn Pre Exam 2013 (Agricultural Engineering)

Q. 93 Whenever there exists a sudden drop (say 1 or 2 m) on a channel bed, what kind of soil conservation structure is usually preferred at this point

(A) Graded bund (B) Deep trench

(C) Drop inlet structure (D) Drop structure

Q. 94 The two broad categories of gullies predominantly found in India have cross sections of

(A) rectangular shaped and parabolic shaped

(B) Straight sections and curved sections

(C) U-shaped and V-shaped cross sections

(D) None of the above

Q. 95 Usually, the highest volume of soil loss per unit area occurs during

(A) Rill erosion (B) Sheet erosion

(C) Stream bank erosion (D) Land slides on hills

Q. 96 Two prominent grasses used for conserving soil and water offering effective control of soil erosion under Indian conditions are

(A) Doob grass and buffalo grass

(B) Vetiver (Khus) and Sacrum Munja (Moonji)

(C) Napier grass and stylo grass

(D) Both (A) and (C)

Q. 97 In terracing in the field is divided into

(A) Large number of small channels

(B) Scrics of strips levelled independently

(C) Small check basins

(D) All of these

Q. 98 Interception loss is

(A) More towards end of a storm (B) More at the middle of a storm

(C) More at beginning of a storm (D) Uniform throughout the storm

Q. 99 In a chute spillway the flow is usually

(A) Uniform (B) Subcritical

(C) Critical (D) Supercritical

Q. 100 The highly wind erosion affected state in India is

(A) M.P. (B) Gujrat

(C) A.P. (D) Rajasthan

Q. 101 The term "middle third" rule is generally used while designing the check dams. What does it really deal in such designing process ?

(A) Crushing (B) Sliding

(C) Breaking (D) Thrust

Q. 102 Which of the following is not the component of a water harvesting pond ?

(A) Emergency outlet (B) Inlet conduit/channel/weir

(C) Earthen embankment (D) Drainage well

Q. 103 Which equation is most commonly used for designing drainage system in an agricultural catchment specifically the distance and configurations of drainage tiles etc.

(A) USDA equation (B) USLE

(C) Hooghoutt's equation (D) All of the above

Q. 104 There are two different watersheds "P" and "Q" both have same size, same location, same soils, same shape and all other factors same, except that P is flat and Q is sloppy having multidirectional slope. During a same intense rainfall event, on which watershed the drainage will be difficult and delayed ?

(A) P

(B) Q

(C) Both will have same, if the rain is same

(D) Nothing can be said

OCS AEn Exam 2011 (Pre) – II (Agricultural Engineering)

Q. 105 Saltation is a type of soil movement in ……… erosion.

(A) Wind (B) Water

(C) Stream bank (D) Anthropogenic

Q. 106 In case of field strip cropping, the width of strips are

(A) Unequal (B) Non uniform

(C) Uniform (D) Reducing

Q. 107 Contouring for erosion control affect the .... factor.

(A) Erosivity (B) Erodibility

(C) Slope (D) Conservation practice

Q. 108 The length of standard experimental plots to measure soil erosion is .... Meter.

(A) 20 (B) 8

(C) 22 (D) 25

Q. 109 Vegetated waterways with Bermuda grass having bed slope of 5 to 10% in erosion resistance soil need to have permissible velocity of flow .... m/sec.

(A) 1.8 (B) 2.1

(C) 2.4 (D) 2.7

Q. 110 To control elevation changes up to 6 m in a channel, ... spillway is used.

(A) Drop (B) Pipe

(C) Chute (D) Box

Q. 111 Watershed is a basic .... unit, which can be used for all development of activities.

(A) Geological (B) Geophysical

(C) Physical (D) Hydrological

Q. 112 Watershed management for soil conservation involves prioritization and treatment of critical areas of watershed with respect to ....

(A) Rainfall (B) Sediment

(C) Erosivity (D) Vegetation

Q. 113 .... is one of the major activity associated with rainwater harvesting and reuse.

(A) Recharge (B) Irrigation

(C) Drainage (D) Flood control

CGPSC State Engineering Services Exam (Agriculture Engineering) Part - II

Q. 114 Usually, area lost due to side and lateral bund is taken as of main bund

(A) 20 % (B) 30 %
(C) 40 % (D) 60 %
(E) 50 %

Q. 115 The type of bunds formed at extreme ends of the contour bund

(A) Supplemental bunds (B) Side bunds
(C) Shoulder bunds (D) Marginal bunds
(E) Both (A) and (B)

Q. 116 The type of soil having the lower soil erosion problem is

(A) Sandy (B) Loamy
(C) Clay (D) Both (B) and (C)
(E) Both (A) and (B)

Q. 117 In modified USLE, the new factor introduced is

(A) Rainfall factor (B) Soil factor
(C) Slope factor (D) Runoff factor
(E) Temperature factor

Q. 118 Contour trenching is followed when land slope exceeds

(A) 16 % (B) 25 %
(C) 33 % (D) 40 %
(E) 50 %

Q. 119 A Series of closed contours with higher values inside indicate

(A) Depression (B) Valley
(C) Hill (D) Level land
(E) None of these

Q. 120 The order of erosion with respect to configuration

(A) Convex > Concave > Complex (B) Convex > Complex > Concave

(C) Concave > Complex > Convex (D) Concave > Convex > Complex

(E) None of these

Q. 121 The farm ponds formed by excavated soil at the site is

(A) Spring (B) Dugout

(C) Off-stream (D) Levee

(E) None of these

Q. 122 The type of bed load sample having high trap efficiency is

(A) Basket (B) Tray

(C) Pressure difference (D) Both (B) and (C)

(E) Both (A) and (B)

Q. 123 Contour bunds and terraces are

(A) Agronomical measures (B) Mechanical measures

(C) Biological measures (D) Both (B) and (C)

(E) Both (A) and (B)

Q. 124 Generally length of staggered contour trnches is

(A) 2 m (B) 4 m

(C) 5 m (D) 6 m

(E) 8 m

Q. 125 The most harmful type of soil erosion is

(A) Splash (B) Sheet

(C) Rill (D) Land slide

(E) Both (A) and (B)

Q. 126 The cross-sectional area of conservation ditch is

(A) 0.583 m^2 (B) 1.583 m^2

(C) 2.583 m^2 (D) 3.583 m^2

(E) 4.583 m^2

Q. 127 For design of permanent structures, number of years of rainfall data generally considered for the design.

(A) 10 – 20 (B) 20 – 30

(C) 25 – 30 (D) 30 – 60

(E) 40 – 80

Q. 128 The movement of soil particles having sizes in range of 0.05 to 2.5 mm through a series of benches is known as

(A) Surface creep (B) Surface transformulation

(C) Saltation (D) Suspension

(E) Both (A) and (B)

Q. 129 Soil detachability increases as the

(A) Size of particle reduces

(B) Size of particle increases

(C) Impact angle of raindrop reduces

(D) Length of slop reduces

(E) No change

UKPSC A.En. Exam 2007 Paper – II

Q. 130 The wind erosion movement of the soil particles can be carried out by

(A) Suspension only

(B) Suspension, saltation and surface creep

(C) Saltation and surface creep

(D) None of the above

Q. 131 From the erosion point of view, the most active position of the gully is

(A) Gully bed (B) Gully head

(C) Gully tail (C) None of the above

Q. 132 If a land has 30 % slope, the vertical interval for a 3.50 m wide level bench terrace with 1 : 1 riser slope, will be

(A) 1 m (B) 1.5 m

(C) 2.0 m (D) 3.0 m

Q. 133 As per land capability classification, the lands best suited for cultivation belong to

(A) Class-VIII lands (B) Class-I lands

(C) Class-IV lands (D) Class-VI lands

Q. 134 Central soil and Water Conservation Research and Training institute is located in

(A) Shimla (B) Bhubaneswar

(C) New Delhi (D) Dehradun

Q. 135 The ratio of sediment yield to gross soil erosion is known as

(A) Bifurcation ratio (B) Sediment delivery ratio

(C) Erosion generation ratio (D) Sediment deposition ratio

Q. 136 The "Cox" formula is used to determine

(A) Rainfall velocity (B) Rainfall intensity

(C) Vertical interval between contour bunds

(D) Earthwork volume in the pond design

Q. 137 The major portion of soil erosion by wind takes place in the form of

(A) Surface creep (B) Suspension

(C) Saltation (D) None of the above

Q. 138 Stepped spillway is a modified form of a

(A) Drop inlet spillway (B) Inclined spillway

(C) Straight drop spillway (D) Ogee spillway

Q. 139 End-control method is used to determine

(A) Contour interval (B) Average precipitation

(C) Storage capacity of pond (D) Amount of soil-erosion

Q. 140 The elongation ratio of a watershed having area(A) and length(L) would be

(A) $1.128A^{0.50}/L$ (B) $1.128L/A^2$

(C) $1.128L^2/A$ (D) $1.128L^2/A^2$

Q. 141 In a watershed management programme PRA stands for

(A) Public relation agency (B) Public relation authority

(C) Program research analysis (D) Participatory rural appraisal

Q. 142 The general form of relationship between rainfall depth and area of watershed is

(A) $p = p_0 \exp(-K A^n)$ (B) $p = p K A^n$

(C) $p = p + K A^n$ (D) None of the above

Q. 143 Contour bunds are constructed, where

(A) Rainfall is more than 1500 mm (B) Rainfall is less than 600 mm

(C) Water drainage is required (D) Soil is heavy

Q. 144 Which of the following is not a biological methods of soil erosion control ?

(A) Field strip cropping (B) Control bunding

(C) Buffer strip cropping (D) None of the above

Q. 145 Soil erosion due to wind is a problem of mainly

(A) Humid regions (B) Arid regions

(C) Hilly reasons (D) None of the above

ASCO (RPSC) Exam (Engineering) 2011

Q. 146 Temporary gully control structures are structural measures such as

(A) Woven-wire

(B) Brush wood and logs

(C) Loose stone and boulder check dams

(D) All of the above

Q. 147 Materials used for constructing permanent gully control structures are

(A) Earth material, vegetation

(B) Stones and bunches of trees

(C) Masonary, reinforce concrete, earth etc

(D) None of the above

Q. 148 The natural grassed waterways are generally found in the shape of

(A) Square (B) Triangular

(C) Parabolic (D) Rectangular

Q. 149 Doubling the length of the slope increases soil erosion by about

(A) 1.4 times (B) 2 times

(C) 2.4 times (D) 3.0 times

Q. 150 When the grade and dimensions of a channel are known, the terrace channel carrying capacity can be calculated from

(A) Rational method (B) Cook's method

(C) Manning's formula (D) Thiesson method

Q. 151 The height of the graded bunds should not be less than

(A) 1.00 m (B) 0.75 m

(C) 0.45 m (D) 2.00 m

Q. 152 Table top bench terraces are suitable for area of

(A) High rainfall (B) Medium rainfall

(C) Low rainfall (D) Irrespective of total rainfall

Q. 153 A typical sand dune has highest steeper slope to leeward side of

(A) 15 % (B) 33 %

(C) 50 % (D) 65 %

Q. 154 In an area having general slope of 10 %, bench terracing is to be done to bring the land under cultivation, 4.5 m width of land is sufficient for doing necessary agricultural operations. If risers are to be laid on 1 : 1 gradient, the vertical interval between two consecutive benches will be

(A) 0.45 m (B) 0.50 m

(C) 4.00 m (D) 1.00 m

Q. 155 If depth of water over the crest of a spillway is 45 cm. Asuming $(0.45)^{3/2} = 0.3$, the length of the crest of the drop spillway for a discharge of 5.3 m^3/sec will be

(A) 0.45 m (B) 1.77 m

(C) 10.00 m (D) 5.30 m

Q. 156 Soil conservation practices are

(A) Contour farming and terracing (B) Strip farming

(C) All of the above (D) None of the above

Q. 157 Soil erosion is a process of transportation of soil by

(A) Wind (B) Water

(C) Glacier (D) All of the above

Q. 158 Forces opposing the soil uplift and entrainment are

(A) Gravity, friction and cohesion

(B) Wind velocity and particle weight

(C) Friction and cohesion only

(D) None of the above

Q. 159 If the land slope is increased 4 times, the velocity of water flowing over it will increase

(A) Three times (B) Two times

(C) Negligible (D) Four times

Q. 160 Loose rock fill dams are preferred for gully control, where

(A) Gullies are of greater depth (B) Gullies are of lesser depth

(C) Tones are readily available (D) None of the above

Q. 161 What type of material is most likely to be transported as suspended load ?

(A) Clay

(B) Silt

(C) Sand

(D) Depends on the "energy" of the stream

Q. 162 Why is soil erosion more common in areas that lack adequate vegetation ?

(A) Because gravity pulls on the plants and increases erosion

(B) Because plant roots help keep the soil in place

(C) Because soil erosion is stopped by reduced vegetation

(D) Because the lack of adequate vegetation affects the soil pressure and erosion rate

Q. 163 Design of vegetated waterways is based upon

(A) Manning's formula (B) Cook's method

(C) Thiessen method (D) Universal soil loss equation

Q. 164 The minimum wind velocity required to initiate movement of soil particles is known as

(A) Critical velocity (B) Threshold velocity

(C) Dynamic threshold velocity (D) Intrinsic velocity

Q. 165 Sediment is transported in channels by

(A) Saltation (B) Suspension

(C) Bed load movement (D) All of the above three factor

Q. 166 Pick up the correct statement

(A) Sloping outwards bench terraces are effective in low rainfall areas

(B) Sloping inwards bench terraces are effective in low rainfall areas

(C) Table top bench terraces are effective in low rainfall areas

(D) Broad-base terraces are constructed in small and scattered land holdings

Q. 167 Which of the following statement is correct ?

(A) The purpose of the wing wall is to prevent seepage under the structure

(B) Wing wall is constructed to prevent swelling effect of the water

(C) Chute spillway provides temporary storage on up-steam side

(D) Wing wall is constructed at an angle of 60 degree from side wall

Q. 168 Area lost due to contour bunding is given by

(A) $0.305(s/(a + b)))$ (B) $s \times b/VI$

(C) $100\ s \times b/VI$ (D) $W_a \times h/2$

Q. 169 In the stream having depth up to 3 m, the suspended sediment is measured at the depth of

(A) 0.6 d (B) 0.8 d

(C) 1.5 d (D) 0.2 d

Q. 170 Which of the following outlet is not made in drop spillway ?

(A) Straight (B) Flared

(C) Curved (D) Box

Q. 171 Contour bunding is to be done in a land having 4 % gradient. If horizontal interval of the bund is 100 m, the vertical interval between the bunds will be

(A) 25 m (B) 0.25 m

(C) 4 m (D) 2.5 m

Q. 172 The woven wire dams are backed with well temped earth fill slope of at least

(A) 4 : 1 (B) 3 : 1

(C) 2 : 1 (D) 1 : 1

Q. 173 To construct a loose rock dam, a trench is made which forms the base of dam on which the stones are laid in the rows. The approximate depth of the trench should be

(A) 10 cm (B) 30 cm

(C) 50 cm (D) 70 cm

Q. 174 The ratio of the diameter of a circle with same area as the basin to the maximum length is called

(A) Circulatory ratio (B) Elongation ratio

(C) Compactness ratio (D) Stream length ratio

Q. 175 In building the slab dams the posts at an interval of 1.2 m are set in row across the gully to a depth of about

(A) 30 cm (B) 50 cm

(C) 70 cm (D) 100 cm

ASCO (RPSC) Exam (Forest) 2011

Q. 176 An erosion with intensity ranging from 0.5 to 5 m^3/ha/year is considered to be

(A) Insignificant erosion (B) Slight erosion

(C) Severe erosion (D) Geological erosion

Q. 177 Hydrologic soil group-C contains

(A) Clay loam and shallow sandy loam

(B) Heavy plastic clay

(C) Certain saline soil

(D) Deep sand

Q. 178 Which of the following conditions would lead to the most soil erosion ?

(A) Heavy rainfall (from a single storm) in an area of low average annual rainfall

(B) Steep slopes in a tropical rain forest

(C) An agricultural region where soil conservation practices are implemented

(D) An area that has been urbanised for several decades or more

Q. 179 Rill erosion usually begins in the

(A) Lower part of land slope (B) Upper part of land slope

(C) Middle part of land slope (D) Entire length of land slope

Q. 180 The limit of slope length at which soil erosion begins at called as

(A) Optimum slope length (B) Critical slope length

(C) Allowable slope length (D) Unit slope length

Q. 181 Erosion intensity of sheet and wind erosion is expressed as

(A) m^3/ha (B) tons/ha

(C) m^3/sec (D) Both (A) and (B)

Q. 182 Longitudinal slope of gully bed is almost parallel to the adjoining land slope. It is generally found in case of

(A) V-shaped gully (B) U-shaped gully

(C) Inactive gully (D) Active gully

Q. 183 In surface creep process the transportation of soil particles varies from

(A) 3 to 38 % (B) 55 to 72 %

(C) 80 to 90 % (D) 7 to 25 %

Q. 184 In general the contour bunding is practiced in all type of soil except

(A) Clayey and black cotton soil (B) Loamy soil

(C) Brown soil (D) Highly permeable soil

Q. 185 Double row brushwood check dams are economically constructed to control the gullies of

(A) 1.5 m to 2.1 m width (B) 1.5 to 2.1 m depth

(C) 5 to 10 m depth (D) 3 to 5 m depth

Q. 186 Drop inlet and chute spillways consist of

(A) Inlet and outlet (B) Inlet, conduit and outlet

(C) Weir and outlet (D) Earth dam and outlet

Q. 187 In LUCC system, the sub-classes indicates the problem of

(A) Soil type (B) Major land use

(C) Slope (D) Slope length

Q. 188 Strip cropping technique for erosion control is more effective when it is practiced under

(A) Crop rotation (B) Bunding

(C) Terracing (D) Single cropping system

Q. 189 Contour bund length in metre per hectare is given

(A) 10000/HI (B) S×100/HI

(C) VI×100/S (D) HI×100/S

Q. 190 The most suitable structure for drainage line treatment followed in all regions of southern Rajasthan is

(A) Loose stone check dam (B) Graded bund

(C) Contour bund (D) PRT

Q. 191 The structure with masonry wall constructed across a Nala for the purpose of water harvesting is known as

(A) Culvert (B) Anicut

(C) Farm pond (D) Percolation tank

Q. 192 Side slope of medium gullies varies from

(A) 8 to 15 % (B) 2 to 3 %

(C) 5 to 7 % (D) 3 to 6 %

UKPSC A.En. Exam 2012 Paper - II

Q. 193 The Stanford model is used to

(A) Estimate the soil loss (B) Estimate the runoff

(C) Estimate the rainfall depth (D) Estimation of soil formation

Q. 194 The value of 'C' in rational equation is generally

(A) Less than 1 (B) More than 1

(C) Equal to 5 (D) All of the above

Q. 195 For design of grassed waterway the average grade used is

(A) 1 % (B) 5 %

(C) 10 % (D) 15 %

Q. 196 Erosion index is based on rainfall intensity

(A) I_{15} (B) I_{30}

(C) I_{45} (D) I_{60}

Q. 197 The spacing of check dam is kept between

(A) 25 – 75 m (B) 150 – 450 m

(C) 80 – 180 m (D) 75 – 100 m

Q. 198 Check dams are provided with the main objective of

(A) Ponding of water

(B) Causing upstream sedimentation

(C) To reduce channel slope

(D) All the above

Q. 199 Graded bund is used for safe disposal of excess runoff in the area with

(A) High rainfall (B) Relatively impervious soil

(C) Both (A) and (B) (D) None of the above

Q. 200 Those gullies whose dimension does not change with time are called as

(A) Active gullies (B) U-shaped gullies

(C) Inactive gullies (D) V-shaped gullies

Q. 201 Drop inlet spillways are located at the place of

(A) Medium gully slope (B) High depression

(C) Gully face (D) Upstream area of gully

Q. 202 The difference in elevation between top of the dam and normal reservoir level is called as

(A) Minimum free board (B) Maximum free board

(C) Critical free board (D) Normal free board

Q. 203 The Manning's coefficient of roughness for grassed channels is specifically known as

(A) Retardance coefficient (B) Coefficient of discharge

(C) Manning's constant (D) Flow coefficient

Q. 204 Coarse textured soils erode mainly through

(A) Surface creep (B) Saltation

(C) Suspension (D) None of the above

Q. 205 Strip cropping for erosion control is more effective when it is practiced under

(A) Terracing (B) Crop rotation

(C) Bunding (D) Mono-cropping

Q. 206 Batter slope in bench terraces is primarily provided for

(A) Moisture retention (B) Controlling sloughing

(C) Both (A) and (B) (D) Stability to the fill material

Kerala AAO Exam 2017

Q. 207 Soil erosion is a process of

(A) Transportation of H_2O (B) Detachment of soil particles

(C) Transportation of soil particles (D) Both (B) and (C)

Q. 208 Consider the following pairs

I. Windbreaks – obstruct wind flow

II. Shelter breaks – do not obstruct wind flow completely

III. Shelter breaks – planted across the deviation of wind

Which of the pair given is/are correct ?

(A) I only (B) II and III only

(C) I and II only (D) I, II and III only

Graduate Aptitude Test in Engineering - 2007

Q. 209 The soil loss from a field with 5% slope and for crop management factor of 0.25 is 44.80 Mg ha^{-1}. Contouring along with crop management factor of 0.15 is adopted as the soil conservation measure in the field. The changed soil loss from the field is

(A) 1.61 Mg ha^{-1} (B) 2.68 Mg ha^{-1}

(C) 16.12 Mg ha^{-1} (D) 26.87 Mg ha^{-1}

Graduate Aptitude Test in Engineering - 2008

Q. 210 The following design parameters of contour bunds constructed on a land of 4% slope are given:

V.I. = 1.2 m, base width = 2.5 m, top width = 0.5 m, height = 1.0 m. Assuming the length for side and lateral bunds as 30% of the length of contour bunds, the land area lost due to bunding is

(A) 0.156 % (B) 2.50 %

(C) 10.83 % (D) 12.52 %

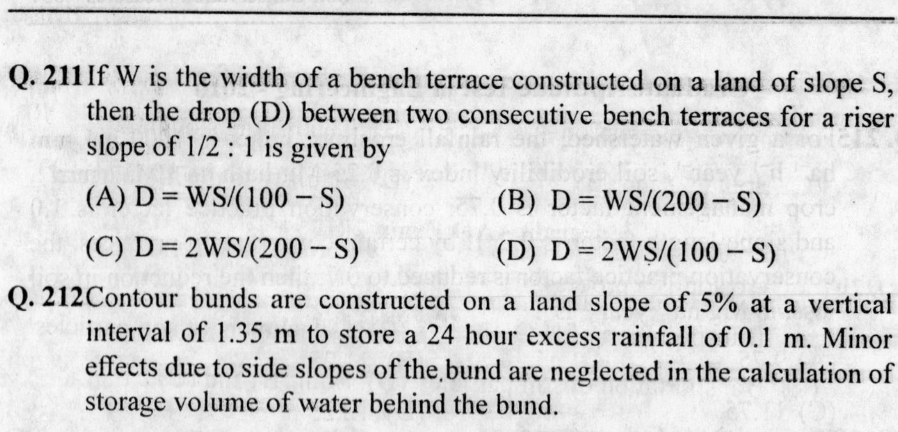

Q. 211 If W is the width of a bench terrace constructed on a land of slope S, then the drop (D) between two consecutive bench terraces for a riser slope of 1/2 : 1 is given by

(A) D = WS/(100 − S) (B) D = WS/(200 − S)

(C) D = 2WS/(200 − S) (D) D = 2WS/(100 − S)

Q. 212 Contour bunds are constructed on a land slope of 5% at a vertical interval of 1.35 m to store a 24 hour excess rainfall of 0.1 m. Minor effects due to side slopes of the bund are neglected in the calculation of storage volume of water behind the bund.

(i) The depth of impounding immediately behind the contour bund is

(A) 0.32 m (B) 0.42 m

(C) 0.52 m (D) 0.62 m

(ii) The water spread length behind the bund is

(A) 12.4 m (B) 10.4 m

(C) 8.4 m (D) 6.4 m

Graduate Aptitude Test in Engineering - 2009

Q. 213 Contour bund is constructed on a land with S per cent slope. If the length of each contour bund is L and vertical interval is d, then the number of contour bunds per hectare of land area is

(A) $\frac{S}{LD}$ (B) $\frac{LD}{S}$

(C) $\frac{100S}{LD}$ (D) $\frac{10000S}{LD}$

Q. 214 The peak runoff volume from the catchment between two contour bunds constructed on a 2.5% slope is 972.5 m^3. The contour bunds have top width of 500 mm, height 600 mm, side slope 2:1, vertical interval 1.0 m and length 250 m. If the crest of the overflow weir is at a height of 300 mm from the ground and time available for the excess water to flow through the weir is 20 minutes, discharge in m^3 per minute through the weir is

(A) 15.0 (B) 20.0

(C) 25.0 (D) 48.6

Graduate Aptitude Test in Engineering - 2010

Q. 215 For a given watershed, the rainfall erosivity index is 1000 mJ mm ha^{-1} h^{-1} $year^{-1}$, soil erodibility index is 0.25 Mg ha h ha^{-1} MJ^{-1} mm^{-1}, crop management factor is 0.75, conservation practice factor is 1.0 and slope length factor is 0.2. If by certain conservation practices, the conservation practice factor is reduced to 0.7, then the reduction in soil loss, in Mg ha^{-1} $year^{-1}$ is

(A) 9.75 (B) 11.25

(C) 11.75 (D) 12.25

Graduate Aptitude Test in Engineering - 2011

Q. 216 In wind erosion process, the rate of soil movement (S) depends on wind speed(v), threshold limit of wind speed required to move the soil (v_m) and average soil particle size (d). This interrelationship is expressed as

(A) S $(v - v_m)^2 d^{1/2}$ (B) S $(v - v_m)^2 d$

(C) S $(v - v_m)^3 d^{1/2}$ (D) S $(v - v_m)^3 d$

Q. 217 For the eastern Himalayan region having hill slope of 16%, a bench terrace system is to be designed. It is found that the depth of cut for constructing the bench terrace cannot be more than 0.40 m due to the limitation of the soil depth. If a batter slope of 1/2:1 is proposed, the maximum possible width of the terrace and the required earthwork per hectare (assuming cut equal to fill) will be

(A) 3.4 m, 680 m^3 (B) 4.6 m, 920 m^3

(C) 4.8 m, 960 m^3 (D) 5.0 m, 1000 m^3

Graduate Aptitude Test in Engineering - 2012

Q. 218 It is proposed to construct bench terraces on a 10% hill slope. If the batter slope is 1/2 H : 1 V, the percentage area that will be lost for cultivation due to bench terracing is

(A) 4.68 (B) 5.47

(C) 6.25 (D) 6.78

Graduate Aptitude Test in Engineering - 2013

Q. 219 A 16 m high wind break is constructed to protect soil from wind erosion due to wind velocity of 18 m s^{-1} at 15 m height. The minimum wind velocity at 15 m height capable of moving the soil fraction is 8 m s^{-1}. The angle of deviation of prevailing wind direction from the perpendicular to the wind break is 30°. The distance of full protection from the wind break in m is

(A) 60.44 (B) 104.69

(C) 306.00 (D) 530.01

Graduate Aptitude Test in Engineering - 2014

Q. 220 A triangular shaped grassed waterway with a longitudinal slope of 2.5% is to carry a discharge of 1.5 m^3 s^{-1} with a permissible velocity of 1.2 m s^{-1}. The side slope of the channel is 1.5:1 (Horizontal : Vertical). Without considering free board, the top width of the channel in m is ...

Q. 221 Bench terraces are to be constructed on a 15% hill slope. The batter slope is 1:1 and vertical interval is 2.5 m. The earth work in cutting is equal to filling. The quantum of earthwork in m^3 ha^{-1} will be

(A) 2656 (B) 2818

(C) 4248 (D) 5312

Graduate Aptitude Test in Engineering - 2016

Q. 222 The most suitable hydraulic structure for conveying water from higher elevation to lower elevation across the earthen bund is

(A) Drop structure (B) Pipe drop structure

(C) Chute spillway (D) Gabion structure

Q. 223 If the width of bench terrace is W, drop D and existing land slope S; then for 150% batter slope, the drop D will be

(A) $\frac{WS}{100}$ (B) $\frac{WS}{100-S}$

(C) $\frac{2WS}{200-S}$ (D) $\frac{3WS}{300-2S}$

Graduate Aptitude Test in Engineering - 2017

Q. 224 Bunds are to be constructed to conserve rainwater in a farm having 6% slope. If horizontal interval between two bunds is 30 m and there is no loss of water, the required height of the bund to store rainwater from an 18 cm rainfall event with 10 years of return period in cm will be ....

Graduate Aptitude Test in Engineering - 2018

Q. 225 Match the following items in Column I with the corresponding items in Column II:

Column I	Column II
P. Flownet	1. Soil erosion
Q. EI30 Index	2. Curve number
R. Groynes	3. Equipotential line
S. Isobath	4. Groundwater
T. Runoff	5. River bank

(A) P-4, Q-2, R-5, S-3, T-1 (B) P-4, Q-3, R-1, S-5, T-2

(C) P-3, Q-1, R-5, S-4, T-2 (D) P-3, Q-1, R-2, S-4, T-5

Q. 226 A rectangular chute spillway is designed for gully erosion control with a drop (H) of 5 m, peak flow of 1.81 m^3/s and maximum inlet water level of 0.81 m. Frictional head loss over the apron is 20% of H. The design height of chute blocks is equal to the depth of flow in the water way. Acceleration due to gravity (g) is 9.81 m s^{-2}. If coefficient of discharge of the weir section is 1.66, height of the chute blocks is .... m.

Graduate Aptitude Test in Engineering - 2019

Q. 227 Match the following items between Column-1 and Column-II with the most appropriate combinations

Column-I	Column-II
(1) Uniformly spaced contour lines	(P) Flat ground
(2) Widely spaced contour lines	(Q) Steep ground
(3) Closely spaced contour lines	(R) Hill
(4) A series of close contours with high value inside	(S) Uniform slope

(A) 1-P, 2-R, 3-S, 4-Q (B) 1-S, 2-P, 3-Q, 4-R

(C) 1-Q, 2-S, 3-P, 4-R (D) 1-S, 2-Q, 3-P, 4-R

Q. 228 A soil conservation structure has an expected life of 10 years and is designed for a flood magnitude of return period 50 years. The risk of this hydrologic design in percentage is ....

Q. 229 A parabolic grassed water channel 8 m wide at the top and 60 cm deep is laid on a slope of 3%. Assuming the value of 'n' in Manning's formula as 0.04 $m^{-1/3}$ s, the discharge capacity (in m^3 s^{-1}) of the channel is ....

Q. 230 On a 4% land slope in a medium rainfall zone, the horizontal spacing of bunds in meter and the length of bunds per hectare in meter, respectively are

(A) 25 and 300 (B) 25 and 400

(C) 30 and 300 (D) 30 and 400

Graduate Aptitude Test in Engineering - 2020

Q. 231 The following data were used for a watershed experiencing soil erosion problem:

Rainfall Erosivity Index= 280 MJ mm/(ha.h.year), Soil Erodibility Index= 0.38 ton ha h MJ^{-1} mm^{-1}, Slope length = 200 m, Average slope of the land= 8%, Slope steepness factor = 0.85, Cropping management factor = 0.35, and Conservation practice factor= 0.60

If the slope length is reduced to half, percentage reduction in soil loss is ....

Q. 232 The Curve Number (CN) of a watershed of 40 ha area under given hydrologic soil group, land use and management practices, and Antecedent Moisture Condition (AMC)-II is 80. The initial abstraction is 20% of maximum retention. For the rainfall event of 40 mm, the direct runoff in mm is ....

Graduate Aptitude Test in Engineering - 2021

Q. 233 While assessing the intensity of agricultural drought, a negative value of aridity index indicates that the area is classified as

(A) Severely arid (B) Moderately arid

(C) Mildly arid (D) Non-arid

Q. 234 It is proposed to develop bench terraces in an area having land slope of 10%. If the vertical interval between the bench terraces is 2.5 m and the batter slope is 100%, working width (in m) and the area lost for cultivation (in per cent), respectively will be

(A) 22.50 and 0.05 (B) 25.00 and 0.50

(C) 22.50 and 10.45 (D) 25.00 and 10.45

Q. 235 A windbreak, 15 m in height and 200 m in length, is established to protect the land from wind erosion in an arid area. The minimum wind velocity at the height of 15 m above the ground required to move the most erodible soil fraction is 9.6 m/s. If 5-year return period wind velocity at 15 m height is 16 m/s and the wind direction deviates 20° from the line perpendicular to the windbreak, the area protected by the windbreak in ha is...........

Q. 236 The hydrologic reservoir routing methods use

(A) Bernoulli's equation only

(B) Hydrologic continuity equation only

(C) Muskingum equation only

(D) Both the hydraulic momentum and hydrologic continuity equations

Answers Key

1	2	3	4	5	6	7	8	9	10
A	D	B	C	C	A	B	A	C	A
11	12	13	14	15	16	17	18	19	20
B	A	C	A	D	C	C	C	B	C
21	22	23	24	25	26	27	28	29	30
D	A	B	B	B	C	B	B	D	B
31	32	33	34	35	36	37	38	39	40
B	B	A	D	C	A	A	C	D	D
41	42	43	44	45	46	47	48	49	50
B	D	A	A	C	A	B	A	D	A
51	52	53	54	55	56	57	58	59	60
C	D	D	C	B	A	D	A	B	B
61	62	63	64	65	66	67	68	69	70
D	D	B	D	D	D	A	A	C	A
71	72	73	74	75	76	77	78	79	80
A	D	D	B	D	B	B	A	B	B
81	82	83	84	85	86	87	88	89	90
B	B	C	D	A	D	C	B	D	C

91	92	93	94	95	96	97	98	99	100
C	D	D	C	D	B	B	C	D	D
101	102	103	104	105	106	107	108	109	110
D	D	C	A	A	C	D	C	B	C
111	112	113	114	115	116	117	118	119	120
D	B	A	B	B	B	D	C	C	A
121	122	123	124	125	126	127	128	129	130
B	B	B	B	C	B	C	C	B	B
131	132	133	134	135	136	137	138	139	140
B	D	B	D	B	C	C	B	C	A
141	142	143	144	145	146	147	148	149	150
D	A	B	B	B	D	C	C	A	C
151	152	153	154	155	156	157	158	159	160
C	B	B	B	C	C	D	A	B	C
161	162	163	164	165	166	167	168	169	170
A	B	A	B	D	A	B	B	A	B
171	172	173	174	175	176	177	178	179	180
C	C	B	B	D	B	A	A	A	B
181	182	183	184	185	186	187	188	189	190
D	B	D	A	B	B	B	A	A	A
191	192	193	194	195	196	197	198	199	200
B	A	A	A	B	B	B	D	C	C
201	202	203	204	205	206	207	208	209	210
B	D	A	A	B	D	D	D	D	C
211	212	213	214	215	216	217	218	219	220
C	C, B	C	C	B	C	B	B	B	2.74
221	222	223	224	225	226	227	228	229	230
A	B	D	80.5	C	0.14	B	18.29	7.45	B
231	232	233	234	235	236				
29.29	8.20	D	C	2.88	B				

Explanations

Q. 18

Bifurcation Ratio: The bifurcation ratio is the ratio between the number of streams in one order and in the next.

Hence, Number of second order streams = 4×4 = 16

Q. 50

Refer explanation for Q. 221.

Q. 51

The following stages of surface gully development are generally recognized:

Stage 1: Formation stage

Stage 2: Development stage

Stage 3: Healing stage

Stage 4: Stabilization stage

Q. 101

The Middle Third Rule states that no tension is developed in a wall or foundation if the resultant force lies within the middle third of the structure.

Q. 104

In flat area water will flow in all the directions(no defined outlets to capture water), therefore watershed drainage will be difficult and delayed.

Q. 132

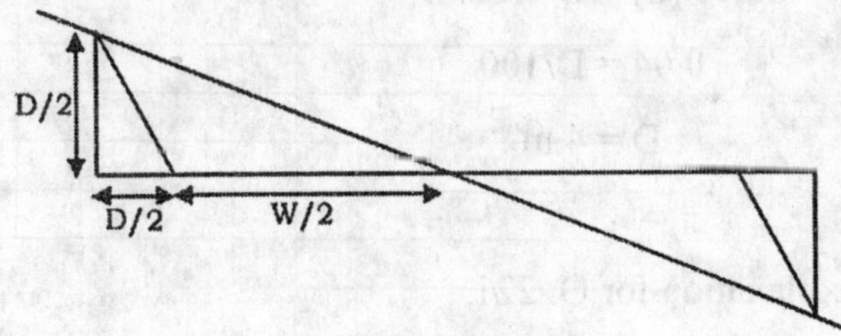

Referring above diagram

$$\text{Land slope(S)} = \frac{\frac{D}{2}}{\frac{D}{2}+\frac{W}{2}} \times 100$$

Putting S = 30 and W = 3.5 m in above formula

We get D = 1.5 m

Q. 149

We know

Soil loss (L) = $(\lambda_e/22.13)^m$

Where, m = 0.5 (taken for steep slopes)

Hence, $L_1/L_2 = (\lambda_{e1}/\lambda_{e2})^{0.5}$

Doubling the slope length will give $L_2 = (2)^{0.5}\ L_1 = 1.4\ L_1$

Q. 154

$$\text{Discharge(Q)} = \frac{2}{3}\ C_d b\sqrt{2g}\ H^{3/2}$$

$$\Rightarrow \quad 5 = \frac{2}{3} \times 0.68 \times b \times \sqrt{2 \times 9.81}\ (0.45)^{3/2}$$

$$\Rightarrow \quad d = 10.00 \text{ m}$$

Q. 159

According to Manning's formula

$$\text{Discharge(Q)} \propto (S)^{0.5}$$

Hence, $Q_1/Q_2 = (S_1/S_2)^{0.5}$

Doubling the slope length will give $S_2 = (4)^{0.5}\ S_1 = 2\ S_1$

Q. 171

We know

$$\text{Slope (S)} = D \times 100/H$$

$$\Rightarrow \quad 0.04 = D/100$$

$$\Rightarrow \quad D = 4 \text{ m}$$

Q. 189

Refer explanation for Q. 221.

Q. 209

We know

$$\text{Soil loss} = RKLSPC_1$$

$$\Rightarrow \quad 44.80 = RKLSP \times 0.25$$

$$\Rightarrow \quad RKLSP = 44.80/0.25 = 179.2 \text{ Mg/ ha}$$

Now, Again

Soil loss = $(RKLSP)C_2$

$$= 179.2 \times 0.15 = 26.88 \text{ Mg/ ha}$$

Q. 210

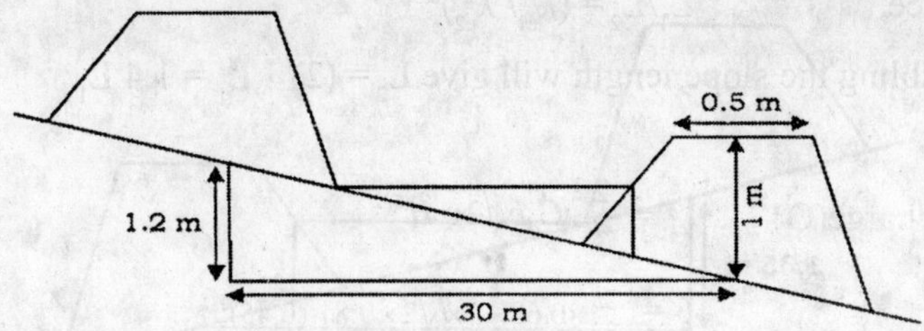

Bottom width of bund (b) = 2.5 m

Horizontal interval between bunds (HI) = $\frac{VI}{S}\times 100$

$$= \frac{1.2}{0.4}\times 100 = 30 \text{ m}$$

Length of bund per hectare (L_B) = 10000/30 = 333.33 m

Adding 30% length for side bunds

$L_B = 333.33\times 1.30 = 433.329$ m

Area lost due to bunding per hectare = $L_B b/10000$

$$= 433.329\times 2.5/10000 = 10.83\ \%$$

Q. 211

Referring below figure

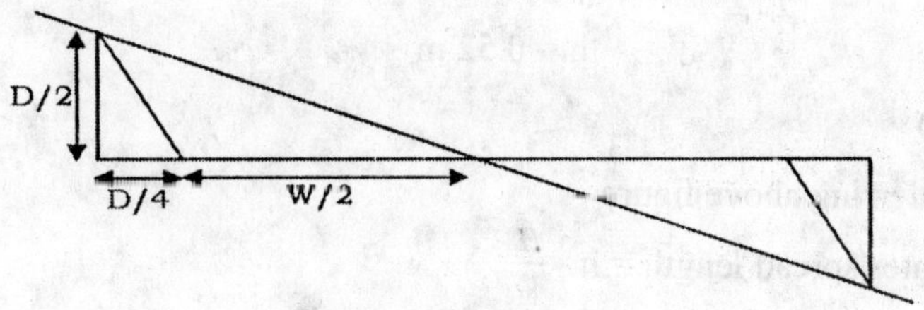

$$\text{Land slope(S)} = \frac{\frac{D}{2}}{\frac{D}{4}+\frac{W}{2}}\times 100$$

$$\Rightarrow \quad DS + 2WS = 200D$$

$$\Rightarrow \quad D = \frac{2WS}{200-s}$$

Q. 212 (i)

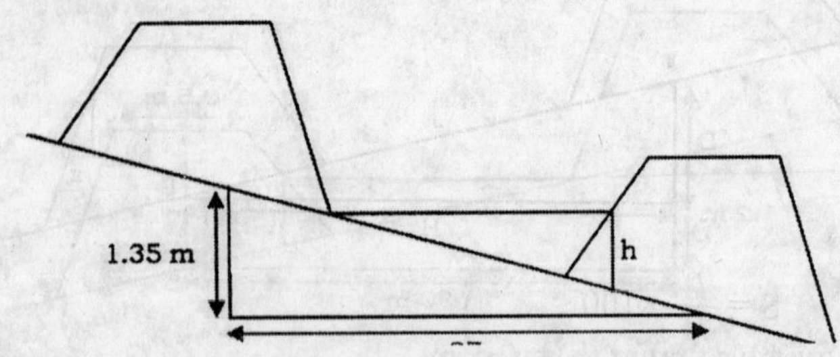

Horizontal interval between bunds (H) = $\frac{V}{S}\times 100$

$= \frac{1.35}{5}\times 100 = 27$ m

Consider length of bund = L_B

Runoff volume = $27\times 0.1\times L_B$

Water stored behind bund = Cross sectional area× L_B

$= \frac{1}{2}\times h\times h\times \frac{H}{V}\times L_B$

$= \frac{1}{2}\times h^2\times 20\times L_B$

∵ Runoff volume = Storage volume of water behind the bund

$\frac{1}{2}\times h^2\times 20\times L_B = 27\times 0.1\times L_B$

⇒ h = 0.52 m

(ii)

Referring above figure

Water spread length = $h\times \frac{H}{V}$

⇒ $= 0.52\times 20 = 10.4$ m

Q. 213

Consider horizontal interval is H.

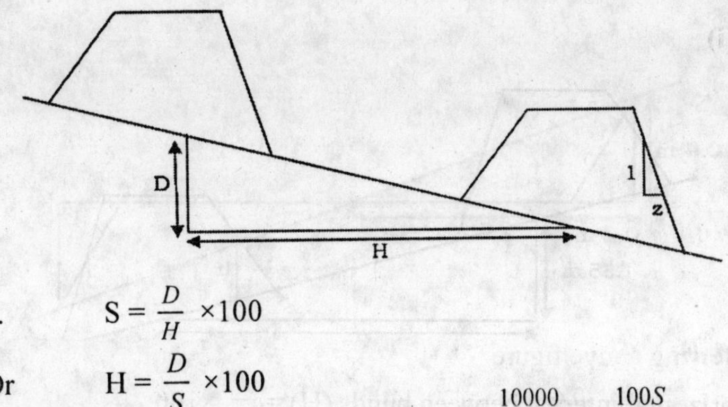

$$\therefore \qquad S = \frac{D}{H} \times 100$$

$$\text{Or} \qquad H = \frac{D}{S} \times 100$$

$$\text{Number of contour bunds per hectare} = \frac{10000}{L \times \frac{D}{S} \times 100} = \frac{100S}{LD}$$

Q. 214

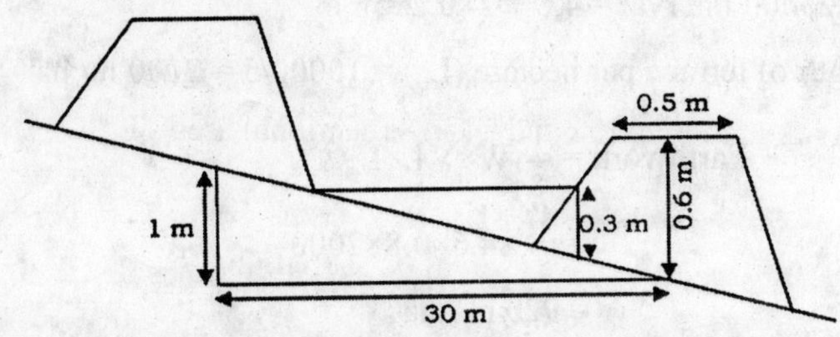

Runoff volume = 972.5 m^3

Storage volume = Cross sectional area of water×L_B

$$= \frac{h^2}{2} \times (Z + 100/S) \times L_B$$

$$= \frac{(0.3)^2}{2} \times (2 + 100/2.5) = 472.5 \text{ m}^3$$

∴ Excess runoff to be disposed off = Runoff volume – Storage volume

= 972.5 – 472.5

= 500 m^3 in 20 minutes

Therefore, Discharge = 25 m^3/ min

Q. 215

$$\text{Soil loss} = RKLSC(P_1 - P_2)$$

$$= 1000 \times 0.25 \times 0.2 \times 0.75 \times (1.0 - 0.7)$$

$$= 11.25 \text{ Mg ha}^{-1} \text{ year}^{-1}$$

Q. 217

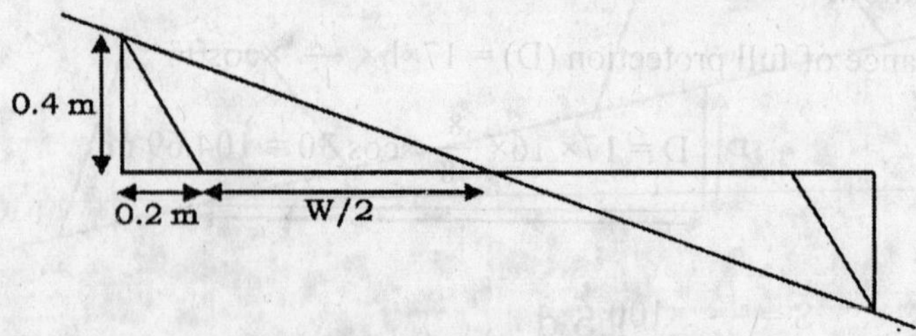

Referring above figure

$$\frac{0.4}{0.2+(w/2)} = 0.16$$

$\Rightarrow$ w = 4.6 m

Horizontal interval = 4.6 + 2×0.2 = 5 m

Length of terrace per hectare (L_B) = 10000/5 = 2000 m/ ha

$$\therefore \quad \text{Earthwork} = \frac{1}{8} W \times V.I. \times L_B$$

$$= \frac{1}{8} \times 4.6 \times 0.8 \times 2000$$

$$= 920 \text{ m}^3/\text{ha}$$

Q. 218

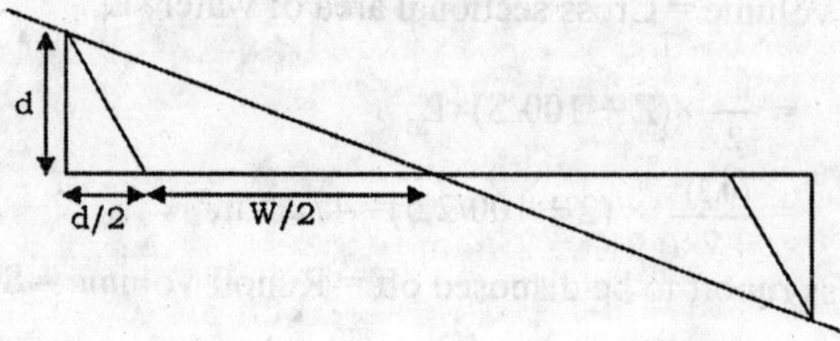

Referring above figure

$$\frac{2d}{d+w} = 0.10$$

$\Rightarrow$ w = 19 d

$$\because \quad \text{Area lost} = \frac{\textit{Older surface length} - \textit{Newly created surface length}}{\textit{Older surface length}} \times 100$$

$$= \frac{\sqrt{(10d)^2 + d^2} - \sqrt{(19d/2)^2}}{\sqrt{(10d)^2 + d^2}} \times 100$$

$$= 5.47\ \%$$

Q. 219

We know,

Distance of full protection (D) = $17 \times h \times \frac{V_m}{V} \times \cos\theta$

$$D = 17 \times 16 \times \frac{8}{18} \times \cos 30 = 104.69 \text{ m}$$

Q. 220

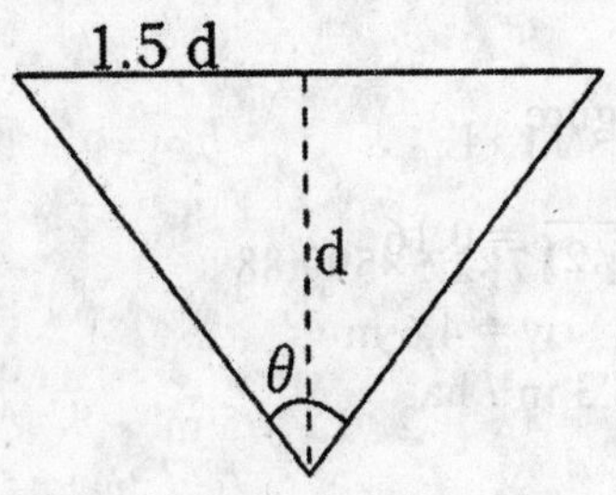

We know

Cross sectional area (A) = $\frac{Dischareg}{Velocity}$

$$= \frac{1.5}{1.2} = 1.25 \text{ m}^2$$

Referring above figure

Cross sectional area (A) = $\frac{1}{2} \times 2 \times 1.5d$

$\Rightarrow$ $1.25 = 1.5d^2$

$\Rightarrow$ $d = 0.913$ m

Therefore, Top width = 3d

$= 3 \times 0.913 = 2.74$ m

Q. 221

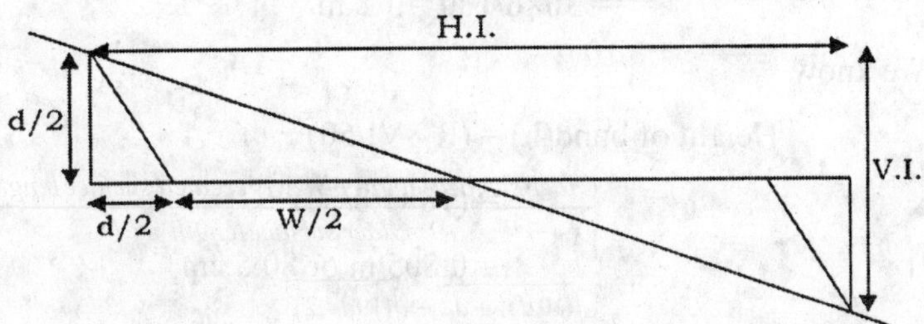

Referring above figure

Horizontal interval (HI) = $\frac{VI}{S} \times 100$

$= \frac{2.5}{0.15} \times 100 = 16.67$ m

Length of terrace per hectare (L_B) = 10000/16.67 = 599.88 m/ ha

Width of cut (W) = 16.67 - 2.5 = 14.17 m

We know

Earthwork = $\frac{1}{8}$ W×V.I.×L_B

$= \frac{1}{8} \times 14.17 \times 2.5 \times 599.88$

= 2656.3 m^3/ ha

Q . 223

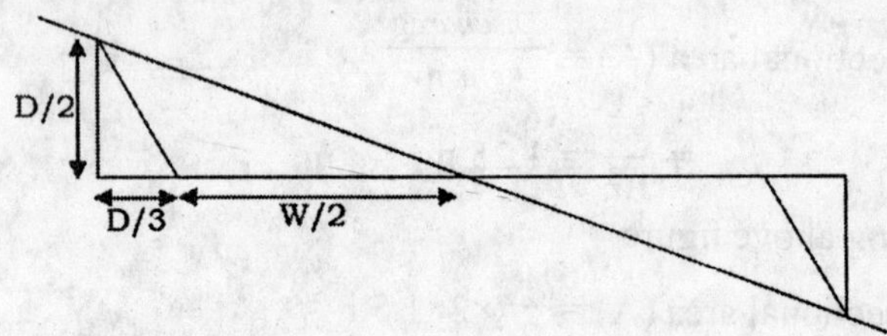

Referring above figure

$$\text{Land slope (S)} = \frac{D/2}{D/3 + W/2} \times 100$$

⇒ D = 3WS/(300 – 2S)

Q. 224

Vertical interval (VI) = HI×100/S

= 30×6/100 = 1.8 m

We know

Height of bund(h) = $(R_e \times VI/50)^{1/2}$

= $(18 \times 1.8/50)^{1/2}$

= 0.805 m or 80.5 cm

Q. 225

Note - A groyne is a rigid hydraulic structure built from an ocean shore or from a bank (in rivers) that interrupts water flow and limits the movement of sediment. Groynes are hard shoreline protection structures which aim to protect the shoreline from coastal erosion.

Q. 226

Peak discharge (Q) = $1.77\ L\ H^{1.5}$

$\Rightarrow \quad 1.81 = 1.77 \times L \times (0.8)^{1.5}$

$\Rightarrow$ Inlet Width (L) = 1.43 m

Velocity at stilling basin (v) = $(2gh_e)^{1/2}$

($\because$ h_e is the head loss at apron

= Drop in guly head – 0.20× Drop in guly head

= 5 – 0.20×5 = 4 m)

Therefore, $v = \sqrt{(2 \times 9.81 \times 4)} = 8.86$ m/s

Since, Peak discharge (Q)

= Area of the water body on the crest × velocity of flow

= L.y×v

$\Rightarrow \quad 1.81 = 1.40 \times y \times 8.86$

$\Rightarrow \quad y = 0.14$ m

Q. 228

The probability of occurrence of event(P) = 1/50 = 0.02

Risk is calculated by means of the binomial distribution given in simplified form as follows:

$R = 1 - (1 - P)^n$

$= 1 - (1 - 0.02)^{10} = 0.1829$ or 18.29 %

Q. 229

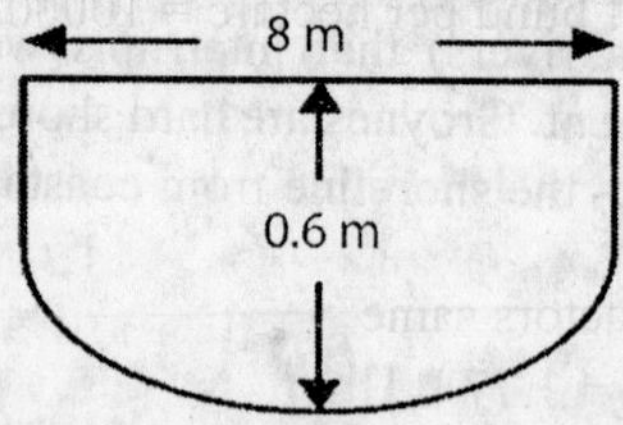

Slope (S) = 0.03

Manning's Value (n) = 0.04

Top width (b) = 8 m

Maximum Depth (d) = 0.6 m

Wetted Perimeter (P) = $b + \frac{8d^2}{3b}$

$= 8 + \frac{8(0.6)^2}{3\times8} = 8.12$ m

Cross Sectional Area (A) = $\frac{2}{3}$ b d

$= \frac{2}{3} \times8\times0.6 = 3.2 \text{ m}^2$

Thus,

Hydraulic Radius (R) = A/P

R = 3.2/8.12 = 0.394 m

Velocity of discharge (V) = $\frac{R^{2/3}S^{1/2}}{n}$

$= \frac{(0.394)^{2/3}(0.03)^{1/2}}{0.04} = 2.327$ m/s

Thus

Discharge (Q) = A×V

$= 3.2\times2.327 = 7.45 \text{ m}^3/\text{s}$

Q. 230

For medium rainfall

Vertical interval (VI) = 0.1 s + 0.6

$= 0.1\times4 + 0.6 = 1$ m

Horizontal interval (HI) = $1\times100/4 = 25$ m

and Length of bund per hectare = $10000/25 = 400$ m

Q. 231

Soil loss = RKLSPC

Keeping all other factors same

Soil loss α L

Where

Slope length (L) = $(\lambda_e/22.13)^m$

Where, m = 0.5 (taken for steep slopes)

Hence, $L_1/L_2 = (\lambda_{e1}/\lambda_{e2})^{0.5}$

Doubling the slope length will give $L_2 = (1/2)^{0.5}\, L_1 = 0.7071L_1$

Hence,

Percent reduction in soil loss = $(L_1 - 0.7071L_1)\times100/ L_1 = 29.29\%$

Alternatively,

Soil loss α LS

$$LS = \frac{\sqrt{L}}{100}\,(0.76 + 0.53S + 0.076S^2)$$

Put, L = 200 and S = 8 ⇒ LS = 1.395

And, L = 100 and S = 8 ⇒ LS = 0.9864

Hence,

Percent reduction in soil loss = $(1.395 - 0.9864)\times100/ 1.395 = 29.29\%$

Q. 232

$$\text{Depth of runoff (S)} = \frac{25400}{CN} - 254$$

$$= \frac{25400}{80} - 254 = 63.5 \text{ mm}$$

Hence, $$\text{Runoff volume (Q)} = \frac{(P-0.2S)^2}{P+0.8S} = 8.20 \text{ mm}$$

Q. 234

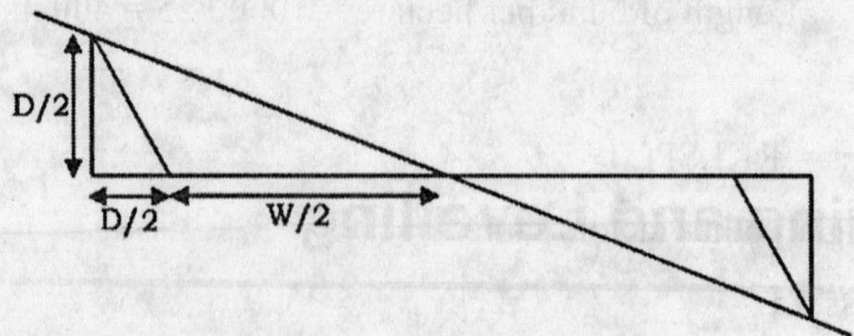

Referring above diagram

$$\text{Land slope(S)} = \frac{\frac{D}{2}}{\frac{D}{2}+\frac{W}{2}} \times 100$$

Putting S = 10 and D = 2.5 m in above formula

We get W = 22.5 m

Again, Referring above figure

Original width = $(12.5^2 + 1.25^2)^{1/2}$ = 12.56 m

Width lost = 12.56 – 11.25 = 1.31 m

Hence, $\quad \text{Area lost} = \frac{Width\,lost}{Original\,Width} \times 100$

$$= \frac{1.31}{12.56} \times 100 = 10.43\ \%$$

Q. 235

We know,

Distance of full protection (D) = $17 \times h \times \frac{V_m}{V} \times \cos\theta$

$$D = 17 \times 15 \times \frac{9.6}{16} \times \cos 20 = 143.77 \text{ m}$$

$\Rightarrow$ Area protected = 143.77×200= 28754 m^2 or 2.88 ha

28

Surveying and Levelling

UKPSC Combined Assistant Engineering Exam 2013 Paper - I

Q. 1 Fore sight is also known as

(A) Positive sight (B) Minus sight

(C) Intermediate sight (D) Vertical sight

Q. 2 Reduced level of a point is its height or depth above or below

(A) The ground surface (B) The assumed datum

(C) The line of collimation (D) Both (A) and (b)

Q. 3 The length of Gunter's chain is

(A) 100 feet (B) 50 feet

(C) 66 feet (D) 75 feet

Q. 4 Length of a metric chain is

(A) 1 m (B) 20 m

(C) 50 m (D) 100 m

Q. 5 The bench mark fixed at the end of a day's of survey work is called

(A) Permanent bench mark (B) Arbitrary bench mark

(C) Temporary bench mark (D) None of the above

Q. 6 The imaginary line joining the geographical North and South is known as

(A) Geographical meridian (B) Magnetic meridian

(C) Arbitrary meridian (D) All of the above

Q. 7 The bearing of line AB is 40° 26’ 10” and bearing of AC is 140° 30’ 40”. What will be angle between lines AB and AC?

(A) 180° 56’ 50” (B) 130° 26’ 10”

(C) 100° 4’ 30” (D) 230° 4’

Q. 8 Alidade is an important instrument used in which survey ?

(A) Plane table survey (B) Chain survey

(C) Theodolite survey (D) Compass survey

Q. 9 Instrument used for measurement of areas of figure having irregular boundaries are

(A) Anemometer (B) Planimeter

(C) Rolling meter (D) All of the above

Q. 10 If a chain is too long, the measured distance will be less than actual, then the correction will be

(A) Can be negative or positive (B) Zero

(C) Negative only (D) Positive only

Q. 11 The type of survey in which the shape of earth is taken into account is called

(A) Geodetic survey (B) Curvature surface

(C) Precise survey (D) All above

Q. 12 Height of instrument method is used for determination of

(A) Reduced level (B) Discharge

(C) Flow velocity (D) Back bearing

Q. 13 Direct ranging in survey is done

(A) When one line overlap into other

(B) When two ends of survey line are inter visible

(C) When two ends are not visible

(D) None of the above

Q. 14 Curvature of the earth is taken into account in surveying when the extent of the area is more than

(A) 50 km² (B) 100 km²

(C) 250 km² (D) None of the above

Q. 15 The chain with 100 links and 100 feet length is known as

(A) Gunter's chain (B) Engineer's chain

(C) Metric chain (D) None of the above

Q. 16 The survey in which curvature of the earth is not considered is known as

(A) Chain survey (B) Aerial survey

(C) Compass survey (D) Plane survey

Q. 17 The longest chain line in chain surveying is called

(A) Check line (B) Tie line

(C) Base line (D) All above three

Q. 18 Errors in surveying are

(A) Instrumental error (B) Personal error

(C) Measurement error (D) All of the above

Q. 19 The bearing of line taken in the direction of opposite to the progress of work is known as

(A) Fore bearing (B) Back bearing

(C) Whole circle bearing (D) None of the above

Q. 20 Amongst various land levelling methods, the most common method is

(A) Plane method (B) Darcy method

(C) Contour adjustment method (D) Plane inspection method

Q. 21 The volume of earthwork in land levelling can be computed by

(A) Trapezoidal rules (B) Prismoidal method

(C) End – area method (D) Both (A) and (B)

Q. 22 The permissible depth of earthwork under land levelling work is about

(A) ± 3 cm (B) 4.5 cm

(C) 5.0 cm (D) 10.0 cm

Q. 23 The prismoidal formula is used to determine ....

(A) Runoff rate (B) Earthwork volume

(C) Land slope (D) Infiltration rate

Q. 24 Which of the following is not a land grading method ?

(A) Profile method (B) Laser method

(C) Trapezoidal method (D) Plane inspection method

Q. 25 Four point method is used to determine

(A) Earthwork volume (B) Cross slope

(C) Longitudinal slope (D) Volume of water

Q. 26 Which of the following is not a system of recording levels in the field book ?

(A) Height of instrument method (B). Collimation method

(C) Rise and fall method (D) Black and white method

APPSC AEES Agriculture Engineering Exam 2016

Q. 27 Length of the surveyor's chain is

(A) 33 ft. (B) 66 ft.

(C) 100 ft. (D) 150 ft.

MPPSC Assistant Agricultural Engineer 2013

Q. 28 The primary objective of surveying is to prepare a ....

(A) Cross-section (B) Drawing

(C) Map (D) Sketch

Q. 29 In face left position of theodolite, the position of vertical circle is on .... of observer.

(A) Left (B) Right

(C) Front (D) Back

Q. 30 The reading taken by level on levelling staff kept at point of known elevation is entered as .... reading.

(A) Back sight (B) Intermediate sight

(C) Fore sight (D) None of these is correct

Q. 31 The bearing 168°13' in WCB system can be expressed as .... in QB system.

(A) N 168°13' E (B) S 191°47' W

(C) S 11°47' E (D) N 11°47' W

RPSC AEn Pre Exam 2013 (Agricultural Engineering)

Q. 32 The angle of intersection of the two plane mirrors of an optical square is

(A) 30° (B) 45°

(C) 60° (D) 90°

Q. 33 For Earth work estimation during terrace construction, we need to refer a standard ratio in which is generally known as

(A) Rise : fall ratio (B) Up : down ratio

(C) Cut : fill ratio (D) All of the above

Q. 34 Size of a theodolite is specified by

(A) The length of the telescope

(B) The diameter of the vertical circle

(C) The diameter of the lower plate

(D) The diameter of the upper plate

Q. 35 The R.L. of the point A which is on the floor is 100 m and back side reading on A is 2.455 m. If the fore sight reading on point B, which is on the ceiling is 2.745 m, the R.L. of point B will be

(A) 94.80 (B) 99.71

(C) 100.29 (D) 105.20

Q. 36 Among below given statements only one statement is correct. Select the correct statement

(A) A contour is not necessarily a closed curve

(B) A contour represents a ridge line if the concave side of lower value contour lies towards the higher value contour

(C) Two contours of different elevations do not cross each other except in case of an overhanging cliff

(D) All of the above statements are correct

OCS AEn Exam 2011 (Pre) – II (Agricultural Engineering)

Q. 37 The volume of earthwork in land leveling is calculated by end area method or prismoidal method. What is the expression for volume of earthwork(V) using prismoidal formula (A_1 = Area of first end plane; A_2 = Area of second end plane; A_m = Area of middle section parallel to end planes; L = Distance between end planes ?

(A) $V = L\times(A_1 + A_2)/2$ (B) $V = L\times A_m$

(C) $V = L\times(A_1 + 4A_m + A_2)/6$ (D) $V = L\times(A_1 + A_m + A_2)/3$

Q. 38 Back sight in surveying is:

(A) Rod reading taken on a point of known elevation

(B) Rod reading on a point, the elevation of which is to be determined

(C) Rod reading on an intermediate point

(D) None of the above

Q. 39 The length of a line measured by 30 metre chain was found to be 330 m. The true length of the line is if the chain was 10 cm too long.

(A) 330.10 m (B) 331.10 m

(C) 331.0 m (D) 333.3 m

Q. 40 Tags in the metric chain are fixed at every length of:

(A) 2 m (B) 4 m

(C) 5 m (D) 6 m

CGPSC State Engineering Services Exam (Agriculture Engineering) Part - II

Q. 41 In land grading, the ratio of cut-fill should be

(A) 0.8 – 1.2 (B) 1.2 – 2

(C) 2 – 2.5 (D) 2.5 – 3.5

(E) 3.5 – 10

Q. 42 When the R.F. is 1/5000 then the scale is

(A) 1 cm = 5 cm (B) 1 cm = 50 cm

(C) 1 cm = 50 m (D) 1 cm = 5 m

(E) None of these

Q. 43 The diagonal scale is used to read ……… dimensions.

(A) One (B) Two

(C) Three (D) Four

(E) Five

Q. 44 In surveyors compass the range of bearing are from

(A) 0° – 90° (B) 0° – 180°

(C) 0° – 270° (D) 0° – 360°

(E) None of these

Q. 45 The bearing of the line in the direction of progress of survey is called as

(A) Reduced bearing (B) Fore bearing

(C) Back bearing (D) Whole circle bearing

(E) None of these

Q. 46 The length of Ranging rod is

(A) 3 m (B) 4 m

(C) 5 m (D) 6 m

(E) 9 m

Q. 47 S 31° 28' W convert the quadrant bearing to whole circle bearing

(A) 212° 28' (B) 211° 28'

(C) 210° 28' (D) 209° 28'

(E) 213° 28'

Q. 48 Simpson's rule is also known as

(A) Parabolic line (B) Hyperbolic line

(C) Chord rule (D) Both (B) and (C)

(E) Both (A) and (B)

Q. 49 Check line is also called as

(A) Base line
(B) Chain line
(C) Tie line
(D) Proof line
(E) Centre line

Q. 50 In case of long sights, the horizontal line is not a level, due to

(A) Curvature
(B) Refraction
(C) Local attraction
(D) Slope
(E) Both (A) and (C)

Q. 51 The instrument used in fast needle method is

(A) Chain
(B) Compass
(C) Theodolite
(D) Cross staff
(E) All of the above

Q. 52 In second quadrant, WCB =

(A) 180° + R.B.
(B) 180° - R.B.
(C) 360° + R.B.
(D) 360° - R.B.
(E) 90° - R.B.

Q. 53 The levelling used for determining the surface undulations on either side of the given line

(A) Cross sectioning
(B) Profile levelling
(C) Check levelling
(D) Reciprocal levelling
(E) Base levelling

Q. 54 The graduated drum of a planimeter is divided into ……… divisions.

(A) 10
(B) 50
(C) 60
(D) 80
(E) 100

Q. 55 Find included angle between lines 64°28' and 190°32'

(A) 136°4'
(B) 126°4'
(C) 116°4'
(D) 106°4'
(E) 102°4'

Q. 56 Apparent movement of the image relative to the cross hairs is known as

(A) Curvature (B) Refraction

(C) Parallax (D) None

(E) Both (A) and (B)

Q. 57 A line makes an angle of 45° to horizontal. The slope of the line is

(A) 1 % (B) 40 %

(C) 70 % (D) 80 %

(E) 100 %

UKPSC A.En. Exam 2007 Paper – I

Q. 58 Contour lines cross Ridge lines and Valley lines are on an angle of

(A) 45 degree (B) 90 degree

(C) 180 degree (D) 60 degree

Q. 59 The elevation of a point can be determined by using the instrument named

(A) Cross staff (B) Optical square

(C) Dumpy level (D) Prismatic compass

Q. 60 Which is the method related to plane table survey ?

(A) Conduction method (B) Radiation method

(C) Multiplication method (D) Subtraction method

Q. 61 The magnetic bearing of a line is $48^0 24'$. Calculate the true bearing of the line if magnetic declination is $5^0 38'$ eastward

(A) $54^0 80'$ (B) $23^0 24'$

(C) $54^0 02'$ (D) $48^0 62'$

Q. 62 The distance between two stations was known to be 1200 m. When measured by a 30 m long chain, it was found to be 1210 m. What was the actual length of the chain ?

(A) 29.50 m (B) 29.75 m

(C) 30.50 m (D) 30.75 m

Q. 63 Orientation is done in

(A) Chain survey (B) Compass survey

(C) Plane table survey (D) Level survey

Q. 64 When both ends of a survey line are not intervisible, then the method of ranging used is

(A) Direct ranging method (B) Indirect ranging method

(C) Improper ranging method (D) None of the above

Q. 65 The area measured with a 30 m long chain was found to be 900 m^2. If the chain was found to be 30 cm too long, the true area will be

(A) 882 m^2 (B) 890 m^2

(C) 918 m^2 (D) 950 m^2

Q. 66 The whole circle bearing of a line is 100 degree. Its reduced bearing will be

(A) S 100^0 W (B) S 80^0 W

(C) S 80^0 E (D) S 100^0 E

Q. 67 If a line makes an angle of 200^0 15' 40" from north direction clockwise, its reduced bearing will be

(A) 120^0 15' 40" (B) N 20^0 15' 40" W

(C) S 20^0 15' 40" W (D) N 120^0 5' 40" E

Q. 68 If the fore bearing of a line is 40^0 25' 30", its back bearing will be

(A) 220^0 25' 30" (B) 140^0 25' 30"

(C) 40^0 25' 30" (D) 130^0 25' 30"

Q. 69 The magnetic bearing of a line is 48^0 20' 30". The true bearing of this line for magnetic declination of 4^0 20' E will be

(A) 44^0 20' 30" (B) 44^0 30'

(C) 52^0 40' 30" (D) 48^0 24' 30"

Q. 70 Contour line is a line joining the points of

(A) Equal elevation (B) Equal slope

(C) Equal bearing (D) None of the above

Q. 71 A pentagraph is used to

(A) Enlarge, reduce and copy a map

(B) Compute area of a map

(C) Draw the contour lines

(D) Measure landslope

Q. 72 One mile is equal to

(A) 80 furlong (B) 50 furlong

(C) 8 furlong (D) 20 furlong

Q. 73 For high degree of precision measurement which type of tape is more suitable

(A) Cloth tape (B) Metallic tape

(C) Steel tape (D) Inwar tape

Q. 74 A line passing through true North and South is called

(A) Magnetic meridian (B) True meridian

(C) Magnetic bearing (D) True bearing

Q. 75 In quadrantal bearing system, the bearing of a line varies from

(A) 0 – 360 degree (B) 0 – 180 degree

(C) 0 – 90 degree (D) None of the above

Q. 76 Magnetic declination at a point is the horizontal angle between

(A) True bearing and reduced level

(B) Magnetic bearing and reduced level

(C) True meridian and magnetic meridian

(D) True level and magnetic level

Q. 77 If the difference between fore bearing of a line and its back bearing is not 180 degree, then the compass is suffering from

(A) Wind velocity (B) Sunlight

(C) Local attraction (D) None of the above

Q. 78 Closed contour lines with the higher elevation value inside represents

(A) Flat surface (B) Hillock

(C) Valley (D) Undulations

Q. 79 The vertical distance between any two consecutive contours is called

(A) Vertical equivalent (B) Horizontal equivalent

(C) Contour interval (D) Vertical sight

Q. 80 Foresight is also known as

(A) Positive sight (B) Minus sight

(C) Horizontal sight (D) Vertical sight

Q. 81 The process of depicting ground elevations at a vertical cross section along a survey line is called

(A) Fly levelling (B) Check leveling

(C) Reciprocal levelling (D) Profile leveling

Q. 82 If contour lines are straight, parallel and equally spaced, then they represent

(A) An irregular surface

(B) A plane uniformly sloppy surface

(C) A hilly surface

(D) A depression

Q. 83 If one method to obtain reduced level of a point is height of instrument method, the other method is

(A) Rise and fall method (B) Quadrantal method

(C) Left and right method (D) Differential method

Q. 84 Which of the following method is not used in plane table survey ?

(A) Radiation method (B) Traversing method

(C) Intersection method (D) Conduction method

ASCO (RPSC) Exam (Engineering) 2011

Q. 85 The length of a link in a chain is

(A) 20 cm (B) 30 cm

(C) 35 cm (D) 40 cm

Q. 86 For ranging a line, the number of ranging rods required is

(A) At least two (B) At least three

(C) At least four (D) At least five

Q. 87 The preliminary inspection of the area to be surveyed is known as

(A) Rough survey (B) Primary survey

(C) Route survey (D) Reconnaissance survey

Q. 88 The closing error in a closed traverse is adjusted by

(A) Lenmanns rule (B) Slide rule

(C) Bowditch's rule (D) Simpson's rule

Q. 89 The working edge of the alidade is known as the

(A) Parallel edge (B) Beveled edge

(C) Parallax edge (D) Fiducial edge

Q. 90 Spirit level in plane table is used for

(A) Centering (B) Sighting

(C) Marking north (D) Levelling

Q. 91 A datum surface in levelling is a

(A) Horizontal surface (B) Vertical surface

(C) Level surface (D) None of the above

Q. 92 Which of the following method of contouring is most suitable for hilly terrain ?

(A) Direct method (B) Square method

(C) Cross section method (D) Techo metric method

Q. 93 A relatively permanent point of reference whose elevation is known as

(A) Reduced level (B) Benchmark

(C) Level surface (D) Datum point

Q. 94 The maximum tolerance in a 20 m chain is

(A) ± 2 mm (B) ± 3 mm

(C) ± 5 mm (D) ± 8 mm

Q. 95 Height of instrument method of levelling is

(A) More accurate than rise and fall method

(B) Less accurate than rise and fall method

(C) Quicker and less tedious for large number of intermediate sights

(D) None of the above

Q. 96 Electronic theodolite of various ranges in which measured angles are displayed originally on display board are based on which one of the following ?

(A) Special optical technology

(B) Introduction of microprocessor technology

(C) Electro optical technology

(D) Special gearing

Q. 97 Surveys which depict the natural features of a country are known as

(A) Cadastral survey (B) Engineering survey

(C) Topographic survey (D) Geological survey

Q. 98 Perpendicular offsets may be taken by setting the right angle in the ratio

(A) 3 : 4 : 5 (B) 2 : 4 : 9

(C) 3 : 6 : 9 (D) 2 : 4: 5

Q. 99 A series of closely spaced contour lines represents a

(A) Steep slope (B) Gentle slope

(C) Uniform shop (D) Plane surface

Q. 100 In plane table survey, the operation which must be carried out is

(A) Resection (B) Orientation

(C) Radiation (D) Intersection

Q. 101 The U-fork and plumb bob are required for

(A) Leveling (B) Orientation

(C) Centering (D) Marking north line

Q. 102 The benchmark (BM) established by the survey of India is known as

(A) GTS BM (B) Arbitrary BM

(C) Permanent BM (D) Levelled BM

Q. 103 For topographical survey of the area vertical angle is taken in

(A) Chain survey (B) Plane table survey

(C) Theodolite survey (D) Dumpy level survey

Q. 104 Pick up the correct statement

(A) A line joining the point of same elevation is termed as contour line

(B) A line joining the points of different reduced level is known as contour line

(C) A line joining the equal slope is termed as contour line

(D) A line joining the equal length is termed as contour line

Q. 105 At the valley line the contour line form curve of

(A) U-shape (B) V-shape

(C) L-shape (D) R-shape

ASCO (RPSC) Exam (Forest) 2011

Q. 106 Odd out from the following which does not relate to chain survey

(A) Metric chain (B) Jacob's chain

(C) Gunter's chain (D) Engineer's chain

Q. 107 Which of the following field is largest ?

(A) Field-A : 10 square kilometre (B) Field B : 100 hectare

(C) Field C : 625 acre (D) Field D : 2500 vigha

Q. 108 Which one of the following is the latest survey equipment ?

(A) DGPS (B) Theodolite

(C) Total servey station (D) Dumpy level

UKPSC A.En. Exam 2012 Paper – I

Q. 109 In chain surveying, following lines are used :

(A) Base line, check line, contour line

(B) Base line, contour line, tie line

(C) Check line, tie line, contour line

(D) Base line, check line, tie line

Q. 110 In plane surveying

(A) The curvature of the earth is considered

(B) The surveys extent over small areas

(C) The surveys extent over large areas

(D) None of the above

Q. 111 The area of the plot can be calculated by

(A) Mid ordinate rule or average ordinate rule

(B) Trapezoidal rule or Simpson rule

(C) Both (A) and (B)

(D) None of the above

Q. 112 The sum of interior angle's of a closed traverse is

(A) (2n – 4) × 90° (B) (2n + 4) × 90°

(C) (n – 4) × 90° (D) (n + 4) × 90°

Q. 113 The cross slope of the road is called

(A) Road crown (B) Road gradient

(C) Road camber (D) None of the above

Q. 114 The quadrantal bearing in quadrantal system never exceeds

(A) 90° (B) 180°

(C) 270° (D) None of the above

Q. 115 Increasing contours towards centre represents a

(A) Ridge (B) Valley

(C) Hillock (D) Pond

Q. 116 A 1 km long straight line on 60° slope will measure on a 1 : 10000 scale as .... cm

(A) 10 (B) 5

(C) 1 (D) 0.5

Q. 117 Contour lines on a map cut one another only in case of

(A) Over hanging cliff (B) A ridge line

(C) A valley line (D) Flat land

Q. 118 When the chain is found to be too long, the error in measuring distance will be

(A) Positive (B) Negative

(C) Compensating (D) None of the above

Q. 119 The correction to be applied to each 30 m chain for a line measured along 60° slope could be .... m.

(A) 3 (B) 10

(C) 3.75 (D) 45

Q. 120 The operation of levelling to determine the elevation between two points is known as

(A) Simple levelling (B) Fly levelling

(C) Differential levelling (D) None of the above

Q. 121 Contour lines will always be

(A) Parallel to general slope of the field

(B) Across the general slope of the field

(C) They can be in any direction

(D) None of the above

Q. 122 The unit cost of any item of work is determined by using

(A) Specification (B) Volume analysis

(C) Rate analysis (D) Economic analysis

Q. 123 The object of a survey is to prepare plan or map so that it may represent the area on a

(A) Vertical plane (B) Horizontal plane

(C) Circular plane (D) None of the above

Q. 124 The method of plane table surveying which is performed by having a single station only

(A) Radiation method (B) Intersection method

(C) Traversing method (D) None of the above

Q. 125 Dip of magnetic needle at equator is .... degree

(A) Zero (B) 45

(C) 90 (D) 180

Q. 126 A theodolite can measure

(A) Difference in level (B) Bearing of line

(C) Zenith angle (D) All the above

Q. 127 The vertical distance between any two consecutive contours is called

(A) Vertical equivalent (B) Horizontal equivalent

(C) Contour interval (D) Contour gradient

Q. 128 Method of surveying in which the field work and plotting are done simultaneously is called

(A) Plane tabling (B) Mapping

(C) Compass surveying (D) Drawing

Q. 129 Only linear measurements are made by

(A) Chain survey (B) Plane table survey

(C) Compass survey (D) Theodolite survey

UKPSC A.En. Exam 2012 Paper - II

Q. 130 If 'θ' be the curvature angle of an open ditch the radius of curvature (r) of the open ditch in metres can be calculated using the formula

(A) $R = 15/\sin(\theta/4)$ (B) $R = 15/\sin(\theta/2)$

(C) $R = 15/\cos(\theta/4)$ (D) $R = 15/\cos(\theta/2)$

Graduate Aptitude Test in Engineering - 2007

Q. 131 The areas within the contour lines at the site of a proposed reservoir and dam are as follows:

Contour, m	20	22	24	26	28	30	32
Area, m^2	100	220	600	1800	4500	10000	25000

If 20 m R.L. represents the bottom of the reservoir and 32 m R.L. represents the water surface, the volume of water in the reservoir obtained by the trapezoidal formula is

(A) 21110 m^3 (B) 32220 m^3

(C) 42220 m^3 (D) 59340 m^3

Q. 132 The angle between the lines AB and BC whose respective bearings are 35 and 140 is

(A) 75° (B) 115°

(C) 175° (D) 185°

Graduate Aptitude Test in Engineering - 2008

Q. 133 The area of a map plotted to a scale of 1:3000 measures 9069.37 mm^2. The 20 m chain used for this survey was short by 0.2 m. The true land area it represents is

(A) 83281 m^2 (B) 82449 m^2

(C) 80808 m^2 (D) 80000 m^2

Q. 134 To measure the different in level precisely between two points with a leveling instrument having collimation error, the method to be used is

(A) Reciprocal leveling (B) Check leveling

(C) Compound leveling (D) Profile leveling

Graduate Aptitude Test in Engineering - 2009

Q. 135 Line of sight through the leveling instrument is called

(A) Backsight (B) Foresight

(C) Line of collimation (D) Sight of collimation

Q. 136 In a reciprocal leveling, the level set up close to point P gave readings of 1.6 m and 0.8 m at stations P and Q respectively. The reading obtained by setting up the level close to point Q were 1.4 m and 0.5 m on stations P and Q respectively. Total error of collimation, curvature and refraction in m is

(A) 0.05 (B) 0.10

(C) 0.55 (D) 0.85

Graduate Aptitude Test in Engineering - 2010

Q. 137 A land survey is conducted on 40 m×40 m grids and the elevations of grids in m from mean sea level are as follows.

	A	B	C
1	102.3	103.0	103.7
2	101.5	102.4	102.3
3	101.2	103.5	102.6

Assuming that cut is equal to fill, the volume of earthwork required to level the area in m^3 is

(A) 4480 (B) 4840

(C) 5480 (D) 6480

Graduate Aptitude Test in Engineering - 2011

Q. 138 The magnetic bearing of a line at a station point is found to be 182°. The magnetic declination at the station is 3°E and the local attraction is − 1° for correction. The true bearing of the line is

(A) 186° (B) 184°

(C) 180° (D) 178°

Graduate Aptitude Test in Engineering - 2012

Q. 139 The difference between Fore Bearing and Back Bearing of a traverse line is

(A) Exactly 90° (B) Less than 180°

(C) Exactly 180° (D) Greater than 180°

Graduate Aptitude Test in Engineering - 2014

Q. 140 A 20 m chain used for surveying is found to be actually 19.7 m. If the actual distance is 1200 m, the chain distance in m will be ....

Q. 141 The process of determining the elevation of different points in a vertical plane is known as

(A) Levelling (B) Surveying

(C) Contouring (D) Tacheometry

Graduate Aptitude Test in Engineering - 2016

Q. 142 The back sight of 1.258 m was observed for the bench mark (BM) at reduced level (RL) of 48 m. The corresponding fore sight on the staff held vertically inverted to the underside of a bridge beam is 4.645 m. The RL at the underside of the bridge beam in m is

(A) 44.613 (B) 46.936

(C) 51.581 (D) 53.903

Graduate Aptitude Test in Engineering - 2018

Q. 143 If the departure and latitude of a line during Theodolite survey are 70.03 m and - 100 m, respectively, the whole circle bearing of this line in degree is

(A) 55 (B) 125

(C) 135 (D) 145

Graduate Aptitude Test in Engineering - 2020

Q. 144 Area enclosed by different contours of a pond is given in the following Table. Using trapezoidal formula, the total estimated capacity of pond in m^3 is _

Contour Value (m)	100	101	102	103	104	105	106
Area Enclosed (m^2)	410	580	640	740	960	1100	1240

Q. 145 The underside beam of a railway bridge, marked as permanent Bench Mark (BM) with Reduced Level (RL) of 85.168 m, is taken as reference for leveling operation. The Back Sight (BS) on the staff held vertically inverted to the BM is 3.645 m. For the Fore Sight (FS) of 1.523 m at a point in the construction site, the RL or elevation in m is ....

Graduate Aptitude Test in Engineering - 2021

Q. 146 While carrying out a traverse survey ABCDA′ using a theodolite with the originating station A, the departures and latitudes of the lines, as obtained, are shown in the following figure (not drawn to scale). It is seen that, due to the observational errors, the originating station A and its computed station A′ are not the same. For this survey, the 'closing error' in m is

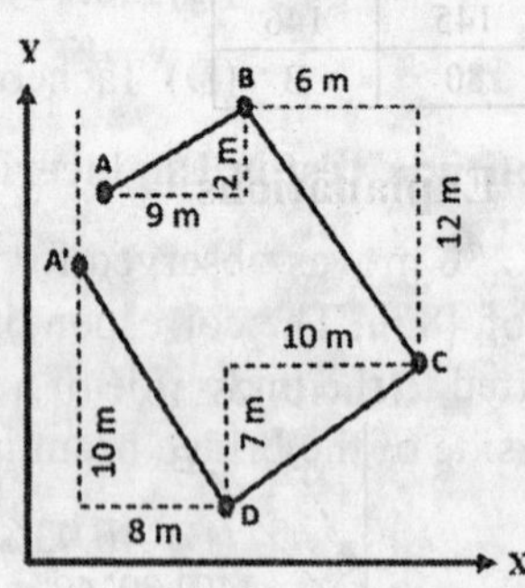

(A) 6.33 (B) 7.62
(C) 33.73 (D) 35.21

Answers Key

1	2	3	4	5	6	7	8	9	10
A	B	C	B	C	A	C	A	B	D
11	12	13	14	15	16	17	18	19	20
A	A	B	C	B	D	C	D	B	A
21	22	23	24	25	26	27	28	29	30
D	A	B	C	A	D	A	C	A	A
31	32	33	34	35	36	37	38	39	40
C	B	C	C	D	C	C	A	B	C
41	42	43	44	45	46	47	48	49	50
B	C	C	A	B	A	B	A	D	A
51	52	53	54	55	56	57	58	59	60
C	B	A	E	B	C	E	B	C	B
61	62	63	64	65	66	67	68	69	70
C	B	C	B	C	C	C	A	C	A
71	72	73	74	75	76	77	78	79	80
A	C	D	B	C	C	C	B	C	B
81	82	83	84	85	86	87	88	89	90
D	B	A	D	A	B	D	D	D	D
91	92	93	94	95	96	97	98	99	100
C	D	B	C	C	C	C	A	B	B
101	102	103	104	105	106	107	108	109	110
C	A	C	A	B	B	A	A	D	B
111	112	113	114	115	116	117	118	119	120
C	A	C	A	C	B	A	B	C	C
121	122	123	124	125	126	127	128	129	130
B	C	B	A	A	D	C	A	A	B

131	132	133	134	135	136	137	138	139	140
D	A	D	A	C	A	A	B	C	1218.27
141	142	143	144	145	146				
A	D	D	4860	80	B				

Explanations

Q. 7

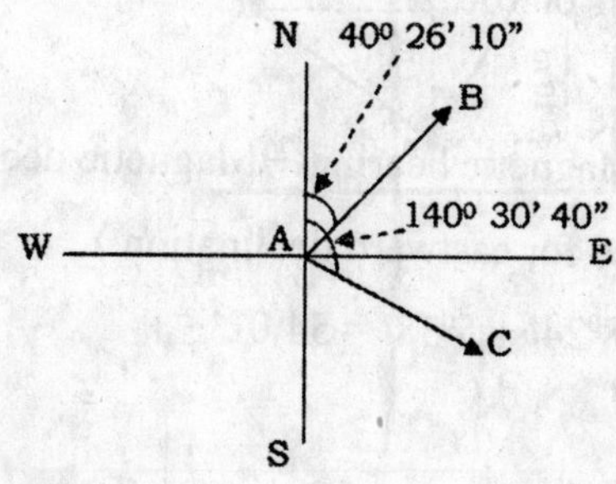

Referring above figure

Angle between lines AB and AC = 140° 30' 40" – 40° 26' 10" = 100° 4' 30"

Q. 35

Refer explanation for Q. 142

Q. 39

In every chainage it measures 0.1 m less. For measuring 330 m length there will be 330/30 = 11 chainages.

Therefore, true length will be = 330 + 11×0.1 = 331.10 m

Q. 47

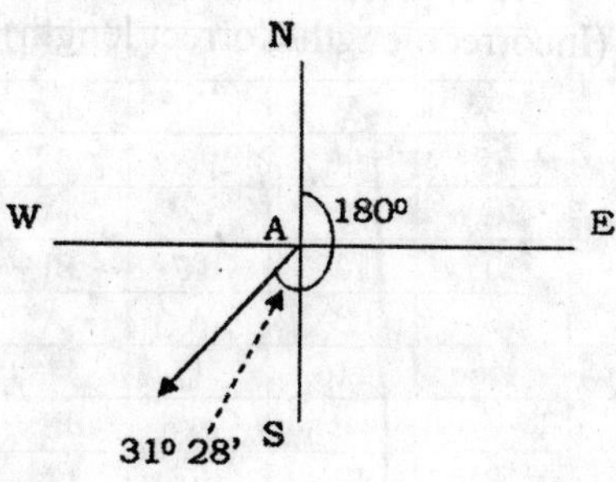

Referring above figure

Whole circle bearing = $180^0 + 31^\circ 28' = 211^\circ 28'$

Q. 55

Refer explanation for Q. 7.

Q. 57

Slope = $\tan 45^0 = 1$ or 100 %

Q. 61

True bearing = Magnetic bearing + Magnetic declination

("+" sign is used for eastward declination.)

$= 48^0 24' + 5^0 38' = 54^0 02'$

Q. 62

In every chainage it measures 10 m more. For measuring 1200 m length there will be 1200/30 = 40 chainages. In every chainage, extra length measured is 10/40 = 0.25 Therefore, true length of chain will be = 30 - 0.25 = 29.75 m

Q. 65

Method I

We know

True area = $(\text{Incorrect length}/\text{Correct length})^2 \times \text{Measured area}$

$= (30.3/30) \times 900 = 918 \text{ m}^2$

Note:

True distance = (Incorrect length/Correct length)×Measured length

True volume = $(\text{Incorrect length}/\text{Correct length})^3 \times \text{Measured area}$

Method II

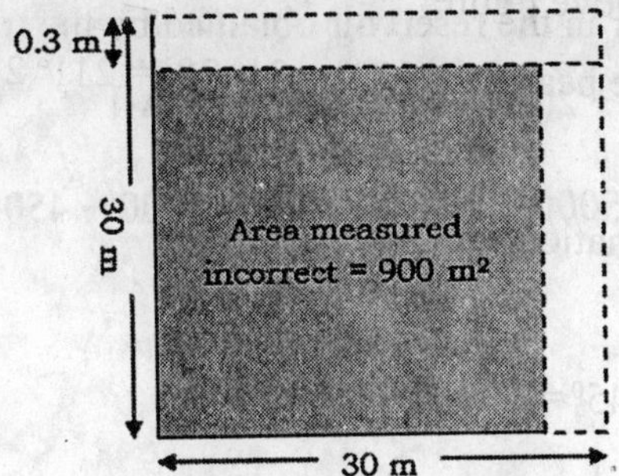

Area measure due to incorrect length = 900 m^2

But, actual area will be more that will be left as shown in figure. Therefore actual area of the land = 900 + area left

$$= 900 + 0.3\times29.70 + 0.3\times30 = 918\ m^2$$

Q. 67

Refer explanation for Q. 47.

Q. 69

Refer explanation for Q. 61.

Q. 107

Field A: 10 km^2 = 1000 ha

Field B: 100 ha

Field C: 625 acre = 625×0.405 = 253.13 ha

Field D: 2500 vigha = 2500/4.01 = 623.44 ha

Hence, field A is largest.

Q. 108

DGPS- Differential Global Positioning System.

Q. 119

Slope = 60^0, Therefore H = 30×cos60 = 15 m

Correction = $H^2/2L$

$$= (15)^2/2\times30 = 3.75\ m$$

Q. 131

Volume of water in the reservoir obtained by the trapezoidal formula

$$V = \frac{h}{2}\,[A_1 + A_7 + 2(A_2 + A_3 + A_4 + A_5 + A_6]$$

$$= \frac{2}{2}\,[100 + 25000 + 2(220 + 600 + 1800 + 4500 + 10000]$$

$$= 59340\ m^3$$

Q. 132

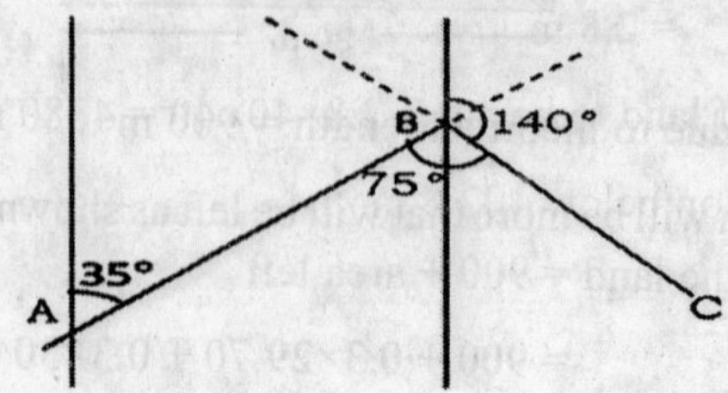

As shown in above figure, angle between AB and BC is 75°.

Q. 133

$$\text{Area on field} = \frac{9069.37}{1000\times1000}\times(3000)^2$$

$$= 81624.33\ m^2$$

Now, $\text{Actual area} = 81624.33\times\frac{19.8}{20} = 80808\ m^2$

Q. 135

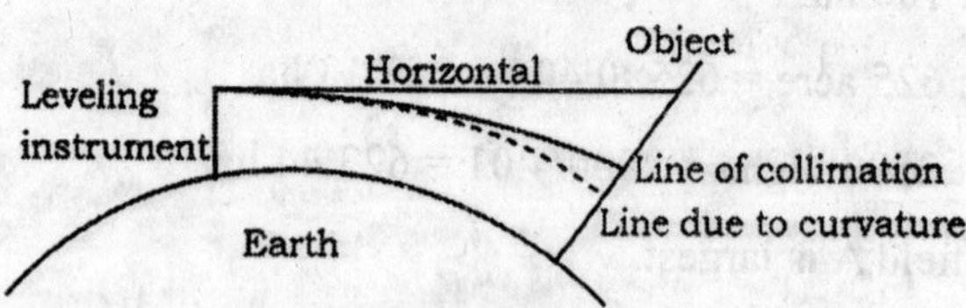

Q. 136

In reciprocal leveling

Total error of collimation, curvature and refraction

$$= \frac{\textit{Differnce of ellivations at first point} - \textit{Differnce of ellivations at second point}}{2}$$

$$\text{Error} = \frac{(1.6-1.4)-(0.8-0.5)}{2} = -0.05$$

Q. 137

Mean level of grids = (102.3 + 103 + 103.7 + 101.5 + 102.4 + 102.3 + 101.2 + 103.5 + 102.6)/9 = 102.5 m

Land elevation above 102.5 m will be cut and below 102.5m will be filled.

For grid elevation above 102.5m.

Level differences = (103.7 − 102.5) + (103 − 102.5) + (103.5 − 102.5) + (102.6 − 102.5)

= 2.8 m

Thus volume of land to be cut = 2.8×40×40 = 4880 m^3

For grid elevation below 102.5 m.

Level differences = (102.5 − 102.3) + (102.5 − 101.5) + (102.5 − 102.4) + (102.5 − 102.3) + (102.5 − 101.2) = 2.8 m

Thus volume of land to be filled = 2.8×40×40 = 4480 m^3

Q. 138

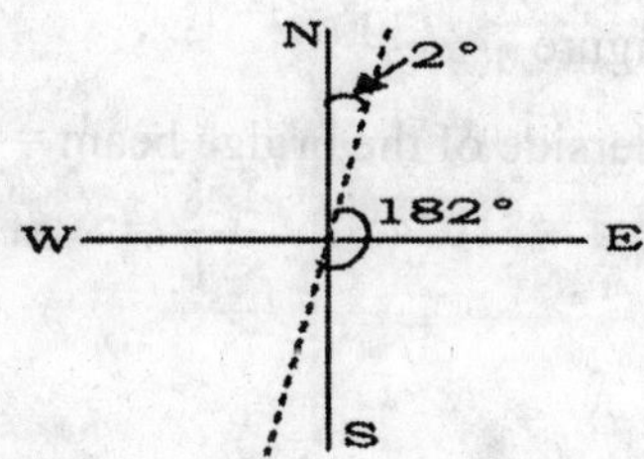

True bearing = Magnetic bearing ± magnetic declination

= 182° + (3° − 1°)

= 184°

Q. 139

Bearings are measured with respect to North Direction in Clock-wise direction.

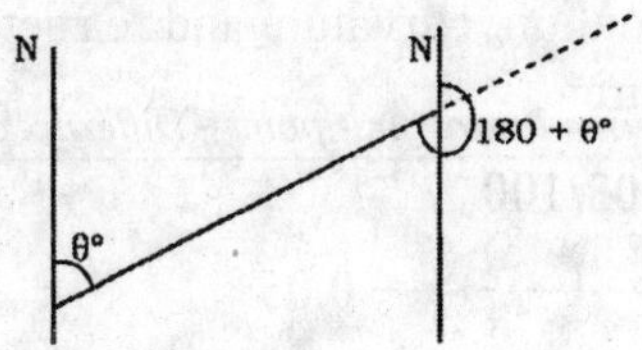

Thus, Back Bearing – Fore Bearing = (180°+ θ°)- (θ°) = 180°

Q. 140

We know

$$\frac{\textit{Actual Distance}}{\textit{Incorrect Chain Distance}} = \frac{\textit{Incorrect Chain Length}}{\textit{Correct Chain Length}}$$

$$\frac{1200}{\textit{Incorrect Chain Distance}} = \frac{19.7}{20}$$

$$\Rightarrow \quad \text{Incorrect Chain Distance} = \frac{20 \times 1200}{19.7} = 1218.27 \text{ m}$$

Q. 142

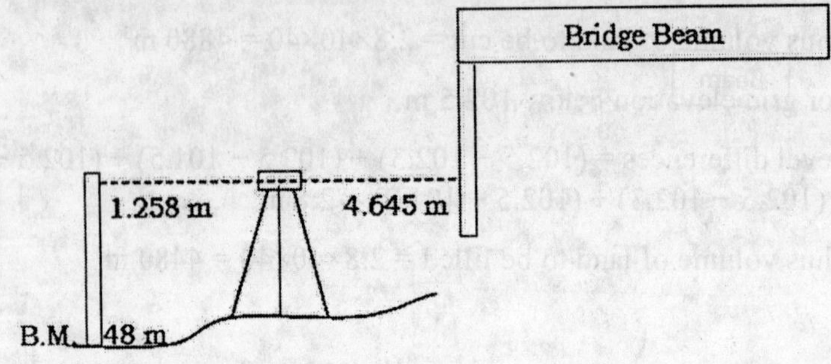

Referring above figure

The RL at the underside of the bridge beam = 48 + 1.258 + 4.645

= 53.903 m

Q. 143

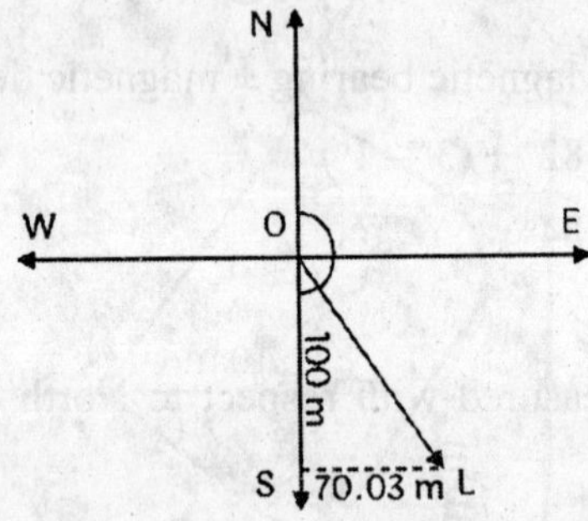

Referring the figure

$\tan(\angle SOL) = 70.03/100$

$\angle SOL = 35^{\circ}$

Hence, Whole circle bearing of the line is

$\angle NOL = 180^\circ - 35^\circ = 145^\circ$

Q. 144

Volume of water in the pond obtained by the trapezoidal formula is

$$V = \frac{h}{2}\ [A_1 + A_7 + 2(A_2 + A_3 + A_4 + A_5 + A_6\ ]$$

$$= \frac{1}{2}\ [\ 440 + 1240 + 2(\ 580 + 640 + 740 + 960 + 1100]$$

$$= 4860\ m^3$$

Q. 145

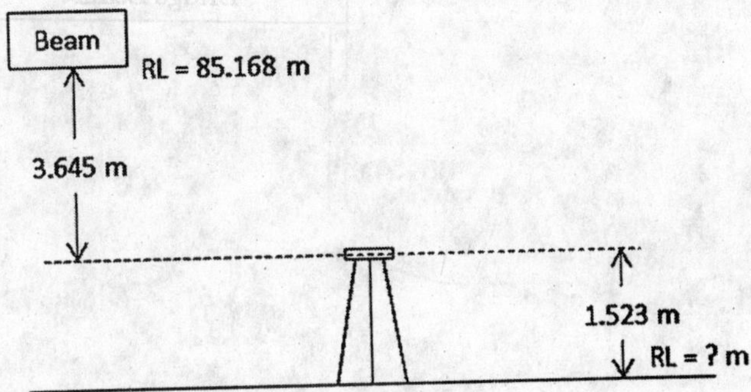

Referring above figure

RL of point = 85.168 – 3.645 – 1.523 = 80 m

Q. 146

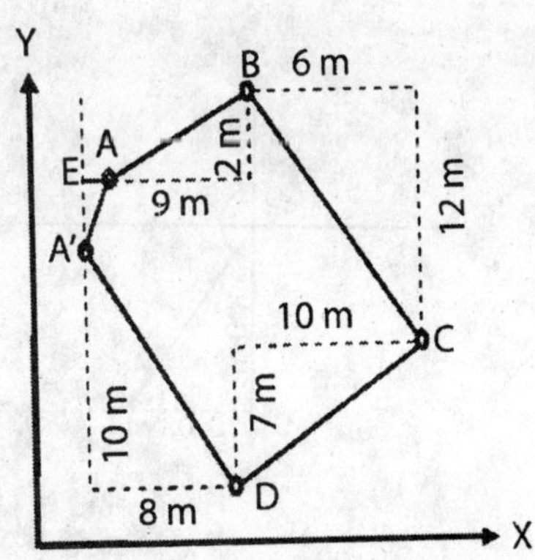

Reffering above figure

AE =3 m and A'E = 7 m

$AA' = (3^2 + 7^2)^{1/2} = 7.62$ m

V. Agricultural Process Engineering

29

Heat Transfer in Food Processing Steady State

APPSC AEES Agriculture Engineering Exam 2016

Q. 1 Psychrometric chart is a graphical representation of which properties of air

(A) Chemical (B) Aerodynamic

(C) Hygroscopic (D) Thermodynamic

Q. 2 At air 400° C and 50 % Rh has a wet bulb depression of 100° C. If the relative humidity decreases to 40 %, the wet bulb depression will

(A) Increase (B) Decrease

(C) Remain constant (D) Follow no definite trend

Q. 3 Regeneration is economical when product is

(A) Heated (B) Cooled

(C) Heated and cooled (D) None of the above

Q. 4 The dimensions of thermal diffusivity are:

(A) MLT^{-2} (B) M^0LT

(C) $M^{-1}LT$ (D) None of the above

Q. 5 In a heat exchanger, the rate of heat transfer from the hot fluid to the cold fluid

(A) Varies as square of the area

(B) Varies directly as the area and its LMTD

(C) Directly proportional to LMTD and inversely as the area

(D) None of the above

Q. 6 Milk and fruit juice are deaerated before they are allowed to flow through pasteurizer. This is done in order to:

(A) Reduce fouling of pasteurizer(B) Increase rate of heat transfer

(C) Reduce oxidative deterioration(D)Reduce microbial load

Q. 7 The specific gravity of skim milk is

(A) Lower than whole milk (B) Same as whole milk

(C) Higher than whole milk (D) Same as water

Q. 8 Metals are good conductors of heat because

(A) They contains free electrons

(B) They have high density

(C) Their atoms collide frequently

(D) Their atoms are relatively far apart

Q. 9 Thermal vapour compression is used in

(A) Evaporator (B) Homogenizer

(C) Pasteurizer (D) None of the above

RPSC AEn Pre Exam 2013 (Agricultural Engineering)

Q. 10 Sensible heating or cooling process of air-vapor mixture on psychometric chart is represented by

(A) Horizontal line (B) Vertical line

(C) Inclined Line (D) None of the above

OCS AEn Exam 2011 (Pre) – II (Agricultural Engineering)

Q. 11 The difference between dry bulb and dew point temperature is known as

(A) Dew point depression (B) Wet bulb depression

(C) Dry bulb depression (D) None of the above

Q. 12 Air at 60 °C and 60 % relative humidity has a wet bulb depression of 10 °C. If at the same temperature the relative humidity decreases to 50%, the wet bulb depression will

(A) Increase (B) Decrease

(C) Remain constant (D) Follow no definite trend

Q. 13 Sensible Heat Factor (SHF) is defined as the ratio of sensible heat to

(A) Latent heat (B) Total heat

(C) Specific heat (D) None of the above

Q. 14 Horizontal and uniformly spaced lines on a psychometric chart indicate

(A) Dry bulb temperature (B) Wet bulb temperature

(C) Dew point temperature (D) Specific humidity

Q. 15 LMTD for a heat exchanger is given

(A) $\frac{\Delta t_2 - \Delta t_1}{log \frac{\Delta t_2}{\Delta t_1}}$ (B) $\frac{\Delta t_2 - \Delta t_1}{log \frac{\Delta t_1}{\Delta t_2}}$

(C) $\frac{\Delta t_2 - \Delta t_1}{\frac{\Delta t_2}{\Delta t_1}}$ (D) $\log \frac{\Delta t_2 - \Delta t_1}{\frac{\Delta t_2}{\Delta t_1}}$

Q. 16 The relationship between Heat Utilization Factor (HUF) and Coefficient of Performance of a drier is

(A) HUF = 1 – COP (B) HUF = 1 + COP

(C) HUF = 1/COP (D) HUF = COP

CGPSC State Engineering Services Exam Paper – I

Q. 17 The amount of heat required to raise the temperature of a given mass by unit volumc is known as

(A) Specific heat (B) Thermal conductivity

(C) Heat capacity (D) Specific gravity

(E) Enthalpy

Q. 18 SI unit of thermal conductivity

(A) Wm/K (B) W/Km

(C) Km/W (D) W MK

(E) WmK

Q. 19 The process of breakdown of complex molecules in organic compound under the influence of yeast, bacteria, enzyme is known as

(A) Gasification (B) Pyrolysis

(C) Fermentation (D) Digestion

(E) Liquefaction

Q. 20 Specific heat of saturated water at 100 ^{0}C is

(A) Negative (B) Positive

(C) Zero (D) 1

(E) 2

Q. 21 The diffusivity has the same dimension as

(A) Kinematic viscosity (B) Concentration

(C) Density (D) Specific gravity

(E) Thermal conductivity

Q. 22 The material having thermal conductivity irrespective of direction is called

(A) Isotropic (B) Anisotropic

(C) Amorphous (D) Homogeneous

(E) Heterogeneous

Q. 23 Calculate the specific heat of orange juice concentrate having a solids content of 35 %

(A) 0.24 BTU/(1b R^0 F) (B) 0.48 BTU/(1b R^0 F)

(C) 0.72 BTU/(1b R^0 F) (D) 0.90 BTU/(1b R^0 F)

(E) 0.97 BTU/(1b R^0 F)

UKPSC A.En. Exam 2007 Paper – I

Q. 24 Which is not the type of heat exchanger

(A) Parallel flow (B) Counter flow

(C) Cross flow (D) Indirect flow

Q. 25 Fick's law relates to

(A) Surveying (B) Diffusion

(C) Transmission (D) Conduction

Q. 26 The heat required to raise the temperature of 1 kg of dry air and its accompanying water vapour by 1 ^{0}C is known as

(A) Specific heat (B) Humid heat

(C) Heat co-efficient (D) None of the above

Q. 27 As pressure is reduced, the latent heat value

(A) Increases (B) Decreases

(C) Remains the same (D) None of the above

Q. 28 The increase in boiling point of a solution over boiling point of water is known as

(A) Boiling point elevation (B) Boiling point depression

(C) Boiling point constant (D) None of the above

UKPSC A.En. Exam 2007 Paper – II

Q. 29 Internal energy of an ideal gas is a function of

(A) Temperature and volume (B) Pressure and volume

(C) Pressure and temperature (D) Temperature alone

Q. 30 To convert 1 gram of ice to one gram of water, the amount of heat required is

(A) 540 calories (B) 80 calories

(C) 240 calories (D) 100 calories

Q. 31 An isothermal process occurs when

(A) Pressure is constant (B) Temperature is constant

(C) Volume is constant (D) None of the above

Q. 32 A thermodynamic process occurring at constant volume is called

(A) Isochoric (B) Isobaric

(C) Adiabatic (D) Isenthalpic

Q. 33 Universal gas constant of a perfect gas

(A) Increases with temperature

(B) Decreases with temperature

(C) Increases with molecular weight

(D) Is always constant

Q. 34 Enthalpy is the heat supplied to a system at

(A) Constant temperature (B) Constant pressure

(C) Constant volume (D) None of the above

Q. 35 Hygrometer is used to measure

(A) Humidity (B) Heat conductivity

(C) Work done (D) Relative density

UKPSC A.En. Exam 2012 Paper – I

Q. 36 For steady state processes, accumulation of both mass and energy will be

(A) Less than zero (B) More than zero

(C) Zero (D) Infinite

Q. 37 Wet bulb temperature lines on psychometric chart are

(A) Horizontal (B) Vertical

(C) Inclined (D) Curved

Q. 38 Celsius and Fahrenheit temperature scales have the identical numerical values for a temperature of

(A) 0 °C (B) – 40 °C

(C) – 212 °C (D) 32 °C

Graduate Aptitude Test in Engineering - 2007

Q. 39 Effectiveness of countercurrent heat exchanger is given by

$$\epsilon = \frac{1-exp\left[-NTU\left(1-\frac{C_{min}}{C_{max}}\right)\right]}{1-\frac{C_{min}}{C_{max}}exp\left[-NTU\left(1-\frac{C_{min}}{C_{max}}\right)\right]}$$

If same liquid at the same flow rate is used as heating and cooling media through a countercurrent double tube heat exchanger then effectiveness is given by.

(A) $\frac{NTU-1}{NTU}$ (B) $\frac{NTU}{NTU+1}$

(C) $\frac{NTU-1}{NTU+1}$ (D) $\frac{NTU-1}{NTU+2}$

Q. 40 A rectangular fin of length 12 cm, width 22 cm and thickness 1.5 cm is connected to a tube at temperature of 0°C. The thermal conductivity of the fin material is 150 $Wm^{-1}K^{-1}$. The tip of the fin is not insulated. Air at a temperature of 5°C is in contact with the fin. The heat transfer coefficient between the fin and the air is 25 $W\,m^{-2}\,K^{-1}$. The rate of heat transfer is

(A) 3.33W (B) 6.63 W

(C) 9.13 W (D) 15.23 W

Q. 41 In order to reduce heat loss, a steam line with a tube diameter of 1.0 cm is insulated with a material having thermal conductivity of 0.108 W m^{-1} K^{-1}. Heat is dissipated from the outer surface of the insulating material by natural convection with a heat transfer coefficient of 12 W m^{-2} K^{-1} into the ambient at a constant temperature. The heat loss becomes maximum when the thickness of insulation is

(A) 0.5 mm (B) 2 mm

(C) 4 mm (D) 6.5 mm

Graduate Aptitude Test in Engineering - 2008

Q. 42 The thermal conductivity of a common metal used in fabrication of food processing equipment is given as 120 BTU ft^{-1} h^{-1} °F^{-1}. This value in J m^{-1} $s^{-1}K^{-1}$ will be

(A) 2.08 (B) 20.8

(C) 208 (D) 280

Q. 43 For foods whose composition is known, the following equation holds good

$C_p = 1.424\ m_c + 1.5594\ m_p + 1.675\ m_f + 0.837\ m_a + 4.187\ m_m$

Where, C_p is specific heat in kJ kg^{-1} K^{-1}, and m_c, m_p, m_f, m_a & m_m are mass fractions of carbohydrates, proteins, fats, ash & moisture, respectively.

The specific heat of a food containing 40% carbohydrates, 20% protein, 10% fat, 5%ash and 25% moisture will be

(A) 1.42 (B) 2.14

(C) 4.21 (D) 6.41

Q. 44 A cork slab of 100 mm thickness has one face at −12 °C and the other face at 21 °C. If the mean thermal conductivity (k) of the cork is 0.042 J m^{-1} s^{-1} K^{-1}, the rate of heat transfer (J s^{-1}) through one m^2 of the wall will be

(A) 13.9 (B) 9.3

(C) 5.0 (D) 2.5

Q. 45 Peas which have an average diameter of 6 mm are blanched to give a temperature of 85°C at the centre. The initial temperature of the peas is 15°C and the temperature of the blancher water is 95°C. Assuming that the heat transfer coefficient is 1200 Wm^{-2} K^{-1} and, for peas, the thermal conductivity is 0.35 $Wm^{-1}K^{-1}$, the specific heat is 3.3 KJ kg^{-1} K^{-1}, and the density is 980 kgm^{-3}. Fourier number (F_o) is 0.32, the time of blanching will be

(A) 26.6 s (B) 26.0 s

(C) 20.6 s (D) 20.0 s

Graduate Aptitude Test in Engineering - 2009

Q. 46 Fruit juice flowing at the rate of 600 kg h^{-1} is to be heated using same flow rate of hot fruit juice in a countercurrent regenerator. The specific heat capacity of the fruit juice is 3.9 kJ kg^{-1} K^{-1}. The overall heat transfer coefficient of the regenerator is 512 W m^{-2} K^{-1} and the area of the regenerator is 3.5 m^2. The effectiveness of the regenerator is

(A) 0.547 (B) 0.734

(C) 0.837 (D) 0.943

Q. 47 An insulating material has a thermal conductivity of 0.03 W m^{-1} K^{-1}. If 60 mm of this material is applied as insulation on a heat transfer surface, the r-value of the insulation in m^2 K W^{-1} is

(A) 1 (B) 2

(C) 3 (D) 6

Graduate Aptitude Test in Engineering - 2010

Q. 48 20% sucrose solution is boiled and frozen separately. If latent heat of vaporization at 100°C and the latent heat of crystallization at 0°C are 2257 and 334 kJ kg^{-1} respectively, then the ratio of freezing point depression to boiling point elevation is

(A) 6.3 (B) 4.2

(C) 3.6 (D) 2.4

Graduate Aptitude Test in Engineering - 2011

Q. 49 A material having thermal conductivity k insulates a spherical object of diameter d. The heat transfer coefficient between the insulating material and the environment is h_o. The critical thickness of insulation for maximum heat transfer rate is

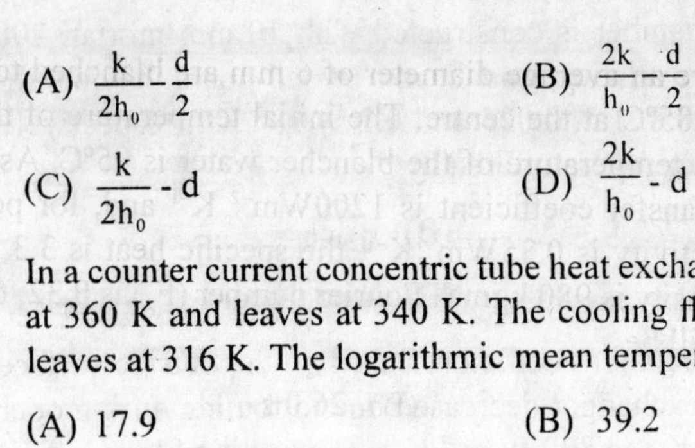

(A) $\frac{k}{2h_0} - \frac{d}{2}$ (B) $\frac{2k}{h_0} - \frac{d}{2}$

(C) $\frac{k}{2h_0} - d$ (D) $\frac{2k}{h_0} - d$

Q. 50 In a counter current concentric tube heat exchanger, the hot fluid enters at 360 K and leaves at 340 K. The cooling fluid enters at 300 K and leaves at 316 K. The logarithmic mean temperature difference in K is

(A) 17.9 (B) 39.2

(C) 41.9 (D) 57.3

Q. 51 A constant heat flux of 500 W m^{-2} is supplied to one face of a food material having a plate like structure with a thickness of 10 mm. The thermal conductivity of the food material is 1.5 W m^{-1} °C^{-1}. From the other face of the food material, heat is dissipated by convection into a fluid of 40 °C temperature. The heat transfer coefficient of the fluid is 100 W m^{-2} °C^{-1}. The temperature in °C of the surface to which the heat flux is supplied will be

(A) 43.3 (B) 45.3

(C) 48.3 (D) 54.3

Graduate Aptitude Test in Engineering - 2012

Q. 52 Two small parallel plane square surfaces, each measuring 4 mm × 4 mm are placed 0.5 m apart (centre to centre) with 30° angle between the radial distance and both the surface normals. The view factor between the two surfaces is

(A) 1.53×10^{-5} (B) 1.76×10^{-5}

(C) 3.82×10^{-3} (D) 4.41×10^{-3}

Graduate Aptitude Test in Engineering - 2013

Q. 53 Milk enters into the heating section of a high temperature short time (HTST) pasteurization plant at a temperature of 45°C and leaves at 72°C. Hot water at temperature of 95 °C enters countercurrently into the heat exchanger and leaves at 77°C. The effectiveness of the heat exchanger is

(A) 0.18 (B) 0.36

(C) 0.54 (D) 0.84

Q. 54 A cold storage chamber is constructed with 10 mm mortar, 200 mm brick, 100 mm insulation and 5 mm wood-board having thermal conductivities of 0.8, 1.5, 0.025 and 0.2 W m^{-1} K^{-1}, respectively. The resistance of 4 K m^2 W^{-1} is offered by

(A) Mortar (B) Brick

(C) Insulation (D) Wood-board

Q. 55 The overall heat transfer coefficient based on the outside surface area of a tubular heat exchanger decreased due to fouling during operation from 1000 W m^{-2} K^{-1} to 800 W m^{-2} K^{-1}. The fouling film coefficient of the heat exchanger in W m^{-2} K^{-1} is ....

Graduate Aptitude Test in Engineering - 2014

Q. 56 Select the most appropriate option about boiling and condensation processes as expressed by the statements P, Q and R.

P – The quantities of heat involved in evaporation and condensation of unit mass of fluid are identical

Q – The boiling and condensation of a single compound normally occur isothermally

R – The condensation is achieved at or below dew point and boiling occurs at triple point

(A) All P, Q and R are true (B) Only P and Q are true

(C) Only P is true (D) Only Q is true

Graduate Aptitude Test in Engineering – 2015

Q. 57 The wall of a cold storage is made up of four layers: concrete, brick, cardboard, paint with respective thickness of 5, 60, 8 and 1 mm, and their corresponding thermal conductivities are 0.8, 0.7, 0.04 and 0.15 W m^{-1} K^{-1}. The overall resistance of the wall to conduction heat transfer in m^2 K W^{-1} is ....

Q. 58 Two very large parallel walls (grey bodies) facing each other have emissivities of 0.5 and 0.7. The view factor between these two walls is ....

Q. 59 In a counter-current flow double pipe heat exchanger (DPHE), temperature difference between the hot and cold liquids at all positions is held constant at C. If the effectiveness of heat exchanger is 0.65 and heat capacity ratio of hot and cold liquids is 1, the number of transfer units (NTU) is ....

Q. 60 Two kg mass of air at 40° C with 0.023 kg water vapour per kg dry air to produce 5 kg mass of air at 60% relative humidity at 28°C. Assume all the streams are at normal atmosphere pressure (101.325 kPa). The saturation vapour pressure of water in kPa at 28° C is ....

Graduate Aptitude Test in Engineering – 2016

Q. 61 Hot water at 95°C is used in a plate heat exchanger for heating 2 kg s^{-1} fruit juice from 45°C to 75°C. Specific heat capacity of fruit juice is 3.7 kJ kg^{-1} K^{-1}. Final temperature of the hot water is 70°C. Overall heat transfer coefficient is 1122 W $m^{-2}K^{-1}$. Heat transfer area is 12.75 m^2. The log-mean temperature correction factor is ....

Q. 62 In a drying experiment, the constant rate of drying is found to be 3.6 kg water $m^{-2}h^{-1}$. Dry bulb and wet bulb temperatures of the drying air are 75°C and 37°C, respectively. Latent heats of vapourization at the dry bulb and the wet bulb temperatures are 2321 and 2414 kJ kg^{-1}, respectively. Convective heat transfer coefficient in W m^{-2} K^{-1} for the drying operation is ....

Q. 63 An air – water vapour mixture is at 35°C and normal atmospheric pressure with absolute humidity of 0.02 kg water vapour kg^{-1} dry air. Its humid volume in m^3 kg^{-1} dry air is ....

Q. 64 A 2.5 m long pipe is insulated at both ends. It has ID and OD as 50 mm and 56 mm, respectively. Its log-mean heat transfer area in m^2 is ....

Graduate Aptitude Test in Engineering – 2017

Q. 65 For a psychometric ratio of 1003 J kg^{-1} K^{-1}, the latent heat of vaporization at the wet bulb temperature of 35°C is 2418.90 kJ kg^{-1}. The saturation vapour pressure is 19.7 kPa corresponding to the dry bulb temperature of 60°C. If the relative humidity of air is 20%, the saturation vapour pressure of the air at the wet bulb temperature in kPa will be ....

Q. 66 A sphere (3.5 cm diameter) made of copper (ρ = 8954 kg m^{-3}; C_p = 0.4 kJ kg^{-1} K^{-1}; k = 375 W $m^{-1}K^{-1}$) is initially at uniform temperature of 200°C. It is suddenly placed in an environment of 35°C having convective film coefficient of 12 W $m^{-2}K^{-1}$. After 18 minutes of exposure, the temperature of the sphere in °C will be ....

Q. 67 In a vertical tube single effect evaporator, the boiling film coefficient inside the tubes is 1350 W m^{-2} K^{-1}. Steam condensation film coefficient outside is 7500 W m^{-2} K^{-1}. Thermal conductivity of 20 tubes made of SS 304 is 16 W m^{-1} K^{-1}. The Vertical tubes are 4.3 m long and are of 25 mm ID and 27 mm OD, maintaining a steady 15° temperature difference across the tube walls. Assume no boiling point rise, no heat losses, the feed enters the evaporator at the boiling point and the latent heat of vaporization of water is 2346.3 kJ kg^{-1}. Total water evaporation rate from the evaporator bundle of tubes in kg h^{-1} will be ....

Q. 68 Hot water at 95°C is sent through a countercurrent tube–in-tube heat exchanger with cold water entering at 25°C. Hot/cold water specific heat capacity is 4.2 kJ kg^{-1} K^{-1}. Flow of hot and cold water are 2.7 and 4.1 kg min^{-1} respectively. Overall heat transfer coefficient is 55 W m^{-2} K^{-1} and the area of heat transfer is 5 m^2. Cold water outlet temperature from the heat exchanger in °C will be ....

Q. 69 Steam (h_f = 632.2, h_{fg} = 2113.2, h_g = 2745.4 kJ kg^{-1}) at 150°C is used to sterilize milk by direct steam injection. Milk is initially at 90°C and after sterilization, the blend of resultant milk and condensed steam (h_f = 567.6, h_{fg} = 2159.2, h_g = 2726.7 kJ kg^{-1}) are at 135°C. Specific heat capacity of milk is 3.8 kJ kg^{-1} K^{-1}. Assuming no energy loss, the amount of milk (in kg) sterilized per kg steam supplied is

(A) 21.57 (B) 12.74

(C) 9.73 (D) 4.48

Graduate Aptitude Test in Engineering – 2018

Q. 70 Convective film heat transfer coefficient (h) of air (thermal conductivity = 0.025 W m^{-1} K^{-1}) in W m^{-2} K^{-1}, trapped between two parallel glass panes in a window 1.5 mm apart, is

(A) 3.75 (B) 16.67

(C) 37.52 (D) 60.21

Q. 71 Log mean temperature difference (LMTD) correction factor (F) is applicable to

(A) Spiral tube heat exchanger (B) Steam jacketed kettle

(C) Plate heat exchanger (D) Evaporator tubes

Q. 72 Which one of the following statements related to parboiling of paddy is NOT correct?

(A) Irreversible granule swelling occurs during heating.

(B) Parboiling process is used to salvage wet or damaged paddy.

(C) Parboiled rice takes less time to cook as compared to raw rice due to starch gelatinization.

(D) Milling yield of paddy increases due to starch gelatinization.

Q. 73 In a falling film evaporator, the inside and outside diameters of the tube wall are 25 mm and 27 mm, respectively. The tube is made of SS304 (thermal conductivity = 15 W m^{-1} K^{-1}) and the inside convective film coefficient is 750 W $m^{-2}K^{-1}$. Outside the tube wall, film coefficient of condensing steam is 7175 W m^{-2} K^{-1}. Based on the inside area of the tube, the overall heat transfer coefficient is .... W $m^{-2}K^{-1}$.

Q. 74 Mass flow rate of outside air through a duct is 735 kg h^{-1}. Absolute humidity of air is 0.025 kg water vapor per kg dry air at 40°C. This air is mixed with equal quantity of moist exhaust air coming out of a counter current dryer at 55°C. The mixed air is heated up to 75°C to dry barley, fed at 174 kg h^{-1} , at an initial moisture content of 31% (wet basis) to a final value of 7% (dry basis). The absolute humidity of the exhaust air from the dryer in 'kg water vapor per kg dry air will be ...

Graduate Aptitude Test in Engineering – 2019

Q. 75 A batch of 100 kg grain at 32% moisture content(wet basis) is being dried using hot air at 70 °C and 30% RH. The values of Henderson equation's constants c and n for the grain are 8.5x10^{-6} and 2.07, respectively. Considering the maximum possible drying of the batch, the quantity of moisture removed in kg is

(A) 10.20 (B) 17.05

(C) 21.98 (D) 25.07

Q. 76 Two streams of air with the following conditions are adiabatically mixed

Stream	Flow rate, kg dry air/h	Dry bulb temperature, °C	Absolute humidity, g water vapour /(kg dry air)
Fresh air	727	35	27
Recycled air	1020	55	40

Latent heat of vapourization of water at 0 °C = 2501 kJ/kg

Specific heat capacity of dry air = 1.005 kJ/(kg K)

Specific heat capacity of water vapour = 1.880 kJ (kg K)

Using above values, the dry bulb temperature and the absolute humidity of the mixed air in °C and g water vapour/(kg dry air), respectively are

(A) 43 and 30 (B) 44 and 31

(C) 45 and 33 (D) 46 and 35

Q. 77 A tube-in-tube counter-flow heat exchanger is heating oil from 35 °C to 77 °C by circulating hot water at 100 °C. The outlet temperature of water is 70 °C. The log-mean-temperature difference (LMTD) is

(A) exactly equal to the mean arithmetic temperature difference

(B) significantly greater than the mean arithmetic temperature difference

(C) significantly smaller than the mean arithmetic temperature difference

(D) very nearly equal to the mean arithmetic temperature difference

Q. 78 Air-water vapour mixture at 1 atmosphere pressure has 0.035 kg water vapour/(kg dry air) and dry bulb temperature of 37 °C. The value of Universal gas constant is 8.314 kJ/(kg mole K). The humid volume of this air-water vapour mixture in m^3 (kg dry air)$^{-1}$ is ....

Q. 79 In an air blast freezing operation, a flat tray of 1.0 m x 1.0 m x 0.02 m dimensions is used to freeze filled depodded peas. Bulk density and moisture content of peas are 550 kg/m^3 and 85% (wb.), respectively. Latent heat of freezing from. water to ice at –1 °C is 335 kJ/kg and heat transfer occurs identically from the top and the bottom surfaces of the tray. Convective film heat transfer coefficient on the heat transfer surfaces of the tray is 30 $Wm^{-2}K^{-1}$ and the thermal conductivity of frozen peas is 0.54 W m^{-1} K^{-1}. Assuming the tray to be a semi-infinite slab, the freezing time (in minutes) to completely freeze the product is ....

Graduate Aptitude Test in Engineering – 2020

Q. 80 Fresh tomato juice containing 6% (w/w) total solids enters in a single effect evaporator at a feed rate of 500 kg/h to concentrate up to 36% (w/w) solids. In this process, the rate of water removal in kg/h is......

Q. 81 Heat gain is occurring through a composite cold storage wall, made of brick and polyurethane foam insulation (thickness and thermal conductivity values are given below). If the exposed surfaces of brick and insulation are at 45 °C and 10 °C, respectively, the temperature at the interface of brick and insulation in °C is

Material	Thickness (mm)	Thermal conductivity (W m $^{-1}$ °C $^{-1}$)
Brick	100	0.30
Insulation	150	0.05

Q. 82 Saturated steam at 121 °C is used to sterilize pineapple juice by direct steam injection. The initial temperature of the juice is 80 °C and after sterilization, the blend of diluted juice exits the sterilizer at 95 °C. The enthalpy values of steam and condensate water are given in Table below. Specific heat capacity of the juice is 3.9 kJ/(kg °C)• Assuming no energy loss to the surroundings in the process of sterilization, the ratio of juice sterilized to steam utilized is

Temperature (°C)	Condensate water h_f(kJ/kg)	Saturated steam h_g(kJ/kg)	Phase change h_{fg}(kJ/kg)
95	397.9	2668.1	2270.2
121	507.7	2707.1	2205.2

(A) 34.79 (B) 37.94

(C) 39.47 (D) 43.97

Q. 83 Hot refined oil at 120 °C enters a concentric tube-in-tube heat exchanger (HE) at the rate of 20 kg/min. The oil is cooled by water entering at a temperature of 30 °C from the other end of the HE at the rate of 50 kg/min. Specific heat capacities of oil and water are 1.9 and 4.2 kJ/ (kg °C), respectively. The effectiveness of the HE may be taken as 0.7. Assuming no heat loss to the surrounding under steady-state condition, the exit temperature of water from the HE in °C is

(A) 39.9 (B) 41.4

(C) 57.7 (D) 63.0

Q. 84 Air water vapour mixture at 30 °C DBT and 40% RH is heated to 65 °C DBT and 30 °C WBT and is used as drying medium under the constant rate period drying of spinach leaves. Specific heat capacities of dry air and pure water vapour are 1.005 and 1.88 kJ/(kg °C), respectively. Using the properties given in Table below, the value of absolute humidity in kg water vapour per kg dry air for the dryer exit air at DBT of 45 °C is ....

Temperature (°C)	30	45	65
Saturation Vapour Pressure (kPa)	4.16	9.48	24.92
Latent Heat of Vapourisation (kJ/kg)	2430.7	2394.9	2346.3

Q. 85 Hydrothermal treatment of paddy makes

(A) Shelling more difficult

(B) Polishing of parboiled rice easier

(C) Higher retention of vitamins and minerals

(D) Kernel soft, resulting in faster cooking

Q. 86 Both particle formation and drying process are carried out by

(A) Flash dryer (B) Fluidized bed dryer

(C) Pneumatic conveyor dryer (D) Spray dryer

Q. 87 A contact plate freezer extracts thermal energy from a 24 mm thick slab of boneless meat containing 85% (w/w) water. Initially the slab is at the freezing point of meat, that is 272.5 K and corresponding latent heat of freezing is 335 kJ/kg water. The plate temperature of the freezer is assumed steady at 247.5 K. Bulk density of the slab is 750 kg/m^3. The thermal conductivity value for frozen meat is 1.5 W/(m K)• The minimum duration required for complete freezing of the slab in seconds is ....

Graduate Aptitude Test in Engineering – 2021

Q. 88 One-dimensional generalized heat conduction equation representing temperature distribution in a sphere, based on thermal conductivity k, specific heat capacity Cp, density ρ, and energy generation E, can be written as, $\frac{1}{r^n}\frac{\partial}{\partial r}\left(r^n k\frac{\partial T}{\partial r}\right) + E = \rho C_p \frac{\partial T}{\partial t}$ where the value of n is

(A) 1 (B) 2

(C) 3 (D) 4

Q. 89 Fresh potatoes of mass 1000 kg are dried from 14% to 93% total solids. If 7% of original potatoes is lost in peeling, the product yield from fresh potatoes in percent is ....

Q. 90 Match the correct items in column 1 with column 2

Column 1		Column 2	
P	Pipe-in-pipe heat exchanger	1	Cooling of air
Q	Shell and tube heat exchanger	2	Simultaneous co-current and counter current heat exchange
R	1-2 shell and tube heat exchanger	3	Large flow rate
S	Cross flow heat exchanger	4	Small heat exchange area

(A) P-1, Q-2, R-4, S-3 (B) P-2, Q-3, R-4, S-1

(C) P-3, Q-4, R-2, S-1 (D) P-4, Q-3, R-2, S-1.

Q. 91 Molecular masses of water and air are 18.02 and 28.97 kg/(kg mol), respectively. Air in a room is at 40 °C under a total pressure of 101.3 kPa absolute and contains water vapour at a partial pressure of 4.0 kPa. If saturated vapour pressure of water at 40 °C is 7.37 kPa, the relative humidity of this air in per cent is..........

Q. 92 One side of a solid food block of 10 cm thickness is subjected to a heating medium having a film heat transfer coefficient of 70 W/(m^2°C). The other side of the food block is being cooled by a medium having a film heat transfer coefficient of 100 W/(m^2 °C). The food block is having a thermal conductivity of 0.2 W/(m °C). and the contact area of the block available for heat transfer is 1 m^2. Heat transfer rate in the block at steady state is 100 J/s. The temperature difference between the two sides of the block in °C is ....

Q. 93 Pitting is a process of

(A) Mixing of pulses with red earth

(B) Mixing of pulses with edible oil

(C) Scratching of pulses by emery roller during its milling

(D) Beating of oil seeds for oil extraction

Answers Key

1	2	3	4	5	6	7	8	9	10
D	A	C	D	B	C	B	C	A	A
11	12	13	14	15	16	17	18	19	20
A	A	B	D	A	A	C	D	C	A
21	22	23	24	25	26	27	28	29	30
A	A	C	D	B	B	A	A	D	B
31	32	33	34	35	36	37	38	39	40
B	A	D	B	A	C	C	B	B	B
41	42	43	44	45	46	47	48	49	50
C	C	B	A	A	B	B	C	B	C
51	52	53	54	55	56	57	58	59	60
C	A	C	C	4000	B	0.299	0.412	1.857	3.72
61	62	63	64	65	66	67	68	69	70
0.69	63.53	0.90	0.4145	5.47	123.73	Bonus	55.10	12.74	B
71	72	73	74	75	76	77	78	79	80
C	C	655.09	Bonus	D	D	D	0.927	Bonus	416.67

81	82	83	84	85	86	87	88	89	90
41.5	C	B	0.018	C	D	410.04	B	14	D
91	92	93							
54.27	52.43	C							

Explanations

Q. 12

At 100 % Relative humidity

Wet bulb depression = 0 ($\because$ Wet bulb temperature = Dry bulb temperature)

Decreasing relative humidity, wet bulb temperature will increase. Therefore Wet bulb depression will increase.

Q. 30

1 calorie will raise 1 gram of water 1 degree Celsius. If 1 g of water is given 30 calories, its temperature will go up 30 degrees.

Melting ice at 0 °C: It takes 80 calories to melt 1 gram of ice. If 1 g of ice (at 0 °C) is given 80 calories, it will melt and the final temperature of the water will be 0°C.

Freezing water at 0 °C: It also takes 80 calories to freeze 1 gram of water. If 1 g of water (at 0 °C) gives away 80 calories, it will freeze and its final temperature will be 0 °C.

Boiling water at 100 °C: 540 calories are needed to turn 1 gram (at 100 °C) of water to steam.

Q. 39

$$\text{Effectiveness of a double tube heat exchanger} = \frac{NTU}{1+NTU}$$

Q. 40

Width of fin (W) = 22 cm = 0.22 m; Length of fin (l) = 12 cm = 0.12 m

Thickness of fin (t) = 1.5 cm = 0.015 m; Initial temperature of fin (T_i) = 0°C

Air temperature (T_a) = 5°C

Heat transfer coefficient between fin and air (h) = 25 W m^{-2} K^{-1}

Thermal conductivity of fin material (K) = 150 W m^{-1} K^{-1}

We know

$$\text{Heat transfer rate(Q)} = \sqrt{hPKA}\ .(T_a - T_i).\tanh(ml)$$

Where

P = fin perimeter = 2(w + t) = 2(0.22 + 0.015) = 0.47 m

A = cross sectional area = w.t = 0.22×0.015 = 0.0033 m^2

$$m = \sqrt{\frac{hP}{KA}} \sim \sqrt{\frac{2P}{Kt}} = \sqrt{\frac{50 \times 0.47}{150 \times 0.0033}} = 6.89$$

ml = 6.89×0.12 = 0.827

Putting above values

$$Q = \sqrt{25 \times 0.47 \times 150 \times 0.0033} \times (5 - 0)\tanh(0.827)$$

$$= 8.186 \text{ W}$$

$$\text{Fin efficiency } ((\eta) = \frac{tanh(ml)}{ml} = \frac{0.6789}{0.827} = 82.09\ \%$$

Therefore,

Actual heat transfer rate = 0.8209×8.186 = 6.7 W

Q. 41

Given that

Tube diameter (D) = 1 cm, Radius of tube (R) = 0.005 m

Thermal conductivity of insulation (K) = 0.108 W m^{-1} K^{-1}

Convective heat transfer coefficient of insulation (h) = 12 W m^{-1} K^{-1}

Heat loss becomes maximum at critical thickness of insulation.

We know that

Critical thickness of insulation for cylinder (t_c)

$$= \frac{K}{h} - R$$

$$= \frac{0.108}{12} - 0.005 = 0.004 \text{ m} = 4 \text{ mm}$$

Note :- Critical thickness of insulation for spherical object = $\frac{2K}{h} - R$

Q. 42

1 BTU = 1055 J; 1 ft = 0.3048 m; 1°F = 5/9 K

Therefore

120 BTU $ft^{-1}h^{-1}°F^{-1}$ = 120 × 1055/(0.3048×3600 × $\frac{5}{9}$) = 207.68 J/ m s K = 208 J/ m s K

Q. 43

Given

Carbohydrate mass fraction (m_c) = 0.40; Protein mass fraction (m_p) = 0.20

Fat mass fraction (m_f) = 0.10; Ash mass fraction (m_a) = 0.05

Moisture mass fraction (m_m) = 0.25

∴ Specific heat (C_p) = $1.424m_c + 1.594m_p + 1.675m_f + 0.837m_a + 4.187m_m$ kJ kg^{-1} K^{-1}

= 1.424×0.40 + 1.594×0.20 + 1.675×0.10 + 0.837×0.05 + 4.187×0.25

= 2.14 kJ kg^{-1} K^{-1}

Q. 44

T_1 = 21°C; T_2 = -12°C; T = 33°C

A = 1 m^2; k = 0.042 J m^{-1} s^{-1} C^{-1}; x = 0.1 m

$$\text{Heat transfer rate (q)} = \frac{0.42 \times 1 \times 33}{0.1}$$

= 13.9 J s^{-1}

Q. 45

Heat transfer coefficient (k) = 0.35 W m^{-1} K^{-1} Specific heat (C_p) = 3.3 kJ kg^{-1} K^{-1}

Fourier number (F_o) = 0.32; Density of peas (ρ) = 980 kg/ m^3

Average diameter of peas (R_p) = 0.003 m

We know

$$\text{Fourier number } (F_o) = \frac{kt}{\rho C_p (D_p)^2}$$

$$\Rightarrow \quad 0.32 = \frac{0.35 \times t}{980 \times 3.3 \times 1000 \times (0.003)^2}$$

$$\Rightarrow \quad t = 26.61 \text{ seconds}$$

Q. 46

Given that

Overall heat transfer coefficient (U) = 512 W $m^{-2}K^{-1}$

Specific heat capacity of fruit juice (C_p) = 3.9×1000 J kg^{-1} K^{-1}

Effectiveness of regenerator (ξ) = $\frac{NTU}{NTU+1}$

Where

$$NTU = \text{Number of transfer units} = \frac{UA}{m(C_p)_{min}}$$

$$NTU = \frac{512 \times 3.5 \times 3600}{600 \times 3.9 \times 1000} = 2.76$$

$$\because \quad \xi = \frac{2.76}{2.76+1} = 0.734$$

Q. 47

Thermal conductivity (k) = 0.03 W m^{-1} K^{-1}; Insulation thickness (L) = 0.06 m

$$R - \text{value of insulation} = \frac{L}{k}$$

$$= \frac{0.06}{0.03} = 2$$

Q. 48

Boiling point elevation (ΔT_b) = mK_b

Where m = molality (mol particles/ Kg H_2O)

K_b = 0.52 °C.Kg H_2O/mol particles

Freezing point depression (ΔT_f) = mK_f

Where m = molality (mol particles/ Kg H_2O)

K_b = 1.86 °C.Kg H_2O/mol particles

Here, molality is same for both conditions thus

Freezing point depression/ Boiling point elevation) = $\frac{1.86}{0.52}$ = 3.6

Q. 50

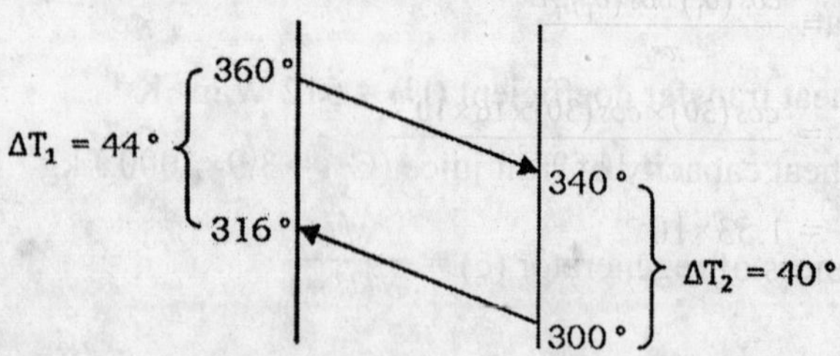

Logarithmic mean temperature difference (ΔT_{lmtd}) = $\dfrac{\Delta T_1 - \Delta T_2}{ln(\Delta T_1 / \Delta T_2)}$

$$= \frac{44-40}{ln(44/40)} = 41.9 \text{ K}$$

Q. 51

Heat transfer (Q) = 500×A W

(∵ Consider, surface area of food material = A)

Overall heat transfer coefficient (U) = $\dfrac{1}{\dfrac{0.01}{1.5}+\dfrac{1}{100}} = 59.88 \text{ W m}^{-2}\text{ K}^{-1}$

We know

$$UA\,\Delta T = 500 \times A$$

Temperature of surface = T (Assume)

$$\Rightarrow \quad 59.88 \times A \times (T-40) = 500 \times A$$

$$\Rightarrow \quad T = 48.35°C$$

Q. 52

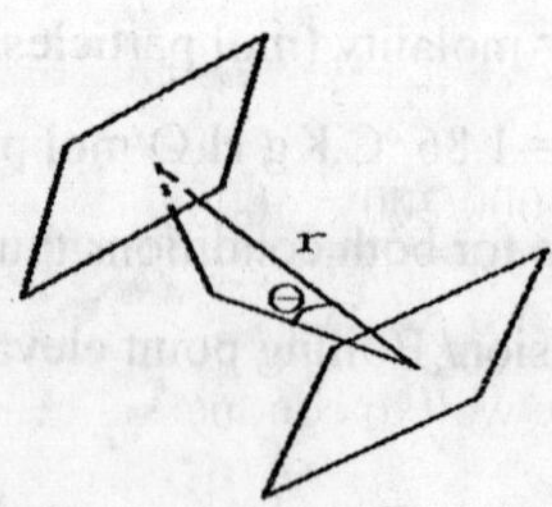

Area of plate (A) = 4×4 = 16 mm² or 16×10^{-6} m²

Centre to centre distance (r) = 0.5 m

Angle between the radial distance and both the surface = $\theta_1 = \theta_2 = 30°$

We know

$$\text{View factor} = \frac{cos(\theta_1).cos(\theta_1).A}{\pi r^2}$$

$$= \frac{cos(30)\times cos(30)\times 16\times 10^{-6}}{3.14\times 0.5^2}$$

$$= 1.53\times 10^{-5}$$

Q. 53

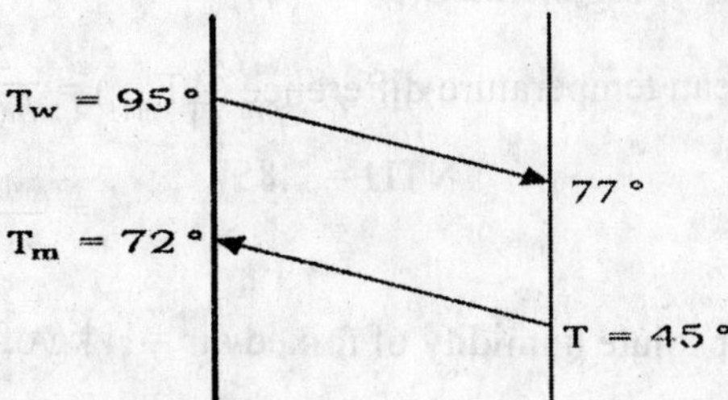

Temperature of milk (T_m) = 72°C ; Temperature of hot water (T_w) = 95°C;

Temperature in pasteurization plant (T) = 45°C;

We know

$$\text{Effectiveness of heat exchanger } (\varepsilon) = \frac{T_m - T}{T_w - T} = \frac{72-45}{95-45} = 0.54$$

Q. 54

By checking L/K,

Option (C) gives resistance of 4 K m^2 W^{-1}

L/K = 0.1/0.025 = 4 K m^2 W^{-1}

Q. 55

1/Fouling Factor = $1/U_1 - 1/U_2$

Fouling Factor = 800×1000/200 = 4000

Q. 57

Overall resistance of the wall to conduction heat transfer

$R = 1/U = L_1/K_1 + L_2/K_2 + L_3/K_3 + L_4/K_4$

= 5/800 + 60/700 + 8/40 + 1/150

= 0.299 m^2 K W^{-1}

Q. 58

View factor for infinite parallel gray bodies (planes)

$$= 1/(1/\epsilon_1 + 1/\epsilon_2 - 1)$$

$$= 1/(1/0.5 + 1/0.7 - 1)$$

$$= 0.412$$

Q. 59

Effectiveness of regenerator (ξ) = $\frac{NTU}{NTU + 1}$

$\Rightarrow$ $0.65 = \frac{NTU}{NTU + 1}$

$\Rightarrow$ NTU = 1.857

Q. 60

Consider, Absolute humidity of mixed air = H kg of water vapour/ kg of dry air

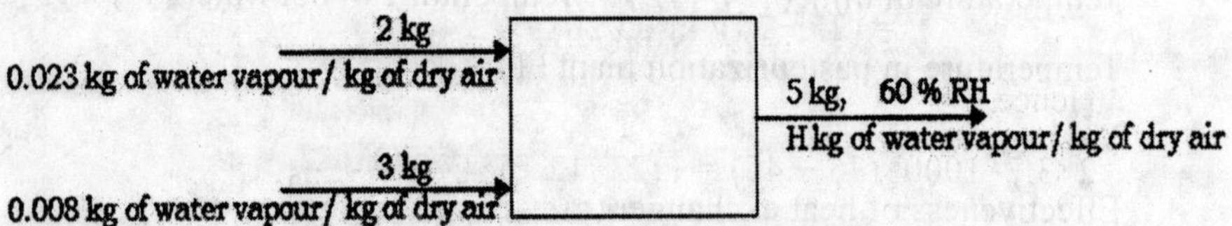

Applying water balance

$2 \times 0.023 + 3 \times 0.008 = H \times 5$

$\Rightarrow$ H = 0.014 kg of water vapour/ kg of dry air

We know

Absolute humidity (H) = $0.622 \frac{P_v}{P - P_v}$

P = atmospheric pressure = 101.325 kPa

$\Rightarrow$ $0.014 = 0.622 \frac{P_v}{101.325 - P_v}$

$\Rightarrow$ Water vapour pressure (P_v) = 2.23 kPa

Again,

Relative humidity (∅) = 0.60 = P_v / P_s

$\Rightarrow$ Saturation water vapour pressure (P_s) = 3.72 kPa

Q. 61

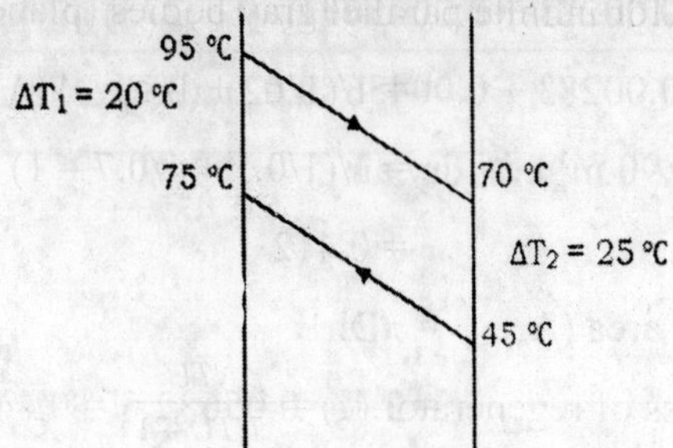

We know

Rate of heat transfer (q) = $mC_p \Delta T$

$= UAF\Delta T_L$

Where, $\Delta T_L = (\Delta T_2 - \Delta T_1) / \ln(\Delta T_2 / \Delta T_1)$

$\Delta T_L = (\Delta T_2 - \Delta T_1) / \ln(\Delta T_2 / \Delta T_1)$

$= (25 - 20) / \ln(25/20) = 22.41$ °C

Hence,

$2 \times 3.7 \times 1000 \times (75 - 45) = 1122 \times 12.75 \times F \times 22.14$

$\Rightarrow$ $F = 0.69$

Q. 62

Rate of drying (R_c) = 3.6 kg water $m^{-2}h^{-1}$

= 3.6/3600 kg water $m^{-2}s^{-1}$

Latent heats of vapourization wet bulb temperature (λ_w) = 2414 kJ kg^{-1}

We know

Rate of constant drying(R_c) = $\frac{h(T - T_w)}{\lambda_w}$

$\Rightarrow$ $3.6/3600 = h(75 - 37)/(2400 \times 10^3)$

$\Rightarrow$ $h = 63.53$ W m^{-2} K^{-1}

Q. 63

We know

Humid volume = total volume of 1 kg of dry air plus its accompanying water vapour.

Mathematically,

$\therefore V_H = (0.00283 + 0.00456\ H)\times(t + 273)\ m^3/kg$

$= (0.00283 + 0.00456\times0.02)\times(35 + 273)$

$= 0.90\ m^3\ kg^{-1}$ dry air

Q. 64

Outer surface area (A_2) $= \pi Dh$

$= 3.14\times0.056\times2.5 = 0.44\ m^2$

Outer surface area (A_1) $= \pi dh$

$= 3.14\times0.050\times2.5 = 0.39\ m^2$

$$\text{Log mean heat transfer area} = \frac{A_2 - A_1}{ln(A_2/A_1)}$$

$$= \frac{0.44 - 0.39}{ln(0.44/0.39)}$$

$$= 0.4145\ m^2$$

Q. 65

We know

Relative humidity = P_w/P_s

$\Rightarrow \quad 0.20 = P_w/19.7$

$\Rightarrow \quad P_w = 3.94$ k Pa

Atmospheric pressure(P_a) = 101.33 k Pa

Again

$$P_w = P_{wb} - \frac{(P_a - P_{wb})(T_a - T_{wb})}{1555.56 - 0.722\times T_{wb}}$$

$$\Rightarrow \quad 3.94 = P_{wb} - \frac{(101.33 - P_{wb})(60 - 35)}{1555.56 - 0.722\times 35}$$

$\Rightarrow \quad P_{wb} = 5.47$ kPa

Q. 66

For sphere, Volume(V) /Surface Area(A) = d/6 = 0.035/6 m

We know,

$$(T - T_w)/(T_0 - T_w) = e^{\left(\frac{-hA}{\rho C_p V}\right)t}$$

$$\Rightarrow \quad (T - 35)/(200 - 35) = e^{\left(\frac{-12\times 6}{8954\times 0.4\times 1000\times 0.035}\right)\times 18\times 60}$$

$$\Rightarrow \quad T = 123.73\ ^0C$$

Q. 67

We know

Heat transfer rate due to conduction and convection is given by

$$q = \frac{T_{in} - T_o}{\frac{1}{h_i A_i} + \frac{R-r}{k}\frac{D_{lm}}{D_o} + \frac{1}{h_o A_o}}$$

Where,

$T_{in} - T_o = 15\ ^0C$ or 288 K

$h_i = 1350$ W m^{-2} K^{-1}; $h_o = 7500$ W m^{-2} K^{-1}

$A_1 = \pi D_i L = 3.14\times 0.025\times 4.3 = 0.3376$ m^2

$A_0 = \pi$ Do $L = 3.14\times 0.027\times 4.3 = 0.3646$ m^2

$R - r = (0.027 - 0.025)/2 = 0.001$ m

And $D_{lm} = (D_0 - D_i)/\ln(D_0 / D_i)$

$= (27 - 25)/\ln(27/25) = 0.026$ m

Therefore

$$q = \frac{288}{\frac{1}{0.3376\times 1350} + \frac{0.001\times 0.026}{16\times 0.027} + \frac{1}{0.3646\times 7500}}$$

= 109908.50 W or 109.91 kW

∵ There are 20 tubes in evaporator system.

Hence, total heat transferred = 20×109.91 = 2198.17 kW

Again, consider water evaporation rate = x kg/sec

Applying heat balance

$$x \times 2346.30 = 2198.17$$

$$\Rightarrow \quad x = 0.9369 \text{ kg/s or } 3372.84 \text{ kg/h}$$

Q. 68

Heat capacity for hot water(C_h) = $(mC_p)_h$ = 2.7×4200/60 = 189 W/K

Heat capacity for cold water(C_h) = $(mC_p)_c$ = 4.1×4200/60 = 287 W/K

∵ Number of transfer units(NTU) = UA/C_{min}

= 55×5/189 = 1.455

And C_{min}/C_{max} = 189/287 = 0.659

Therefore, Effectiveness of heat exchanger

$$(\epsilon) = \frac{1 - exp\left[-NTU\left(1 - \frac{C_{min}}{C_{max}}\right)\right]}{1 - \frac{C_{min}}{C_{max}} exp\left[-NTU\left(1 - \frac{C_{min}}{C_{max}}\right)\right]}$$

$$= \frac{1 - exp\left[-1.455(1 - 0.659)\right]}{1 - 0.659 \times exp\left[-1.455(1 - 0.659)\right]} = 0.653$$

Hence,

Heat extracted by hot fluid = $\epsilon \times C_h \times (T_{hi} - T_{ci})$

= 0.653×189×(95 – 25) = 8639.19 W

Heat absorbed by cold fluid = $C_c \times (T_{co} - T_{ci})$

= $287 \times (T_{co} - 25)$ W

Applying heat balance

$287 \times (T_{co} - 25) = 8639.19$

⇒ $T_{co} = 55.10\ ^0C$

Q. 69

Heat lost by steam to sterilize the milk = 2745.40 – 567.60 = 2177.80 kJ/kg

Heat gained by milk = $mC_p\Delta t$

= m×3.8×(135 – 90)

Applying heat balance

m×3.8×(135 – 90) = 2177.80

m = 12.74 kg

Q. 70

Convective heat transfer coefficient(h)

= Thermal conductivity/ Thicness of air film

$= 0.025/0.0015 = 16.67\ W\ m^{-2}\ K^{-1}$

Q. 73

For conduction and convection heat transfer

$$\frac{1}{UAi} = \frac{1}{hiAi} + \frac{\ln(Do/Di)}{2\pi KL} + \frac{!}{hoAo}$$

$$\Rightarrow \frac{1}{U \times \pi DiL} = \frac{1}{hi \times \pi DiL} + \frac{\ln(Do/Di)}{2\pi KL} + \frac{1}{ho \times \pi DoL}$$

$$\Rightarrow \frac{1}{U \times Di} = \frac{1}{hi \times Di} + \frac{\ln(Do/Di)}{2K} + \frac{1}{ho \times Do}$$

$$\Rightarrow \frac{1}{U \times 0.025} = \frac{1}{750 \times 0.025} + \frac{\ln(0.027/0.025)}{2 \times 15} + \frac{1}{7175 \times 0.027}$$

Overall heat transfer coefficient based on inner area(U)

$= 655.09\ W\ m^{-2}\ K^{-1}$

Q. 74

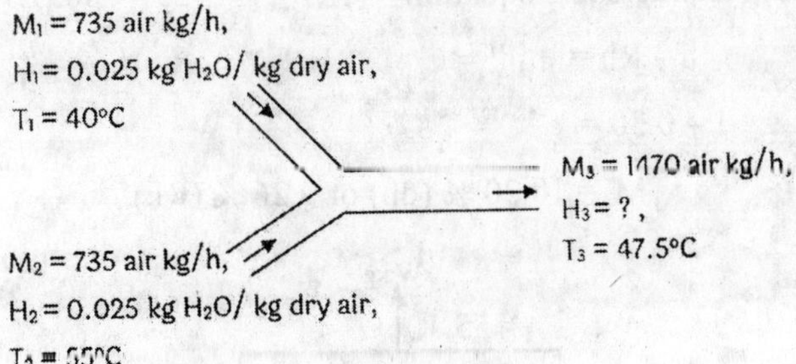

Temperature of mixed air = (40 + 55)/2 = 47 °C

For mixing of two air streams

$$M_1H_1 + M_2H_2 = (M_1 + M_2)H_3$$

$$\Rightarrow \quad 735 \times 0.025 + 735 \times H_2 = (735 + 735) \times H_3 \qquad \text{(i)}$$

And

$$T_1H_1 + T_2H_2 = (T_1 + T_2)H_3$$

$$\Rightarrow \quad 40 \times 0.025 + 55 \times H_2 = (40 + 55) \times H_3 \qquad \text{(ii)}$$

Solving above equations $H_2 = 0.025$ kg H_2O/ kg dry air And $H_3 = 0.025$ kg H_2O/ kg dry air

Absolute humidity of air dose not change with increase in temperature. Hence, Absolute humidity of mixed air at 75 °C (H_3) = 0.025 kg H_2O/ kg dry air

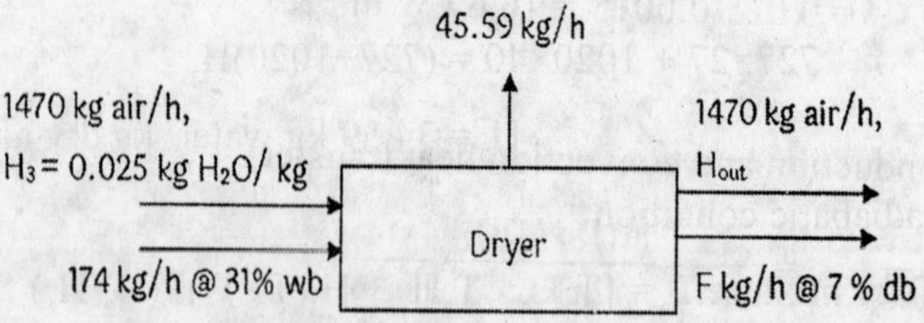

Referring above figure 174×(1 – 0.31) = F×(1 – 0.065)

(∵ 7 % db = 0.07/1.07 = 6.5 % wb)

⇒ F = 128.41 kg Rate of Water removed = 174 – 128.41 = 45.59 kg/h

We know Water removal rate = $(H_{out} - H_3)\times(M_1 + M_2)$

$$\Rightarrow \quad 45.59 = (H_{out} - 0.025)1470$$

$$\Rightarrow \quad H_{out} = 0.031 + 0.025 = 0.056 \text{ kg H2O/ kg dry air}$$

Q. 75

Applying Handerson's equation

$$1 - Rh = e^{-CTM_e^n}$$

$$\Rightarrow \quad 1 - 0.30 = e^{-8.5\times10^{-6}\times343\times M_e^{2.07}}$$

$$\Rightarrow \quad M_e = 10.20\ \%\ (db)\ \text{or}\ 9.26\ \%\ (wb)$$

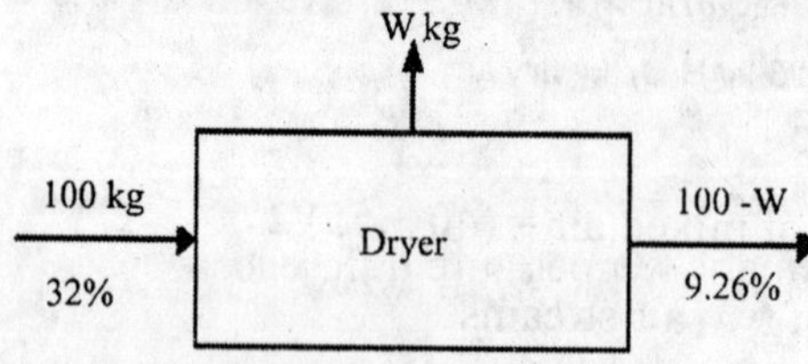

Applying Mass balance equation

$$0.32\times100 = (100 - W)\times0.0926 + W$$

$$\Rightarrow \quad W = 25.06 \text{ kg}$$

Q. 76

Applying mass balance equation

$$m_1H_1 + m_2H_2 = (m_1 + m_2)H_3$$

$$\Rightarrow \quad 727\times27 + 1020\times40 = (727+1020)H_3$$

$$\Rightarrow \quad H_3 = 34.59 \text{ kg water/ kg dry air}$$

For adiabatic condition

TBT for mixture $T_3 = [T_1H_1 - T_2H_1 + H_3(T_2\text{-}T_1)]/(H_2\text{-}H_1)$

$= [35\times40 - 55\times27 + 34.59(55 - 35)]/(40 - 27)$

$= 46.68\ ^{O}C$

Q. 77

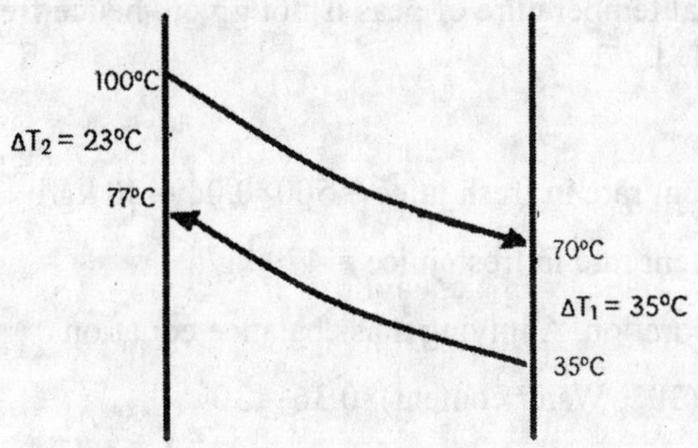

Logarithmic mean temperature difference $(\Delta T_{lmtd}) = \dfrac{\Delta T_1 - \Delta T_2}{ln(\Delta T_1 / \Delta T_2)}$

$= \dfrac{35-23}{ln(35/23)} = 28.58°C$

Mean Airthematic temperature difference$(\Delta T_A) = \dfrac{\Delta T_1 + \Delta T_2}{2}$

$= \dfrac{35+23}{2} = 29\,°C$

Q. 78

Humid volume = (0.00283 + 0.00456H)(t+273)

= (0.00283 + 0.00456×0.035)(37+273)

= 0.927 m^3/ kg dry air

Q. 79

Given

Tray thickness (a) = 0.02 m; Density of peas (ρ) = 550 kg/ m^3

Latent heat of crystallization (λ) = 335 kJ/ kg = 335000 J/ kg

Convective heat transfer coefficient (h) = 30 W/ (m^2K)

Thermal conductivity of meat (k) = 0.54 W m^{-1} k^{-1}

Freezing Temperature (T_f) = - 1°C, Initial Temperature (T_i) = Data Missing

We know

Freezing Time (t) = $\dfrac{\lambda\rho}{T_f - T_i}\left(\dfrac{a}{2h} + \dfrac{a^2}{8k}\right)$

Here, Initial temperature of peas is not given, hence freezing time can't be calculated.

Q. 80

Solid content rate in fresh juice = 500×0.06 = 30 kg/h

Water content rate in fresh juice = 470 kg/h

After evaporation, Applying mass balance equation

⇒ (30 + Water content)×0.36 = 30

⇒ Water content = 53.33 kg/h

Hence, Rate of water removal = 470 – 53.33 = 416.67 kg/h

Q. 81

We know that,

$Q = \Delta T/ \sum R_{th}$

Here (ΔT) overall = $T_1 - T_3$ = 35°C and ΣR t_h = $R_{th1} + R_{th2}$

$R_{th1} = L_1/k_1A$

$= 0.1/0.30A$ K/W

$R_{th2} = 0.15/0.05A$ K/W

Hence, $Q = 35/(0.1/0.30A + 0.15/0.05A) = 10.51/A$ W

Again, To find temperature drop across

$\Rightarrow$ $Q = \Delta T/\sum R_{th}$

$\Rightarrow$ $10.51/A = (45 - T)/0.1/0.30A$

$\Rightarrow$ $T = 41.5\ ^{o}C$

Q. 82

Consider,

Juice Sterilized = J kg

Steam utilized = S kg

Heat utilized to dilute the juice= Enthalpy of saturated steam – Enthalpy of condensate water

$\Rightarrow$ $J \times 3.9 \times (95 - 80) = 2707.1 \times S - 397.9 \times S$

$\Rightarrow$ $J/S = 39.47$

Q. 83

$C_{hot} = mC_p = 20 \times 1.9 = 38\ kJ/^{0}C$

$C_{cold} = mC_p = 50 \times 4.2 = 210\ kJ/^{0}C$

$C_{min} = mC_p = 20 \times 1.9 = 38\ kJ/^{0}C$

We know

$$\text{Effectiveness of heat exchanger} = \frac{C_{cold}(T_{cold\ out} - T_{cold\ in})}{C_{min}(T_{hot\ in} - T_{cold\ in})}$$

$$\Rightarrow \quad 0.7 = \frac{210(T_{cold\ out} - 30)}{38(120 - 30)}$$

$\Rightarrow$ $T_{cold\ out} = 41.4\ ^{0}C$

Q. 84

We know,

Relative humidity ($\varnothing$) = $P_v / P_s = 0.40$

$\Rightarrow$ Water vapour pressure (P_v) = $0.40 \times 4.16 = 1.664$kPa

Absolute humidity (H_{30}) = $0.622\ \dfrac{P_v}{P - P_v}$

(P = atmospheric pressure = 101.325 kPa)

$$\Rightarrow \qquad H_{30} = 0.622\left(\frac{1.664}{101.325-1.664}\right)$$

$$= 0.0104 \text{ kg of water vapour/ kg of dry air}$$

In a system by heating a mixture its relative humidity.

Hence, Absolute humidity (H_{65}) = 0.0104kg of water vapour/ kg of dry air

Applying Energy Balance

Heat consumed) = Latent heat of mixture(at 45°C) + Latent heat of mixture(at 65°C)

(During drying littlewater mixes with the heated water vapour.)

$$\Rightarrow \qquad 20\times(1.005 + 1.88\times0.0104) = 2346.3\times H_2 - 2394.9\times0.0104$$

$$\Rightarrow \qquad H_2 = 0.018 \text{ kg of water vapour/ kg of dry air}$$

Q. 87

Meat block thickness (a) = 24 mm = 0.024 m; Density of meat (ρ) = 750 kg/ m^3

Latent heat of crystallization (λ) = 335 kJ/ kg = 335000 J/ kg

Thermal conductivity of meat (k) = 1.5 W m^{-1} K^{-1}

Freezing Temperature (T_f) = – 0.5°C; Initial Temperature (T_i) = –25.5°C

We know

$$\text{Freezing Time (t)} = \frac{\lambda\rho}{T_f - T_i}\left(\frac{a}{2h} + \frac{a^2}{8k}\right)$$

Here, $\frac{a}{2h}$ is very small in respect to $\frac{a^2}{8k}$. Therefore, $\frac{a}{2h}$ is neglected.

$$\Rightarrow \qquad t = \frac{0.85\times335000\times750}{-0.5-(-25.5)}\left(\frac{0.024^2}{8\times1.5}\right)$$

$$\Rightarrow \qquad t = 410.04 \text{ seconds}$$

Q. 89

Amount of lost potatoes in peeling =1000x7/100=70 kg

Amount of peeled potatoes = 1000 – 70 = 930 kg

Let W = Water content and P = Product Amount

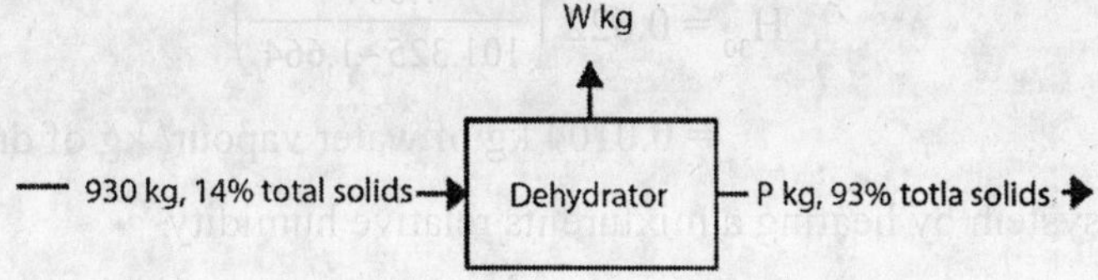

Mass balance:

930 = W + P

And, Total solid balance: 930x0.14 = Px 0.93

⇒ P = 140 kg

⇒ % Yield from fresh potatoes = (140x100)/1000 = 14 %

Q. 91

$$\text{Relative humidity } (\varnothing) = \frac{P_v}{P_s} \times 100 = \frac{4}{7.37} \times 100 = 54.27\ \%$$

Q. 92

Film heat transfer coefficient on one side (h_i) = 70 W.(m². °C)$^{-1}$

Film heat transfer coefficient on other side (h_o) = 100 W.(m². °C)$^{-1}$

Heat transfer area (A) = 1 m²

Heat transfer rate = 100 J/s

Thermal conductivity = 0.2 W. (m². °C)$^{-1}$

$$\text{Heat transfer rate} = \frac{\Delta T}{\frac{1}{A}\left(\frac{1}{h_1}+\frac{x}{k}+\frac{1}{h_o}\right)}$$

$$\Rightarrow \quad 100 = \frac{\Delta T}{\frac{1}{1}\left(\frac{1}{70}+\frac{0.1}{0.2}+\frac{1}{100}\right)}$$

Temperature difference between the two sides of the block (ΔT) = 52.43 °C

30

Drying, Evaporation and Separation of Food Components

UKPSC Combined Assistant Engineering Exam 2013 Paper - I

Q. 1 LSU dryer was developed at

(A) London State University (B) Ludhiana State University

(C) Lucknow State University (D) Lusiana State University

Q. 2 Maximum recommended air temperature for drying of wheat is:-

(A) 20°C (B) 40°C

(C) 60°C (D) 90°C

Q. 3 Hullers are used to

(A) Boiling of rice (B) Remove husk from paddy

(C) Grinding of rice (D) Storage of rice

Q. 4 Handerson's equation is used to determine

(A) Critical duration of storage (B) Critical depth of flow

(C) Critical temperature (D) Equilibrium moisture content

Q. 5 Separation of grains by screening is done on the basis of

(A) Size (B) Colour

(C) Texture (D) Weight

Q. 6 If the moisture content is 20 % on wet basis. The moisture content on dry basis will be

(A) 5 % (B) 15 %

(C) 35 % (D) 25 %

Q. 7 In thin layer drying the grain depth is limited up to

(A) 20 cm (B) 30 cm

(C) 40 cm (D) 50 cm

Q. 8 Which of the following is not an EMC model ?

(A) Kelvin equation (B) Harkins-Jura equation

(C) Rankine equation (D) Henderson equation

Q. 9 In commercial rotary dryers the length of drum is:-

(A) From 3 to 6 m (B) From 8 to 11 m

(C) From 13 to 16 m (D) From 18 to 21 m

Q. 10 The drier which uses the principle of thin layer drying is known as:-

(A) Baffle type dryer (B) Sack dryer

(C) Continuous flow dryer (D) Batch type dryer

Q. 11 Post harvest losses of fruits and vegetables in handling is about:-

(A) 35 % (B) 40 %

(C) 45 % (D) 50 %

Q. 12 In a batch type pasteurizer, milk is heated at

(A) 61 °C for 30 minutes (B) 61 °C for 30 seconds

(C) 40 °C for 15 minutes (D) 40 °C for 15 seconds

Q. 13 Dryers utilizing high gas temperature of 500°C or more. But for short exposure time are called

(A) Fluidized bed dryers (B) Flash dryers

(C) Turbo dryers (D) Drum dryers

Q. 14 The temperature range for ultra high temperature (UHT) sterilization of milk is

(A) 90 – 100 °C (B) 105 – 115 °C

(C) 135 – 150 °C (D) None of the above

Q. 15 Which of the following process is used to separate liquids having different boiling points

(A) Crystallization (B) Fractional distillation

(C) Evaporation (D) Fractional crystallization

Q. 16 The shape of rotary and gyratory screen are generally

(A) Triangular (B) Circular or rectangular

(C) Trapezoidal (D) None of the above

APPSC AEES Agriculture Engineering Exam 2016

Q. 17 In case of intermittent drying, total drying time reduces as exposure time

(A) Increases (B) Decreases

(C) No effect (D) None of the above.

Q. 18 Generally temperature in multiple effect evaporators will .... in subsequent effects

(A) Decrease (B) Increase

(C) Remains constant (D) None of the above

Q. 19 Theoretically number of ways to increase cyclone efficiency when particle size reduces from 10 micron to 5 micron diameter are

(A) By increasing entrance velocity

(B) By increasing length of cyclone

(C) Temperature in the cyclone

(D) 1 and 2

Q. 20 The removal of few large particles in an initial process is

(A) Scalping (B) Cleaning

(C) Grading (D) Sorting

Q. 21 Disc separator separates materials on the basis of

(A) Length (B) Width

(C) Weight (D) Shape

Q. 22 For biological materials, the relationship between EMC and RH was given by

(A) Janssen (B) Rankine

(C) Henderson (D) Chung Fast

MPPSC Assistant Agricultural Engineer 2013

Q. 23 If the moisture content at dry basis is 17.67 %, the moisture content on wet basis is ....

(A) 21.46 % (B) 17.67 %

(C) 15.02 % (D) None of these is correct

Q. 24 Removal of moisture from any product to bone dry level is called as ....

(A) Drying (B) Evaporation

(C) Dehydration (D) None of these is correct

Q. 25 The temperature to which air and its water vapour must be cooled till saturation at constant humidity is known as ....

(A) Dry bulb temperature (B) Wet bulb temperature

(C) Dew point temperature (D) Ambient temperature

Q. 26 Unit of specific heat in MKS system is ....

(A) Kcal/kg-°C (B) Kcal/kg-m-°C

(C) Kcal/kg (D) Kcal/m-°C

Q. 27 The design and development of cleaning and grading machinery is done considering which of the following important physical property of material ?

(A) Porosity (B) Size

(C) Specific gravity (D) True density

Q. 28 True density of granular material can be determined by ....

(A) Standard charts (B) Pycnometer

(C) Water displacement (D) None of these is correct

Q. 29 When rate of moisture loss from a product to the surrounding is equal to rate of moisture gain by the product from surrounding then the product is in .... with surrounding atmosphere.

(A) Inequilibrium (B) Equilibrium

(C) Tune (D) None of these is correct

Q. 30 The relation $1 - R_h = e^{-CTM_e^n}$ for equilibrium moisture content was suggested by ....

(A) Dalton (B) Bond

(C) Henderson (D) Rittinger

Q. 31 As compared to pure water the freezing point of water solution will be ....

(A) Same (B) Lower

(C) Higher (D) None of these is correct

Q. 32 The latent heat of vapourization of water is ....

(A) 55 Kcal (B) 540 Kcal

(C) 55 cal (D) 540 cal

Q. 33 The air velocity at which a particle remains suspended in a long vertical pipe is known as .... of the particle.

(A) Drag coefficient (B) Critical velocity

(C) Specific velocity (D) Terminal velocity

Q. 34 Hull and germ are removed during the .... operation of cereals.

(A) Hulling (B) Milling

(C) Cutting (D) None of these is correct

Q. 35 .... is the unit of thermal conductivity in SI system.

(A) J/°C (B) J/s-°C

(C) J/s-m (D) J/s-m-°C

Q. 36 Cutting, Crushing, Compressing are the methods used for ....

(A) Agricultural Process Engineering

(B) Drying

(C) Size reduction

(D) Slicing

Q. 37 In order to separate ...., fluidized bed separator is used.

(A) Heavier seeds (B) Medium seeds

(C) Lighter seeds (D) All seeds

Q. 38 In batch type dryers, the air temperature is usually kept at ....

(A) 20 °C (B) 45 °C

(C) 55 °C (D) 60 °C

Q. 39 Crushing law was proposed by ....

(A) Bond (B) Kick

(C) Rittinger (D) Stokes

UKPSC Combined Junior Engineering Exam 2013 Agricultural Engineering Paper-I

Q. 40 The amount of water vapour present in a unit weight of dry air under a given set of conditions is known as

(A) Humidity (B) Absolute humidity

(C) Relative humidity (D) None of the above

Q. 41 The temperature at which a mixture of air and water vapour has to be cooled at constant humidity to make it saturated is known as

(A) Dry bulb temperature (B) Wet bulb temperature

(C) Dew point temperature (D) Ambient temperature

Q. 42 The first phase of seed processing consists of

(A) Scalping (B) Debearding

(C) Hulling (D) All of the above

Q. 43 The second phase of processing in which removal of weed seeds, broken seeds and inert materials are involve, is done by

(A) Air screen (B) Air screen cleaner

(C) Aspirator (D) Scalper

Q. 44 Pneumatic separation of fruits by grader depends on

(A) Size and shape of the fruit (B) Density of the fruit

(C) Surface resistance of the fruits (D) All of the above

Q. 45 Equivalent moisture content of food grains is measured in

(A) Dry weight basis (B) Wet weight basis

(C) Measuring instruments (D) All of the above

Q. 46 Design of a grain dryer aims at

(A) Finding the dimensions of the equipment

(B) Drying agent and heat requirement

(C) Both of above (A) & (B)

(D) None of the above

Q. 47 Gur (Jaggery) making process from sugarcane involves

(A) Crushing of sugarcane

(B) Heating of the sugarcane juice to make it semi-solid

(C) Cooling down and make jaggery

(D) All of the above process

Q. 48 Sucrose content in sugarcane is about

(A) 5 – 10 percent (B) 10 – 15 percent

(C) 15 – 20 percent (D) None of the above

Q. 49 The largest sugarcane producing country in the world is

(A) Canada (B) India

(C) Russia (D) America

Q. 50 To separate mustard seed from wheat, the best type of separator is

(A) Specific gravity separator (B) Spiral separator

(C) Centrifugal separator (D) All of the above

Q. 51 A scalper is used for

(A) Grading the material (B) Removing of stones

(C) Fine separation of materials (D) Rough separation of materials

Q. 52 The number of holes per square inch in a 20 mesh screen will be

(A) 800 (B) 400

(C) 200 (D) 20

Q. 53 During seed processing, the moisture content of seed should be

(A) < 15 percent (B) 15 – 20 percent

(C) 20 – 25 percent (D) > 25 percent

Q. 54 Gravity separator work on the principle of

(A) Gravity (B) Drag

(C) Both (A) and (B) (D) None of the above

Q. 55 Fruits and vegetables are preferably dried in a

(A) Spray drier (B) Fluidized bed drier

(C) Freeze drier (D) Cabinet drier

Q. 56 Soyabean is mostly used in India for production of

(A) Edible oil (B) Pulses

(C) Milk substitutes (D) Processed foods

Q. 57 LSU type dryer was developed in

(A) India (B) England

(C) USA (D) Canada

Q. 58 The main cause of quality deterioration of food is

(A) Climate (B) Microorganisms

(C) Rats (D) All of the above

Q. 59 The process of removing moisture from the food product is called

(A) Canning (B) Sterilization

(C) Pasteurization (D) Dehydration

Q. 60 Parboiling of rice includes the action of

(A) Water and temperature (B) Water and chemical

(C) Oil and chemical (D) None of the above

Q. 61 Respiration of food release the gas

(A) Oxygen (B) Carbon monoxide

(C) Carbon dioxide (D) Nitrogen

Q. 62 The maximum produced vegetable crop in India is

(A) Potato (B) Tomato

(C) Brinjal (D) Okra

Q. 63 In India parboiled rice is mostly produced and consumed in

(A) Northern region (B) Southern region

(C) Eastern region (D) Western region

Q. 64 Bio ethanol is primarily produced from fermentation of

(A) Rice grain (B) Sugarcane molasses

(C) Grapes (D) None of the above

Q. 65 Most cereal food contain mainly

(A) Fat (B) Vitamin

(C) Carbohydrate (D) Protein

RPSC AEn Pre Exam 2013 (Agricultural Engineering)

Q. 66 Which of the following is NOT an advantage of dehydration under controlled conditions over sun drying ?

(A) Cost (B) Time of drying

(C) Quality of drying (D) None of the above

Q. 67 When evaporators are compared with dryers, they are

(A) Less efficient (B) More efficient

(C) Effect is same (D) None of the above

Q. 68 Explosion Puffing is a process of

(A) Drying (B) Extrusion

(C) Size reduction (D) All the above

OCS AEn Exam 2011 (Pre) – II (Agricultural Engineering)

Q. 69 Accuracy of moisture content determined by direct method is .... the indirect method.

(A) More than (B) Less than

(C) Equal to (D) None of the above

Q. 70 100 kg paddy is dried from 25 % moisture content (dry basis) to 15 % moisture content (dry basis). The amount of moisture removal will be

(A) 8 kg (B) 10 kg

(C) 20 kg (D) 37.5 kg

Q. 71 Osaw moisture meter measure the moisture on the basis of

(A) Thermal conductivity (B) Electrical capacitance

(C) Electrical resistance (D) Thermal diffusivity

Q. 72 The product quality is poorer in which type of dryer under same operating conditions ?

(A) Fluid bed dryer (B) Spray dryer

(C) Tray dryer (D) Freeze dryer

Q. 73 Constant rate of drying is directly proportional to

(A) Wet bulb temperature

(B) Latent heat of vaporization

(C) Convective heat transfer coefficient

(D) None of the above

Q. 74 Thin layer drying equation is given by

(A) $\frac{M - M_e}{M_0 - M_e} = e^{-k\theta}$ (B) $\frac{M - M_0}{M_e - M_0} = e^{-k\theta}$

(C) $\frac{M_e - M}{M_0 - M_e} = e^{k\theta}$ (D) $\frac{M_e - M_0}{M - M_e} = e^{k\theta}$

Q. 75 Steam economy of single effect evaporator without vapour recompression is

(A) Always greater than one (B) Less than one

(C) About one fourth (D) About half

CGPSC State Engineering Services Exam Paper – I

Q. 76 A desorption isotherm is a plot of equilibrium moisture content versus relative humidity at a given temperature for a biological material which has been subjected to a

(A) Drying environment (B) Wetting environment

(C) EMC (D) ERH

(E) Packaging

Q. 77 Dielectric meters are used for measurement of moisture content of grains by recording the dial reading of

(A) Resistance (B) Capacitance

(C) Emf (D) Voltage

(E) Current

Q. 78 Knowledge of moisture content is necessary to decide welter the grain should be

(A) Dried before storage (B) Soaked before storage

(C) Milled before storage (D) Parboiled before storage

(E) Consumed (E) 45 cm

Q. 79 Safe storage moisture content of paddy ranges

(A) 12 - 14 % (w.b.) (B) 12 - 14 % (d.b.)

(C) 16 – 18 % (w.b.) (D) 16 – 18 (d.b.)

(E) 20 % (w.b.)

Q. 80 The relationship between EMC and RH for biological materials has been given by

(A) Perry (B) Rankine

(C) Henderson (D) Jannsen

(E) Stock

Q. 81 The critical moisture content of agricultural produce is

(A) In between constant rate and falling rate period

(B) Above the falling rate period

(C) Equivalent to initial moisture content

(D) Equivalent to final moisture content

(E) Equivalent to final moisture content

Q. 82 Stokes's law is used to find out

(A) Terminal velocity (B) Drag coefficient

(C) Surface tension (D) Specific gravity

(E) Viscosity

Q. 83 In deep dryer drying system if heated air at a temperature of 450 °C is employed, each layer thickness recommended is

(A) ≤ 45 cm (B) 50 cm to 1.0 cm

(C) 1.0 cm to 1.25 cm (D) 1.25 to 1.50 cm

(E) ≥ 1.55 cm

Q. 84 The vacuum filters are limited to maximum filtering pressure of

(A) One (B) Two

(C) Three (D) Five

(E) Seven

Q. 85 Use of filter aid is to

(A) Increase the filtering efficiency

(B) Decrease the filtering efficiency

(C) To give body to the filtrate

(D) To increase the mass of cake

(E) To decrease the mass of cake

Q. 86 The space between the individual wires of a wire mesh screen is

(A) Mesh number (B) Screen aperture

(C) Wire per linear inch (D) Screen capacity

(E) Screen size

Q. 87 Primary aim of blanching of fruit is

(A) To destroy surface microorganisms

(B) To increase softness of the fruits

(C) To destroy the enzymes

(D) To increase the palatability of the fruit

(E) To increase the texture

Q. 88 A cyclone separator is used for separating

(A) Particle from liquids (B) Liquid droplets from gases

(C) Fine particles from solids (D) Both (A) & (B)

(E) Both (B) & (C)

Q. 89 Separation of liquids from solids by the application of pressure is known as

(A) Extraction (B) Expression

(C) Filtration (D) Leaching

(E) Both (A) & (D)

Q. 90 For acid delinting of cotton seed, the required concentration of sulphuric acid is

(A) 75 ml per kg of seed (B) 125 ml per kg of seed

(C) 175 ml per kg of seed (D) 250 ml per kg of seed

(E) 275 ml per kg of seed

Q. 91 Extraction of soluble constituent from a solid by means of solvent is known as

(A) Distillation (B) Leaching

(C) Evaporation (D) Sublimation

(E) Extraction

UKPSC A.En. Exam 2007 Paper – I

Q. 92 Type of moisture that can be removed by drying is

(A) Equilibrium moisture (B) Total moisture

(C) Free moisture (D) None of the above

Q. 93 With increase in temperature, equilibrium moisture content of grains

(A) Decreases (B) Increases

(C) Remains constant (D) None of the above

Q. 94 With increase in moisture content of grain, the angle of repose

(A) Increases (B) Decreases

(C) Does not change (D) None of the above

Q. 95 At constant temperature, with increase in relative humidity drying capacity of air

(A) Increases (B) Decreases

(C) Remains constant (D) None of the above

Q. 96 Drying is a process of

(A) Heat transfer only (B) Mass transfer only

(C) Heat and mass transfer both (D) None of the above

Q. 97 Separation of grain by screening is done on the basis of

(A) Size (B) Colour

(C) Texture (D) None of the above

Q. 98 The amount of water to be removed from 10 kg of food material in order to reduce the moisture content from 80 % (wet basis) to 25 % (dry basis) is

(A) 7.5 kg (B) 5 kg

(C) 8 kg (D) 6 kg

Q. 99 The difference between initial and equilibrium moisture content of a food is known as

(A) Unbound moisture content (B) Bound moisture content

(C) Critical moisture content (D) Free moisture content

Q. 100 Recommended air temperature for a grain drying is nearly

(A) 45 ^{0}C (B) 90 ^{0}C

(C) 60 ^{0}C (D) 100 ^{0}C

UKPSC A.En. Exam 2012 Paper – I

Q. 101 Rice polishing involves

(A) Removal of dust particles

(B) Removal of husk from paddy

(C) Removal of bran layer from brown rice

(D) Addition of white coating

Q. 102 Expression is separation of liquid from solids by

(A) Drying (B) Evaporation

(C) Condensation (D) Applied pressure

Q. 103 Most commonly used solvent in solvent extraction of oil is

(A) Carbon tetrachloride (B) Petroleum spirit

(C) Kerosene (D) Hexane

Q. 104 Air temperature in solar drying in comparison to sun drying is kept

(A) Lower (B) Equal

(C) Higher (D) Multiple

Q. 105 Operations like cleaning, grading drying, shelling, hulling, polishing and grinding of cereals are known as

(A) Pre-harvest technology of cereals

(B) Post-harvest technology of cereals

(C) Food technology

(D) All of the above

Q. 106 Polishing is generally done in case of

(A) Wheat milling (B) Rice milling

(C) Dal milling (D) Gram milling

Q. 107 Oil from oil cake is removed by

(A) Expeller (B) Ghani

(C) Rotary mill (D) Solvent extraction

Q. 108 In grain separation the angle of inclination of screens from horizontal generally ranges as ……… degree

(A) 20 – 25 (B) 15 – 18

(C) 4 – 12 (D) 0 – 3

Q. 109 Classification of cleaned products into various quality fractions depending upon various commercial values and other usages is known as

(A) Packaging (B) Grading

(C) Mixing (D) Fixing

Q. 110 The moisture content attained by a grain with respect to a set of atmospheric temperature and relative humidity is known as

(A) Normal moisture content (B) Equilibrium moisture content

(C) Critical moisture content (D) Average moisture content

Q. 111 In a solvent extraction process of oil, pelletization is the process common to the preparation of

(A) Soyabean (B) Groundnut

(C) Rice bran (D) Rapeseed

Q. 112 In falling film evaporator, the boiling point is influenced by

(A) Hydrostatic pressure

(B) External pressure

(C) Both external pressure and hydrostatic head

(D) None of the above

Kerala AAO Exam 2017

Q. 113 Match the following

List-I	List-II
P. Spiral separator	1. Length of seed
Q. Disc separator	2. Shape and surface texture
R. Magnetic separator	3. Shape or rolling ability
S. Cylinder separator	4. Surface texture and stickiness of seed

(A) P-1, Q-2, R-3, S-4 (B) P-3, Q-1, R-4, S-2

(C) P-2, Q-3, R-1, S-4 (D) P-4, Q-1, R-3, S-2

Graduate Aptitude Test in Engineering - 2007

Q. 114 A heater is placed in front of a continuous countercurrent dryer. Air at 40°C and 70% RH is fed into the heater from which the air exits at 65°C. If saturation vapour pressure at 40°C and 65°C are 0.074 bar and 0.250 bar respectively, then relative humidity of the air coming out of the heater and entering the dryer is

(A) 21 % (B) 27 %

(C) 32 % (D) 38 %

Q. 115 At 65°C, Henderson constant C and n are 7.4×10^{-4} K^{-1} and 0.56 respectively. The equilibrium moisture content corresponding to 40% relative humidity is

(A) 38 % (wet basis) (B) 78 % (dry basis)

(C) 87 % (wet basis) (D) 358 % (dry basis)

Graduate Aptitude Test in Engineering - 2008

Q. 116 Water activity (a_w) is a ratio of

(A) Vapour pressure of water to partial pressure of water in the product

(B) Partial pressure of water in air to partial pressure of air at saturation

(C) Vapour pressure of water in equilibrium with the food to vapour pressure of pure water at the same temperature

(D) Vapour pressure of pure water to vapour pressure of water in equilibrium with food

Q. 117 Heated air at 50°C and 10% relative humidity is used to dry rice in a bin dryer. The air leaves the bin under saturated condition. The corresponding data for humidity ratio as read from the psychometric chart are 0.0078 and 0.019 kg water per kg dry air. The amount of water removed per kg of dry air will be

(A) 0.0112 kg (B) 0.021 kg

(C) 0.112 kg (D) 0.121 kg

Q. 118 Potatoes are dried from 14% to 93% total solids. Considering 8% peeling losses, the product yield from one tonne of raw potato will be

(A) 10.56 % (B) 13.85 %

(C) 15.25 % (D) 20.58 %

Q. 119 One hundred kilogram of a food grain is dried from 18% wb to 13% wb moisture content. The total amount of water removed from the grain is

(A) 6.82 kg (B) 6.28 kg

(C) 5.75 kg (D) 5.57 kg

Q. 120 Mechanical separation is divided into

(A) Cleaning, sorting, sieving and filtration

(B) Grading, weighing, sieving and filtration

(C) Sedimentation, centrifugation, filtration and sieving

(D) Sedimentation, centrifugation, cleaning and sieving

Q. 121 In order to freeze a fruit juice its thermodynamic temperature is

(A) Higher than the freezing point of water

(B) Below the freezing point of water

(C) Equal to the freezing point of water

(D) Dependent upon the water content of the fruit juice

Q. 122 Following two groups of equipment and their working principles of purpose are given

GROUP I	GROUP II
(i) Pneumatic conveyor	(a) Air blowing or suction
(ii) Hammer mill	(b) Feed grinding
(iii) Cyclone separator	(c) Centrifugal force
(iv) Pycnometer	(d) Stress / strain measurement

Identify the incorrect pair

(A) i - a (B) ii - b

(C) iii - c (D) iv – d

Graduate Aptitude Test in Engineering - 2009

Q. 123 Vegetable seeds are stored at absolute temperature of 320 K and relative humidity of 20%. If Henderson equation for equilibrium relationship is valid for this case where the values of constants C and n are 6.5×10^{-6} and 1.8 respectively, the equilibrium moisture content of the seeds will be

(A) 5.6 % (B) 10.2 %

(C) 13.4 % (D) 20.5 %

Q. 124 Particles having average diameter of 20 μm and particle density of 1000 kg m^{-3} enter a cyclone of 500 mm diameter at a linear velocity of 20 m s^{-1}. The separation factor of the cyclone is

(A) 136 (B) 163

(C) 316 (D) 613

Q. 125 If absolute humidity at saturation and percentage humidity of air are 0.075 kg water vapour (kg dry air)$^{-1}$ and 60% respectively, the relative humidity of air is

(A) 57.4 % (B) 60.0 %

(C) 62.7 % (D) 74.5 %

Q. 126 Bulk density of paddy with 28% moisture content on wet basis is 650 kg m^{-3}. The dry solid bulk density of paddy in kg m^{-3} is

(A) 468 (B) 508

(C) 832 (D) 904

Q. 127 Wheat weighing 4900 N at moisture content of 25% on wet basis is to be dried to moisture content of 10% on dry basis. The amount of moisture evaporated from the wheat in kg is

(A) 87 (B) 103

(C) 116 (D) 156

Graduate Aptitude Test in Engineering - 2010

Q. 128 Air at 40°C temperature has partial vapour pressure of 2.4 kPa. If universal gas constant is 8.314 kJ kg mole^{-1} K^{-1} and total pressure is 101.325 kPa then humid volume of air in m^3 (kg dry air)$^{-1}$ is

(A) 0.809 (B) 0.908

(C) 1.089 (D) 1.098

Graduate Aptitude Test in Engineering - 2011

Q. 129 300 kg of parboiled and wet paddy having an initial moisture content of 28% (dry basis) was dried in a batch type dryer. During drying operation, grain sample was drawn at regular time intervals to measure its moisture content (dry basis). It was found that the moisture content of the grain borne a linear relationship with drying time till 30 minutes of drying operation at which moisture content was measured to be 20%. The drying rate (kg h^{-1}) during this interval was

(A) 22.3 (B) 37.5

(C) 41.7 (D) 48.1

Q. 130 The average interstitial velocity in a packed bed is 2.5 m s^{-1} and the superficial velocity under minimum fluidized condition based on the cross-section of the empty container is 1 m s^{-1}. Solid particles having density of 1200 kg m^{-3} and size of 0.15 mm are to be fluidized using air at 3 atm absolute pressure and 30 °C temperature having a density of 3.5 kg m^{-3}. The empty bed cross section is 0.4 m^2 and the bed contains 400 kg of solid. Pressure drop at minimum fluidized condition in Pa is

(A) 3.5×10^3 (B) 6.5×10^3

(C) 9.7×10^3 (D) 17.1×10^3

Graduate Aptitude Test in Engineering - 2012

Q. 131 A packed bed of 480 kg solid particles having particle size of 0.15 mm and density of 800 kg m^{-3} is fluidized using air at 25 °C and 1 atmospheric pressure. If the cross section of the empty bed is 0.45 m^2 and voidage at minimum fluidizing condition is 0.5, then the minimum height of the fluidized bed, in m is

(A) 7.4 (B) 5.4

(C) 2.7 (D) 1.0

Q. 132 Final mass flow rate of osmotically dehydrated cherries after finish drying from 18% dry basis moisture content to 11.5% wet basis moisture content is 5000 kg per hour. The dryer efficiency is 70%, latent heat of vapourization is 2345 kJ kg^{-1}, specific heat of air is 1.005 kJ kg^{-1} K^{-1}, drying temperature is 50 °C and the specific volume of ambient air at 25 °C is 0.866 m^3 kg^{-1}. The necessary air flow requirement for the drying system in m^3 min^{-1} is

(A) 477 (B) 587

(C) 625 (D) 702

Q. 133 A single effect vacuum evaporator has 100 tubes of 25 mm diameter. One thousand kg feed of milk per hour with 15% TS is concentrated to 20% TS in the evaporator. Film heat transfer coefficients on either sides of the tube are 5000 and 800 W m^{-2} K^{-1}. Thermal conductivity of 1.5 mm thick SS tubes is 15 W m^{-1} K^{-1}. Latent heat of vapourization under vacuum is 2309 kJ kg^{-1}. For 10 °C temperature difference across the tube wall, the height of each tube, in m is

(A) 1.36 (B) 2.13

(C) 2.56 (D) 3.17

Graduate Aptitude Test in Engineering - 2013

Q. 134 Slope and intercept of the BET equation relating $[a_w/x(1-a_w)]$ and a_w, where a_w is water activity and x is the moisture content on dry basis, are 18 and 2, respectively. The values of thermodynamic constant C and BET monolayer moisture content x_m (%) are respectively

(A) 40 and 20 (B) 30 and 15

(C) 20 and 10 (D) 10 and 5

Q. 135 A fish fillet of 5 mm thickness having 85% moisture (wet basis) is to be frozen using a plate freezer. The plates are at –35°C and the heat transfer coefficient between the fillet and the freezer plates can be assumed to be 2.0 W $m^{-2}K^{-1}$. The initial freezing temperature of fish is –2.5°C, latent heat of fusion is 330 kJ kg^{-1}, density of fish is 1100 kg m^{-3} and thermal conductivity of frozen fish is 1.5 W $m^{-1}K^{-1}$. The time required to freeze the fillet from the initial freezing temperature in hour(s) is ....

Q. 136 A rotary dryer is used to dry 1200 kg/h of paddy containing 30% moisture (wet basis) to give a product containing 15% moisture (wet basis). Alternately, a portion of the dry product may be recycled and mixed with the fresh feed such that the mixed feed enters the dryer with moisture content of 20% (wet basis). The moisture evaporation rate without recycle and the paddy recycle rate in kg/h respectively in the dryer are

(A) 211.76 and 2400 (B) 211.76 and 600

(C) 256.5 and 2400 (D) 256.5 and 600

Q. 137 In a screen separator, the mass flow rates of feed, overflow and underflow are 150, 140 and 10 kg h^{-1}, respectively. Mass fraction of material in the feed and overflow are 0.9 and 0.96, respectively. The effectiveness of separation in percentage is

(A) 32 (B) 42

(C) 52 (D) 62

Graduate Aptitude Test in Engineering - 2014

Q. 138 A centrifuge with rotational speed of 1440 rpm is used for separating oil from a dispersion in which oil is present in the form of spherical globules of 47 μm diameter. The density of oil is 886 kg m^{-3}. Separation occurs at an effective radius of 4 cm. Viscosity and density of water are 0.705 cP and 1000 kg m^{-3}, respectively. The velocity of oil through water, in mm s^{-1}, is ....

Q. 139 A tray type paddy separator is used to separate paddy from a binary mixture of paddy and brown rice fed at the rate of 1600 kg h^{-1}. Mass fractions of paddy in feed, separated paddy and separated rice are 0.2, 0.7 and 0.02, respectively. The mass of 1000 paddy grains is 24.5 g. Separation effectiveness and number of paddy grains recycled per second, respectively are

(A) 0.35 and 7373 (B) 0.93 and 3733

(C) 0.76 and 3773 (D) 0.83 and 3361

Q. 140 In a tray drying under adiabatic process, hot air enters the dryer at 50°C dry bulb temperature (DBT) and 10% relative humidity (RH) with enthalpy of 70.2 kJ per kg dry air. The air leaving the dryer gains 20% moisture. Considering saturated water vapour pressure as 12.349 kPa at 50°C DBT, the enthalpy of air leaving the dryer in kJ per kg dry air is ....

Q. 141 A fat rich food product remains most stable in the water activity (a_w) range of

(A) $a_w < 0.1$ (B) $0.1 < a_w < 0.2$

(C) $0.3 < a_w < 0.4$ (D) $0.5 < a_w < 0.6$

Graduate Aptitude Test in Engineering - 2015

Q. 142 The efficiency of a cyclone separator can be increased by

(A) Decreasing the size of the particles

(B) Increasing the velocity of inlet air

(C) Reducing the length of the separator

(D) Reducing the diameter of air outlet

Q. 143 The correct statement in respect of rice parboiling process is

(A) Kernel structure becomes soft and it cooks easily

(B) Heat treatment during parboiling preserves some antioxidants

(C) Parboiled rice retains more proteins, vitamins and minerals

(D) Shelling of parboiled rice becomes more difficult

Q. 144 The values of C and n for raisin are 1.283×10^{-4} K^{-1} and 1.02 respectively in the Henderson equation. The equilibrium moisture content of raisin in percent dry basis corresponding to 70% relative humidity at 40°C is ....

Q. 145 Match the following items between **Column-I** and **Column-II** with the most appropriate combinations.

Column - I	Column - II
(i) Microwave dryer	(1) Cyclone separation
(ii) Spray dryer	(2) Sublimation of water
(iii) Freeze dryer	(3) Dielectric drying
(iv) Drum dryer	(4) Drying of fruit pulp

(A) 1-Q, 2-S, 3-R, 4-P (B) 1-R, 2-P, 3-Q, 4-S

(C) 1-R, 2-S, 3-Q, 4-P (B) 1-Q, 2-S, 3-P, 4-R

Q. 146 A single effect long-tube evaporator has ten tubes each of 2.5 cm diameter and 6 m length. It concentrates pineapple juice from 18° Brix to 23° Brix. The feed rate into the evaporator is 557 kg/h at the boiling point of 70°C (latent heat of vaporization = 2333.82 kJ/kg). Neglecting boiling point rise, the overall heat transfer coefficient in W m^{-2} K^{-1} for 12°C temperature gradient across the tube walls is ....

Q. 147 A fat globule of 1.5 μm diameter is rising up in a stagnant skim milk medium of 1005 kg/m^3 density and 1.5 cP viscosity. If the density of the fat globule is 915 kg/m^3, the steady rising velocity of the globule in μm/s is ....

Graduate Aptitude Test in Engineering - 2016

Q. 148 One hundred kilogram spice is extracted for essential oil using twice the amount of a pure organic solvent. The extracted solid mass contains 5% residual oil (oil-free solid mass basis). The liquid extracted mass contains 20% oil. Assume no solvent is retained by the extracted solid mass. Initial mass of the oil in the spice in kg is ....

Answers Key

1	2	3	4	5	6	7	8	9	10
D	C	B	D	A	D	A	C	A	C
11	12	13	14	15	16	17	18	19	20
A	A	B	C	B	B	B	A	D	A
21	22	23	24	25	26	27	28	29	30
A	C	C	C	B	A	C	B	B	C
31	32	33	34	35	36	37	38	39	40
B	B	D	B	D	C	C	B	C	B
41	42	43	44	45	46	47	48	49	50
C	D	B	D	D	C	D	B	B	B
51	52	53	54	55	56	57	58	59	60
D	B	A	C	D	A	C	D	D	A
61	62	63	64	65	66	67	68	69	70
C	A	C	B	C	A	B	A	A	A
71	72	73	74	75	76	77	78	79	80
C	C	C	A	B	A	B	A	A	C
81	82	83	84	85	86	87	88	89	90
A	A	E	A	A	B	C	C	B	B
91	92	93	94	95	96	97	98	99	100
B	C	A	A	B	C	A	A	D	A
101	102	103	104	105	106	107	108	109	110
C	D	D	C	B	B	D	C	B	B
111	112	113	114	115	116	117	118	119	120
C	B	B	A	D	C	A	B	C	C
121	122	123	124	125	126	127	128	129	130
B	D	C	B	C	A	A	B	B	C
131	132	133	134	135	136	137	138	139	140
C	BONUS	D	D	3.30	A	D	18	D	70.18
141	142	143	144	145	146	147	148		
C	B	C	28.05	B	1388.54	0.0736	52.5		

Explanations

Q. 6

Moisture content at dry basis = M. C. wet basis×100/(1 – M. C. wet basis)

$$= 0.20 \times 100/(1 - 0.20) = 25\ \%$$

Q. 23

Refer explanation for Q.6.

Q. 66

Mass of paddy = 100 kg

Initial moisture content = 25 % db or 20 % wb

Final moisture content = 15 % db or 13.04 % wb [$\because M_{wb} = \frac{M_{db}}{M_{db}+1}$]

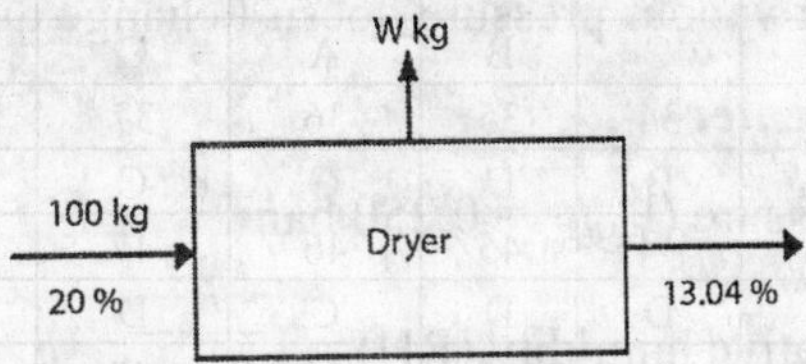

Applying Mass balance sequation

$$0.20 \times 100 = (100 - W) \times 0.1304 + W$$

$$\Rightarrow \quad W = 8 \text{ kg}$$

Q. 98

Initial moisture content = 80 % wb

Final moisture content = 25 % db or 20 % wb [$\because M_{wb} = \frac{M_{db}}{M_{db}+1}$]

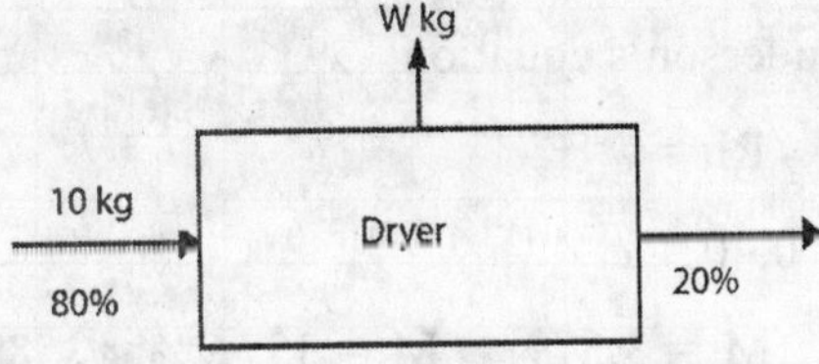

Applying Mass balance equation

$$0.80 \times 10 = (10 - W) \times 0.20 + W$$

$$\Rightarrow \quad W = 7.5 \text{ kg}$$

Q. 114

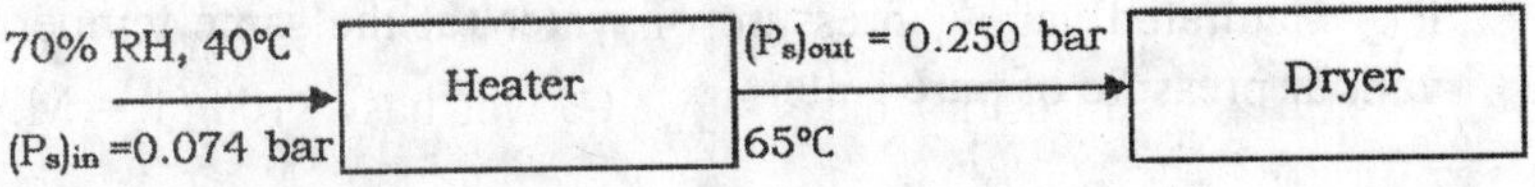

At inlet of heater

Saturation pressure $(P_s)_{in}$ = 0.074 bar

$\therefore$ Relative humidity (RH) = $\frac{P_v}{(P_s)in}$

$\Rightarrow$ $0.70 = \frac{P_v}{0.074}$

$\Rightarrow$ $P_v = 0.0518$ bar

∵ Partial water vapour pressure doesn't change during heating of air.

At outlet of heater

Saturation pressure $(P_{ss})_{out} = 0.250$ bar

$\therefore$ Relative humidity (RH) = $\frac{P_v}{(P_s)out}$

$$RH = \frac{0.0518}{0.259} = 21\%$$

Q. 115

Given that

Relative humidity (∅) = 40 %

Temperature (T) = 338 K

Constants: C = 7.410^{-4} K^{-1} and n = 0.56

Applying Handerson's equation

$$1 - Rh = e^{-CTM_e^n}$$

$$\Rightarrow \quad 1 - 0.40 = e^{-7.4\times10^{-4}\times338\times M_e^{0.56}}$$

$$\Rightarrow \quad M_e = 3.579 \Rightarrow M_e = 357.9\% \approx 358\% \text{ (db)}$$

Q. 116

Water Activity $(a_w) = P_v/P_s$

Where

P_v = Partial pressure of water in the headspace of a product

P_s = Saturated vapour pressure of water (at the same temperature) or vapour pressure of pure water

Q. 117

Water removed = $H_2 - H_1$

$$= 0.019 - 0.0078 = 0.0112 \text{ kg}$$

Q. 118

Basis 1000 kg potato entering

As 8% of potatoes are lost in peeling, potatoes to drying are 920 kg, solids 129 kg.

Mass in	(kg)	Mass out	(kg)
Raw Potatoes:		Dried Product:	
Potato solids	140 kg	Potato solids	129 kg
Water	860 kg	Associated water	10 kg
		Total product	139 kg
		Losses:	
		Peelings	
		- solids	11 kg
		- water	69 kg
		Water evaporated	781 kg
		Total losses	861 kg
Total	1000 kg	Total	1000 kg

$$\text{Product yield} = \frac{Product \times 100}{Raw\,material} = \frac{139 \times 100}{1000} = 13.9\ \%$$

Q. 119

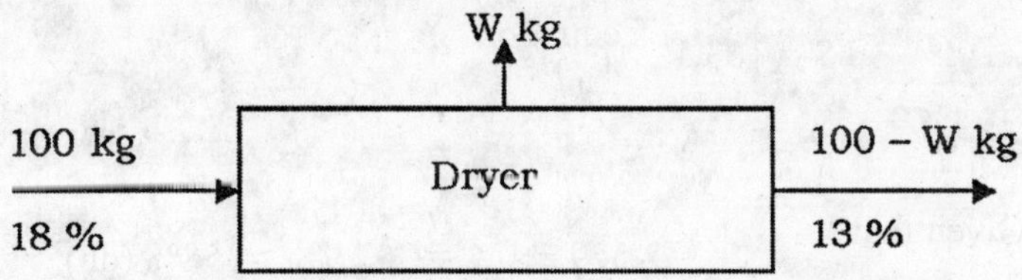

Applying Mass balance equation

$$0.18 \times 100 = (100 - W) \times 0.13 + W$$

$$\Rightarrow \quad W = 5.75 \text{ kg}$$

Q. 120

Mechanical separations can be divided into four groups - sedimentation, centrifugal separation, filtration and sieving.

Q. 123

Given that

Relative humidity ($\varnothing$) = 20 %

Temperature (T) = 320 K

Constants: C = 6.510^{-6} and n = 1.8

Applying Handerson's equation

$$1 - Rh = e^{-CTM_e^n}$$

$$\Rightarrow 1 - 0.20 = e^{-6.5\times10^{-6}\times320\times M_e^{1.8}}$$

$$\Rightarrow \ln(0.80) = -6.5\times10^{-6}\times320\times M_e^{1.8}$$

$$\Rightarrow M_e = 13.42\ \%$$

Q. 124

Given that

Radius of cyclone(R) = 500 mm = 0.5 m

Entrance velocity (V) = 20 m/s

∵ Seperation factor (performance factor) of cyclone separator

$$(S) = \frac{2V^2}{Dg}$$

$$S = \frac{2\times20^2}{0.5\times9.81}$$

$$S = 163.1$$

Q. 125

Given that

Saturation humidity (H_s) = 0.622 $\frac{P_s}{P-P_s}$ = 0.075 kg water vapour / kg dry air

$$P_s = 10.90\ \text{kPa}$$

Percentage humidity (H_p) = $\frac{H}{H_s}$ = 0.60

Where

H = absolute humidity = 0.622 $\frac{P_v}{P-P_v}$

P = atmospheric pressure = 101.3 kPa

Thus, $$H_p = \frac{\frac{P_v}{P - P_v}}{\frac{P_s}{P - P_s}}$$

$$\Rightarrow \quad 0.60 = \frac{\frac{P_v}{P - P_v}}{\frac{0.075}{0.622}}$$

$$\Rightarrow \quad P_v = 6.83 \text{ kPa}$$

Therefore,

Relative humidity (∅) = $\frac{P_v}{P_s} \times 100 = \frac{6.83}{10.90} \times 100 = 62.7\ \%$

Q. 126

Bulk density of paddy (ρ_b) = 650 kg/m³

Moisture content (w) = 28% (wb)

Weight of water (M_w) = 650×0.28 kg

Weight of dry solids (M_d) = 650 (1 – 0.28) = 468 kg

Volume filled by water (V_w) = 1000 m³

Volume of dry solids (V_d) = 1m³ (consider volume of bulk paddy is 1 m³)

∴ Dry density = M_d/ V_d

Dry density = 468/1 = 468 kg/m³

Q. 127

Mass of wheat = 4900/9.81 = 499.5 kg

Final moisture content = 10 % db or 9.09 % wb [$\because M_{wb} - \frac{M_{db}}{M_{db}+1}$]

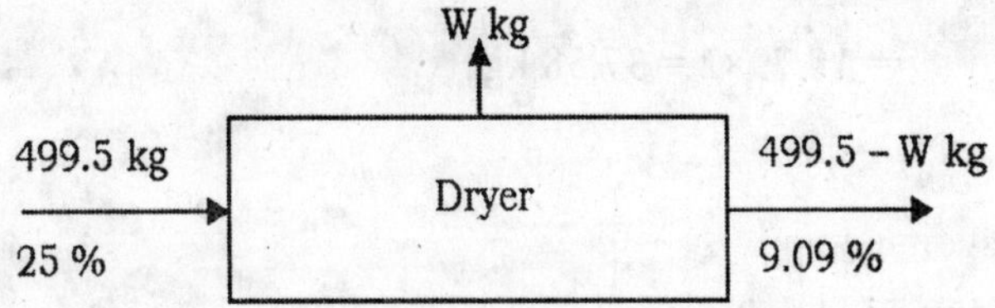

Applying Mass balance equation

$$0.25 \times 499.5 = (499.5 - W) \times 0.0909 + W$$

$$\Rightarrow \quad W = 87.4 \text{ kg} \approx 87 \text{ kg}$$

Q. 128

Humidity of air (H) = $0.622\times(P_v/P-P_v)$

$= 0.622\times(2.4/98.925) = 0.15$

Thus

Humid volume = $(0.00283 + 0.00456H)(t + 273)$

$= (0.00283 + 0.00456\times0.15)\times(40 + 273)$

$= 0.908\ m^3/kg$ dry air

Q. 129

Initial moisture content = 28 % db or 21.89 % wb

$[\because M_{wb} = \frac{M_{db}}{M_{db}+1}]$

Final moisture content = 20% db or 16.67 % wb

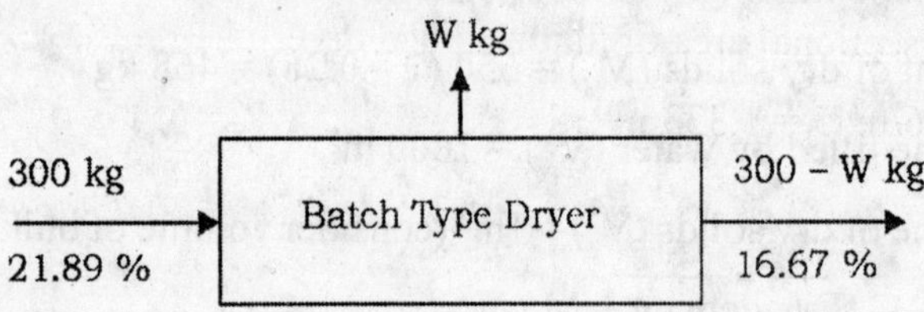

Applying Mass balance equation

$0.2189\times300 = (300 - W)\times0.1667 + W$

$\Rightarrow$ W = 18.79 kg

$\because$ 18.79 kg moisture is removed in 30 min, hence moisture removed per hour

$= 18.79\times2 = 37.58$ kg

Q. 130

Given

M = 400 kg; A = 0.4 m^2

$\rho_d = 1200\ kg/m^3$; $\rho_a = 3.5\ kg/m^3$

As we know porosity of particles is zero. Thus the volume of fluidized container will be equal to the volume of particles.

$\pi r^2 L = M/\rho_d$

$\Rightarrow \quad 0.4L = 400/1200$

$\Rightarrow \quad L = 0.833$ m

Because porosity = 0

Thus $L_{min} = L = 0.833$ m

Pressure Drop $(\Delta P) = L \times (\rho_d - \rho_a) \times g$

$= 0.833 \times (1200 - 3.5) \times 9.81 = 9.7 \times 10^3$ Pa

Q. 131

Volume of solids(V) = mass/density

= 480/800 = 0.6 m

Consider solid is filled upto L m.

Now, Volume of fluidized bed dryer = Volume of solids

Cross sectional area of fluidized bed dryer ×height = 0.6

L = 0.6/0.45 = 1.333 m

Now,

$$\frac{L_{min}}{L} = \frac{1-n}{1-n_{min}}$$

$$\Rightarrow \quad \frac{L_{min}}{1.333} = \frac{1}{1-0.5}$$ (Pororsity at no fluidization (n) = 0)

= 2.67 m

Q. 132

Initial TS = 18 % db = 15.25% wb

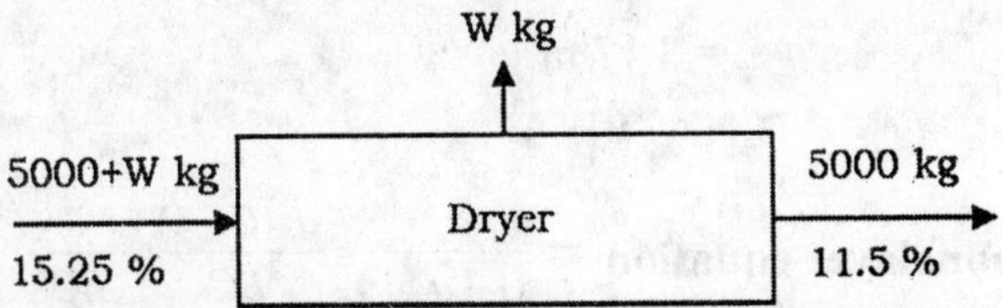

Applying Mass balance equation

$0.1525 \times (5000 + W) = 0.115 \times 5000 + W$

$\Rightarrow \quad W = 221.24$ kg/ h

Heat supplied = 221.24×2345 = 518807.8 kJ/ h

Actual heat supplied (Q_s) = 518807.8/0.70 = 741154 kJ/ h

We know that

$$Q_s = \frac{V}{V_H} C_p (T_i - T_i)$$

$$741154 = \frac{V}{0.866} \times 1.005 \times (50 - 25)$$

⇒ V = 25545.84 m³/ h or 425.76 m³/ min

Q. 133

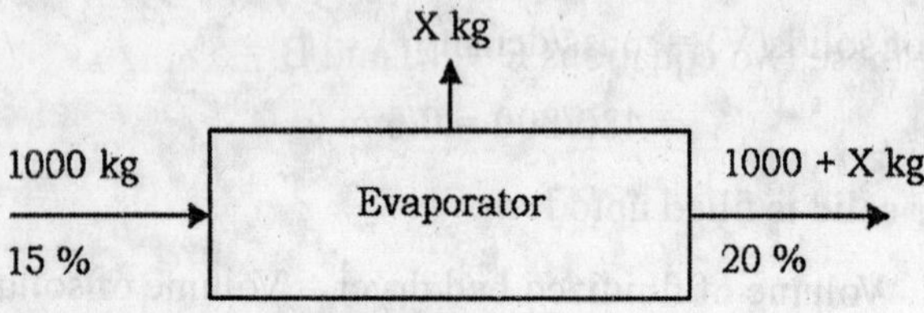

Applying Mass balance equation

$$0.15 \times 1000 = 0.20 \times (1000 - X)$$

⇒ X = 250 kg/ h

Heat supplied (Q_s) = 250×2309 = 577250 kJ/ h = 160347.22 W

Overall heat transfer coefficient (U) = $\dfrac{1}{\dfrac{1}{5000} + \dfrac{0.0015}{15} + \dfrac{1}{800}}$ = 645.16 W m^{-2} K^{-1}

∵ Heat supplied (Q_s) = n.U A ΔT

= n U (πDL) ΔT (∵ Tube surface area(A) = πDL)

⇒ 160347.22 = 100×645.16×3.14×0.25×L×10

⇒ L = 3.17 m

Q. 134

BET monolayer equation ⇒ $\dfrac{a_w}{(1 - a_{w})B} = \dfrac{a_w (C-1)}{BC} + \dfrac{1}{BC}$

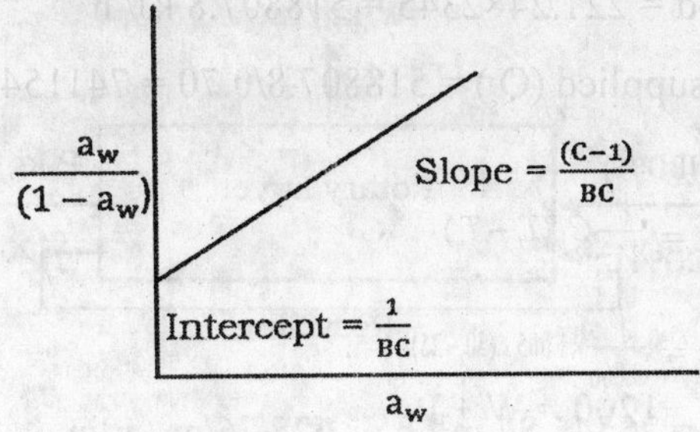

As given $\frac{1}{BC} = 2$ and $\frac{(C-1)}{BC} = 18$

Solving these two equations C = 10 and B = 5%

Q. 135

Given

Fish fillet thickness (a) = 5 mm = 0.005 m; Density of fish (ρ) = 1100 kg/ m^3

Latent heat of fusion (λ) = 330 kJ/ kg = 330000 J/ kg

Convective heat transfer coefficient (h) = 2.0 W m^{-2} k^{-1}

Thermal conductivity (k) = 1.5 W m^{-1} k^{-1}

Freezing Temperature (T_f) = – 35°C; Initial Temperature (T_i) = – 2.55°C

We know

$$\text{Freezing Time (t)} = \frac{\lambda\rho}{T_f - T_i}\left(\frac{a}{2h} - \frac{a^2}{8k}\right)$$

$$t = \frac{330000 \times 1100}{-2.5-(-35)}\left(\frac{0.005}{2\times 2} - \frac{0.005^2}{8\times 1.5}\right)$$

$$t = 3.30 \text{ hrs}$$

Q. 136

Without Recycling

Mass Balance Equation

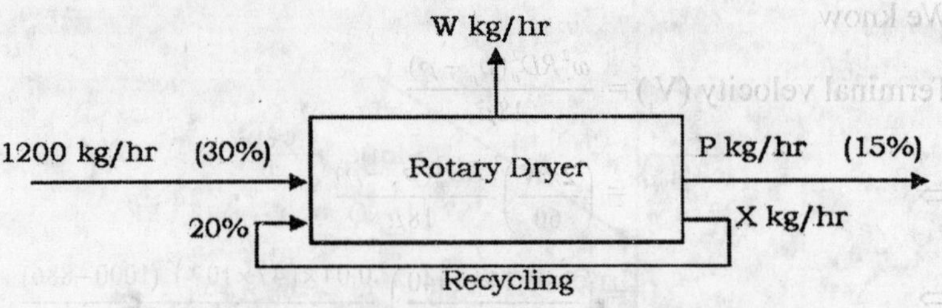

$$1200 = W + P \tag{i}$$

$$\Rightarrow \quad 1200\times0.30 = W + P\times0.15 \tag{ii}$$

By solving above equations

$W = 211.76$ kg/h

And $P = 988.24$ kg/h

After Recycling

Consider portion recycled is = X kg/h

At mixing of this X kg/hr (15 % M.C.) and 1200 kg/hr (30% M.C.) gives feed at 20 % M.C.) .

Thus $X\times0.15 + 1200\times0.30 = 0.20\times (1200 + X)$

$\Rightarrow \quad X = 2400$ Kg/hr

Q. 137

Mass fraction of underflow (m_f) = 0.06;

Mass fraction of overflow (m_o) = 0.96

Mass fraction of feed (m_f) = 0.90

$$\text{Effectiveness of screen} = \frac{m_0(m_o - m_f)(m_f - m_u)(1-m_u)}{m_f(1-m_f)(m_0 - m_u)^2}$$

$$= \frac{0.96(0.06)(0.84)(0.94)}{0.9(0.1)(0.9)^2} = 62\ \%$$

Q. 138

Rotational speed (N) = 1440 rpm; Effective separation radius (R) = 0.04 m

Diameter of particle (D_p) = 47×10^{-6} m;

Viscosity of oil (μ) = 0.705 cP or 705×10^{-6} Pa s

Density of oil (ρ_p) = 886 kg/ m^3; Density of water (ρ) = 1000 kg/ m^3

We know

$$\text{Terminal velocity } (V_t) = \frac{\omega^2 R D_p^2 (\rho_p - \rho)}{18\mu}$$

$$\Rightarrow \quad = \left(\frac{2\pi N}{60}\right)^2 \frac{R D_p^2 (\rho_p - \rho)}{18\mu}$$

$$\Rightarrow \quad = \left(\frac{2 \times 3.14 \times 1440}{60}\right)^2 \frac{0.04 \times \left(47 \times 10^{-6}\right)^2 (1000 - 886)}{18 \times 705 \times 10^{-6}}$$

$$\Rightarrow \quad = 0.018 \text{ m/ s} = 18 \text{ mm/ s}$$

Q. 139

The mass flow rates

$$\Rightarrow \quad F.X_f = P.Xp + R.Xr$$

$$1600 \times 0.2 = P \times 0.7 + (1600 - P) \times 0.02$$

$$P = 423.5294 \text{ Kg/h and } R = 1176.4706 \text{ Kg/h}$$

The effectiveness of the screen is

$$E = \frac{PR}{(F)^2}\left[\frac{X_p(1 - X_r)}{X_f(1 - X_f)}\right]$$

$$\Rightarrow \quad E = \frac{423.5294 \times 1176.4706}{(1600)^2}\left[\frac{0.7 \times (1 - 0.02)}{0.2 \times (1 - 0.2)}\right] = 0.83$$

$$\text{Number of paddy grain recycled per Second} = \frac{423.5294 \times 0.7 \times 1000 \times 1000}{3600 \times 24.5}$$

$$= 3361 \text{ paddy grains}$$

Q. 140

We know

$$\text{Relative humidity } (R_h) = \frac{P_v}{P_s}$$

$$P_v = 0.10 \times 12.349 = 1.2349 \text{ k Pa}$$

$$\text{Absolute humidity (H)} = 0.622 \frac{P_v}{P - P_v}$$

$$= 0.622\left(\frac{1.2349}{101.3 - 1.2349}\right)$$

$$= 0.0077 \text{ kg } H_2O/ \text{ kg dry air}$$

For Adiabatic drying

$$\frac{H-H_s}{T-T_s}=\frac{1.005+1.88H}{\lambda_s}$$

$$\Rightarrow \quad \frac{70.2-H_s}{50-0}=\frac{1.005+1.88\times0.0077}{2501}$$

$$\Rightarrow \quad H_s = 70.18 \text{ kJ/ kg dry air}$$

Q. 141

At Water Activity below 0.6, almost all microbial activity is inhibited in food product. And, at water activity 0.3 to 0.4 there is Minimum oxidation velocity of food product. Therefore, To prevent moisture uptake packaging is required on this level of water activity.

Q. 144

Given that

Relative humidity (∅) = 70 %

Temperature (T) = 313 K

Constants: C = 1.28310^{-4} K^{-1} and n = 1.02

Applying Handerson's equation

$$1 - Rh = e^{-CTM_e^n}$$

$$\Rightarrow \quad 1-0.70 = e^{-1.283\times10^{-4}\times313\times M_e^{1.02}}$$

$$\Rightarrow \quad \ln(0.30) = -1.283\times10^{-4}\times313\times M_e^{1.02}$$

$$\Rightarrow \quad M_e = 28.05\% \text{ (db)}$$

Q. 146

Overall evaporation = Feed rate×(1 – Concentration of product / Concentration of feed)

= 557×(1 – 18/23) = 121.09 kg/h

Thus, total heat used (Q) = 121.09×2333.82/3600 = 78.48 kW

(∵ Heat of vaporization Q =mλ)

∵ Surface area of tubes = 10×2πrh

= 10×2×3.14×0.0125×6 = 4.71 m^2

We know that

$$\text{Total heat consumed in evaporator (Q)} = \text{U A T}$$

$$78.48\times1000 = \text{U}\times4.71\times12$$

$$\Rightarrow \quad \text{U} = 1388.54 \text{ W m}^{-2}\text{ K}^{-1}$$

Q. 147

Applying Stoke's Law

$$V_t = \frac{D_p^2.g.(\rho_p - \rho_a)}{18.\mu}$$

$$= \frac{(1.5\times10^{-6})^2 \times 9.81\times(1005-915)}{18\times1.5\times10^{-2}} = 7.36\times10^{-10} \text{ m/sec}$$

$$= 0.0736 \ \mu \text{ m/s}$$

Q. 148

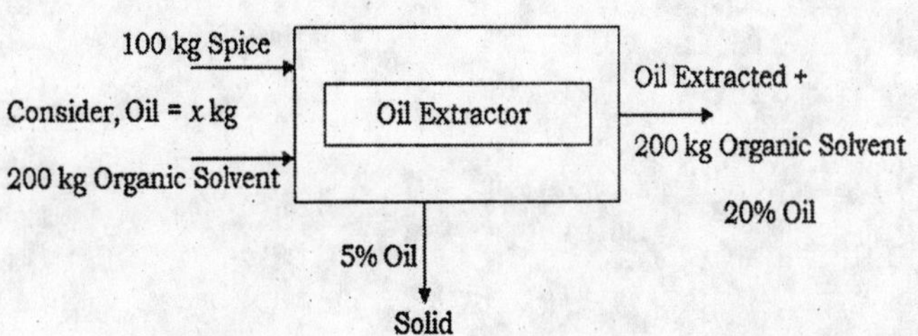

Referring above figure

$\because$ (Oil Extracted + 200 kg Organic Solvent)×(20/100) = Oil Extracted

$\Rightarrow$ Oil extracted = 50 kg

Hence, Mass of solid = 100 – 50 = 50 kg

Mass balance for oil,

$x = 50\times0.05 + (50+200)\times0.20 = 52.5$ kg

31

Size Reduction Milling and Homogenization

UKPSC Combined Assistant Engineering Exam 2013 Paper- I

Q. 1 Which of the following is not a type of grinder?

31

Size Reduction Milling and Homogenization

UKPSC Combined Assistant Engineering Exam 2013 Paper - I

Q. 1 Which of the following is not a type of grinder ?

(A) Attrition mill (B) Burr mill

(C) Hammer mill (D) Solvent extraction mill

Q. 2 Which of the following is not a part of a hammer mill?

(A) Rotor (B) Cast iron plates

(C) Hammer (D) Screen

Q. 3 Parboiling of rice before milling is done to

(A) Decrease nutritional value (B) No change in nutritional value

(C) Decrease of yield (D) Increase nutritional value

Q. 4 The size of the food grains in attrition mill is reduced by

(A) Shear and crushing (B) Impact and shear

(C) Impact and crushing (D) Impact only

Q. 5 In a ball mill or pebble mill, most of the size reduction is done by

(A) Shearing (B) Cutting

(C) Impact (D) Crushing

Q. 6 In rice polishing is done for the following job

(A) A coating is applied on the outer surface of brown rice

(B) A layer of bran is removed form brown rice

(C) A layer of starch is removed

(D) Only husk is removed

Q. 7 Types of liquid – liquid emulsion is

(A) Milk (B) Water

(C) Sugarcane juice (D) Oil

APPSC AEES Agriculture Engineering Exam 2016

Q. 8 In size reduction of fine powders, which of the following laws in more applicable:

(A) Kick' Law (B) Rittinger's Law

(C) Bond's Law (D) All of the above

Q. 9 In a rubber roll paddy sheller, the direction or rotation and peripheral speeds of the Rollers are respectively

(A) Same and equal (B) Same and different

(C) Opposite and equal (D) Opposite and different

Q. 10 In dry milling process of pulses, prior to treatment with oil, the following operation isDone

(A) Conditioning (B) Grading

(C) Pitting (D) Scalping

UKPSC Combined Junior Engineering Exam 2013 Agricultural Engineering Paper-I

Q. 11 A good rice huller does the following work :

(A) Removes the husk from the paddy

(B) Does not break the rice

(C) Produce good quality rice

(D) All of the above

Q. 12 There is high percentage of breakage of rice grain in

(A) Disc type hullers (B) Roller type hullers

(C) Both (A) & (B) (D) None of the above

Q. 13 The process of husking or dehusking is also known as

(A) Shelling (B) Bruising

(C) Scalping (D) None of the above

Q. 14 When the paddy is shelled, brown colored rice is obtained, which is known as

(A) Par boiled rice (B) Brown rice

(C) Both (A) & (B) (D) None of the above

Q. 15 Average weight of 1000 wheat grains ranges between

(A) 30 – 35 gms (B) 35 – 45 gms

(C) 45 – 50 gms (D) None of the above

Q. 16 By processing soyabean seeds, various products which are prepared can help in

(A) Certain cancer diseases (B) Cardio-vascular diseases

(C) Both (A) & (B) (D) None of the above

Q. 17 In order to extract juices from sugarcane crushers, the king roller is fitted in

(A) Bullock drawn crusher at the top of the roller gear.

(B) Power driven cane-crusher, it is fitted separately above the axle of the roller

(C) None of these

(D) Both (A) & (B)

Q. 18 Crushing capacity of power operated cane crusher is

(A) < 60 % (B) 70 %

(C) > 80 % (D) None of the above

Q. 19 Grinding is a process of

(A) Drying (B) Polishing

(C) Size reduction (D) Packaging

Q. 20 Modern flour mills for wheat milling are

(A) Disc type (B) Attrition type

(C) Under runner type (D) Roller type

Q. 21 Energy required to grind any material is expressed by

(A) Fick's law (B) Kick's law

(C) Newton's law (D) Stokes law

Q. 22 Angle of repose for wheat grains is about

(A) 15° (B) 20°

(C) 25° (D) 30°

Q. 23 Shellers are used in the processing of

(A) Wheat (B) Maize

(C) Soyabean (D) Paddy

RPSC AEn Pre Exam 2013 (Agricultural Engineering)

Q. 24 In attrition mill, the size of food grain is reduced by

(A) Impact (B) Impact and shear

(C) Impact and crushing (D) Shear and crushing

Q. 25 Paddy consists rice husk

(A) 18 to 20 % (B) 20 to 24 %

(C) 20 to 30 % (D) 20 to 28 %

Q. 26 In general, during milling operation of cereals .... are removed.

(A) Hull and endosperm (B) Germ and endosperm

(C) Hull and germ (D) Seed coat and endosperm

OCS AEn Exam 2011 (Pre) – II (Agricultural Engineering)

Q. 27 In rubber roll sheller, the size of the grain and the clearance are fixed, then the length of husking zone increases with .... in diameter of the roll.

(A) Decrease (B) Increase

(C) Either increase or decrease (D) Neither increase nor decrease

CGPSC State Engineering Services Exam Paper – I

Q. 28 The process of breaking down of solid material through the application of mechanical force is called

(A) Centrifugation (B) Filtering

(C) Size reduction (D) Agitation

(E) Mixing

Q. 29 Size reduction of grains is caused by impact in

(A) Gyratory crusher (B) Crushing roll

(C) Hammer mill (D) Jaw crusher

(E) Ball mill

Q. 30 The ratio of surface energy created by crushing to the energy absorbed by the solid is called

(A) Working index (B) Crushing efficiency

(C) Capacity of crusher (D) Both (A) & (B)

(E) Both (A) & (C)

Q. 31 In grinding machines

(A) Residence time is larger and throughput is smaller

(B) Residence time is smaller and throughput is larger

(C) Both residence time and throughput is larger

(D) Both residence time and throughput is smaller

(E) Only throughput is larger

Q. 32 If 'd_1' and 'd_2' are average initial and final sizes of material and K is a constant then E, energy required for size reduction of unit mass is given by

(A) $E = K [(d_1 - d_2) / (d_1 \times d_2)]$ (B) $E = K [(d_1 \times d_2) / (d_1 - d_2)]$

(C) $E = K [(d_2 - d_1) / (d_1 \times d_2)]$ (D) $E = K [(d_1 \times d_2) / (d_2 \times d_2)]$

(E) $E = K [(d_1 - d_2) - (d_1 \times d_2)]$

UKPSC A.En. Exam 2007 Paper

Q. 33 Emery roller is used in

(A) Wheat milling (B) Corn milling

(C) Pulse milling (D) None of the above

Q. 34 In a vertical bullock driven crusher, the position of king roller is

(A) Movable (B) Fixed

(C) Fluctuating (D) None of the above

Q. 35 Hullers are not recommended for paddy milling because

(A) They are costly (B) The bran quality is poor

(C) The breakage is more (D) They are heavy

Q. 36 Which of the following is not related to size reduction ?

(A) Hammer mill (B) Parboiler

(C) Crusher (D) Attrition mill

UKPSC A.En. Exam 2012 Paper – I

Q. 37 Rittinger's and Kick's law are related with

(A) Size reduction (B) Heat transfer

(C) Mass transfer (D) Momentum

Q. 38 Which is not the size reduction process ?

(A) Crushing (B) Impact

(C) Shearing (D) Mixing

Q. 39 Wheat milling involves

(A) Break rolls and emery rolls

(B) Reduction rolls and emery rolls

(C) Break rolls and reduction rolls

(D) None of the above

Q. 40 If n_1 is the amount of husk, n_2 the amount of unhulled grains and n_3 is the total quantity of grains, the effectiveness of hulling is

(A) $(n_3 - n_2)/n_1$ (B) $(n_1 - n_2)/n_1$

(C) $(n_3 - n_2)/n_3$ (D) $(n_1 - n_3)/n_1$

Graduate Aptitude Test in Engineering - 2007

Q. 41 A pulse mill grinds Bengal gram of 2 mm volume-surface mean diameter to powder of 100 μm volume-surface mean diameter. The ratio of Rittinger's to Kick's constant in the grinding operation is

(A) 0.317 kWh kg^{-1} (B) 3.15 mm

(C) 315.34 μm (D) 152.793 $kWhton^{-1}$

Graduate Aptitude Test in Engineering - 2008

Q. 42 The percentage of husk, bran and bran oil received from rice milling are respectively

(A) 20, 5 and 25 (B) 5, 10 and 30

(C) 20, 5 and 40 (D) 20, 10 and 20

Q. 43 The results of sieve analysis of a food powder are presented in the following two tables.

Table 1:

Sieve aperture (μm)	Mass retained (%)
12.5	13.8
7.5	33.6
4.0	35.2
2.5	12.8
0.75	4.6

Table 2:

Average diameter of particles (μm)	Mass retained on the sieve x (%)	(d) (x) = dx
0.375	4.6	1.725
1.625	12.8	20.800
3.250	35.2	114.4
5.750	33.6	193.2
10.000	13.8	138.0
Total	100	468.125

The mass mean diameter of the sample will be

(A) 8.46 μm (B) 6.48 μm

(C) 4.86 μm (D) 4.68 μm

Q. 44 A material consisting of 20 mm particles is crushed to an average size of 5 mm and requires 18 kJ kg^{-1} energy for this size reduction. If other conditions are similar, the energy required (kJ kg^{-1}) to crush the material from 25 mm to 3 mm needs to be calculated.

(i) The energy requirement calculated using Rittinger's law will be

(A) 61.53 (B) 35.16

(C) 16.43 (D) 5.82

(ii) The energy requirement calculated using Kick's law will be

(A) 72.39 (B) 52.76

(C) 27.55 (D) 14.85

(iii) The energy requirement calculated using Bond's law will be

(A) 57.34 (B) 30.57

(C) 15.79 (D) 11.25

Graduate Aptitude Test in Engineering - 2009

Q. 45 A ball mill of 1.8 m diameter is charged with balls each having diameter of 40 mm for grinding solid material. The rotational speed of the balls is 80% of the critical speed. The operating speed of rotation in revolution per minute is

(A) 18 (B) 22

(C) 26 (D) 30

Q. 46 If the length, breadth and thickness of a rice grain are 7 mm, 3 mm and 2 mm respectively, the sphericity of the grain is

(A) 0.33 (B) 0.50

(C) 0.67 (D) 0.75

Graduate Aptitude Test in Engineering - 2011

Q. 47 Sphericity of a particle is defined as: (surface to volume ratio for a sphere having identical volume as that of the particle) / (the surface to volume ratio of the particle). The measure of sphericity of a cube shaped sugar crystal is

(A) $\left(\frac{\pi}{6}\right)^{1/3}$ (B) $\left(\frac{\pi}{6}\right)^{2/3}$

(C) $\left(\frac{\pi}{4}\right)^{1/3}$ (D) $\left(\frac{\pi}{4}\right)^{2/3}$

Graduate Aptitude Test in Engineering - 2012

Q. 48 Work index in size reduction can be obtained by multiplying Bond's energy constant with

(A) 10 (B) $\sqrt{10}$

(C) $\sqrt[3]{10}$ (D) $\sqrt[4]{10}$

Graduate Aptitude Test in Engineering - 2013

Q. 49 A high pressure dairy homogenizer operates under upstream and downstream pressures of 200 and 40 bar respectively homogenizing 30 L of whole milk per hour. Density and specific heat capacity of whole milk are 1030 kg m^{-3} and 3.8 kJ kg^{-1} K^{-1}, respectively. Assuming complete energy conservation, the temperature rise of whole milk in degree Celsius is ....

Q. 50 The work index of a material is 6.25. If 80% of the feed and 80% of the product pass through IS Sieve No. 340 (3.25 mm opening) and IS Sieve No. 40 (0.42 mm opening), respectively, the power consumed in kW to crush 5000 kg h^{-1} of sorghum is

(A) 9.77 (B) 20.49

(C) 26.29 (D) 32.29

Q. 51 A milk fat globule of 2 μm diameter is rising in whole milk of density 1030 kg m^{-3} and coefficient of viscosity 10^{-3} Poise. If the fat density is 950 kg m^{-3}, the time needed to rise 10 mm for this fat globule in min is

(A) 0.57 (B) 34.57

(C) 35.57 (D) 95.57

Graduate Aptitude Test in Engineering - 2014

Q. 52 Energy required to grind a given mass of particles from a mean diameter of 12 mm to 4 mm is 12 kJ kg^{-1}. If energy consumed to grind the same mass of particles of 2 mm mean diameter to x mm mean diameter is 252 kJ kg^{-1}, the value of x using Rittinger's Law will be..........

Q. 53 The observations recorded in a pulse de-husking operation are:

S.No.	Parameters	Before de-husking	After de-husking
1	Whole(split) kernel content, %	0.5	72.3
2	Broken kernel content, %	0.7	11.2
3	Mealy waste content, %	1.1	16.5

For this operation the effectiveness of wholeness (in decimal) of kernels will be..........

Graduate Aptitude Test in Engineering – 2015

Q. 54 A ball mill of 1.8 m diameter is loaded with steel balls each having a diameter of 6 cm. The rotational speed of the ball is kept at 75% of the critical speed. The operational speed of the ball mill in rpm is ....

Graduate Aptitude Test in Engineering – 2016

Q.55 In a particle size analysis, the following results are obtained:

Mass of particles, g	2	5	7	4	1
Mean size of particles, μm	350	240	200	150	100

Volume-surface mean diameter of the particles in μm is ....

Q. 56 A high speed tubular ultracentrifuge with bowl radius of 100 mm and height 500 mm rotates at 20000 rpm and settles starch particles (average diameter of 20 μm) on the wall. The ratio of centrifugal force to the gravitational force acting on the particle is ....

Q. 57 Match the following

(P) Wheat milling — (1) Rubber rolls

(Q) Paddy dehusking — (2) Abrasive emery roll cylinder

(R) Pulse dehusking — (3) break and reduction rolls

(S) Spice grinding — (4) Hammer mills

(A) P-2, Q-1, R-3, S-4 (B) P-1, Q-2, R-4, S-3

(C) P-3, Q-1, R-2, S-4 (D) P-1, Q-3, R-4, S-2

Graduate Aptitude Test in Engineering – 2017

Q. 58 A batch of 1000 kg of apples containing 6.2% bruised apples is sorted by an electronic color sorter. The sorted apples contain 927.3 kg red apples and 17.4 kg bruised apples. Remaining red and bruised apples are delivered at the rejection outlet of the sorter. Overall effectiveness of the sorter is ....

Q. 59 Spherical dust particles of 50 μm are settling under gravity in air at 21°C and normal atmospheric pressure. Density of partiles is 1250 kg m^{-3} and density of air is 1.2 kg m^{-3}. Considering viscosity of air as 1.81×10^{-5} Pa s, the settling velocity of dust in mm s^{-1} will be ....

Graduate Aptitude Test in Engineering – 2019

Q. 60 The clean paddy production per annum is 160 million tonnes. Average milling quality analysis indicates the husk content, total yield and degree of polish as 22%, 73.32% and 6%, respectively. For an average bran oil yield of 20%, the annual rice bran oil potential in million tonnes is

(A) 1.268 (B) 1.498

(C) 1.617 (D) 1.945

Q. 61 In a rubber roll sheller, 250 mm diameter rolls are set at a clearance of 1 mm. If the mean thickness of paddy grains being shelled is 2 mm, the length of husking zone in mm is ....

Q. 62 A cream separator has discharge radii of 6 cm and 9 cm and the density of cream and skim milk are 860 and 1035 kg/m^3, respectively. The ideal radius (in meter) for placing the feed inlet is

(A) 0.085 (B) 0.098

(C) 0.113 (D) 0.174

Q. 63 Head rice contents in the samples collected at feed inlet, head rice outlet and broken rice outlet of an indented cylinder grader are 82%, 94% and 15%, respectively. If the grader receives the feed at 1200 kg/h, the flow rate (in kg/h) of head rice in the broken rice stream is

(A) 20.17 (B) 27.34

(C) 182.28 (D) 1017.72

Graduate Aptitude Test in Engineering – 2020

Q. 64 Choose the correct combination of process (Column I) performed by corresponding machine component(s) (Column II)

Column I	Column II
P. Paddy dehusking	I. Abrasive emery roll cylinder
Q . Wheat milling	2. Attrition disk
R. Pulse dehusking	3. Break roll and reduction roll
S. Spice grinding	4. Rubber rolls

(A) P-4, Q-1, R-3, S-2 (B) P-4, Q-3, R-1, S-2

(C) P-1. Q-4, R-3, S-2 (D) P-2, Q-1, R-4, S-3

Q. 65 Tray type paddy separator is employed to separate paddy from a binary mixture of paddy and brown rice at a feed rate of 1200 kg/h. Mass fractions of paddy in feed, separated paddy and brown rice streams are 0.2, 0.75 and 0.02, respectively. The amount of paddy in separated paddy stream in kg/h is ....

Graduate Aptitude Test in Engineering – 2021

Q. 66 A ball mill of 200 cm diameter grinds solid materials while operating with 10 cm size balls. If the same ball mill is used for wet grinding, charged with 20 cm diameter balls, the change in the operating speed in rpm is ……… (Take $\pi = 3.14$ and $g = 9.81\ m/s^2$)

Q. 67 Match the following hulling mechanism in column 1 with the corresponding machine in column 2

Column 1		Column 2	
P	Shear and compression	1	Blade type emery scourer
Q	Friction and abrasion	2	Horizontal *Gota* machine
R	Shear, compression and friction	3	Rubber roll dehusker
S	Impact, abrasion and friction	4	Under runner disc sheller

(A) P-3, Q-2, R-4, S-1 (B) P-3, Q-1, R-2, S-4

(C) P-3, Q-1, R-4, S-2 (D) P-4, Q-3, R-1, S-2

Q. 68 In a size reduction operation, the power required to crush 2 ton of feed material per hour is 7.2 kW. Eighty per cent of the feed and product material pass through 4.75 mm and 0.5 mm sieve openings, respectively. The work index of the material is

(A) 6.5 (B) 7.4

(C) 11.9 (D) 14.8

Q. 69 An air screen grain cleaner unit of capacity one ton/h with two screens was evaluated with a feed containing 8.5% impurities. During the operation, the clean grain at blower outlet, overflow of 1st screen and underflow of second screen were found to be 0.3%, 1.2% and 0.8%, respectively. If the clean grain contains 0.6% of impurities, the cleaning efficiency of the cleaner unit in per cent would be. ………..

Q. 70 A 30 μm thick membrane having 3 m^2 surface area is used to separate NaCl from a solution at steady state condition. The mass transfer coefficient of NaCl at the solution side is 1×10^{-6} m/s and that at the other side of the membrane is 3×10^{-7} m/s. Concentration of NaCl in the solution is 0.03 g/(100 mL) and, that on the other side of the membrane is assumed to be zero. Permeability of the membrane is 9×10^{-6} m/s. The rate of removal of the NaCl by the membrane in g/h is

(A) 0.73 (B) 0.81

(C) 0.86 (D) 0.93

Answers Key

1	2	3	4	5	6	7	8	9	10
D	B	D	A	C	B	A	B	D	C
11	12	13	14	15	16	17	18	19	20
D	C	B	B	B	C	D	C	C	D
21	22	23	24	25	26	27	28	29	30
B	C	D	D	B	C	B	C	C	B
31	32	33	34	35	36	37	38	39	40
A	A	C	B	C	B	A	D	C	C
41	42	43	44	45	46	47	48	49	50
C	D	D	B, C, B	C	B	A	B	4.09	A
51	52	53	54	55	56	57	58	59	60
D	0.25	0.735	24.06	193.5	44715	C	0.71	94	B
61	62	63	64	65	66	67	68	69	70
22.27	D	B	B	222	4.05	A	C	93.32	A

Explanations

Q. 41

Feed mean diameter (D_f) = 2 mm or 200 μm

Product mean diameter (D_p) = 100 μm

We know

$$\frac{K_r \ (J \ \mu m / kg)}{K_k \ (J / kg)} = \frac{ln(\frac{D_f}{D_p})}{\left(\frac{1}{D_p} - \frac{1}{D_f}\right)}$$

$$= \frac{ln(\frac{2000}{100})}{\left(\frac{1}{100} - \frac{1}{2000}\right)} = 315.34 \ \mu m$$

Q. 42

Rice contains 20% husk, 8 - 10 % bran and 12 - 22% bran oil.

Q. 43

Given

$\sum(d)(x) = 468.125$ μm

Total mass retained on the sieve ($\sum x$) = 100%

$\therefore$ Mass mean diameter (D_m) = $\dfrac{\sum(d)(x)}{\sum x}$

$= \dfrac{468.125}{100} = 4.68\ \mu m$

Q. 44 (i)

Applying Rittinger's law

Energy Requirement (E) $= K_R F_C \left[\dfrac{1}{L_2} - \dfrac{1}{L_1}\right]$

Therefore, $$\frac{E}{E'} = \frac{\frac{1}{L_2} - \frac{1}{L_1}}{\frac{1}{L_2'} - \frac{1}{L_1'}}$$

$\Rightarrow$ $$\frac{18}{E'} = \frac{\frac{1}{5} - \frac{1}{20}}{\frac{1}{3} - \frac{1}{25}}$$

$\Rightarrow$ E' = 35.19 kJ/ kg

(ii)

Applying Kick's law

Energy Requirement(E) = $K_k F_C \ln(L_1/L_2)$

Therefore, $$\frac{18}{E'} = \frac{ln(L_1 / L_2)}{ln(L'_1 / L'_2)} = \frac{ln(20/5)}{ln(25/3)}$$

$\Rightarrow$ E' = 27.53 kJ/kg

(iii)

Applying Bond's law

Energy Requirement(E) = $K_B F_C \dfrac{\frac{1}{\sqrt{L_2}} - \frac{1}{\sqrt{L_1}}}{\frac{1}{\sqrt{L'_2}} - \frac{1}{\sqrt{L'_1}}}$

Therefore, $$\frac{18}{E'} = \frac{\frac{1}{\sqrt{5}} - \frac{1}{\sqrt{20}}}{\frac{1}{\sqrt{3}} - \frac{1}{\sqrt{25}}}$$

$\Rightarrow$ E' = 30.38 kJ/kg

Q. 45

Given that

Diameter of ball mill (D) = 1.8 m, R = 0.9 m

Diameter of ball (d) = 40 mm, r = 0.02 m

For ball mill

$$\text{Speed of ball mill (N)} = \frac{1}{2\pi}\sqrt{\frac{g}{R-r}}$$

$$= \frac{1}{2\pi}\sqrt{\frac{9.81}{0.9-0.02}} = 0.5314 \text{ rev/s} = 31.88 \text{ rpm}$$

$\therefore$ Operating speed of rotation = N× η

$$= 31.88 \times 0.80 = 25.5 \text{ rpm}$$

Note:-

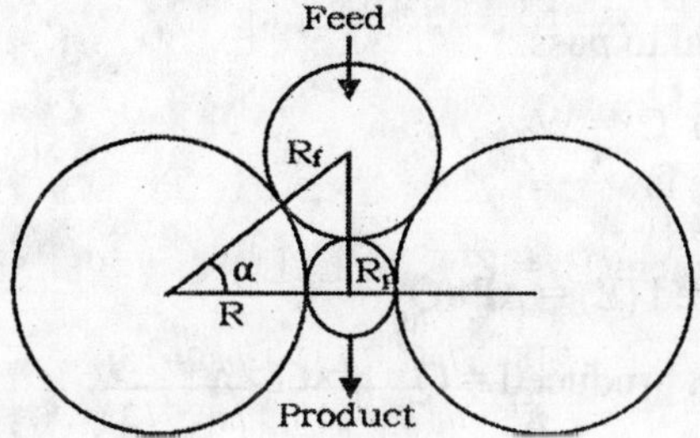

R = Radius of roller

R_f = Radius of feed grain

R_p = Radius of grain product

Refering above figure

$$\cos\alpha = \frac{R+R_p}{R+R_f}$$

Q. 46

We know

$$\text{Sphericity} = \frac{(l.b.t)^{1/3}}{l}$$

$$= \frac{(7\times 3\times 2)^{1/3}}{7} = 0.50$$

Q. 48

Bond's law is used to calculate the energy required for size reduction from:-

$$E = C_b.\sqrt{\frac{1000}{d_2}} - \sqrt{\frac{1000}{d_1}}$$

Or

$$\frac{E}{W} = \sqrt{\frac{100}{d_2}} - \sqrt{\frac{100}{d_1}}$$

Where

W (J kg[1]) = the Bond Work Index (40000–80000 J kg^{-1} for hard foods such as sugar or grain.

d_1 (m) = diameter of sieve aperture that allows 80% of the mass of the feed to pass and

d_2 (m) = diameter of sieve aperture that allows 80% of the mass of the ground material to pass.

Therefore, $\sqrt{10}\ C_b = W$

Q. 49

Energy Required (E) = ΔP×Q

And Heat produced = Q× ρ ×C_p×ΔT

Thus Q××C_p×ΔT = ΔP×Q

⇒ $\Delta T = 160\times10^5/(1030\times3.8\times10^3) = 4.09°C$

Q. 50

Applying Bond's Law

$$\frac{P}{f} = 0.3162 W_i \left[\frac{1}{\sqrt{D_p}} - \frac{1}{\sqrt{D_f}}\right]$$

$$P = 0.3162\times4.62\times5\times\left[\frac{1}{\sqrt{0.42}} - \frac{1}{\sqrt{3.25}}\right] = 9.77 \text{ kW}$$

Q. 51

Applying Stoke's Law

$$V_t = \frac{D_p^2.g.(\rho_p - \rho_a)}{18.\mu}$$

$$= \frac{(2\times10^{-6})^2\times9.81\times(1030-950)}{18\times10^{-3}} = 174.4\times10^{-9} \text{ m/sec}$$

Thus, t = D/V_t = 10×10^{-3}/(/(174.4×10^{-9}) = 5734.2 sec = 95.57 min

Q. 52

According to Rittinger's Law

$$E = K_R f_c \left(\frac{1}{L_p} - \frac{1}{L_f}\right)$$

$$\therefore \quad \frac{252}{12} = \frac{\left(\frac{1}{x} - \frac{1}{2}\right)}{\left(\frac{1}{4} - \frac{1}{12}\right)}$$

$$\Rightarrow \quad x = 0.25 \text{ mm}$$

Q. 53

Effectiveness of wholeness of kernels ($\in_{wk}$)

$$= \frac{k_2 - k_1}{(k_2 - k_1) + (d_2 - d_1) + (m_2 - m_1)}$$

$$= \frac{72.3 - 0.5}{(72.3 - 0.5) + (11.2 - 0.7) + (16.5 - 1.1)}$$

$$= 0.735$$

Q. 54

Given that

Diameter of ball mill (D) = 1.8 m, R = 0.9 m

Diameter of ball (d) = 60 mm, r = 0.03 m

For ball mill

$$\text{Speed of ball mill (N)} = \frac{1}{2\pi}\sqrt{\frac{g}{R - r}}$$

$$= \frac{1}{2\pi}\sqrt{\frac{9.81}{0.9 - 0.03}} = 0.5347 \text{ rev/s} = 32.08 \text{ rpm}$$

∴ Operating speed of rotation = N× η

= 32.08×0.75 = 24.06 rpm

Q. 55

Mean Diameter (D_{pi}) m	Mass of particles, g	Mass Fraction (X_i)
350	2	2/19 = 0.1053
240	5	5/19 = 0.2632
200	7	7/19 = 0.3684
150	4	4/19 = 0.2105
100	1	1/19 = 0.0526

$\because$ Volume surface mean diameter $(S_p) = \dfrac{1}{\sum_{i=1}\left(\dfrac{X_i}{D_{pi}}\right)}$

$\sum_{i=1}\left(\dfrac{X_i}{D_{pi}}\right)$ $= 0.1053/350 + 0.2632/240 + 0.3684/200 + 0.2105/150 + 0.0526/100$
$= 0.005169$

$\therefore \quad S_p = 1/0.005169 = 193.5\ \mu m$

Q. 56

Centrifugal force $(F_c) = mrw^2 = mv^2/r$

$= mr(2\pi N/60)^2$

Gravitational force $(F_g) = mg$

Hence, $F_c/F_g = r(2\pi N/60)^2/g$

$= 0.1\times(2\times3.14\times20000/60)^2/9.81$

$= 44715$

Q. 58

It can be easily seen through the below flow chart, how the process is flowing.

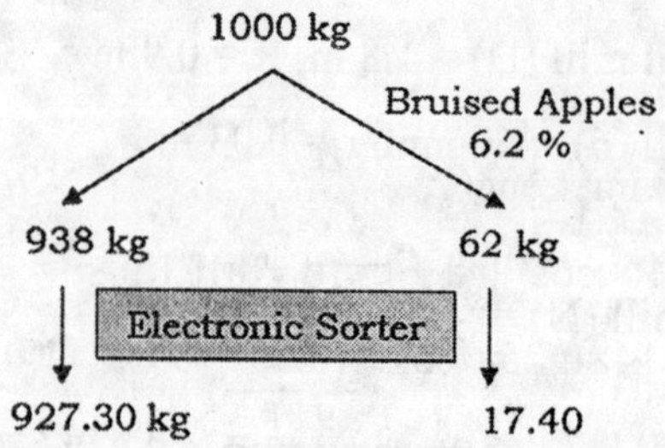

$\therefore$ Overall effectiveness = (927.30/938)[(62 – 17.4)/62]

= 0.71

Q. 59

Applying Stoke's law

Diameter of particle $(D_p) = 50\times10^{-6}$ m;

Viscosity of oil $(\mu) = 1.81\times10^{-5}$ Pa s

Density of oil $(\rho_p) = 1250$ kg/ m³; Density of water $(\rho) = 1.2$ kg/ m³

We know

$$\text{Settling velocity } (V_t) = \frac{D_p^2 g(\rho_p - \rho)}{18\mu}$$

$$= \frac{D_p^2 g(\rho_p - \rho)}{18\mu}$$

$$= \frac{9.81 \times \left(50 \times 10^{-6}\right)^2 (1250 - 1.2)}{18 \times 1.81 \times 10^{-5}}$$

$$= 0.094 \text{ m/ s} = 94.00 \text{ mm/ s}$$

Q. 60

Percentage of bran = 100 – 22 – 73.32 = 4.68 %

Bran yield = 160×4.68/100 = 7.488 million tonnes

Hence, Bran oil yield = 7.488×20/100 = 1.498 million tonnes

Q. 61

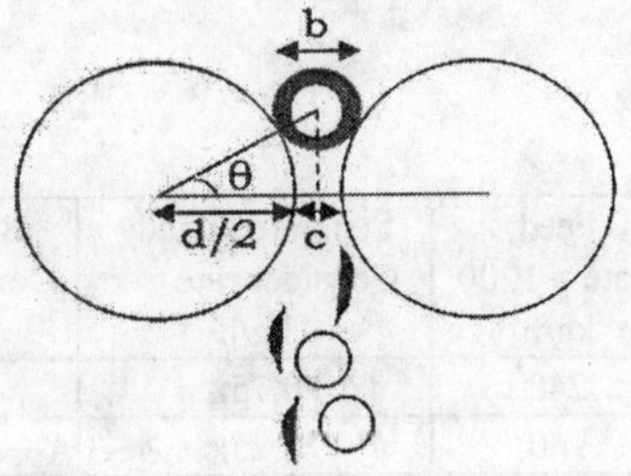

Length of husking zone(L)

$$= \pi d \times \cos^{-1}[(d + c)/(d + b)]/180$$

$$= 3.14 \times 0.25 \times \cos^{-1}[(0.25 + 0.001)/(0.25 + 0.002)]/180$$

$$= 0.02227 \text{ m or } 22.27 \text{ mm}$$

Q. 62

We know

$$r_n^2 = \frac{\rho_A r_1^2 - \rho_B r_2^2}{\rho_A - \rho_B}$$

$$= \frac{1030 \times 0.09^2 - 860 \times 0.06^2}{1030 - 860}$$

$$\Rightarrow \quad r_n = 0.176 \text{ m}$$

Q. 63

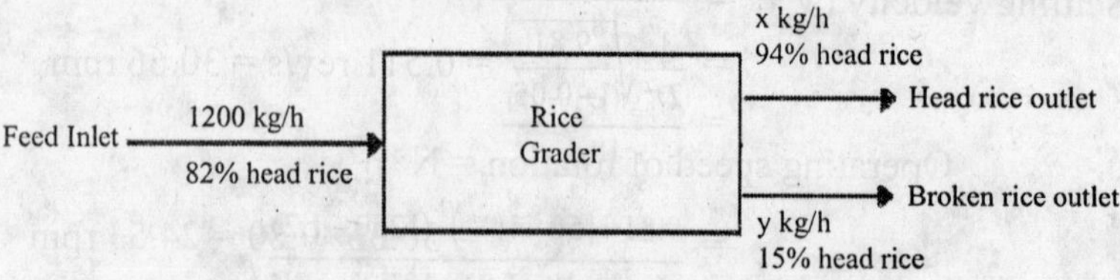

Applying Mass balance equation

$$x + y = 1200 \quad (1)$$

Again $$0.94x + 0.15y = 1200 \times 0.82 \quad (2)$$

Solving equation (1) and (2)

$x = 1017.72$ kg/h and $y = 182.28$ kg/h

Hence, Flow rate of head rice in the broken rice stream $= 182.28 \times 0.15$

$= 27.34$ kg/h

Q. 65

Rate / Component	Feed Rate = 1200 kg/h	Separated Paddy Consider rate = x kg/h	Brown Rice Consider rate = y kg/h
Paddy	240	0.75x	0.02y
Brown Rice	960	0.25x	0.98y

Hence,

$$0.75x + 0.02y = 240 \quad (i)$$

$$0.25x + 0.98y = 960 \quad (ii)$$

Solving (i) and (ii)

$x = 295.89$ kg/h and $y = 904.11$ kg/h

Hence, Amount of paddy in separated Paddy $= 0.75x = 222$ kg/h

Q. 66

Situation I :-

Radius of ball mill (R) = 1 m

Radius of ball (r) = 5 cm = 0.05 m

Speed of ball mill (N) = $\frac{1}{2\pi}\sqrt{\frac{g}{R-r}}$

$$= \frac{1}{2\pi}\sqrt{\frac{9.81}{1-0.05}} = 0.511 \text{ rev/s} = 30.66 \text{ rpm}$$

∴ Operating speed of rotation = N×η

$$= 30.66\times0.80 = 24.53 \text{ rpm}$$

Situation II :-

Radius of ball mill (R) = 1 m

Radius of ball (r) = 10 cm = 0.10 m

Speed of ball mill (N) = $\frac{1}{2\pi}\sqrt{\frac{g}{R-r}}$

$$= \frac{1}{2\pi}\sqrt{\frac{9.81}{1-0.10}} = 0.525 \text{ rev/s} = 31.50 \text{ rpm}$$

∴ Operating speed of rotation = N× η

$$= 31.50\times0.65 = 20.48 \text{ rpm}$$

Change in operating speed = 24.53 – 20.48 = 4.05 rpm

Q. 68

Applying Bond's Law

$$\frac{P}{f} = 0.3162W_i\left[\frac{1}{\sqrt{D_p}} - \frac{1}{\sqrt{D_f}}\right]$$

⇒ $7200 = 0.3162\times Wi\times2000\times\left[\frac{1}{\sqrt{0.5}} - \frac{1}{\sqrt{4.75}}\right]$

⇒ Wi = 11.98

Q. 69

Fraction of clean seed at clean seed outlet (E) = 1 – 0.006 = 0.994

Fraction of clean seed in feed (F) = 1 – 0.085 = 0.915

Fraction of clean seed at foreign matter outlets

(G) = 0.003 + 0.012 + 0.008 = 0.023

We know

$$\text{Cleaning efficiency(C)} = \frac{E(F-G)(E-F)(1-G)}{F(E-G)^2(1-F)}$$

Putting above values in the above formula, we get, C = 0.9332 or 93.32 %

Q. 70

Concentration of NaCl in in one side of membrane, C_1 = 0.03 g/(100 ml) =300 g/m^3

Concentration of NaCl in in one side of membrane, C_2 = 0 g/m^3

Mass transfer coefficient of NaCl at the solution side (K_1) = 1 x 10^{-6} m/s

Mass transfer coefficient of NaCl at the other side (K_2) = 3 x 10^{-7} m/s

Permeability of the membrane (P) = 9 x 10^{-6} m/s

We know

$$\text{Mass flux(M)} = \frac{C_1 - C_2}{\left(\frac{1}{K_1} + \frac{1}{K_2} + \frac{1}{P}\right)}$$

Putting above values in the above formula, we get, M = 67.5×10^{-6} g/(s.m^2)

∴ Rate of removal of the NaCl = Mass flux x Membrane surface area

= 67.5×10^{-6}×3 = 202.5 g/s or 0.73 g/h

32

Material Handling and Food Storage

UKPSC Combined Assistant Engineering Exam 2013 Paper - I

Q. 1 The high moisture grain should be kept in

(A) Air tight structure (B) Aerated structure

(C) Pucca kothi (D) Gunny bags in closed room

Q. 2 Most commonly used storage structure by Food Corporation of India in India is

(A) Pusa bins (B) CAP storage

(C) Hapur bins (D) None of the above

Q. 3 In the cold storage, the storage temperature is normally

(A) – 1 °C to 10 °C (B) 15 °C to 25 °C

(C) 30 °C to 40 °C (D) 45 °C to 55 °C

Q. 4 Generally the frozen food are stored at a temperature of

(A) 0 °C (B) –4 °C

(C) –8 °C (D) –18 °C

Q. 5 The moisture accumulation in a grain mass stored for a long time when outside environment is hot, taken place at

(A) Top surface (B) Bottom surface

(C) Middle surface (D) Near the walls

Q. 6 Which standard is applied in International trade of agri-food commodities

(A) B.I.S. (B) Codex-Alimenta

(C) Agmark (D) None of the above

Q. 7 Brine is solution of

(A) Sodium chloride in water

(B) Sugar + Magnesium chloride in water

(C) Calcium chloride in water

(D) None of the above

Q. 8 Necessary hopper slope angles to achieve reliable mass flow is

(A) 30°–40° (B) 40°–50°

(C) 60°–70° (D) None of the above

Q. 9 To prevent spoilage, silage should be removed at the rate of

(A) 5 cm/day (B) 10 cm/day

(C) 15 cm/day (D) 20 cm/day

Q. 10 Pressure drop in fluid flow through granular material is best estimated by

(A) Black - Kozney equation (B) Burke - Plummer equation

(C) Ergun equation (D) Fourrier

Q. 11 In cold storage, apples can be safely stored for the following period

(A) 5 months (B) 10 months

(C) 15 months (D) 20 months

APPSC AEES Agriculture Engineering Exam 2016

Q. 12 Centrifugal discharge is used in

(A) Belt conveyor (B) Chain conveyor

(C) Screw (auger) conveyor (D) Bucket elevator

Q. 13 In a godown extra space for alleyways for inspection and disinfection of stacks is provided which is generally about

(A) 30 % (B) 20 %

(C) 5 % (D) 1 %

MPPSC Assistant Agricultural Engineer 2013

Q. 14 Food spoilage occurs due to ....

(A) Bacteria
(B) Fungi
(C) Yeast
(D) All options are correct

UKPSC Combined Junior Engineering Exam 2013 Agricultural Engineering Paper-I

Q. 15 In bucket elevator, the material discharge chute is situated at

(A) Bottom
(B) Top
(C) Lower half but not at bottom
(D) Upper half but not at top

Q. 16 The shape of trough in screw conveyor is used as

(A) U-shaped
(B) L-shaped
(C) V-shaped
(D) J-shaped

RPSC AEn Pre Exam 2013 (Agricultural Engineering)

Q. 17 Which one of the following bucket elevator is used for grain handling ?

(A) Positive discharge
(B) Centrifugal discharge
(C) Continuous bucket type discharge
(D) All of the above

OCS AEn Exam 2011 (Pre) – II (Agricultural Engineering)

Q. 18 Food grain silos are used to store the grain in

(A) Bags
(B) Crates
(C) Bulk
(D) Both (A) and (B)

Q. 19 The process of moving air through stored grain at low flow rates to maintain or improve its quality is called

(A) Aeration
(B) Fumigation
(C) Ventilation
(D) Infiltration

Q. 20 In which regions of India, Morai type storage structures are used

(A) Eastern and Southern (B) Western and Northern

(C) Eastern and Western (D) Southern and Northern

Q. 21 Asbestos sheets of bag storage Godown are applied .... to prevent mould growth.

(A) Sodium Sulphate (B) Magnesium Sulphate

(C) Zinc Sulphate (D) Copper Sulphate

Q. 22 To remove field heat from vegetables immediately after harvesting should be

(A) Pre cooled (B) Refrigerated

(C) Stored at ambient temperature (D) Kept open

Q. 23 The chemical preservative recommended by FPO is/are

(A) KMS and KNO_3 (B) SO_2 and Sodium Benzoate

(C) Benzoic acid (D) None of the above

Q. 24 In dilute solution of Sodium Hydroxide is heated to 100 – 129 °C. Food is passed through this solution which softens the skin, and high pressure water sprays then remove the skin. This process is known as

(A) Flame peeling (B) Lye peeling

(C) Abrasion peeling (D) Flash peeling

Q. 25 The minimum concentration of soluble solids in Jam is

(A) 45 % (B) 55 %

(C) 65 % (D) 75%

CGPSC State Engineering Services Exam Paper – I

Q. 26 Canned food should be stored in

(A) Kitchen (B) Dry food store

(C) Refrigerator (D) Deep freezer

(E) Insulated room

Q. 27 ……… Conveyor moves granular materials in a closed duct by a high velocity air stream

(A) Chain (B) Pneumatic

(C) Screw (D) Belt

(E) Spiral

Q. 28 Capacity of bucket elevators may vary from ………

(A) 2 – 1000 t/h (B) 2 – 100 t/h

(C) 2 – 50 t/h (D) 2 – 30 t/h

(E) 2 – 10 t/h

Q. 29 Glass bottles are examples for

(A) Primary Package (B) Secondary package

(C) Tertiary package (D) Quaternary package

(E) Both (A) & (D)

Q. 30 For pack ageing beer and canned vegetables the preferred internal enamel coat is

(A) Oleoresins (B) 'S' resistance resins

(C) Epoxy phenolic (D) Both (B) & (C)

(E) Both (A) & (B)

Q. 31 OTR in plastic films stands for

(A) Ozone transfer rate (B) Oxygen transmission rate

(C) Oxygen transfer rate (D) Ozone transmission rate

(E) Oxygen testing rate

Q. 32 Which of the following packaging material has perfect barrier properties

(A) Paper and board (B) Metal

(C) Glass (D) Plastics

(E) Both (A) & (B)

Q. 33 To inhibit the growth of bacteria in the milk, the storage temperature in degree celcius should not exceed

(A) 1 (B) 2

(C) 3 (D) 4

(E) 5

Q. 34 The most common fumigant for storage of cereals is

(A) Zinc Phosphide (B) Ethylene dibromide

(C) Aluminium Phosphide (D) DDT

(E) Both (A) & (B)

Q. 35 Two piece cans can be manufactured by

(A) Drawn & ironed (B) Drawn & redrawn

(C) Ironed & reironed (D) Both (A) & (B)

(E Both (B) & (C)

Q. 36 The following is one of the paper based packing material that act as odour barrier and it is used for foods and beverages due to

(A) Grease proof paper (B) Tissue paper

(C) Wax paper (D) Glassine paper

(E) Vegetable parchment paper

Q. 37 Glass bottles can be manufactured by two processes known as ....

(A) Blow & blow (B) Press & blow

(C) Press & press (D) Both (A) & (B)

(E) Both (A) & (C)

UKPSC A.En. Exam 2007 Paper – I

Q. 38 Silo pits structures primarily are used for

(A) Sanitation (B) Animal fodder storage

(C) Grain storage (D) Manures and fertilizer storage

Q. 39 Which of the following agent is not the major enemy of grains during storage ?

(A) Rodents (B) Insects

(C) Sunlight (D) Microorganisms

Q. 40 Janssen's equation is related to

(A) Storage silo design (B) Size reduction of particles

(C) Grain transportation systems (D) Size separation of grains

Q. 41 At which place the grain remains warmer in the bin during winter ?

(A) Near the wall

(B) Centre of bin below the upper surface

(C) Centre of the bin near the bottom

(D) None of the above

Q. 42 The grain silos are used to store grains in

(A) Bulk (B) Bags

(C) Bulk and bags (D) None of the above

Q. 43 Bukhari type structures are generally used to store

(A) Fodder (B) Water

(C) Grain (D) Vegetables

Q. 44 With the increase in grain moisture content, the thermal conductivity of grain

(A) Increases (B) Decreases

(C) Does not changes (D) Decreases initially then increases

Q. 45 Morai type storage structures are made on a

(A) Raised platform (B) Underground platform

(C) Perforated platform (D) Cotton platform

Q. 46 The factor which is not important for storage of food grains, vegetables and fruits is

(A) Temperature (B) Oxygen

(C) Moisture (D) Carbon dioxide

Q. 47 Dynamic pressure exerted on the bin wall due to loading and unloading of grains is usually taken as

(A) Equal to static load

(B) 2 – 4 times that of static load

(C) 5 – 8 times that of static load

(D) 0.5 times that of static load

Q. 48 Optimum relative humidity for cold storage of grapes are recommended by "CFTRI" is

(A) 20 % (B) 40 %

(C) 60 % (D) 80 %

Q. 49 Rankine's formula is used to determine

(A) Temperature (B) Pressure

(C) Relative humidity (D) None of the above

Q. 50 Constant levels of oxygen and carbon dioxide are maintained in

(A) Modified atmosphere storage (B) Controlled atmosphere storage

(C) Hypobaric storage (D) Cold storage

Q. 51 If true and bulk density of a grain are 1.2 and 0.8 g/cc respectively, its porosity would be nearly

(A) 33 % (B) 50 %

(C) 67 % (D) 5 %

UKPSC A.En. Exam 2012 Paper – I

Q. 52 In which region of the country, Kothar type storage structures are used

(A) Eastern (B) Western

(C) Northern (D) Southern

Q. 53 Safe moisture content for storing grains is .... %

(A) 15 – 20 (B) 10 – 13

(C) 4 – 8 (D) 20 – 27

Q. 54 In a controlled atmosphere, storage of fruits, apart from temperature it is the level of the following which is controlled :

(A) Oxygen alone

(B) Nitrogen alone

(C) Carbon-di-oxide

(D) Both carbon-di-oxide and oxygen

Q. 55 Respiration of grains generates

(A) Heat, oxygen and moisture

(B) Heat, moisture and carbon-di-oxide

(C) Heat, moisture and nitrogen

(D) None of the above

Q. 56 The sequence of growth phases of micro-organism is

(A) Lag-Logarithmic-decay-stationary

(B) Logarithmic-lag-stationary-decay

(C) Logarithmic-lag-decay-stationary

(D) Lag-logarithmic-stationary-decay

Q. 57 The main factors responsible for the losses in quality and quantity of food grains are

(A) Insects (B) Rodents

(C) Dampness (D) All of the above

Q. 58 Bukhari type structures are generally used to store

(A) Fodder (B) Water

(C) Grain (D) Vegetables

Q. 59 Maximum moisture content (w.b.) for safe storage of wheat is .... %

(A) 20 (B) 18

(C) 13 (D) 8

Q. 60 In a bag storage structure the clear distance between the stacks is kept as ……… m.

(A) 1 (B) 2

(C) 3 (D) 4

Q. 61 Janssen's equation is used to design a

(A) Shallow bin (B) Deep bin

(C) Bukhari (D) Kothar

Q. 62 Which is not the structure used to store grain in rural areas ?

(A) Dahej (B) Morai

(C) Kothar (D) Kanaj

Graduate Aptitude Test in Engineering - 2007

Q. 63 A vegetable oil is flowing through a vertical wall as a film. The density and viscosity of the oil are 920 kg m^{-3} and 0.28 Pa s respectively. If the average velocity of the film is 0.05 m s^{-1}, the thickness of the film is

(A) 0.14 mm (B) 0.36 mm

(C) 1.76 mm (D) 2.16 mm

Q. 64 Angle of internal friction for rice grain is 27°, bulk density of rice at 14% moisture content is 833 kg m^{-3} and coefficient of friction between rice and concrete wall is 0.5. For a silo of 5 m diameter and 20 m height, the ratio between the lateral pressures at the bottom of the silo obtained by Rankine and Janssen formulae is

(A) 1.63 (B) 3.16

(C) 6.13 (D) 9.47

Q. 65 Air carrying particles of density of 700 kg m^{-3} and average diameter of 25 μm enters a cyclone of 0.7 m diameter at a tangential velocity of 30 m s^{-1} at 0.35 m. The density and viscosity of air are 1.1614 kg m^{-3} and 1.85×10^{-5} Pa s respectively. The terminal radial velocity of the particle is

(A) 0.17 m s^{-1} (B) 1.69 m s^{-1}

(C) 3.37 m s^{-1} (D) 16.52 m s^{-1}

Graduate Aptitude Test in Engineering - 2009

Q. 66 The theoretical volumetric flow rate of a horizontal screw conveyor is 1500 m^3 h^{-1}. The conveyor screw diameter is 1.2 m and the shaft diameter is 0.6 m. The rotational speed of the screw conveyor is 50 rpm. The pitch of the screw in mm is

(A) 150 (B) 340

(C) 590 (D) 950

Q. 67 The diameter of a grain storage bin is 4 m and the depth is 16 m. It is completely filled with wheat having bulk density of 800 kg m^{-3}. The angle of friction between wheat and wall is 24°C. The ratio of lateral and vertical pressure intensity is 0.4. The lateral pressure intensity of wheat is kPa on the bin wall at 2 m depth is

(A) 2.85 (B) 5.28

(C) 8.25 (D) 8.52

Graduate Aptitude Test in Engineering - 2011

Q. 68 For a mass of grain stored in a bin, the angle of internal friction of the grain is 30°. The ratio of normal pressure to the applied pressure within the bin is

(A) 0.25 (B) 0.33

(C) 0.50 (D) 1.00

Graduate Aptitude Test in Engineering - 2012

Q. 69 A tall silo having height to diameter ratio of 2 is holding 480 tons wheat of bulk density 960 kg m^{-3}. The angle of internal friction for wheat is 25° and for wheat and wall surface is 24°. Applying Airy formula, the maximum lateral pressure in kPa at the bottom of the bin section is

(A) 40.24 (B) 41.79

(C) 42.83 (D) 42.92

Graduate Aptitude Test in Engineering - 2014

Q. 70 With increasing grain height in a deep cylindrical grain bin, the pressure at its base will

(A) Decrease initially and then increase

(B) Increase initially and then decrease

(C) Decrease initially and then remain constant

(D) Increase initially and then remain constant

IFOS - 2011

Q. 71 A horizontal screw conveyer of length 2 m coveys wheat grain having bulk density of 680 kg / m³. The screw diameter, shaft diameter and pitch length of screw are 0.50 m; 0.30 m and 0.45 m respectively. If the screw is completely filled with the grain and rotates at 60 rpm, determine the capacity of the screw conveyer in kg/h.

Graduate Aptitude Test in Engineering – 2015

Q. 72 A horizontal screw conveyer of length 2.4 m coveys wheat grain having bulk density of 680 kg/ m³ and material factor 1.2. The screw diameter, shaft diameter and pitch length of screw are 0.50 m; 0.15 m and 0.4 m respectively. If the screw is completely filled with the grains and rotates at 60 rpm, the capacity of the screw conveyer in m³/h and the actual power required in hp (approximately) are .... and ...., respectively.

(A) 257, 2.8 (B) 258, 2.5

(C) 275, 1.9 (D) 396, 2.9

Graduate Aptitude Test in Engineering – 2016

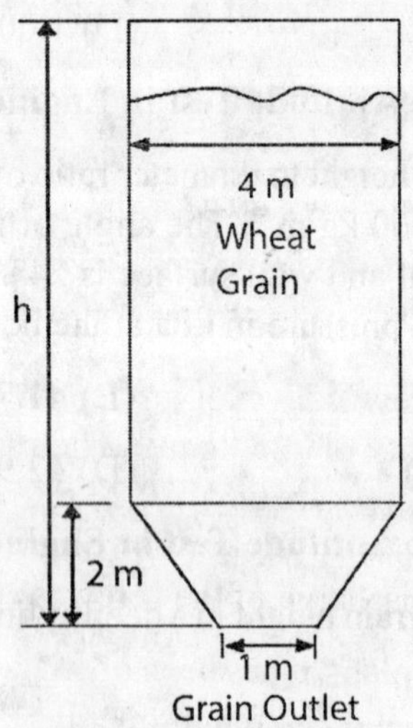

Q. 73 A circular grain silo with conical bottom, as shown in figure, is filled with wheat (true density 1200 kg m^{-3}) with porosity of 0.6. Five hundred metric tonne of wheat fills 80% of its capacity (by volume). The total height(h) of the silo from its grain outlet end in m is ....

Graduate Aptitude Test in Engineering – 2017

Q. 74 Sphericity of a cube with each side as L is ....

Q. 75 A cylindrical shallow bin is filled with grains having angle of repose of 33°. The limiting height to diameter ratio of the bin is.........

Q. 76 Length of the husking zone in a rubber roll paddy dehusker, having *d* as roll diameter, *c* as clearance between the rolls, and *b* as grain thickness, is

(A) $\frac{\pi d}{360} cos^{-1}\left(\frac{d+c}{d+b}\right)$ (B) $\frac{\pi d}{180} cos^{-1}\left(\frac{d+c}{d+b}\right)$

(C) $\frac{\pi d}{360} sin^{-1}\left(\frac{d+c}{d+b}\right)$ (D) $\frac{\pi d}{180} sin^{-1}\left(\frac{d+c}{d+b}\right)$

Q. 77 A cylindrical silo, 3 m in diameter and 20 m high, is filled with barley having bulk density of 625 kg m^{-3}. Coefficient of friction between grain and the bin wall is 0.45 and the ratio of lateral pressure to vertical pressure is 0.4. The lateral pressure at the base of the bin in kPa will be ....

Q. 78 India's annual paddy production is 160 million ton (clean paddy basis). Average husk content of paddy is 22.4% and milled rice yield is 70%. Considering 18% oil content in bran fraction and calorific value of 12 MJ kg^{-1} of husk, the oil potential of bran and energy potential of husk will respectively be

(A) 2.19×10^6 ton; 430080×10^{12} J (B) 2.19×10^6 ton; 210450×10^{12} J

(C) 4.76×10^6 ton; 210450×10^{12} J (D) 4.76×10^6 ton; 430080×10^{12} J

Graduate Aptitude Test in Engineering – 2018

Q. 79 A horizontal screw conveyor of length 3.2 m conveys solid grain having bulk density of 725 kg m^{-3}. The screw diameter, shaft diameter and pitch of the screw are 0.6 m, 0.24 m and 0.48 m, respectively. If the screw is operating with 80% of its volumetric capacity at a speed of 50 rpm, the actual discharge of the screw is .... ton h^{-1}.

Q. 80 In a feed milling plant, it has been observed that 80% of the feed passes through IS sieve 340 (3.25 mm opening) and 80% of the ground feed passes through IS sieve 40 (0.42 mm opening). The power requirement to crush the material with a feed rate of 3 ton h^{-1} is 6 kW. Power requirement to crush 2 ton h^{-1} of the same feed using the above system so that 80% of the ground feed pass through IS sieve 15 (opening 0.157 mm) is .... kW.

Q. 81 Solid food particles having nominal size of 0.2 mm with shape factor of 0.8 and density of 1040 kg m^{-3} are to be fluidized using air at 28°C. The density and pressure of air at the above-mentioned condition are 1.175 kg m^{-3} and 1.013×10^5 Pa, respectively. The voidage at minimum fluidizing condition is 0.48. Use the value of 'g' as 9.81 m s^{-2}. If cross section of the empty bed is 0.5 m^2 and contains 520 kg of solids, the pressure drop at minimum fluidization condition is .... kPa.

Q. 82 A single strength fruit juice is concentrated from 6% total solids (TS) to 24% by ultra filtration. The feed stream has a flow rate of 12 kg min^{-1}. The ultra filtration membrane tube has an inside diameter of 80 mm and the pressure difference applied across the membrane is 2 MPa. If the permeability constant is 4×10^{-5} kg water m^{-2} kPa^{-1} s^{-1}, the length of the membrane tube is .... m.

Graduate Aptitude Test in Engineering – 2019

Q. 83 Angle of internal friction of a certain grain (bulk density 650 kg/m^3) is 30°. A bin filled with this grain experiences a pressure of 60 kPa at its base. Ignoring the factor of safety, the safe height (in meter) to which water (density = 1000 kg/m^3) can be filled in this bin is ....

Graduate Aptitude Test in Engineering – 2020

Q. 84 A cylindrical silo with 3.0 m diameter and height to diameter ratio of 5: 1 is filled with 60 metric ton wheat grains having bulk density of 725 kg/m^3. The coefficient of friction between grain and silo wall is 0.42 and the ratio of lateral pressure to vertical pressure is 0.5. The vertical pressure at the bottom of silo in kPa is .... [use 1 kgf= 9.81 N]

Q. 85 A bucket elevator for lifting parboiled paddy (bulk density = 840 kg/m^3) is operated at a linear speed of 2 m/s• The width of the bucket is 25.4 cm and its cross section is making a subtending angle of 75° at the centre of a circle having 12.7 cm radius. The space between two adjacent buckets on the elevator belt is 40 cm. If the buckets are filled to 80% of their volumetric capacity, the lifting capacity of elevator in kg/min is ....

Graduate Aptitude Test in Engineering – 2021

Q. 86 A cylindrical storage bin with an internal diameter of 4 m and a height of 16 m is completely filled with paddy having bulk density of 640 kg/

m³. The angle of internal friction between grain and bin wall is 30° and the ratio of horizontal to vertical pressures is 0.4. When the grain filling rises from 4 m to 16 m in height, the lateral pressure increases by a multiple of..........

Answers Key

1	2	3	4	5	6	7	8	9	10
B	B	A	D	B	B	A	C	B	C
11	12	13	14	15	16	17	18	19	20
A	D	A	D	B	A	B	C	A	A
21	22	23	24	25	26	27	28	29	30
D	A	B	B	C	B	B	A	A	D
31	32	33	34	35	36	37	38	39	40
B	C	D	C	D	C	D	B	C	A
41	42	43	44	45	46	47	48	49	50
B	A	C	A	A	D	B	D	B	B
51	52	53	54	55	56	57	58	59	60
A	C	B	D	B	D	D	C	C	B
61	62	63	64	65	66	67	68	69	70
B	A	D	B	C	C	B	B	B	D
71	72	73	74	75	76	77	78	79	80
138361	A	105.02	0.806	1.84	B	10.13	A	198.26	7.97
81	82	83	84	85	86				
10.17	7.5	2.04	24.44	540	1.62				

Explanations

Q. 51

Porosity = 1 – Bulk density/True Density

= 1 – 0.8/1.2 – 0.33 or 33%

Q. 63

Given

Density of vegetative oil (ρ) = 920 kg/m³

Dynamic viscosity (μ) = 0.28 Pa s

Average velocity of falling film (v_{av}) = 0.05 m/s

We know that

$$v_{av} = \frac{g\delta^2}{3\nu} = \frac{\rho g\delta^2}{3\mu}$$

$$\delta = \sqrt{\frac{3\mu v_{av}}{\rho g}}$$

$$\delta = \sqrt{\frac{3\times0.28\times0.05}{920\times9.81}} = 0.00216 \text{ m or } 2.16 \text{ mm}$$

Q. 64

Bulk density of rice grain (ρ) = 833 kg/m³

Height of silo (h) = 20 m

Angle of internal friction (∅) = 27°

Coefficient of active grain pressure (K_a) = $\frac{1-sin\varnothing}{1+sin\varnothing} = \frac{1-sin27}{1+sin27} = 0.376$

Applying Rankine formula

∴ Intensity of lateral pressure (P_{lR})

$$= K_a \rho gh$$

$$= 0.376\times833\times20\times9.81 = 61451.41 \text{ N/ m}^2$$

According to Janseen Formula

$$\text{Lateral Pressure } (P_{lj}) = \frac{\rho gR}{\mu}(1-e^{-\mu hK_a/R})$$

Here

Hydraulic radius (R) = D/4 = 5/4 = 1.25

Thus,

$$\text{Lateral Pressure } (P_{lj}) = \frac{833\times9.81\times1.25}{0.5}(1-e^{-0.5\times0.376\times20/1.25})$$

$$= 19420.31 \text{ N/ m}^2$$

Therefore,

$$\text{Required ratio} = \frac{P_{lR}}{P_{lJ}} = \frac{61451.41}{19420.31} = 3.16$$

Q. 65

Density of particles (ρ_p) = 700 kg/m³

Particle diameter (D_p) = 25 μm = 25×10⁻⁶ m

Cyclone diameter (r) = 0.35 m

Tangential velocity (v) = 30 m/s

Density of air (ρ_p) = 1.1614 kg/m^3

Viscosity of air (μ) = 1.85×10^{-5} Pa s

$$\text{Terminal velocity } (V_t) = \frac{D_p^2 . a . (\rho_p - \rho_a)}{18.\mu} = \frac{D_p^2 . \left(\frac{v}{r}\right)^2 . r . (\rho_p - \rho_a)}{18.\mu}$$

$$(\because a = \omega^2 r \Rightarrow a = (\frac{V}{r})^2 r)$$

$$V_t = \frac{(25 \times 10^{-6})^2 . \left(\frac{30}{0.35}\right)^2 \times 0.35 \times (700 - 1.1614)}{18 \times 1.85 \times 10^{-5}} = 3.37 \text{ m/s}$$

Q. 66

Given that

Volumetric flow rate of horizontal screw conveyor = 1500 m^3/h = 25 m^3/min

Consider pitch of screw conveyor = P

N = 50 rev/min

Diameter of screw conveyor (D) = 1.2 m

Diameter of shaft (d) = 0.6 m

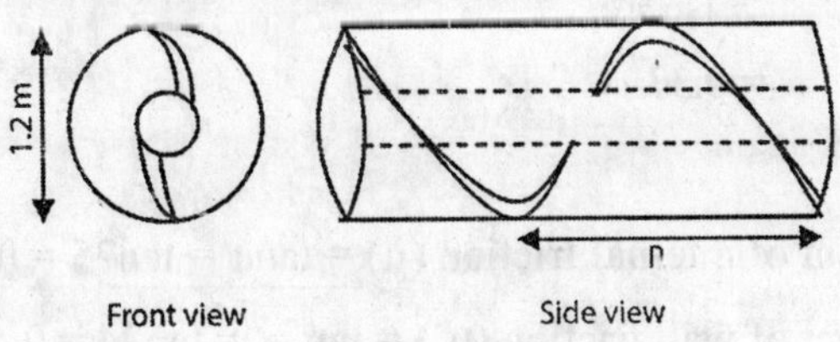

We know that

$$Q = AV$$

$$\therefore \quad Q = 2\int_0^{2\pi} \sqrt{(a - a\cos\theta)^2 + (a\sin\theta)^2}\, d\theta \quad PN$$

$$\Rightarrow \quad 25 = 3.14 \times 2\sqrt{2}a \int_0^{2\pi} \sqrt{(1 - \cos\theta)}\, d\theta \times P \times 50$$

$$\Rightarrow \quad P = 0.590 \text{ m} = 590 \text{ mm}$$

Q. 67

Given that

Diameter of grain bin (D) = 4 m, Height of grain bin (H) = 16m,

Density of grain (ρ)= 800 kg/m³

Ratio of lateral and vertical pressure intensity (K) = P_l/P_v = 0.4,

μ = tan24 = 0.445

Hydraulic radius of grain bin(R) = D/4 = 4/4 = 1 m

Depth of grain bin (h) = 2 m

Here h>D ⇒ Deep bin

∴ Lateral Pressure $(P_l) = \frac{\rho g R}{\mu}(1-e^{-\mu hK/R})$

$$\Rightarrow \quad P_l = \frac{800\times 9.81\times 1}{0.445}(1-e^{-0.445\times 0.4\times 2/1})$$

= 5282.45 N/ m² = 5.28 kPa

Q. 68

We know

$$\frac{P_l}{P_v} = \frac{1-sin\varnothing}{1+sin\varnothing}$$

$$= \frac{1-sin30}{1+sin30}$$

= 0.33

Q. 69

Coefficient of internal friction (μ) = tanφ = tan25 = 0.466

Coefficient of wall friction (μ') = tan φ' = tan24 = 0.445

We know

$$\text{Volume} = \frac{Mass}{Density}$$

$$\Rightarrow \quad \frac{\pi}{4}D^2L = \frac{480\times 1000}{960}$$

$$\Rightarrow \quad \frac{\pi}{4}(L/2)^2L = \frac{480\times 1000}{960}$$

⇒ L = 13.66 m

Applying Airy's Formula

$$P_{max} = gL\left[\frac{}{\sqrt{\mu(\mu+\mu')}+\sqrt{1+\mu^2}}\right]$$

$$= 960\times9.81\times13.66\times\left[\frac{1}{\sqrt{0.466(0.466+0.445)}+\sqrt{1+0.466^2}}\right]^2$$

$$= 41.76 \text{ N}$$

Q. 71

Consider pitch of screw conveyor = 0.45 m

N = 60 rev/min = 3600 rev./ h

Diameter of screw conveyor (D) = 0.5 m

Diameter of shaft (d) = 0.3 m

We know

$$Q = \pi\frac{D^2-d^2}{4}\ PN$$

$$\Rightarrow \quad Q = 3.14\times\left(\frac{(0.5)^2-(0.3)^2}{4}\right)\times0.45\times3600 = 203.472 \text{ m}^3/\text{ h}$$

Therefore,

Capacity of screw conveyer = 203.472×680 = 138361 kg/ h

Q. 72

Given that

Pitch of screw conveyor = 0.4 m

N = 60 rev/min = 3600 rev./h

Diameter of screw conveyor (D) = 0.5 m

Diameter of shaft (d) = 0.15 m

We know that

$$Q = AV$$

$$\therefore \quad Q = \pi\frac{D^2-d^2}{4}\ PN$$

$$\Rightarrow \quad Q = 3.14\times\left(\frac{(0.5)^2-(0.15)^2}{4}\right)\times0.4\times3600$$

$$\Rightarrow \quad Q = 257.17 \text{ m}^3/\text{h}$$

Again,

Power (P) = ρgLQF

$$P = 680\times9.81\times(257.17/3600)\times2.4\times1.2 = 1372.42\ W$$

Hence, power required in hp = 1372.42/746 = 1.84 hp ($\because$ I hp = 746 W)

But, In actual when P < 1 ⇒ multiply P by 1.5.

Actual power required = 1.5×1.84 = 2.76 = 2.8

Q. 73

We know

$$\text{Porosity (n)} = 1 - \frac{\textit{Bulk density}}{\textit{True Density}}$$

Bulk Density = (n – 1)×True density

$$= (1 - 0.6)\times1200 = 480\ kg/m^{-3}$$

Volume of wheat filled = 500000/480

$$= 1041.67\ m^3$$

Volume of cylinder = $\Pi R^2(h - 2)$

$$= 3.14\times(2)^2\times(h - 2)$$

$$= 12.56\times(h - 2)\ m^3$$

Volume of conical section = $\pi R^2X/3 - \pi R^2x/3$

$$= 3.14\times(2^2)\times2/3 - 3.14\times(0.5^2)\times1/3$$

$$= 8.108\ m^3$$

∴ Volume(Capacity) of the bin = Vol. of cylinder + Vol. of conical section

$$= 12.56\times(h - 2) + 8.108$$

∵ Wheat filled its 80% (by volume) capacity.

∴ $(12.56\times(h - 2) + 8.108)\times0.80 = 1041.67$

$$h = 105.02\ m$$

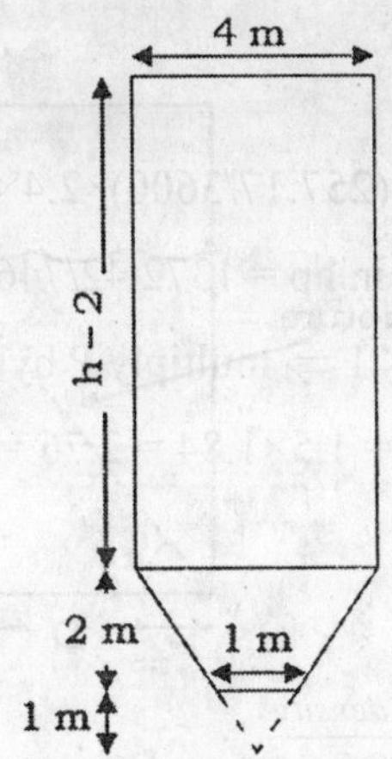

Q. 74

Sphericity of an object = the ratio of the surface area of a sphere (with the same volume as the given particle) to the surface area of particle.

The volume of spherical particle (Vp) = $\pi d^3/6$

$\Rightarrow \quad d = (6V_p/\pi)^{1/3}$

Surface area of a sphere (A_s) = πd^2

or $\quad A_s = \pi\times(6V_p/\pi)^{2/3}$

Thus, $\quad$ Sphericity = $A_s/A_p = \pi\times(6V_p/\pi)^{2/3}/\ A_p$

Where, A_p = Surface area of particle, V_p = Volume of particle, d = Diameter of particle

For cubical particle

Consider side of cube = L

Hence, Volume of cube(V_p) = L^3

And, Surface area (A_p) = $6L^2$

Therefore, Sphericity = $\pi\times(6\times L^3/\pi)^{2/3}/\ 6L^2 = (\pi/6)^{1/3}$ or 0.806

Note : - Sphericity of cylindrical object = 0.874

Q. 75

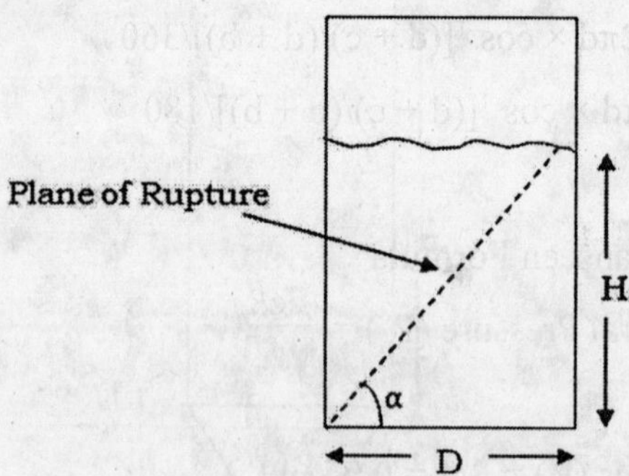

We know

Angle of plane of rupture (α) = $45 + \theta/2$

Where, θ = angle of repose = 33^0

Referring above figure

Limiting Height to Diameter ratio (H/D) = tan α

$\Rightarrow$ $H/D = \tan(45 + 33/2) = 1.84$

Q. 76

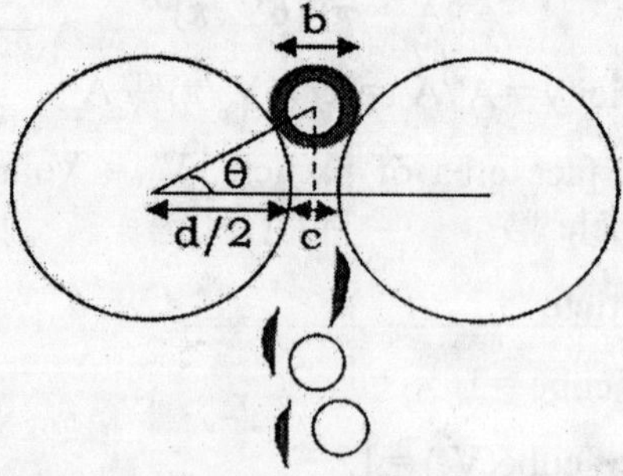

Referring above figure

$$\cos\theta = (d/2 + c/2)/(d/2 + b/2)$$

$$\theta = \cos^{-1}[(d + c)/(d + b)]$$

∵ Length of husking zone is the length of arc(L) on roller, from the point at which the grain is caught by the rollers to the point where the grain is no longer in contact with the rollers.

Hence,

$$L = 2\pi d \times \cos^{-1}[(d + c)/(d + b)]/360$$

$$= \pi d \times \cos^{-1}[(d + c)/(d + b)]/180$$

Q. 77

According to Janseen Formula

$$\text{Lateral Pressure } (P_{lj}) = \frac{\rho g R}{\mu}(1 - e^{-\mu h K_a / R})$$

Here

Hydraulic radius (R) = 3/4 = 0.75 m

$g = 9.81\ m/s^2$; K = lateral to vertical pressure ratio(Janseen's constant) = 0.4

h = depth of grain = 20 m

μ = coefficient of wall friction = 0.45

ρ = bulk density = 625 kg/m³

Thus,

$$\text{Lateral Pressure } (P_{lj}) = \frac{625 \times 9.81 \times 0.75}{0.45}(1 - e^{-0.45 \times 0.4 \times 20/0.75})$$

$$= 10134.65 \text{ Pa or } 10.13 \text{ kPa}$$

Q. 78

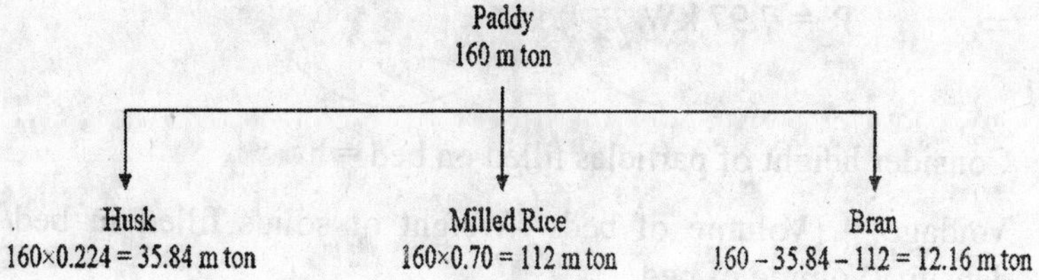

∵ Oil content in bran = 12.16×0.18 = 2.19 m ton or 2.19×10^6 kg

Calorific value of husk = 12 MJ/kg

Hence, Energy potential of husk = $12\times10^6\times35.84\times10^9 = 430080\times10^{12}$ J

Q. 79

Consider pitch of screw conveyor = 0.48 m

N = 50 rev/min = 3000 rev./ h

Diameter of screw conveyor (D) = 0.6 m

Diameter of shaft (d) = 0.24 m

We know

$$Q = [\pi(D^2 - d^2)/4]\ PN$$

$$\Rightarrow \quad Q = [3.14\times((0.6)^2 - (0.24)^2)/4]\times 0.48\times 3000 = 341.83\ m^3/h$$

∵ Volumetric efficiency = 80%

Actual discharge = 341.83×0.80 = 273.46 m^3/ h

Therefore,

Capacity of screw conveyer = 273.46 ×725

= 198258.5 kg/ h or 198.26 ton/h

Q. 80

$$\frac{\dot{P_1}}{P_2} = \frac{0.3162 W_1 f_1 \left[\frac{1}{\sqrt{D_{pl}}} - \frac{1}{\sqrt{D_{fl}}}\right]}{0.3162 W_1 f_2 \left[\frac{1}{\sqrt{D_{p2}}} - \frac{1}{\sqrt{D_{f2}}}\right]}$$

$$\Rightarrow \quad \frac{6}{P_2} = \frac{0.3162\times W_l \times 3\left[\frac{1}{\sqrt{0.42}} - \frac{1}{\sqrt{3.25}}\right]}{0.3162\times W_l\times 2\left[\frac{1}{\sqrt{0.157}} - \frac{1}{\sqrt{3.25}}\right]}$$

$$\Rightarrow \quad P_2 = 7.97\ kW$$

Q. 81

Consider height of particles filled on bed = h

Voidage = (Volume of bed - Weight of solids filled in bed/Particle density)/Volume of bed

$$\Rightarrow \quad 0.48 = 0.5\times h - 520/1040\times 0.5\times h$$

$$\Rightarrow \quad h = 1.92\ m$$

Hence, Pressure drop $= (1-\varepsilon)\times(\rho_d - \rho_a)\times g\times h$

$= (1 - 0.48)\times(1040 - 1.175)\times 9.81\times 1.92$

$= 10174.57$ Pa or 10.17 kPa

Q. 82

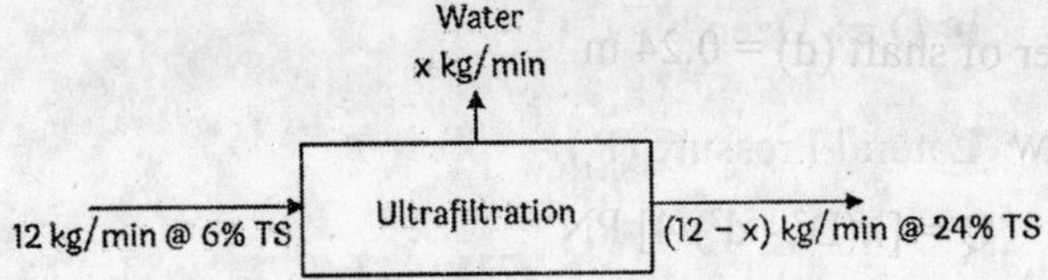

Applying mass balance $12\times0.06 = (12 - x)\times0.24$

$\Rightarrow \quad x = 9$ kg/min or 0.15 kg/s

Consider, length of membrane = L m

Surface area of membrane tube $= \pi dL = 3.14\times0.08\times L = 0.25\,L$

We know Water removed $= k\,A\,\Delta P$

$= 4\times10^{-5}\times0.25\,L\times2000$

$\Rightarrow \quad L = 7.5$ m

Q. 83

Rankine's Coefficient $(k) = (1-\sin\theta)/(1+\sin\theta)$

$= (1-\sin30)/(1+\sin30) = 0.333$

Again $\quad k = P_{lateral}/P_v = \rho gH/P_v$

$\Rightarrow \quad H = 0.333\times60\times10^3/(1000\times9.81) = 2.04$ m

Q. 84

Given that

Diameter of grain bin (D) = 3 m, Height of grain bin (H) = 15 m,

Density of grain (ρ) = 725 kg/m^3

Ratio of lateral and vertical pressure intensity $(K) = P_l/P_v = 0.5$, $\mu = 0.42$

Hydraulic radius of grain bin(R) = D/4 = 3/4 = 0.75 m

We know

$$\text{Volume} = \frac{Mass}{Density}$$

$$\Rightarrow \quad \frac{\pi}{4}D^2h = \frac{60\times1000}{725}$$

$$\Rightarrow \quad \frac{\pi}{4}(3)^2h = 82.76$$

$$\Rightarrow \quad h = 11.70 \text{ m}$$

Depth of grain bin (h) = 11.70 m

Here h>D ⇒ Deep bin

$\therefore$ Lateral Pressure $(P_l) = \frac{\rho g R}{\mu}(1-e^{-\mu hK/R})$

$$= \frac{725\times 9.81\times 0.75}{0.42}(1-e^{-0.42\times 0.5\times 11.70/0.75})$$

$$= 12220.65 \text{ N/ m}^2 = 12.22 \text{ kPa}$$

Therefore,

Vertical pressure(P_v) = 12.22/0.5 = 24.44 kPa

Q. 85

Volume of one bucket = $3.14\times 12.7^2\times 25.4\times 75/360$

$= 2680 \text{ cm}^3$

Bucket fill volume = $2680\times 0.80 = 2144 \text{ cm}^3$

Bucket fill weight = $840\times 2144\times 10^{-6} = 1.80$ kg

Here,

Time taken by one bucket to fill = $0.40/(2\times 60)$

$= 3.33\times 10^{-3}$ min

Hence, Capacity of elevator = $1.8/3.33\times 10^{-3} = 540$ kg/min

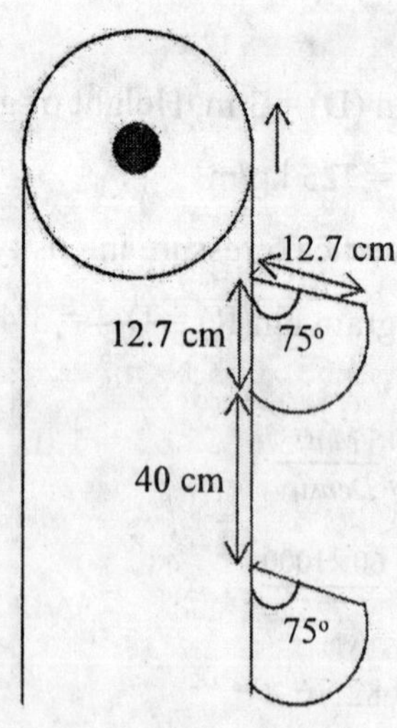

Q. 86

Given that

Diameter of grain bin (D) = 4 m, Height of grain bin (H) = 16 m,

Density of grain (ρ) = 640 kg/m^3

Ratio of lateral and vertical pressure intensity (K) = P_l/P_v = 0.4,

μ = tan30 = 0.577

Hydraulic radius 0f grain bin(R) = D/4 = 4/4 = 1 m

Case I:-

Depth of grain bin (h) = 16 m

Here h>D $\Rightarrow$ Deep bin

According to Janseen Formula

$\therefore$ Lateral Pressure (P_l) = $\frac{\rho g R}{\mu}$ $(1 - e^{-\mu hK/R})$

$\Rightarrow$ $P_l = \frac{640 \times 9.81 \times 1}{0.577}$ $(1 - e^{-0.577 \times 0.4 \times 16/1})$

= 10610.15 N/ m^2

Case II:-

Depth of grain bin (h) = 4 m

Here h=D $\Rightarrow$ Deep bin

Again, According to Janseen Formula

$\therefore$ Lateral Pressure (P_l) = $\frac{\rho g R}{\mu}(1 - e^{-\mu hK/R})$

$\Rightarrow$ $P_l = \frac{640 \times 9.81 \times 1}{0.577}(1 - e^{-0.577 \times 0.4 \times 4/1})$

= 6558.63 N/ m^2

$\therefore$ Required ratio = 10610.15/6558.63 = 1.62

33

Refrigeration

UKPSC Combined Assistant Engineering Exam 2013 Paper - I

Q. 1 Unit of refrigeration is:-

(A) Quintal (B) Gram

(C) kg (D) Ton

APPSC AEES Agriculture Engineering Exam 2016

Q. 2 The part of a refrigeration unit in which the refrigerant changes from vapour to liquid is called:

(A) Evaporator (B) Compressor

(C) Condenser (D) Throttle valve

Q. 3 Freezing temperature of brine is :

(A) Lower than pure water (B) Higher than pure water

(C) Same (D) None of the Above

Q. 4 A refrigerant vapour is to be condensed inside a shell and tube heat exchanger using water as the cooling medium. The water should flow through

(A) Shell side of the heat exchanger

(B) Tube side of the heat exchanger

(C) Any of (A) and (B) above

(D) None of the above

Q. 5 One ton of refrigeration is equivalent to the heat required to melt one tonne of ice in

(A) 6h (B) 12 h

(C) 18 h (D) 24 h

Punjab Mechanic Agricultural Machinery Instructor 2014

Q. 6 Refrigeration is a process in which the decomposition of food is

(A) Retarded (B) Stopped

(C) Speeded up (D) Delayed

UKPSC A.En. Exam 2007 Paper – I

Q. 7 The performance of a heat engine is described by it's

(A) Heating temperature (B) Thermal efficiency

(C) Vapour efficiency (D) None of the above

Q. 8 Heat insulating materials have

(A) High thermal conductivity (B) Low thermal conductivity

(C) Medium thermal conductivity (D) None of the above

Q. 9 Which of the following is not a type of condenser ?

(A) Air cooled (B) Water cooled

(C) Evaporative (D) Freezer

Q. 10 Which of the following is not the type of compressor

(A) Reciprocating (B) Centrifugal

(C) Gravitational (D) Rotary

Q. 11 Reciprocating compressor are

(A) Positive displacement type (B) Negative displacement type

(C) Circular displacement type (D) None of the above

Q. 12 Which of the following in not a part of refrigeration system

(A) Compressor (B) Evaporator

(C) Cooler (D) Condenser

UKPSC A.En. Exam 2012 Paper – I

Q. 13 One ton refrigeration is equivalent to .... K cal/min

(A) 50 (B) 500

(C) 1000 (D) 2000

Graduate Aptitude Test in Engineering - 2007

Q. 14 A refrigerator with a COP of 3.2 uses 2.4 kg min^{-1} refrigerant extracting 150 kJ kg^{-1} heat in the evaporator. Assuming compressor efficiency of 85% the minimum size of the motor is

(A) 0.5 hp (B) 1.5 hp

(C) 2.0 hp (D) 3.0 hp

Q. 15 In a cold store of 30 m×150 m×15 m size, 4000 tonnes of potato having the specific heat of 3.62 kJ kg^{-1} K^{-1} and heat of respiration of 20 W m^{-3} is kept at 30°C. Potato is required to be cooled to 2°C in 30 days. Neglecting other sources of heat, the capacity of the refrigeration plant required is

(A) 6 TR (B) 38 TR

(C) 44 TR (D) 83 TR

Graduate Aptitude Test in Engineering - 2008

Q. 16 The higher and lower temperatures in a refrigerator working on reverse Carnot cycle are 35°C and −15°C respectively. The capacity of the machine is 35.16 kW. The power required will be

(A) 81.6 kW (B) 68.1 kW

(C) 8.61 kW (D) 6.81 kW

Graduate Aptitude Test in Engineering - 2009

Q. 17 Milk weighing 98000 N having specific heat capacity 3.8 kJ kg^{-1} K^{-1} is to be chilled from 40°C to 5°C in one hour in a chilling plant using a refrigerant whose coefficient of performance is 4.7. The total compressor power consumption assuming 100% efficiency is

(A) 79 kW (B) 105 kW

(C) 79 hp (D) 105 hp

Graduate Aptitude Test in Engineering - 2010

Q. 18 Let m, n and p be the numbers of carbon, hydrogen and fluorine atoms in a refrigerant. The identification number of the refrigerant is

(A) R (m + 1) (n − 1) p (B) R (m − 1) (n + 1)p

(C) R(m − 1)(n − 1) p (D) R(m+1)(n + 1) p

Q. 19 A household refrigerator of 1 TR capacity operates half the time during 13-hour long day and 30% time during the nights. If coefficient of performance is 4.7 then at Rs. 3 per kWh, monthly (30 days) electricity bill in Rupees for the refrigerator is

(A) 110 (B) 220

(C) 440 (D) 660

Graduate Aptitude Test in Engineering - 2012

Q. 20 One thousand units of mixed fruit bar, each weighing 100 g with a surface area of 0.01 m², are frozen from 70°C molten mass condition to −20 °C frozen storage condition within 3 hours. The specific heat capacity values of the bar are 3.6 kJ kg^{-1} K^{-1} and 1.97 kJ kg^{-1} K^{-1} before and after freezing point (0°C) respectively. If the latent heat of crystallization is 250 kJ kg^{-1}, the cooling capacity of the refrigeration unit required in tons of refrigeration is

(A) 0.77 (B) 1.43

(C) 1.66 (D) 4.32

Q. 21 A 1 hp motor is used for running a dual cylinder reciprocating compressor of a refrigeration system based on R-134a refrigerant having 185 kJ kg^{-1} cooling capacity. COP of the system is 4.2 and overall efficiency of the compressor is 80%. Specific volume of the refrigerant vapour at suction temperature is 0.15 m^3 kg^{-1}. The compressor with bore diameters of 40 mm each runs at 1440 rpm.

(i) The mass flow rate of the refrigerant in kg min^{-1} is

(A) 1.634 (B) 1.090

(C) 0.813 (D) 0.240

(ii) The compressor stroke length in mm is

(A) 16.8 (B) 33.7

(C) 50.5 (D) 67.4

Graduate Aptitude Test in Engineering - 2013

Q. 22 The horse power of the motor running the compressor of a refrigerator having COP of 4.5 and extracting 200 kJ kg^{-1} of evaporating heat with 1.5 kg min^{-1} refrigerant flow rate is

(A) 0.5 (B) 1.0

(C) 1.5 (D) 2.0

Graduate Aptitude Test in Engineering – 2015

Q. 23 In a cascade refrigeration system, the COPs of the coling cycle and cascade are 3.7 and 4.2 respectively. If tonnage (1 TR = 3.52 kW) of the cooling cycle is 15 and the cascade removes 70% of the total heat rejected in the liquid receiver of the cooling cycle, then the powers required by the cooling and cascade compressors in hp are ………. and………, respectively.

(A) 67.07, 46.95 (B) 52.08, 39.35

(C) 19.13, 14.99 (D) 14.27, 11.18

Q. 24 In a plate freezer, the plate temperature is maintained at – 25°C. Latent heat of crystallization is 335 kJ/kg, and the thermal conductivity and density of frozen meat are 1.25 W m^{-1} K^{-1} and 1060 kg/m^3, respectively. Assuming freezing point of deboned meat at 85% water content on wet basis to be 0°C, the freezing time in minutes for 2 cm thick block of meat, kept between a pair of freezing plates, is ……….

Graduate Aptitude Test in Engineering – 2017

Q. 25 In a cold storage, 10 metric ton of potato is to e brought down from 30 to 8°C storage temperature in 6 hours of air blast at the evaporator temperature of – 10°C. Specific heat capacity of potato is 3.2 kJ kg^{-1} K^{-1}. COP of the refrigeration cycle deployed is 4.2 with evaporator load extraction capacity of 210 kJ kg^{-1}. Neglecting the respiration load of potato, the refrigerant flow rate and compressor power requirement will be

(A) 3.9 kg min^{-1}; 7.76 kW (B) 9.3 kg min^{-1}; 8.50 kW

(C) 9.3 kg min^{-1}; 7.76 kW (D) 3.9 kg min^{-1}; 8.50 kW

Graduate Aptitude Test in Engineering – 2018

Q. 26 One ton of marine fishery products are to be brought down from 32°C temperature to -18°C in half an hour time using a plate freezer. The freezing point of 85% water (based on total mass of the product) present in the fish is -0.3°C. Specific heat capacities of fresh and frozen fish solids (15% of total mass) are 3.2 and 1.8 kJ kg^{-1} K^{-1}, respectively. Specific heat capacity of fresh water is 4.2 kJ kg^{-1} K^{-1} and that of ice is 2.2 kJ kg^{-1} K^{-1}. Latent heat of crystallization of water is 335 kJ kg^{-1}. For freezing the products, the compressor power consumption with a vapor compression refrigeration cycle (coefficient of performance, 3.66) is ………. kW.

Graduate Aptitude Test in Engineering – 2019

Q. 27 A chiller working on mechanical vapour compression refrigeration system (COP = 4.5) is used for cooling 12500 kg of fresh cow milk (c_p = 3.8 kJ/(kg K)) from 30 °C to 4 °C in 3 hours. Assuming ideal compression process, the power consumed by the electric motor in kW and the tonnage of refrigeration (TR), respectively are

(A) 25.4 and 32.5 (B) 25.4 and 114.3

(C) 32.5 and 25.4 (D) 114.3 and 25.4

Graduate Aptitude Test in Engineering – 2020

Q. 28 A cold storage takes 5 hours to bring down the temperature of 100 metric tons of potato from 35 °C to 8 °C. The specific heat capacity of potato is 3.1 kJ/(kg °C). The coefficient of performance (COP) and the latent heat of vapourisation of the refrigerant (R-22)a t an evaporation temperature of - 10 °C are 3.66 and 230 kJ/kg, respectively. Neglecting respiration heat load of potato, and assuming no power loss, the values of refrigerant flow rate and the power input to the compressor are

(A) 121.3 kg/min and 127.1 kW (B) 124.7 kg/min and 121.3 kW

(C) 127.1 kg/min and 121.3 kW (D) 124.7 kg/min and 127.1 kW

Answers Key

1	2	3	4	5	6	7	8	9	10
D	C	A	B	D	D	B	B	C	C
11	12	13	14	15	16	17	18	19	20
A	C	A	D	D	D	D	B	D	B
21	22	23	24	25	26	27	28		
C, B	C	19.13, 14.99	4.025	C	68.02	A	A		

Explanations

Q. 13

We know that to produce 1 TR refrigeration effect 3.5 kW heat is required.

Therefore,

1 TR = 3.5 kW or 3.5 kJ/sec or 210 kJ/min

Again, we know 4.18 kJ = 1 k cal

Hence, 1 TR = 210/4.18 = 50.24 k cal/min

Q. 14

COP of refrigerator = 3.2

Mass flow rate of refrigerant (m) = 2.4 kg/min = 0.04 kg/s

Heat extracted by refrigerant (Q) = 150 kJ/kg = 150×0.04 = 6 kJ/s = 6 kW

Compressor efficiency (η) = 85%

Work done by the compressor (W) = Heat extracted by refrigerant/ COP

$$W = \frac{6}{3.2} = 1.875 \text{ kW}$$

Size of the motor to run the compressor = 1.875/0.85 = 2.21 kW

= 2.21/0.746 = 2.96 hp ≈ 3 hp

Q. 15

Size of cold storage (S) = 30×15×15 = 6750 m^3

Mass of potato (m) = 4000 tonnes = 4000×10^3 kg

Specific heat capacity (C_p) = 3.62 kJ/kg K

Heat of respiration of potato (Q_r) = 20 W/m^3 = 20×6750 = 135 kW

Heat released by the potato when cooled

$$Q = mC_p\Delta T$$

$$Q = 4000\times10^3\times3.62\times(30-2) = 405440\times10^3 \text{ kJ}$$

This much amount of heat is released in 30 days, thus heat released per second

$$Q = \frac{405440\times10^3\times10^3}{30\times24\times3600} = 156.42 \text{ kJ/s} = 156.42 \text{ kW}$$

Capacity of refrigerating plant = Q_r + Q

= 135 + 156.42 = 291.42 kW

As we know that, for producing 1TR (Tonne of refrigeration) effect, 3.5 kW heat is removed from the substance to be cooled.

∴ Capacity of refrigerating plant = 291.42/3.5 = 83.26 TR

Q. 16

COP of reversed Carnot cycle = $\frac{T_2}{T_1 - T_2}$

$$= \frac{273-15}{35+15} = 5.16$$

The capacity of machine (Q) = 35.16 kW

Hence,

Work done on system or power required (W) = $\frac{Q}{COP}$

$$= \frac{35.16}{5.16} = 6.81 \text{ kW}$$

Q. 17

Given that

Mass of milk (m) = 98000/9.8 = 10000 kg

Specific heat capacity (C_p) = 3.8 kJ/kg K

Initial temperature (T_1) = 40°C and final temperature (T_2)= 5°C

Coefficient of performance (C.O.P.) = 4.7

Heat absorbed (Q) = $mC_p\Delta T$

= 10000×3.8×35= 1330000 kJ/ h or 369.44 kW

Thus, total power consumed by compressor (W)

$$= \frac{Q}{C.O.P.}$$

$$= \frac{369.44}{4.7} = 78.605 \text{ kW or } 105.37 \text{ hp}$$

($\because$ 1 hp = 0.746 kW)

Q. 18

For $C_mH_nF_pCl_q$ compounds; in which

$$n + p + q = 2m + 2$$

The designation is $R_{(m-1)(n+1)(p)}$.

Q. 19

We know that to produce 1 TR refrigeration effect 3.5 kW heat is required, coefficient of performance is 4.7 thus work done requirement = 3.5/4.7 =0.745 kW

Thus, one day, energy consumed = 0.745×(13/2 + 11×0.30) = 7.301 kWh

For one month, energy consumed = 7.30×30 = 219.03 kWh

Thus total electricity bill = 3×219.03 = 657.09 Rs.

Q. 20

From 70°C to 0°C (Freezing)

Specific heat $(C_p) = \dfrac{1}{\left(\dfrac{1}{1.97} - \dfrac{1}{3.6}\right)} = 4.35 \text{ kJ kg}^{-1} \text{ K}^{-1}$

Temperature reduction (ΔT) = 70°C

Heat extracted= 100×4.35×70 = 30450 kJ

From 0°C to −20°C (Frozen)

Heat extracted = 100×250 = 25000 kJ

Thus, total heat extracted = 25000 + 30450 = 55450 kJ in 3 hours

Q = 55450/(3×3600) = 5.13 kW

∴ Tons of refrigeration = 5.13/3.5 = 1.47 TR (∵ 3.5 kW = 1 T.R.)

Q. 21 (i)

Motor size = I hp = 0.746 kW

∵ COP of system = 4.2 and compressor efficiency = 80%

Power available for compressor = 0.746×4.2×0.80

= 2.51 kJ/s

Mass flow rate of refrigerant = 2.51/185 = 0.01357 kg/ s = 0.814 kg/ min

(ii)

Mass flow rate of refrigerant = 0.814×0.15 = 0.1221 m³/min

Theoretical volumetric flow rate of refrigerant through both cylinders

$= 2 \times \frac{\pi}{4} \times d^2 \times L \times N$

$= 2 \times \frac{\pi}{4} \times 0.04^2 \times L \times 1440$

= 3.619L m³/min

Therefore,

$3.619L = 0.1221$

$L = 0.0337 \text{ m} = 33.7 \text{ mm}$

Q. 22

Power consumed by compressor = Heat extracting × Refrigerent flow rate

$= 200 \times 1.5/60 = 5 \text{ kW} = 5/0.746 = 6.70 \text{ hp}$

∵ COP of refrigerator = 4.5

Thus, Horse power of motor = 6.70/4.5 = 1.5 hp

Q. 23

Cooing effect = 15×3.52 = 52.8 kW

∵ COP of cooling cycle = 3.7

Therefore, cooling load on compressor due to cooling cycle

= 52.8/3.7 = 14.27 kW

= 14.27/0.746 = 19.13 hp

(∵ 1 hp = 0.746 kW)

Thus, net heat extracted or work load on Cascade compressor

= 52.8 + 14.27

= 67.07 kW

∵ Cascade compressor removes 70% of the total heat reflected in the liquid receiver or by the refrigerant

Therefore, Actual heat extracted or work done by the compressor

= 67.07×0.70

= 46.95 kW

∴ Work done by the compressor = net heat extracted by compressor/ COP of compressor

= 46.95/4.2 = 11.18 kW

= 11.18 /0.746 = 14.99 hp

(∵ 1 hp = 0.746 kW)

Q. 24

Given

Meat block thickness (a) = 2 cm = 0.02 m; Density of meat (ρ) = 1060 kg/ m³

Latent heat of crystallization (λ) = 335 kJ/ kg = 335000 J/ kg

Thermal conductivity of meat (k) = 1.25 W m^{-1} k^{-1}

Freezing Temperature (T_f) = – 25°C; Initial Temperature (T_i) = 0°C

We know

$$\text{Freezing Time (t)} = \frac{\lambda\rho}{T_f - T_i}\left(\frac{a}{2h} + \frac{a^2}{8k}\right)$$

Here, $\frac{a}{2h}$ is very small in respect to $\frac{a^2}{8k}$. Therefore, $\frac{a}{2h}$ is neglecting.

$$t = \frac{0.85 \times 335000 \times 1060}{0 - (-25)}\left(\frac{0.02}{8 \times 1.25}\right)$$

t = 482.94 seconds

t = 8.05 minutes

∵ Block of meat is kept between pair of freezing plates i.e. covered from both sides.

Hence, freezing time will be half.

Freezing time (t) = 8.05/2 = 4.025 minutes

Q. 25

Rate of heat being extracted from potato

= $mC_p\Delta t$

= 10×10³×3.2×(30 – 8) = 704000 kJ

or Work done by refrigerator to remove the heat = 704000/6×3600 = 32.59 kW

Consider, Refrigerant flow rate = x kg/sec

Hence, cooling effect produced = x×210 kW

Applying heat balance

$$210\,x = 32.59$$

$$\Rightarrow \quad x = 9.31 \text{ kg/min}$$

Work done by the compressor (W) = Heat extracted by refrigerant/ COP
= 32.59/4 = 7.76 kW

Q. 26

Total weight of marine fishery = 1000 kg

Water content in marine fishery = 1000×0.85 = 850 kg

Total frozen fish solids = 1000×0.15 = 150 kg

Total refrigeration effect required to freeze the marine fishery = 850×4.2×(32 – 0) + 850×2.2×(0 - (- 18)) + 850×335 + 150×3.2×(3.2 – 0) + 150×1.8×(0 - (- 0.3)) = 448091 kJ in half of hour or 248.94 kJ/s or 248.94 kW

∵ Coeefficient of performance = 3.66

Hence, Compressor power consumption = 248.94/3.66 = 68.02 kW

Q. 27

Cooling effect generated

$= mC_p\Delta t$

= 12500×3.8×(30 – 4)/(3×3600) = 114.35 kW/h

We know

I tonnage of refrigeration = 3.517 kW

Hence, T.R. = 114.35/3.517 = 32.51

And power consumed by electric motor = 114.35/4.5 = 25.41 kW

Q. 28

Refrigeration effect required = 3.1×100×1000×(35 – 8)/(5×3600) = 465 kW

Consider, Refrigerant flow rate = R kg/min

Refrigeration effect produced by evaporator = 230×(R/60) = 3.83R kW

Hence,

$\Rightarrow$ $3.83\ R = 465$

$\Rightarrow$ $R = 121.30$ kg/min

$\because$ COP = Refrigeration effect / Power input to compressor

$\Rightarrow$ Power input to compressor = 465/3.66 = 127.05 kW

34

Thermal Processing

UKPSC Combined Assistant Engineering Exam 2013 Paper - I

Q. 1 The amount of heat required to raise the temperature of 1gm milk by one degree celsius in comparison to water is

(A) 85 % (B) 93 %

(C) 107 % (D) 100 %

Q. 2 The thermal destruction of micro organisms can be expressed as following where:-

N = Viable organism at any time

N_O= Organism present initially

t, k = Constants

(A) $dN/dT = -kN^2$ (B) $dN/dT = 2kN$

(C) $dN/dT = k$ (D) $t = \frac{1}{k} \ln(N_o/N)$

Q. 3 The bimetal thermometers are used at temperature range of

(A) 500 °C to – 12 °C (B) 500 °C to – 185 °C

(C) 538 °C to – 185 °C (D) 538 °C to – 285 °C

CGPSC State Engineering Services Exam Paper – I

Q. 4 Sterilization of milk is carried out to extent of destroying

(A) Only pathogenic organisms (B) Only spoilage organisms

(C) All types of organisms (D) Only heat resistant organisms

(E) All Pathogens only

UKPSC A.En. Exam 2007 Paper – I

Q. 5 The destruction of all microorganism in food by thermal processing is known as

(A) Pasteurization (B) Sterilization

(C) Blanching (D) Scalding

Q. 6 In batch type pasteurizer milk is heated for 30 minutes at a temperature of

(A) 51 °C (B) 61 °C

(C) 71 °C (D) 81 °C

Q. 7 Pasteurization of milk is generally practiced to destroy microorganism which are

(A) Pathogenic (B) Non-pathogenic

(C) Spoilage causing (D) None of the above

Q. 8 In an HTST pasteurizer booster pump is located just before

(A) Heating (B) Regeneration

(C) Cooling (D) Holding

Q. 9 The temperature range for Ultra High Temperature (UHT) sterilization of milk is

(A) 95 – 110 °C (B) 115 – 130 °C

(C) 135 – 150 °C (D) 155 – 170 °C

UKPSC A.En. Exam 2012 Paper – I

Q. 10 Holding time under HTST pasteurizer is

(A) 15 seconds (B) 30 seconds

(C) 45 seconds (D) 1 minute

Q. 11 Method used for cleaning of HTST pasteurizer is called

(A) SIP (B) VIP

(C) CIP (D) DIP

Q. 12 In HTST pasteurizer, milk is heated to at least ………. °C.

(A) 52 (B) 62

(C) 72 (D) 82

Q. 13 Flow diversion valve is a part of

(A) Pasteurizer (B) Grader

(C) Dryer (D) Huller

Q. 14 The booster pump is used in

(A) LTLT Pasteurization (B) HTST Pasteurization

(C) Homogenization (D) Centrifugal separation

Graduate Aptitude Test in Engineering - 2007

Q. 15 Bacillus stearothermophilus has a z value of 10.20°C at a reference temperature of 121°C.

(i) The activation energy for the destruction of Bacillus stearothermophilus is

(A) 327.56 MJ kg $mole^{-1}$ (B) 298.95 MJ kg $mole^{-1}$

(C) 208.35 MJ kg $mole^{-1}$ (D) 75.62 MJ kg $mole^{-1}$

(ii) The z value of the same organism at a reference temperature of 135° C is

(A) 9.73°C (B) 10.20°C

(C) 10.95°C (D) 11.15°C

Q. 16 Bacterial population in milk increases 200 times in 18 hours of storage at 20°C. The increase in population in 3 hours of storage at the same temperature is

(A) 1.34 times (B) 2.42 times

(C) 7.02 times (D) 14.14 times

Graduate Aptitude Test in Engineering - 2008

Q. 17 When suspension of microorganism is heated at constant temperature, the reaction kinetics of decrease in the number of the organism is

(A) Linear (B) Exponential

(C) Parabolic (D) Hyperbolic

Graduate Aptitude Test in Engineering - 2010

Q. 18 Eight log cycle reduction of Clostridium botulinum having z-value of 9°C needs a process time of 1.5 minute at 121°C temperature. The same degree of reduction at 130°C temperature will require a process time of

(A) 72 s (B) 54 s

(C) 18 s (D) 9 s

Graduate Aptitude Test in Engineering - 2011

Q. 19 A mixture of microorganisms, P, Q, R, and S are to be inactivated by autoclaving. Number of spores of the above organisms is 6×10^5, 4×10^6, 2×10^4, and 9×10^5 respectively. Further, their decimal reduction time (D) values in minute are 2.2, 1.8, 0.8, and 1.6 respectively. It is desired to microbiologically inactivate the mixture at 121 °C to obtain a probability of spoilage of 10000^{-1}. The microorganisms that will be inactivated first and that at the last are respectively

(A) P and Q (B) R and P

(C) P and S (D) P and R

Graduate Aptitude Test in Engineering - 2012

Q. 20 Decimal reduction times for Bacillus subtilis are 37 s and 12 s at temperatures of 120 °C and 125 °C, respectively. The temperature rise, in °C, necessary to reduce the first value of decimal reduction time at 120 °C by a factor of 10 is

(A) 7.18 (B) 10.36

(C) 13.06 (D) 16.07

Graduate Aptitude Test in Engineering - 2013

Q. 21 The equation representing the heat of respiration (q) of fruits and vegetables as a function of temperature (Θ) with positive constants a and b is

(A) $q = ae^{b\Theta}$ (B) $q = ae^{-b\Theta}$

(C) $q = a\ln(b\Theta)$ (D) $q = a + b\Theta$

Q. 22 For an initial spore load equal to 25 spores per container inoculated with Clostridium botulinum having $D_{121} = 0.25$ min, the spoilage probability of the container subjected to $F_{121} = 1.5$ min is

(A) 10^{-5} (B) 10^{-6}

(C) 10^{-8} (D) 10^{-9}

IFOS - 2013

Q. 23 Shelf life of food stored at 30°C is 7 days. Assuming that Q_{10} value of deteriorative reactions occurring in the food is 1.8, Shelf life of food when it is stored at 10°C

(A) 847 days (B) 22.68 days

(C) 21 days (D) 49 days

Graduate Aptitude Test in Engineering – 2015

Q. 24 A suspension contains 3.6×10^5 spores of *C.botulinum* having a D-value of 1.5 minute at 121.1°C and 8.5×10^6 spores of *B.* subtilis having a D-value of 0.9 minute at the same temperature. The suspension is heated at a constant temperature of 121.1°C. The heating time needed in minutes for the suspension to obtain a survival probability of 10^{-3} for the most heat resistant organism in it is ....

Graduate Aptitude Test in Engineering – 2016

Q. 25 Milk sterilization kinetics is based on inactivation of index microorganism, *Bacillus stearothermophilus.* The D-values at 121.1°C and 139.1°C are 1.2 min and 0.019 min, respectively. For 12 log-cycle reduction of this microorganism at 130°C, the processing time in second is ....

Graduate Aptitude Test in Engineering – 2017

Q. 26 The reaction rate for destruction of Clostridium botulimum increases 11 times for temperature rise of 10°C from 121.10°C. The decimal reduction time of this organism is 5.7 s at 121.10°C. The minimum sterilization time in seconds at 135°C, for eight log cycle reduction of the organism, is

(A) 0.20 (B) 0.31

(C) 1.63 (D) 2.49

Graduate Aptitude Test in Engineering – 2018

Q. 27 *Mycobacterium tuberculosis* is having decimal reduction time of 13.5 s at 72°C. The count of the organism is to be reduced by eight logarithmic cycles in milk pasteurization with adequate holding at a temperature of 85°C. If Q10 value for the organism is 7.5, the minimum holding time for the desired sterility of this particular organism at 85°C is .... seconds.

Graduate Aptitude Test in Engineering – 2019

Q. 28 A batch of 10000 L milk is to be sterilized and thereafter packed in 20000 packets of 500 ml each. The mean Standard Plate Count (SPC) of *Bacillus subtilis* in 100 samples of fresh milk was found to be 50 per ml. The milk is to be sterilized such that each 500 ml packet is completely devoid of the same organism. Minimum number of log cycle reduction for sterilization of this batch is

(A) 8 (B) 9

(C) 10 (D) 12

Graduate Aptitude Test in Engineering – 2020

Q. 29 A milk processing plant pasteurizes a batch of 12500 L whole milk to inactivate the pathogen *Coxiel!a burnetit,* (decimal reduction time of 14 seconds at 72 °C) prior to packing in 500 milliliter pouches. The initial count of the noted organism is 10 per milliliter. For this batch pasteurization process at 72 °C, resulting in no survivor in any of the packages, the process F-value in seconds is ....

Graduate Aptitude Test in Engineering – 2021

Q. 30 Food cans are sterilized in a retort to inactivate *Clostridium botulinum.* Process time (F_o) of this food material is 150 s and the z value is 10 °C. Temperatures at the slowest heating region of the food can are measured and the average temperature during time periods 0 to 20 min, 20 to 40 min and 40 to 70 min are 71.1 °C, 98.9 °C and 110 °C, respectively. The actual process time in minutes that is required for equivalent sterilization at 121.1 °C is..........

Answers Key

1	2	3	4	5	6	7	8	9	10
B	D	C	C	B	B	A	B	C	A
11	12	13	14	15	16	17	18	19	20
C	C	A	B	B, C	B	B	D	B	B
21	22	23	24	25	26	27	28	29	30
A	B	B	12.83	111.24	C	7.88	B	126	2.45

Explanations

Q. 1

Specific heat of milk = 3.9 kJ/kg K that is 93 % of specific heat of water = 4.2 kJ/kg K.

Q. 15 (i)

Given

Z value of Bacillus stearothermophilus = 10.20°C

$(T_1) = 121°C \Rightarrow T_1 = 394$ K

As we know

$$\frac{D_2}{D_1} = 10^{(T1 - T2)/z}$$

$$\Rightarrow \quad \frac{1}{10} = 10^{(121-T_2)/10.20}$$

$$\Rightarrow \quad T_2 = 131.20°C \Rightarrow T_2 = 404.20 \text{ K}$$

Alternate

$T_2 = z + T_1$

$= 10.20 + 121 = 131.20°C$

We also know

$$\frac{D_2}{D_1} = e^{\frac{E_a\left(\frac{1}{T_2}-\frac{1}{T_1}\right)}{R}}$$

$$\Rightarrow \quad \frac{1}{10} = e^{\frac{E_a\left(\frac{1}{394}-\frac{1}{404.20}\right)}{8.31\times10^3}} \quad (\because R = 8.314\times10^3 \text{ J/ kg mol K})$$

$$\Rightarrow \quad E_a = 298.95 \text{ MJ kg/mole}$$

(ii)

If reference temperature $(T_1) = 135°C \Rightarrow T_1 = 408$ K

We know

$$\frac{D_2}{D_1} = e^{\frac{E_a\left(\frac{1}{T_2}-\frac{1}{T_1}\right)}{R}}$$

$$\Rightarrow \quad \frac{1}{10} = e^{\frac{E_a\left(\frac{1}{T_2}-\frac{1}{408}\right)}{8.31\times10^3}}$$

$$\Rightarrow \quad T_2 = 418.95 \text{ K} = 145.95°C$$

Also, $\quad T_2 = z + T_1$

$$\Rightarrow \quad z = 145.95 - 135 = 10.95°C$$

Q. 16

We know that

$$t = D \log(\frac{N}{N_0})$$

Case I

$$18 = D \log(\frac{1}{200}) \quad \text{(i)}$$

Case II

$$3 = D \log(\frac{1}{x}) \quad \text{(ii)}$$

($\because$ Consider increase in population in 3 hours of storage = x times)

Dividing equation (i) by (ii)

$$\frac{18}{3} = \frac{D\log(\frac{1}{200})}{D\log(\frac{1}{x})} \Rightarrow x = 2.42 \text{ times}$$

Q. 18

We know that

$$D_2/D_1 = 10^{(T_1 - T_2)/Z}$$

Thus, $D_2 = 1.5 \times 10^{(121 - 130)/10}$

$= 9$ seconds

Q. 19

We know

$$t = D \log(\frac{N}{N_0})$$

Putting different-different values of D, N and N_0 corresponding to P, Q, R and S.

For P

$N = 6\times10^5$; $N_0 = 1/10000$; $D = 2.2$ min

$t = 21.51$ min

For Q

$N = 4\times10^6$; $N_0 = 1/10000$; $D = 1.8$ min

$t = 19$ min

For R

$N = 2\times10^4$; $N_0 = 1/10000$; $D = 0.8$ min

$t = 6.64$ min

For S

$N = 9\times10^5$; $N_0 = 1/10000$; $D = 1.6$ min

$t = 15.93$ min

Q. 20

We know that

$$10^{\frac{T_2 - T_1}{z}} = \frac{D_1}{D_2}$$

$$\Rightarrow \quad 10^{\frac{125-120}{z}} = \frac{37}{12}$$

$$\Rightarrow \quad z = 10.22^{\circ}C$$

Q. 22

$$\frac{F_{121}}{D_{121}} = log\frac{N_0}{N}$$

$$\frac{N_0}{N} = 10^{1.5/0.25} = 10^6$$

$\therefore$ Spoilage Probability = $N/N_0 = 10^{-6}$

Q. 23

We know that

$$\frac{k_2}{k_1} = (Q_{10})^{\left(\frac{T_1 - T_2}{10}\right)}$$

$$\Rightarrow \quad k_2 = 7\times(1.8)^{\left(\frac{30-10}{10}\right)}$$

$$\Rightarrow \quad k_2 = 22.68 \text{ days}$$

Q. 24

We know

$$t = D\log\left(\frac{N}{N_0}\right)$$

Putting different-different values of D, N and N_0 corresponding to *C. botulinum* and *B. subtilis*.

For *C. botulinum*

$N = 3.6\times10^5$; $N_0 = 10^{-3}$; $D = 1.5$ min

$t = 1.5\times\log(3.6\times10^5/10^{-3})$

$t = 12.83$ min

For Q *B. subtilis.*

$N = 8.5\times10^6$; $N_0 = 10^{-3}$; $D = 0.9$ min

$t = 0.9\times\log(8.5\times10^6/10^{-3})$

$t = 8.94$ min

Here, ***C. botulinum*** is more heat resistant organism and heating time for given condition is = 12.83 min.

Q. 25

We know that

$$10^{\frac{T_2 - T_1}{z}} = \frac{D_1}{D_2}$$

$$\Rightarrow \quad 10^{\frac{121.1 - 139.1}{z}} = \frac{0.019}{1.2}$$

$$\Rightarrow \quad z = 10°C$$

Again, for D value at 130°C

$$10^{\frac{T_{130} - T_1}{z}} = \frac{D_1}{D_{130}}$$

$$\Rightarrow \quad 10^{\frac{130 - 139.1}{10}} = \frac{0.019}{D_{130}}$$

$$\Rightarrow \quad D_{130} = 0.154 \text{ min} = 9.27 \text{ seconds}$$

∵ Processing time of 12 log-cycle reduction of the microorganism at temperature $T = 12\times D_T$

∴ Processing time = 12×9.27 = 111.24 seconds.

Q. 26

We know

$$D_2/D_1 = 11^{(T_1 - T_2)/10}$$

$$\Rightarrow \quad D_2/5.7 = 11^{(121.10 - 135)/10}$$

$$\Rightarrow \quad D_2 = 0.203 \text{ sec.}$$

Again

Sterilizing time = D $\log(N_0/N)$

$= D \times$ Number of log cycles

$= 0.203 \times 8 = 1.63$ sec.

Q. 27

Time required for eight log cycle reduction $= D \times \log(N/N_0) = 13.5 \times 8$

$= 108$ seconds

Again $k_2/k_1 = (Q_{10})^{\frac{T_1 - T_2}{2}}$

$\Rightarrow \quad k_2 = 108 \times (7.5)^{\frac{72-85}{10}} = 7.88$ seconds

Q. 28

Initial bacterial load in raw milk(N_o) $= 10000 \times 1000 \times 50 = 5 \times 10^8$

After complete sterlization, the bacteria in the milk should be(N) ≤ 1

Hence, Log cycle reduction time $= \log(N_o/N)$

$= \log(5 \times 10^8/1) = 8.699$ or 9

Q. 29

The Decimal reduction time is 14 seconds i.e. It is the time taken by the pasteurization process to reduce the number of spores by 1 log cycle.

Total number of spores in whole milk $= 12500 \times 1000 \times 10 = 125000000$

No. Survival ↓

No. of spores	125000000	12500000	1250000	125000	12500	1250	125	12.5	1.25	0.125
Log cycle reduction	1	2	3	4	5	6	7	8	9	

Hence, F-value $= 9 \times 14 = 126$ seconds

Q. 30

We know that

Process time $(D_T) = D_{121.1} \times 10^{\frac{121.1 - T}{z}}$

Hence,

$$D_{121.1} = D_T \times 10^{\frac{T - 121.1}{z}}$$

For time period 0 to 20 min

$D_{121.1} = 20\times10\ 10^{\frac{71.1-121.1}{10}} = 0.0002$ min

For time period 20 to 40 min

$D_{121.1} = 20\times10\ 10^{\frac{98.9-121.1}{10}} = 0.1205$ min

For time period 40 to 70 min

$D_{121.1} = 20\times10\ 10^{\frac{110-121.1}{10}} = 2.3287$ min

Actual Process Time = 0.0002 + 0.1205 + 2.3287 = 2.4494 min or 2.45 min

35

Dimensionless Numbers

UKPSC Combined Assistant Engineering Exam 2013 Paper - I

Q. 1 Froude number is the ratio between

(A) Inertia force to viscous forces (B) Inertia force to gravity force

(C) Inertia force to shear force (D) Gravity force to inertia force

APPSC AEES Agriculture Engineering Exam 2016

Q. 2 The velocity of a fluid in a pipe of diameter D is V. The pipe is connected to another pipe of diameter 2D. Reynolds number in the pipe of diameter 2D in relation to the pipe of diameter D is

(A) Double (B) Four times

(C) Half (D) Same

Q. 3 Logarithmic mean radius of a pipe in relation to its arithmetic mean radius is always

(A) Smaller (B) Greater

(C) Equal or smaller (D) Equal or greater

Q. 4 Natural convection is associated with

(A) Prandtl number only (B) Reynolds number only

(C) Grashof number only (D) Nusselt number only

UKPSC A.En. Exam 2007 Paper – I

Q. 5 Sherwood number is used in

(A) Conduction of heat transfer (B) Radiation of heat transfer

(C) Mass transfer (D) Freezing of food material

UKPSC A.En. Exam 2012 Paper – I

Q. 6 Grashof number is used in

(A) Conduction heat transfer (B) Radiation heat transfer .

(C) Convection heat transfer (D) Convection mass transfer

Graduate Aptitude Test in Engineering - 2007

Q. 7 Peas of 1.1 cm diameter are dried by air at 65.8°C in a packed bed drier. The void fraction of the bed is 0.35 and the bed has a diameter of 0.5 m and a height of 0.8 m. The flow rater and the viscosity of air are 0.12 kg s^{-1} and 2.03×10^{-5} Pa s respectively. Reynolds number for the packed bed is

(A) 13 (B) 340

(C) 908 (D) 1359

Q. 8 A diatomic, adiabatically compressible fluid having the molecular mass of 16 is flowing through a nozzle at a temperature of 20°C. If the velocity of the fluid is 430 m s^{-1}, the Mach Number is

(A) 0.93 (B) 0.97

(C) 1.03 (D) 1.07

Graduate Aptitude Test in Engineering - 2008

Q. 9 The velocity of fluid in a pipe A of diameter D is V m s^{-1}. This pipe is connected with another pipe B of diameter 2D. Reynold's number in pipe A in relation to pipe B is

(A) Same (B) Half

(C) Double (D) Triple

Q. 10 Milk and rapeseed oil are flowing in pipes of 5 cm diameter with the same flow velocity of 3 m s^{-1}. The densities of milk and rapeseed oil are 1030 and 900 kg m^{-3}, respectively. The viscosity of milk is 2.1×10^{-3} N s m^{-2} and that of rapeseed oil is 118×10^{-3} N s m^{-2}. The values of Reynolds' number for milk and rapeseed oil will be respectively

(A) 73571 and 1144 (B) 1144 and 73571

(C) 73571 and 1144 (D) 1144 and 73571

Q. 11 The viscosity of milk at 21 °C is 2.1×10^{-3} Pa and its density at this temperature is 1029 kg m^{-3}. Milk flows at the rate of 0.12 m^3 min^{-1} in a 205 cm diameter pipe. At 21 °C, the flow of milk will be

(A) Stream line (B) Laminar

(C) Transition (D) Turbulent

Graduate Aptitude Test in Engineering - 2009

Q. 12 Convective heat transfer coefficient outside an ice cream block is 10 W m^{-2} K^{-1}. Thermal conductivity of frozen ice cream is 0.3 W m^{-1} K^{-1}. Convection takes place across a layer of 10 mm of air for 5 minutes. If the density and the specific heat capacity of ice cream are respectively 6000 kg m^{-3} and 2.5 kJ kg^{-1} K^{-1}, then 0.33 is

(A) Biot Number (B) Nusselt Number

(C) Fourier Number (D) Prandtl Number

Graduate Aptitude Test in Engineering - 2010

Q. 13 The dimensionless number in heat transfer corresponding to Sherwood Number is mass transfer is

(A) Biot Number (B) Schmidt Number

(C) Nusselt Number (D) Graetz Number

Graduate Aptitude Test in Engineering - 2011

Q. 14 Milk is agitated in a tank with a rotating impeller. For this system:

Let X = $\frac{P}{\rho N^3 D^5}$, Y = $\frac{D^2 N \rho}{\mu}$, Z = $\frac{DN^2}{g}$

Where, P = power imparted by the impeller to the fluid, N = rate of rotation of the impeller, D = impeller diameter, g = acceleration due to gravity, = fluid density, and μ = fluid viscosity.

The X, Y, Z are

(A) X = Grashof number, Y = Power number, Z = Reynolds number

(B) X = Power number, Y = Reynolds number, Z = Froude number

(C) X = Reynolds number, Y = Froude number, Z = Grashof number

(D) X = Froude number, Y = Grashof number, Z = Power number

Graduate Aptitude Test in Engineering - 2012

Q. 15 If is the kinematic viscosity of air–water vapour mixture and is the mass diffusivity of water vapour in air then the ratio $\frac{\nu}{D_{AB}}$ is known as

(A) Stanton number (B) Prandtl number

(C) Schmidt number (D) Sherwood number

Q. 16 A pair of parallel glass panes, each of 3 mm thickness traps 2 mm layer of stagnant air. Thermal conductivities of glass and air are 0.5 and 0.02 W m^{-1} K^{-1}, respectively. If the film heat transfer coefficient of air is 10 W m^{-2} K^{-1}, then Biot Number is

(A) 1.50 (B) 1.00

(C) 0.06 (D) 0.04

Graduate Aptitude Test in Engineering - 2014

Q. 17 Identify the **INCORRECT** statement about the relevance of various dimensionless numbers in transport processes

(A) Reynolds number is relevant in forced convection and Grashof number is relevant in natural convection

(B) Prandtl number is relevant in heat transfer and Schmidt number is relevant in mass transfer

(C) Biot number is relevant in heat transfer and Froude number is relevant in mass transfer

(D) Nusselt number is relevant in heat transfer and Sherwood number is relevant in mass transfer

Q. 18 For a fully developed laminar flow through a smooth pipe, the relationship between friction factor (f) and Reynolds number (R_e) is

(A) $f \propto (R_e)$ (B) $f \propto (R_e)^{-1}$

(C) $f \propto (R_e)^2$ (D) $f \propto (R_e)^{-2}$

Graduate Aptitude Test in Engineering - 2016

Q. 19 Mass transfer coefficient for equimolar counter- diffusion of water vapour in air is 0.4 m s^{-1} (based on concentration difference). Mass diffusivity of water vapour in air is 3×10^{-4} m^2 s^{-1}. For 100 μm diameter droplet, the Sherwood Number (N_{Sh}) is equal to (A) the mass transfer coefficient (B) diameter divided by mass diffusivity (C) one – third of the mass transfer coefficient (D) three times of the mass diffusivity ....

Graduate Aptitude Test in Engineering - 2018

Q. 20 Molecular diffusivity (DAB) of water vapour in air is 2.6×10^{-5} m^2 s^{-1} over an effective distance of 3 mm. Density and coefficient of viscosity of air are 1.2 kg m^{-3} and 2×10^{-5} kg m^{-1} s^{-1}, respectively. Sherwood Number (NS_h) for water vapour in air is ....

Graduate Aptitude Test in Engineering - 2019

Q. 21 The steady-state mass transfer coefficient (kg) based on water vapour pressure differential (VPD) operating across stagnant, non-diffusing air was estimated to be 0.05 g mole/(s m^2 kPa). If VPD varies from 12 kPa to 7 kPa over a distance of 2 mm, then the mass transfer coefficient (k_y) based on equimolar counter-diffusion in g mole s^{-1} m^{-2} (mole fraction)$^{-1}$ is ....

Answers Key

1	2	3	4	5	6	7	8	9	10
B	C	A	C	C	C	BONUS	A	C	A
11	12	13	14	15	16	17	18	19	20
D	B	C	B	C	C	C	B	C	Bonus
21									
Bonus									

Explanations

Q. 2

Refer Explanation for Q. 9.

Q. 3

We know

Logarithmic mean radius(LMR) = (R - r)/ln(R/r)

or LMR/r = (R/r - 1)/ln(R/r)

Again

Arithmetic mean radius (AMR) = (R + r)/2

or AMR/r = (R/r + 1)/2

Consider, R/r = 1.5

Hence, LMR/r = 1.23 and AMR/r = 1.25 $\Rightarrow$ LMR < AMR

It is valid for all values of R/r. Hence it can be said that the Logarithmic mean radius of a pipe is always smaller than its Arithmetic mean radius.

Q. 7

Diameter of peas (D_p) = 1.1 cm or 0.011 m; Bed diameter (D) = 0.5 m

Bed height (h) = 0.8 m; Flow rate (Q) = 0.12 kg/ s;

Viscosity (μ) = 2.03×10^{-5} Pa s

Void fraction (e) = 0.35

We know

$$Q = AV$$

$$\frac{0.12}{\rho} = \frac{3.14}{4}(0.5)^2 V$$

$$V = \frac{0.6115}{\rho} \text{ m/ s}$$

Superficial air velocity through bed (V') = e.V

$$= 0.35\times \frac{0.6115}{\rho} = \frac{0.214}{\rho} = \text{m/s}$$

∵ Reynolds number (R_e) = $\frac{\rho V' D_p}{(1-e)\mu}$

$$= \frac{\rho \times 0.214 \times 0.011}{(1-0.35)\rho \times 2.03\times10^{-5}}$$

$$= 178$$

Q. 8

Given

Molecular mass of fluid (M) = 16; Specific heat ratio (k) = 1.4×10^3

R = 8.314 J mol^{-1} K^{-1}; Temperature (T) = 20 + 273 = 293 K

Velocity (V) = 430 m/s

∵ Mach number (M_a) = $\frac{V}{\sqrt{kRT/M}}$

$$= \frac{430}{\sqrt{1.4\times10^3 \times 8.314 \times 273/16}} = 0.93$$

Q. 9

For given condition

$$R_e = \frac{\rho VD}{\mu} = \frac{\rho D}{\mu}\left(\frac{4Q}{\pi D^2}\right) \quad [\because Q = AV = (\pi D^2/4)V]$$

$$\Rightarrow \quad R_e \propto \frac{1}{D}$$

Or $\frac{(R_e)_2}{(R_e)_1} = \frac{D_1}{D_2}$

$$\frac{(R_e)_2}{(R_e)_1} = \frac{D}{2D}$$

$\Rightarrow$ $(R_e)_2 = \frac{(R_e)_1}{2}$

Q. 10

Pipe diameter (D) = 0.05 m; Flow velocity (V) = 3 m/ s

For milk

Density of milk (ρ) = 1030 kg/ m³; Viscosity of milk (μ) = 2.1×10^{-3} N s m^{-2}

Reynolds number (R_e) = $\frac{\rho VD}{\mu}$

$$= \frac{1030\times3\times0.05}{2.1\times10^{-3}} = 73571$$

For rapeseed oil

Density of milk (ρ) = 900 kg/ m³; Viscosity of milk (μ) = 118×10^{-3} N s m^{-2}

Reynolds number (R_e) = $\frac{\rho VD}{\mu}$

$$= \frac{900\times3\times0.05}{118\times10^{-3}} = 1144$$

Q. 11

Viscosity of milk at 21°C = 2.10×10^{-3} N s m^{-2}; Diameter of pipe (D) = 0.025 m

Density of milk at 21°C = 1029 kg m^{-3}

Cross-sectional area of pipe = $(\pi/4)D^2$

$= (\pi/4)(0.025)^2$

$= 4.9\times10^{-4}\ m^2$

Rate of flow = 0.12 $m^3\ min^{-1}$ = 0.12/60 $m^3\ s^{-1}$ = 2 x $10^{-3}\ m^3\ s^{-1}$

So,

Velocity of flow = $(2\times10^{-3})/(4.9\times10^{-4})$

$= 4.1\ m\ s^{-1}$

also

$(Re) = (Dv\rho/\mu)$

$= 0.025 \times 4.1 \times 1029/(2.1 \times 10^{-3})$

$= 50{,}230$

and this is greater than 4000 so that the flow is turbulent.

Q. 12

Convective heat transfer coefficient (h) = 10 W m^{-2} K^{-1}

Layer thickness (L) = 0.01 m;

Specific heat capacity (C_p) = 2.5 kJ kg^{-1} K^{-1}

Thermal conductivity (k) = 0.3 W m^{-1} K^{-1}

$\therefore$ Nusselt Number (N_u) = $\frac{10 \times 0.01}{0.3} = 0.33$

Biot Number	Nusselt Number	Fourier Number	Prandtl Number
$B_o = \frac{0.35}{0.60}$ Can be determined	$N_u = \frac{hD}{k}$ D is not given. Can't be determined N_u.	$F_o = \frac{\alpha t}{L^2}$ is not given. Can't be determined F_o.	$P_r = \frac{C_p\mu}{k}$ is not given. Can't be determined P_r.

Q. 13

The equivalent of the Nusselt number in mass transfer is the Sherwood number (S_h).

Nusselt Number

$$N_u = \frac{hD}{k}$$

Where

D is the diameter or characteristic length, h is convective heat transfer coefficient and k is thermal conductivity

Sherwood Number

$$S_h = \frac{k_g D}{D_{wm}}$$

Where

D is the diameter or characteristic length, k_g is mass transfer coefficient and D_{wm} is the diffusivity expressed in kg mole/ (m s).

Q. 15

Schmidt number (Sc) is a dimensionless number defined as the ratio of momentum diffusivity (viscosity) and mass diffusivity, and is used to characterize fluid flows in which there are simultaneous momentum and mass diffusion convection processes.

$$Sc = \frac{\nu}{D}$$

Where

is the kinematic viscosity or (μ/ρ) in units of (m^2/s)

D is the mass diffusivity (m^2/s).; P is the density of the fluid (kg/m^3).

Q. 16

$$\text{Biot number } (N_{BI}) = \frac{hL}{k}$$

$$= \frac{10 \times 0.003}{0.5} = 0.06$$

Q. 19

$$\text{Sherwood Number } (N_{Sh}) = \frac{k_g D}{D_{wm}}$$

$$N_{Sh} = 0.4 \times 100 \times 10^{-6}/3 \times 10^{-4}$$

$$= 0.4/3$$

or $N_{Sh} = k_g/3$ ($\because k_g = 0.4$)

Q. 20

Schmidt number (Sc) = ν/D

$$= \mu/\rho D = 2 \times 10^{-5} / (1.2 \times 2.6 \times 10^{-5}) = 0.641$$

Reynold's number (R_e) = $D\nu\rho/\mu = (1.2 \times 2.6 \times 10^{-5}) / 2 \times 10^{-5} = 1.56$

In the range of $2 \leq R_e \leq 1200$ and $0.6 \leq S_c \leq 2.7$

$\therefore$ Sheerwood number (S_h) = $2 + 0.55 \times R_e^{1/7} \times S_c^{1/3}$

$$= 2 + 0.55 \times (1.56)^{1/7} \times (0.641)^{1/3}$$

$$= 2.505$$

Q. 21

We know

$$K_y = KP_{bm}$$

$$= K \frac{P_2 - P_1}{ln(P_2 / P_1)}$$

$$= 0.05\times[(12-7)/(\ln(12/7)] = 0.464$$

VI. General English, Aptitude and Reasoning

36

General English

Graduate Aptitude Test in Engineering - 2010

Q. 1 Choose the most appropriate word from the options given below to complete the following sentence:

His rather casual remarks on politics .... his lack of seriousness about the subject.

(A) Masked (B) Belied

(C) Betrayed (D) Suppressed

Q. 2 Which of the following options is the closest in meaning to the word below:

Circuitous

(A) Cyclic (B) Indirect

(C) Confusing (D) Crooked

Q. 3 Choose the most appropriate word from the options given below to complete the following sentence:

If we manage to .... our natural resources, we would leave a better planet for our children.

(A) Uphold (B) Restrain

(C) Cherish (D) Conserve

Q. 4 Modern warfare has changed from large scale clashes of armies to suppression of civilian populations. Chemical agents that do their work silently appear to be suited to such warfare; and regretfully, there exist people in military establishments who think that chemical agents are useful tools for their cause.

Which of the following statements best sums up the meaning of the above passage?

(A) Modern warfare has resulted in civil strife.

(B) Chemical agents are useful in modern warfare.

(C) Use of chemical agents in warfare would be undesirable

(D) People in military establishments like to use chemical agents in war.

Graduate Aptitude Test in Engineering - 2011

Q. 5 Choose the most appropriate word from the options given below to complete the following sentence:

Under ethical guidelines recently adopted by the Indian Medical Association, human genes are to be manipulated only to correct diseases for which .... treatments are unsatisfactory.

(A) Similar (B) Most

(C) Uncommon (D) Available

Q. 6 Choose the word from the options given below that is most nearly opposite in meaning to the given word:

Frequency

(A) Periodicity (B) Rarity

(C) Gradualness (D) Persistency

Q. 7 Choose the most appropriate word from the options given below to complete the following sentence:

It was her view that the country's problems had been .... by foreign technocrats, so that to invite them to come back would be counter-productive.

(A) Identified (B) Ascertained

(C) Exacerbated (D) Analyzed

Q. 8 The horse has played a little known but very important role in the field of medicine. Horses were injected with toxins of diseases until their blood built up immunities. Then a serum was made from their blood. Serums to fight with diphtheria and tetanus were developed this way.

It can be inferred from the passage that horses were

(A) Given immunity to diseases

(B) Generally quite immune to diseases

(C) Given medicines to fight toxins

(D) Given diphtheria and tetanus serums

Graduate Aptitude Test in Engineering - 2012

Q. 9 Choose the most appropriate alternative from the options given below to complete the following sentence:

I .... to have bought a diamond ring.

(A) Have a liking (B) Should have liked

(C) Would like (D) May like

Q. 10 Choose the most appropriate alternative from the options given below to complete the following sentence:

Food prices .... again this month.

(A) Have raised (B) Have been raising

(C) Have been rising (D) Have arose

Q. 11 Choose the most appropriate alternative from the options given below to complete the following sentence:

The administrators went on to implement yet another unreasonable measure, arguing that the measures were already .... and one more would hardly make a difference.

(A) Reflective (B) Utopian

(C) Luxuriant (D) Unpopular

Q. 12 Choose the most appropriate alternative from the options given below to complete the following sentence:

To those of us who had always thought him timid, his .... came as a surprise.

(A) Intrepidity (B) Inevitability

(C) Inability (D) Inertness

Q. 13 In the early nineteenth century, theories of social evolution were inspired less by Biology than by the conviction of social scientists that there was a growing improvement in social institutions. Progress was taken for granted and social scientists attempted to discover its laws and phases

Which one of the following inferences may be drawn with the greatest accuracy from the above passage?

Social scientists

(A) Did not question that progress was a fact.

(B) Did not approve of Biology.

(C) Framed the laws of progress.

(D) Emphasized Biology over Social Sciences

Graduate Aptitude Test in Engineering - 2013

Q. 14 The Headmaster .... to speak to you.

Which of the following options is incorrect to complete the above sentence?

(A) Is wanting (B) Wants

(C) Want (D) Was wanting

Q. 15 Mahatama Gandhi was known for his humility as

(A) He played an important role in humiliating exit of British from India.

(B) He worked for humanitarian causes.

(C) He displayed modesty in his interactions.

(D) He was a fine human being.

Q. 16 All engineering students (I)/ should learn mechanics, (II)/ mathematics and (III)/ how to do computation (IV).

Which of the above underlined parts of the sentence is not appropriate?

(A) I (B) II

(C) III (D) IV

Graduate Aptitude Test in Engineering - 2014

Q. 17 Choose the most appropriate word from the options given below to complete the following sentence.

A person suffering from Alzheimer's disease ...... short-term memory loss.

(A) Experienced (B) Has experienced

(C) Is experiencing (D) Experiences

Q. 18 Choose the most appropriate word from the options given below to complete the following sentence .... is the key to their happiness; they are satisfied with what they have.

(A) Contentment (B) Ambition

(C) Perseverance (D) Hunger

Q. 19 Which of the following options is the closest in meaning to the sentence below?

"As a woman, I have no country."

(A) Women have no country.

(B) Women are not citizens of any country.

(C) Women's solidarity knows no national boundaries.

(D) Women of all countries have equal legal rights.

Q. 20 Moving into a world of big data will require us to change our thinking about the merits of exactitude. To apply the conventional mindset of measurement to the digital, connected world of the twenty-first century is to miss a crucial point. As mentioned earlier, the obsession with exactness is an artifact of the information-deprived analog era. When data was sparse, every data point was critical, and thus great care was taken to avoid letting any point bias the analysis.

From "BIG DATA" Viktor Maayer-Schonberger and Kenneth Cukier

The main point of the paragraph is:

(A) The twenty-first century is a digital world.

(B) Big data is obsessed with exactness.

(C) Exactitude is not critical in dealing with big data.

(D) Sparse data leads to a bias in the analysis.

Graduate Aptitude Test in Engineering – 2015

Q. 21 Choose the most appropriate word from the options given below to complete the following sentence:

The principal presented the chief guest with a ...., as token of appreciation.

(A) Momento (B) Memento

(C) Momentum (D) Moment

Q. 22 Choose the appropriate word/phrase from the options given below to complete the following sentence:

Frogs ....

(A) Croak (B) Roar

(C) Hiss (D) Patter

Q. 23 Choose the word most similar in meaning to the given word Educe

(A) Exert (B) Educate

(C) Extract (D) Extend

Q. 24 The following question presents a sentence, part of which is underlined. Beneath the sentence you find four ways of phrasing the underlined part. Following the requirements of the standard written English, elect the answer that produces, the most effective sentence.

Tuberculosis, together with its effect, ranks one of the leading causes of death in India.

(A) Ranks as one of the leading causes of death

(B) Rank as one of the leading causes of death

(C) Has the rank of one of the leading causes of death

(D) Are one of the leading causes of death

Q. 25 Read the paragraph and choose the correct statement

Climate change has reduced human security and threatened human well being. An ignored reality of human progress is that human largely depends upon environmental security. But on the contrary, human progress seems contradictory to environmental security. To keep up both at the required level is a challenge to be addressed by one and all. One of the ways to curb the climate change may be suitable scientific innovations, while the other may be the Gandhian perspective on small scale progress with focus on sustainability.

(A) Human progress and security are positively associated with environmental security.

(B) Human progress is contradictory to environmental security.

(C) Human security is contradictory to environmental security

(D) Human progress depends upon environmental security.

Graduate Aptitude Test in Engineering – 2016

Q. 26 If I were you, I .... that laptop. It's much too expensive.

(A) Won't buy (B) Shan't buy

(C) Wouldn't buy (D) Would buy

Q. 27 He turned a deaf ear to my request. What does the underlined phrasal verb mean?

(A) Ignored (B) Appreciated

(C) Twisted (D) Returned

Q. 28 Choose the most appropriate set of words from the options given below to complete the following sentence. .... is a will, ... is a way.

(A) Wear, there, their (B) Were, their, there

(C) Where, there, there (D) Where, their, their

Graduate Aptitude Test in Engineering – 2017

Q. 29 The ways in which this game can be played .... potentially infinite.

(A) is (B) is being

(C) are (D) are being

Q. 30 If you choose plan P, you will have to .... plan Q, as these two are mutually ......

(A) forgo, exclusive (B) forget, inclusive

(C) accept, exhaustive (D) adopt, intrusive

Graduate Aptitude Test in Engineering – 2018

Q. 31 " When she fell down the ………, she received many ……… but little help."

The words that best fill the blanks in the above sentence are

(A) Stairs, Stares (B) Stairs, Stairs

(C) Stares, Stairs (D) Stares, Stares

Q. 32 " In spite of being warned repeatedly, he failed to correct his ……… behavior."

The word that best fills the blank in the above sentence is

(A) Rational (B) Reasonable

(C) Errant (D) Good

Graduate Aptitude Test in Engineering – 2019

Q. 33 The fishermen,………. the flood victims owed their lives, were rewarded by the government.

(A) whom (B) to which

(C) to whom (D) that

Q. 34 Until Iran came along, India had never been ……… in kabaddi

(A) defeated (B) defeating

(C) defeat (D) defeatist

Q. 35 The nomenclature of Hindustani music has changed over the centuries. Since the medieval period *dhrupad* styles were identified as *baanis*. Terms like *gayaki* and *baaj* were used to refer to vocal and instrumental styles, respectively. With the institutionalization of music education the term *gharana* became acceptable. *Gharana* originally referred to hereditary musicians from a particular lineage, including disciples and grand disciples.

Which one of the following pairings is NOT correct?

(A) *dhrupad, baani* (B) *gayaki,* vocal

(C) *baaj,* institution (D) *gharana*, lineage

Q. 36 "I read somewhere that in ancient times the prestige of a kingdom upon the number of taxes that it was able to levy on its people. It was very much like the prestige of a head-hunter in his own community." Based on the paragraph above, the prestige of a head-hunter depended upon .

(A) the prestige of the kingdom

(B) the prestige of the heads

(C) the number of taxes he could levy

(D) the number of heads he could gather

Graduate Aptitude Test in Engineering – 2020

Q. 37 The untimely loss of life is a cause of serious global concern as thousands of people getkilled .... accidents every year while many other die .... diseases like cardio vascular disease, cancer, etc.

(A) in, of (B) from, of

(C) during, from (D) from, from

Q. 38 He was not only accused of theft .... of conspiracy.

(A) rather (B) but also

(C) but even (D) rather than

Q. 39 Select the word that fits the analogy:

Explicit: Implicit :: Express:

(A) Impress (B) Repress

(C) Compress (D) Suppress

Q. 40 The Canadian constitution requires that equal importance be given to English and French Last year, Air Canada lost a lawsuit, and had to pay a six-figure fine to a French-speaking couple after they filed complaints about formal in-flight announcements in English lasting 15 seconds, as opposed to informal 5 second messages in French.

The French-speaking couple were upset at

(A) the in-flight announcements being made in English.

(B) the English announcements being clearer than the French ones.

(C) the English announcements being longer than the French ones.

(D) equal importance being given to English and French

Graduate Aptitude Test in Engineering – 2021

Q.41 The people .... were at the demonstration were from all sections of society.

(A) whose (B) which

(C) who (D) whom

Answers Key

1	2	3	4	5	6	7	8	9	10
C	B	D	D	D	B	C	B	C	C
11	12	13	14	15	16	17	18	19	20
D	A	A	C	C	D	D	A	C	C
21	22	23	24	25	26	27	28	29	30
B	A	C	A	B	C	A	C	C	A
31	32	33	34	35	36	37	38	39	40
A	C	C	A	C	D	A	B	B	C
41									
C									

Explanations

Q. 1.

(A) Masked – Hide under a false appearance (Opposite word)

(B) Belied – Be in contradiction with (Opposite word)

(C) Betrayed – Reveal unintentionally (Appropriate word)

(D) Suppressed – To put down by force or authority (Irrelevant word)

Q. 2.

Circuitous – Deviating from a straight course (long and Indirect)

(A) Cyclic – Recurring in cycle (Rotary)

(B) Indirect – Not leading by straight line

(C) Confusing – Lacking clarity

(D) Crooked – For shapes (Irregular in shape)

Q. 3.

(A) Uphold – Cause to remain (Not appropriate word)

(B) Restrain – Keep under control (Not appropriate word)

(C) Cherish – Be fond of (Not appropriate word)

(D) Conserve – Keep in safety and protect from harm, decay, loss or destruction. (Appropriate word)

Q. 4.

(A) Modern warfare has resulted in civil strife. (No direct effect of warfare given, not appropriate)

(B) Chemical agents are useful in modern warfare: Passage doesn't mean whether chemical agents are useful or not. (Not appropriate)

(C) Use of chemical agents in warfare would be undesirable: Given that people in military think these are useful. (Not appropriate)

(D) People in military establishments like to use chemical agents in war: Given that people in military think these are useful tools for their work. (Appropriate)

Q. 5.

(A) Similar – Like sb/sth but not exactly the same

(B) Most – Largest in number or amount

(C) Uncommon – Not existing in large numbers or in many places.

(D) Available – Accessible and ready for use or service or convenient for use. (Most Appropriate)

Human genes are to be manipulated only to correct diseases for which available treatments are unsatisfactory.

Q. 6.

(A) Periodicity – The quality of recurring at regular intervals (Synonyms)

(B) Rarity – Infrequency, Something unusual (Most Appropriate)

(C) Gradualness – The quality of being gradual or of coming about by gradual stages

(D) Persistency – Continually recurring to the mind (Not Appropriate)

Q. 7.

(A) Identified – Having the identity known

(B) Ascertained – Found or determined by scientific observations

(C) Exacerbated – Make worse (Appropriate word)

(D) Analysed – Examined carefully and methodically

Q. 11.

(A) Reflective – Thinking deeply (Inappropriate word)

(B) Utopian – An idealistic but impractical (Inappropriate word)

(C) Luxuriant – Growing in extreme abundance (Inappropriate word)

(D) Unpopular – Not liked by a group (Appropriate word)

Q. 12.

(A) Intrepidity – Courageousness (Most appropriate word)

(B) Inevitability – As is certain to happen (Inappropriate word)

(C) Inability – Not being able (Inappropriate word)

(D) Inertness – Immobility by virtue of being inert (Inappropriate word)

Q. 14.

"**want**" is incorrect option , because Verb+s/es is used with III person pronoun.

Q. 15.

He displayed modesty in his interactions.

Q. 16.

Remove "**How**"

Q. 17.

Short-term memory loss is a symptom of Alzheimer's disease. It is a universal truth that should be written in present tense.

Q. 18.

(A) Contentment – Feeling of satisfaction (Appropriate)

(B) Ambition – Desire to do something (Not Appropriate).

(C) Perseverance – Try to achieve aim (Not Appropriate)

(D) Hunger – A strong desire for something (Not Appropriate)

Q. 21

(A) Momento - A misspelling

(B) Memento – A reminder of past events. (Appropriate word)

(C) Momentum – An impelling force or strength, Capacity for further growth. (Not appropriate)

(D) Moment – A particular point in time. (Not appropriate)

Q. 22

(A) Croak – Sound made by a frog.

(B) Roar – Sound made by a lion.

(C) Hiss – Sound made by a snake.

(D) Patter – Sound, as of footsteps.

Q. 23

Educe – Derive, draw out, evoke, Deduce.

(A) Exert – To use power or influence to affect somebody or something. (Not appropriate)

(B) Educate – Prepare, train, teach. (Not appropriate)

(C) Extract – Romove or obtain a substance from something. (Appropriate)

(D) Extend – Make something longer or larger. (Not appropriate)

Q. 24

(A) ranks as one of the leading causes of death

Here, main verb **(ranks)** agrees with the subject (**Tuberculosis** - Singular Noun) of sentence.(Most effective sentence.)

(B) rank as one of the leading causes of death

Here, main verb **(rank)** does not agree with the subject (**Tuberculosis** - Singular Noun) of sentence.(ineffective sentence)

(C) has the rank of one of the leading causes of death

Here, **rank** is not a main verb. Main Verb (**rank**) of sentence has been changed into noun (**the rank**). .(ineffective sentence)

(D) are one of the leading causes of death.(ineffective sentence)

Here, verb (**are**) does not agree with the subject (**Tuberculosis** - Singular Noun) of sentence.

Q. 25

Human progress (innovations) somewhere interferes to the environmental system i.e. damages to the environmental substances. Therefore, it can be concluded, **Human progress is contradictory to environmental security.**

Q. 26

If I were you, I .... that laptop. It's much too expensive.

In these sentences, sentence **If I were you,........** shows supposition for time passed before. Hence "**would**" will be used (Unreal future). Here, it is self understood that the laptop is much too expensive, hard to buy. Hence **'not'** will be used with "**would**".

Q. 27

To turn a deaf ear to somebody/something (**Idiom**) = To ignore or refuse to listen to somebody/something.

Q. 28

Where there is a will, there is a way

It is an old saying. That means, if you really want to do something then you will find a way of it.

Q. 29

Here, Subject " The ways" is plural, it will agree with a plural verb/ helping verb (**are**). If we have "are being" as a verb then sentence will be framed in Passive Voice, that is not our sentence (i.e. given sentence is in Active Voice).

Q. 30

(A) Forgo:- Go without.

Exclusive:- Excluding. (Most appropriate.)

(B) Forget:- Fail to remember.

Inclusive:- Including. (Not appropriate.)

(C) Accept:- Consent to receive.

Exhaustive:- Including. (Not appropriate.)

(D) Adopt:- Legally take.

Intrusive:- Annoyance through being unwelcomed. (Not appropriate.)

Q. 31

Stairs:- A set of steps that lead from one level to another, esp. in a building.

Stares:- Look fixedly at someone or something with one's eyes wide open.

Q. 32

Rational:- Based on or in accordance with reason or logic. (Not Appropriate word.)

Reasonable:- Fair and sensible. (Not Appropriate word.)

Errant:- Erring or straying from the accepted course or standards. (Appropriate word.)

Good:- Advancement of prosperity or well-being. (Not Appropriate word.)

Q. 33

Objective case of who is "*to whom*" .

Q. 34

Here, both the two events occur in past one after another. Hence, the event completing first will take past perfect tense and another event will be expressed in Simple past tense.

Q. 35

As per the data given, correct pairings are:

"dhrupad" is associated with "baani"

"gayaki" is associated with "vocal"

"baaj" is associated with "instrumental"

"gharana" is associated with "musicians from a particular lineage"

Q. 36

Head-hunter refers to the number of heads he could gather.

37

Quantitative Aptitude

Graduate Aptitude Test in Engineering - 2007

Q. 1 An ellipsoidal object has three axes measuring 40 cm, 20 cm and 20 cm respectively. The volume of the object is

(A) 4.23 liters (B) 8.38 liters

(C) 12.63 liters (D) 17.05 liters

Graduate Aptitude Test in Engineering - 2010

Q. 2 5 skilled workers can build a wall in 20 days; 8 semi-skilled workers can build a wall in 25 days; 10 unskilled workers can build a wall in 30 days. If a team has 2 skilled, 6 semi-skilled and 5 unskilled workers, how long will it take to build the wall ?

(A) 20 (B) 18

(C) 16 (D) 15

Q. 3 25 persons are in a room. 15 of them play hockey, 17 of them play football and 10 of them play both hockey and football. Then the number of persons playing neither hockey nor football is:

(A) 2 (B) 17

(C) 13 (D) 3

Graduate Aptitude Test in Engineering - 2011

Q. 4 The sum of n terms of the series 4+ 44 + 444 +.... is

(A) $\frac{4}{81}[10^{n+1} - 9n. - 1]$ (B) $[\frac{4}{81} 10^{n-1} - 9n. - 1]$

(C) $[\frac{4}{81} 10^{n+1} - 9n. - 10]$ (D) $[\frac{4}{81} 10^{n} - 9n. - 1]$

Q. 5 Given that f(y) = | y |/ y, and q is any non-zero real number, the value of | f(q) – f(–q) | is

(A) 0 (B) -1

(C) 1 (D) 2

Q. 6 Three friends, R, S and T shared toffee from a bowl. R took 1/3rd of the toffees, but returned four to the bowl. S took 1/4th of what was left but returned three toffees to the bowl. T took half of the remainder but returned two back into the bowl. If the bowl had 17 toffees left, how many toffees were originally there in the bowl ?

(A) 38 (B) 31

(C) 48 (D) 41

Q. 7 The fuel consumed by a motorcycle during a journey while traveling at various speeds is indicated in the graph below.

The distances covered during four laps of the journey are listed in the table below

Lap	Distance (kilometers)	Average speed (kilometers per hour)
P	15	15
Q	75	45
R	40	75
S	10	10

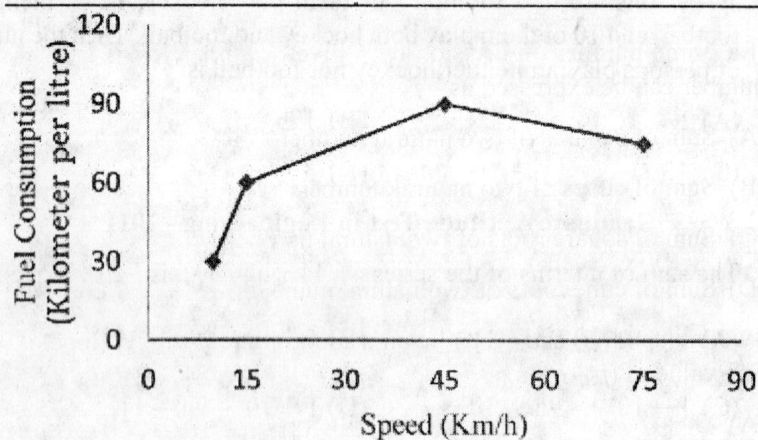

From the given data, we can conclude that the fuel consumed per kilometer was least during the lap

(A) P (B) Q

(C) R (D) S

Graduate Aptitude Test in Engineering - 2012

Q. 8 Two policemen, A and B, fire once each at the same time at an escaping convict. The probability that A hits the convict is three times the probability that B hits the convict. If the probability of the convict not getting injured is 0.5, the probability that B hits the convict is

(A) 0.14 (B) 0.22

(C) 0.33 (D) 0.40

Q. 9 The total runs scored by four cricketers P, Q, R, and S in years 2009 and 2010 are given in the following table:

Player	2009	2010
P	802	1008
Q	765	912
R	429	619
S	501	701

The player with the lowest percentage increase in total runs is

(A) P (B) Q

(C) R (D) S

Q. 10 If a prime number on division by 4 gives a remainder of 1, then that number can be expressed as

(A) Sum of squares of two natural numbers

(B) Sum of cubes of two natural numbers

(C) Sum of square roots of two natural numbers

(D) Sum of cube roots of two natural numbers

Q. 11 Two points $(4, p)$ and $(0, q)$ lie on a straight line having a slope of 3/4. The value of $(p - q)$ is

(A) -3 (B) 0

(C) 3 (D) 4

Q. 12 The arithmetic mean of five different natural numbers is 12. The largest possible value among the numbers is

(A) 12 (B) 40

(C) 50 (D) 60

Graduate Aptitude Test in Engineering - 2013

Q. 13 Velocity of an object fired directly in upward direction is given by V= 80 − 32t, where t(time) is in seconds. When will the velocity be between 32 m/s and 64 m/s?

(A) (1, 3/2) (B) (1/2, 1)

(C) (1/2, 3/2) (D) (1, 3)

Q. 14 In a factory, two machines M1 and M2 manufacture 60% and 40% of the autocomponents respectively. Out of the total production, 2% of M1 and 3% of M2 are found to be defective. If a randomly drawn autocomponent from the combined lot is found defective, what is the probability that it was manufactured by M2 ?

(A) 0.35 (B) 0.45

(C) 0.5 (D) 0.4

Q. 15 Following table gives data on tourists from different countries visiting India in the year 2011.

Country	Number of Tourists
USA	2000
England	3500
Germany	1200
Italy	1100
Japan	2400
Australia	2300
France	1000

Which two countries contributed to the one third of the total number of tourists who visited India in 2011?

(A) USA and Japan

(B) USA and Australia

(C) England and France

(D) Japan and Australia

Q. 16 If $|-2X+9| = 3$ then the possible value of $|-X| - X^2$ would be:

(A) 30 (B) -30

(C) -42 (D) 42

Q. 17 A hemispherical vessel of 300 mm diameter is completely filled with oil and water. If the oil layer is 50 mm deep on the top, the volume of water in the vessel in litres is

(A) 1.27 (B) 3.73

(C) 7.07 (D) 14.14

Graduate Aptitude Test in Engineering - 2014

Q. 18 The total exports and revenues from the exports of a country are given in the two pie charts below. The pie chart for exports shows the quantity of each item as a percentage of the total quantity of exports. The pie chart for the revenues shows the percentage of the total revenue generated through export of each item. The total quantity of exports of all the items is 5 lakh tonnes and the total revenues are 250 crore rupees. What is the ratio of the revenue generated through export of Item 1 per kilogram to the revenue generated through export of Item 4 per kilogram?

(A) 1:2 (B) 2:1

(C) 1:4 (D) 4:1

Exports

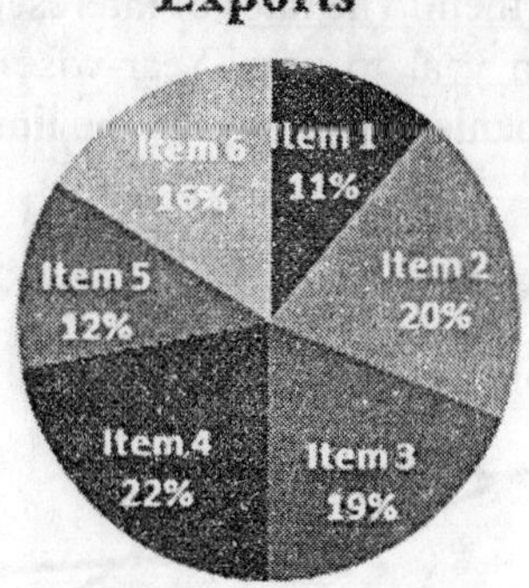

Revenues

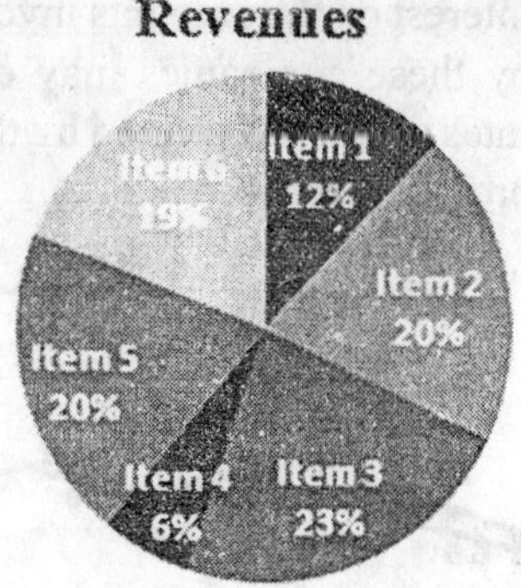

Q. 19 10% of the population in a town is HIV^+. A new diagnostic kit for HIV detection is available; this kit correctly identifies HIV^+ individuals 95% of the time, and HIV^- individuals 89% of the time. A particular patient is tested using this kit and is found to be positive. The probability that the individual is actually positive is ....

Graduate Aptitude Test in Engineering - 2015

Q. 20 Operators □, ⊙ and → are defined by: a □ b = $\frac{a-b}{a+b}$; a ⊙ b = $\frac{a+b}{a-b}$; a → b = ab.

Find the value of (66 □ 6) → (66 ⊙ 6)

(A) – 2 (B) – 1

(C) 1 (D) 2

Q. 21 If $\log_x(5/7) = -1/3$, then the value of x is

(A) 343/125 (B) 125/343

(C) – 25/49 (D) – 49/25

Graduate Aptitude Test in Engineering - 2016

Q. 22 (x % of y) + (y % of x) is equivalent to .

(A) 2 % of xy (B) 2 % of (xy/100)

(C) xy % of 100 (D) 100 % of xy

Q. 23 The sum of the digits of a two digit number is 12. If the new number formed by reversing the digits is greater than the original number by 54, find the original number.

(A) 39 (B) 57

(C) 66 (D) 93

Q. 24 Two finance companies, P and Q, declared fixed annual rates of interest on the amounts invested with them. The rates of interest offered by these companies may differ from year to year. Year-wise annual rates of interest offered by these companies are shown by the line graph provided below.

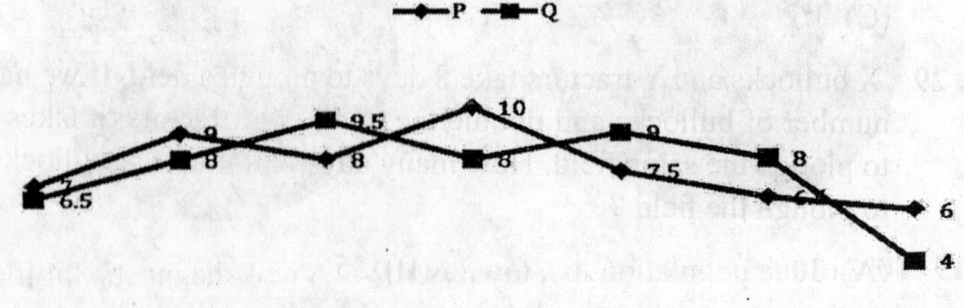

If the amounts invested in the companies, P and Q, in 2006 are in the ratio 8:9, then the amounts received after one year as interests from companies P and Q would be in the ratio

(A) 2:3 (B) 3:4

(C) 6:7 (D) 4:3

Q. 25 A square pyramid has a base perimeter x, and the slant height is half of the perimeter. What is the lateral surface area of the pyramid?

(A) x^2 (B) $0.75\ x^2$

(C) $0.50\ x^2$ (D) $0.25\ x^2$

Q. 26 Ananth takes 6 hours and Bharath takes 4 hours to read a book. Both started readin copies of the book at the same time. After how many hours is the number of pages to be read by Ananth, twice that to be read by Bharath?Assume Ananth and Bharath read all the pages with constant pace.

(A) 1 (B) 2

(C) 3 (D) 4

Graduate Aptitude Test in Engineering - 2017

Q. 27 If *a* and *b* are integers and *a* – *b* is even, which of the following must always be even?

(A) ab (B) $a^2 + b^2 + 1$

(C) $a^2 + b + 1$ (D) ab – b

Q. 28 A couple has 2 children. The probability that both children are boys if the older one is a boy is

(A) 1/4 (B) 1/3

(C) 1/2 (D) 1

Q. 29 X bullocks and Y tractors take 8 days to plough a field. If we halve the number of bullocks and double the number of tractors, it takes 5 days to plough the same field. How many days will it take X bullocks alone to plough the field ?

(A) 30 (B) 35

(C) 40 (D) 45

Q. 30 In the graph below, the concentration of a particular pollutant in a lake is plotted over (alternate) days of a month in winter(average temperature 10 °C) and a month in summer (average temperature 30 °C).

Consider the following statements based on the data shown below

i. Over the given months, the difference between the maximum and minimum pollutant concentrations is the same in both winter and summer.

ii. There are at least four days in the summer month such that the pollutant concentrations on those days are within 1 ppm of the pollutant concentrations on the corresponding days in the winter month.

Which one of the following options is correct?

(A) only I (B) only ii

(C) both I and ii (D) Neither I nor ii

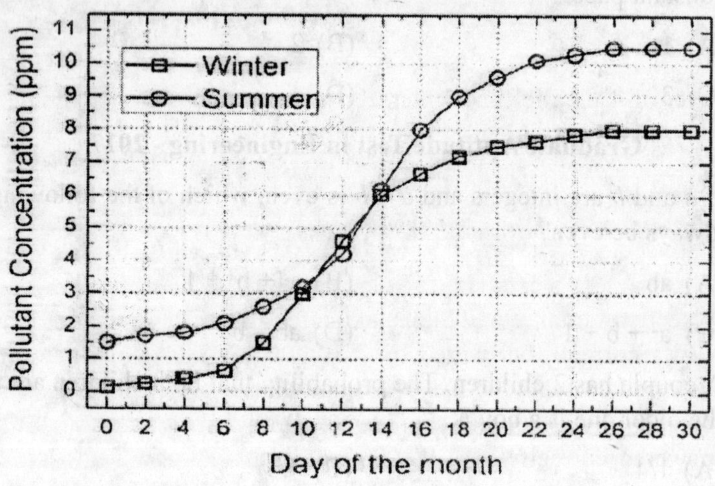

Graduate Aptitude Test in Engineering - 2018

Q. 31 The area of an equilateral triangle is $\sqrt{3}$. What is the perimeter of the triangle ?

(A) 2 (B) 4

(C) 6 (D) 8

Q. 32 Arrange the following three-dimensional objects in the descending order of their volumes:

(i) A cuboid with dimensions 10 cm, 8 cm and 6 cm

(ii) A cube of side 8 cm

(iii) A cylinder with base radius 7 cm and height 7 cm

(iv) A sphere of radius 7 cm

(A) (i), (ii), (iii), (iv) (B) (ii), (i), (iv), (iii)

(C) (iii), (ii), (i), (iv) (D) (iv), (iii), (ii), (i)

Q. 33 An automobile travels from city A to city B and returns to city A by the same route. The speed of the vehicle during the onward and return journeys were constant at 60 km/h and 90 km/h, respectively. What is the average speed in km/h for the entire journey?

(A) 72 (B) 73

(C) 74 (D) 75

Q. 34 A set of 4 parallel lines intersect with another set of 5 parallel lines. How many parallelograms are formed ?

(A) 20 (B) 48

(C) 60 (D) 72

Q. 35 To pass a test, a candidate needs to answer at least 2 out of 3 questions correctly. A total of 6,30,000 candidates appeared for the test. Question A was correctly answered by 3,30,000 candidates. Question B was answered correctly by 2,50,000 candidates. Question C was answered correctly by 2,60,000 candidates. Both questions A and B were answered correctly by 1,00,000 candidates. Both questions B and C were answered correctly by 90,000 candidates. Both questions A and C were answered correctly by 80,000 candidates.

If the number of students answering all questions correctly is the same as the number answering none, how many candidates failed to clear the test ?

(A) 30,000 (B) 2,70,000

(C) 3,90,000 (D) 4,20,000

Q. 36 If $x^2 + x - 1 = 0$ what is the value of $x^4 + 1/x^4$?

(A) 1 (B) 5

(C) 7 (D) 9

Q. 37 In a detailed study of annual crow births in India, it was found that there was relatively no growth during the period 2002 to 2004 and a sudden spike from 2004 to 2005. In another unrelated study, it was found that the revenue from cracker sales in India which remained fairly flat from 2002 to 2004, saw a sudden spike in 2005 before declining again in 2006.

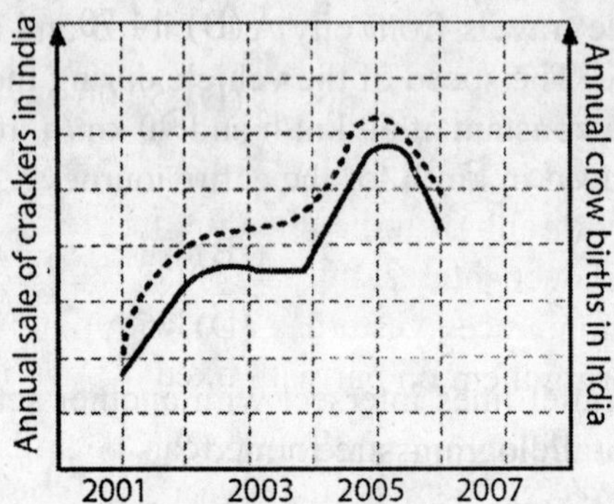

The solid line in the graph below refers to annual sale of crackers and the dashed line refers to the annual crow births in India. Choose the most appropriate inference from the above data.

(A) There is a strong correlation between crow birth and cracker sales.

(B) Cracker usage increases crow birth rate.

(C) If cracker sale declines, crow birth will decline.

(D) Increased birth rate of crows will cause an increase in the sale of crackers.

Graduate Aptitude Test in Engineering - 2019

Q. 38 The radius as well as the height of a circular cone increases by 10%. The percentage increase in its volume is ....

(A) 17.1 (B) 21.0

(C) 33.1 (D) 72.8

Q. 39 Two trains started at 7 AM from the same point. The first train travelled north at a speed of 80 km/h and the second train travelled south at a speed of 100km/h. The time at which they were 540 km apart is ……… AM.

(A) 9 (B) 10

(C) 11 (D) 11.30

Q. 40 In a country of 1400 million population, 70% own mobile phones. Among the mobile phone owners, only 294 million access the internet. Among these users, only half buy goods from e-commerce portals. What is the percentage of these buyers in the country?

(A) 10.50 (B) 14.70

(C) 15.00 (D) 50.00

Q. 41 Since the last one year, after a 125 basis point reduction in repo rate by the Reserve Bank of India, banking institutions have been making a demand to reduce interest rates on small saving schemes. Finally, the government announced yesterday a reduction in interest rates on small schemes to bring them on par with fixed deposit interest rates.

Which of the following statements can be inferred from the given passage?

(A) Whenever the Reserve Bank of India reduces the repo rate, the interest rates on small saving schemes are also reduced.

(B) Interest rates on small saving schemes are always maintained on par with fixed deposit interest rates.

(C) The government sometime takes into consideration the demands of banking institutions before reducing the interest rates on small saving schemes.

(D) A reduction in interest rates on small saving schemes follow only after a reduction in repo by the Reserve Bank of India.

Graduate Aptitude Test in Engineering - 2020

Q. 42 A superadditive function f (.) satisfies the following property

$f(x_1 + x_2) \geq f(x_1) + f(x_2)$

Which of the following functions is a superadditive function for $x > 1$?

(A) e^x (B) $\sqrt{x}$

(C) $1/x$ (D) e^{-x}

Q. 43 The global financial crisis in 2008 is considered to be the most serious world-wide financial crisis, which started with the sub-prime lending crisis in USA in 2007. The sub-prime lending crisis led to the banking crisis in 2008 with the collapse of Lehman Brothers in 2008. The sub -prime lending refers to the provision of loans to those borrowers who may have difficulties in repaying loans, and it arises because of excess liquidity following the East Asian crisis.

Which one of the following sequences shows the correct precedence as per the given passage ?

(A) East Asian crisis → subprime lending crisis→ banking crisis→ global financial crisis.

(B) Subprime lending crisis → global financial crisis→ banking crisis→ East Asian crisis.

(C) Banking crisis→ subprime lending crisis→ global financial crisis → East Asian crisis.

(D) Global financial crisis → East Asian crisis → banking crisis → subprime lending crisis.

Q. 44 It is quarter past three in your watch. The angle between the hour hand and the minute hand is

(A) 0^0 (B) 7.5^0

(C) 15^0 (D) 22.5^0

Q. 45 A circle with centre O is shown in the figure. A rectangle PORS of maximum possible area is inscribed in the circle. If the radius of the circle is a, then the area of the shaded portion is

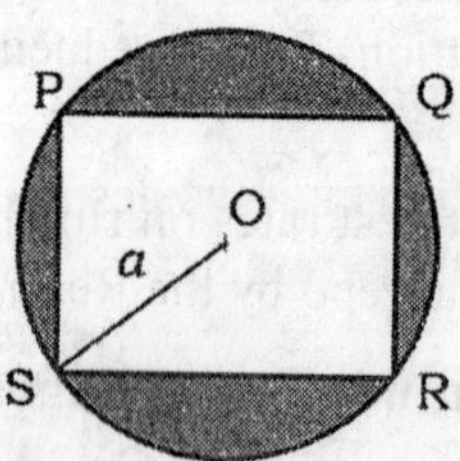

(A) $\pi a^2 - a^2$ (B) $\pi a^2 - \sqrt{2(a)}^2$

(C) $\pi a^2 - 2a^2$ (D) $\pi a^2 - 3\ a^2$

Q. 46 a, b,c are real numbers. The quadratic equation $ax^2 - bx + c = 0$ has equal roots, which is β, then

(A) $\beta = b/a$ (B) $\beta^2 = -ac$

(C) $\beta^3 = bc/(2a)$ (D) $b^2 \neq 4ac$

Q. 47 The following figure shows the data of students enrolled in 5 years (2014 to 2018) for two schools P and Q. During this period, the ratio of the average number of the students enrolled in school P to the average of the difference of the number of students enrolled in schools P and Q is ....

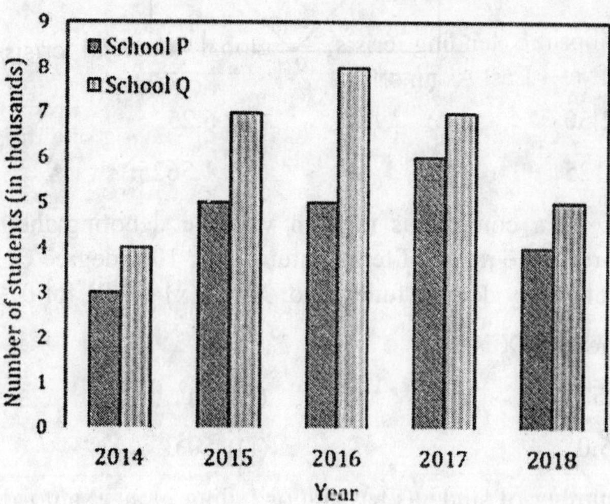

(A) 8:23 (B) 23:8

(C) 23:31 (D) 31:23

Graduate Aptitude Test in Engineering - 2021

Q. 48 For a regular polygon having 10 sides, the interior angle between the sides of the polygon, in degrees, is:

(A) 396 (B) 324

(C) 216 (D) 144

Q. 49 Which one of the following numbers is exactly divisible by $(11^{13}+1)$?

(A) $11^{26}+1$ (B) $11^{33}+1$

(C) $11^{39}-1$ (D) $11^{52}-1$

Q. 50 In the figure shown above, each inside square is formed by joining the midpoints of the sides of the next larger square. The area of the smallest square (shaded) as shown, in cm^2 is:

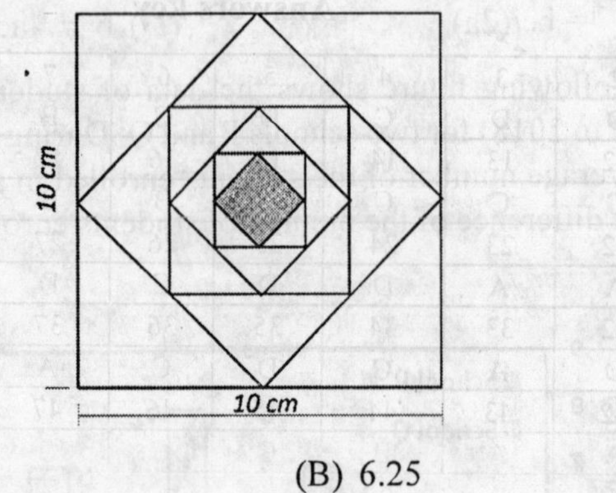

(A) 12.50 (B) 6.25

(C) 3.125 (D) 1.5625

Q. 51 Let X be a continuous random variable denoting the temperature measured. The range of temperature is [0, 100] degree Celsius and let the probability density function of X be $f(x) = 0.01$ for $0 \leq X \leq 100$.

The mean of X is

(A) 2.5 (B) 5.0

(C) 25.0 (D) 50.0

Q. 52 The number of students passing or failing in an exam for a particular subject are presented in the bar chart below. Students who pass the exam cannot appear for the exam again. Students who fail the exam in the first attempt must appear for the exam in the following year. Students always pass the exam in their second attempt.

The number of students who took the exam for the first time in the year 2 and the year 3 respectively, are ....

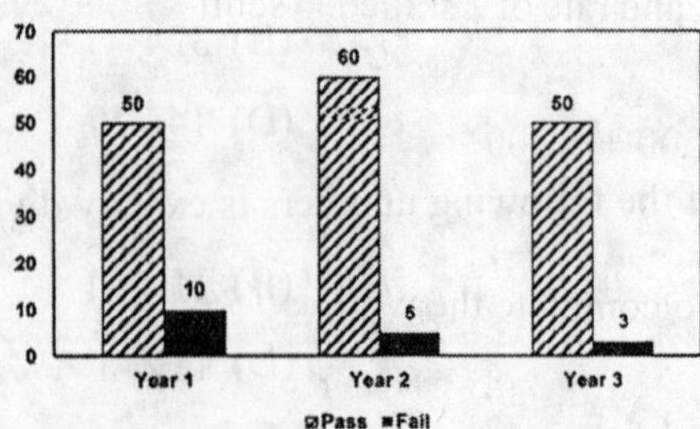

(A) 65 and 53 (B) 60 and 50

(C) 55 and 53 (D) 55 and 48

Answers key

1	2	3	4	5	6	7	8	9	10
B	D	D	C	D	C	B	A	B	A
11	12	13	14	15	16	17	18	19	20
C	C	C	C	C	B	B	D	0.4897	C
21	22	23	24	25	26	27	28	29	30
A	A	A	D	D	C	D	C	A	B
31	32	33	34	35	36	37	38	39	40
C	D	A	C	D	C	A	C	B	A
41	42	43	44	45	46	47	48	49	50
C	A	A	B	C	C	B	D	D	C
51	52								
D	D								

Explanations

Q. 1

Given that a = 40/2 = 20 cm, b = 20/2 = 10 cm , c = 20/2 = 10 cm

Volume of ellipsoid = $\left(\frac{4}{3}\right) \pi$ a b c

$= \frac{4}{3} \times 3.14 \times 20 \times 10 \times 10$

$= 8373.33\ cm^3 = 8.38$ litres

Q. 2

Per day work and rate of one skilled worker = 1/(20×5) = 1/100

Per day work and rate of one semi-skilled worker = 1/(25×8) = 1/200

Per day work and rate of one unskilled worker = 1/(30×10) = 1/300

Per day work and rate of 2 skilled, 6 semi-skilled and 5 unskilled workers is

= 2/100 + 6/200 + 5/300

= 1/15

Thus, Time to complete the work is 15 days.

Q. 3

Refer to Venn diagram:–

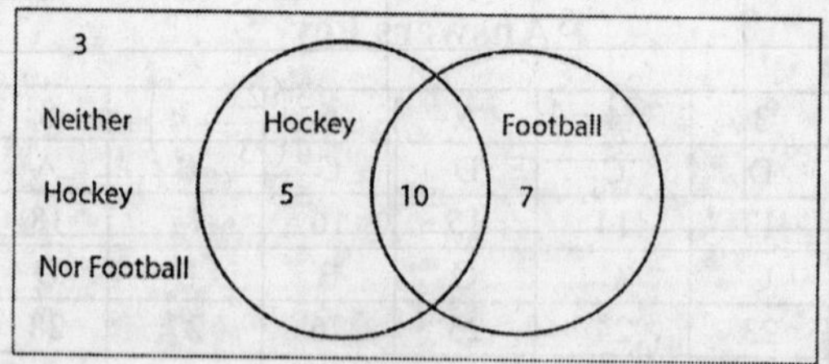

Thus, the number of persons playing neither hockey nor football is 3.

Q. 4

Given that

$$4 + 44 + 444 + 4444 + \ldots\ldots \text{ up to n terms}$$

$\Rightarrow$ $4(1+11+111+1111+\ldots\ldots)$

$\Rightarrow$ $4/9(9+99+999+9999\ldots\ldots)$

It can be written as

$= 4/9[(10 - 1) + (100 - 1) + (1000 - 1) + (10000 - 1) \ldots]$

$= 4/9[(10 + 100 + 1000 + 10000\ldots$ up to n terms) $- (1+11+111+1111 \ldots$ up to n terms)]

By applying geometric progression formula for addition of n terms

$$S_n = \frac{a(r^n - 1)}{r - 1} \text{ for } r > 1$$

$$= 4/9\left[\frac{10(10^n - 1)}{10 - 1} - n\right]$$

$$= 4/81(10^{n+1} - 9n - 10)$$

Q. 5

Give that

$f(y) = \frac{|y|}{y}$ and q is any non zero real number.

$\therefore$ $f(q) = \frac{|q|}{q} = 1$

And $f(-q) = \frac{|-q|}{-q} = -1$

Thus, $f(q) - f(-q)| = |1 - (-1)| = 2$

Q. 6

Consider, total no. of toffees = x

R	S	T
$\frac{x}{3}-4$	$\frac{\frac{2x}{3}-4}{4}-3$	$\frac{\frac{3}{4}\left(\frac{2x}{3}+4\right)+3}{2}-2$

Given that

$$\frac{x}{3}-4+\frac{\frac{2x}{3}-4}{4}-3+\frac{\frac{3}{4}\left(\frac{2x}{3}+4\right)+3}{2}-2+17=x$$

$\Rightarrow$ $x = 48$

Q. 7

Inverse of Fuel consumption gives fuel consumed per kilometer, which is least for Lap Q.

$\therefore$ Fuel consumed per kilometer = 1/75 = 0.013 litre/ km

Q. 8

Probability that B hits = P; Probability that B doesn't hits = 1 – P

Probability that A hits = 3P; Probability that A doesn't hits = 1 – 3P

$\therefore$ Probability of the convict not getting injured = (1 – P)(1 – 3P)

$$0.5 = 3P^2 - 4P + 1$$

$\Rightarrow$ $P = 0.14, 1.19$

$P = 1.19$ is not possible because $0 \le P \le 1$.

Q. 9

Player	2009	2010	Percent increase
P	802	1008	$=\left(\frac{1008-802}{802}\right)\times 100 = 25.69\ \%$
Q	765	912	$=\left(\frac{912-765}{765}\right)\times 100 = 19.22\ \%$
R	429	619	$=\left(\frac{619-429}{429}\right)\times 100 = 44.29\ \%$
S	501	701	$=\left(\frac{701-501}{501}\right)\times 100 = 39.92\ \%$

Q. 10

Such prime numbers are 5, 13, 17, 29, …………

These prime numbers can be expressed

$5 \Rightarrow 1^2 + 2^2$

$13 \Rightarrow 2^2 + 3^2$

$17 \Rightarrow 3^2 + 4^2$

$29 \Rightarrow 2^2 + 5^2$

Q. 11

Equation of line

$$y = \frac{3}{4}x + C$$

Putting, x = 4 and y = p

$$p = 3 + C \quad \text{(i)}$$

Putting, x = 0 and y = q

$$q = 0 + C \quad \text{(ii)}$$

Subtracting equation (ii) from (i)

$$\Rightarrow \quad p - q = 3$$

Q. 12

Given

Sum of natural numbers = 12×5 = 60

Here, Numbers are different and can be 1, 2, 3, 4, and 50.

Q. 13

Case I When V = 32 m/s

$$T = (32 - 80)/(-32) = 3/2$$

Case II When V = 64 m/s

$$t = (64 - 80)/(-32) = 1/2$$

Thus, velocity will be between 32 m/sec and 64 m/sec at (1/2, 3/2).

Q. 14

Consider total component are = 100

Thus no. of component manufactured by M1 = 60

And no. of component manufactured by M2 = 40

Defective component by M1 = 60×2/100 = 1.2

Defective component by M2 = 40×3/100 = 1.2

Thus, probability for both the manufactures = 0.5

Q. 15

Total Number of tourists = 13500

One third of total tourists = 13500/3 = 4500

Thus option (C) only gives the solution.

Q. 16

Taking square of both sides

$$(-2X+9)^2 = 9$$

$\Rightarrow$ $X = 3, 6$

Putting, $X = 6$,

$\Rightarrow$ $-X| - X^2 = |-6| - 6^2 = -30$

Q. 17

$$\text{Volume of spherical cap} = \frac{\Pi h^2}{3}(3R - h)$$

$$= \frac{3.14 \times 10^2}{3} \times (3 \times 15 - 10)$$

$$= 3.73 \text{ litres}$$

Q. 18

Revenue generated through export of Item 1 = $(250\times\frac{12}{100})/(5\times\frac{11}{100})$ crore rs./ lakh tones

Revenue generated through export of Item 4 = $(250\times\frac{6}{100})/(5\times\frac{22}{100})$ lakh tones

Required ratio = $[(250\times\frac{12}{100})/(5\times\frac{11}{100})]/[(250\times\frac{6}{100})/(5\times\frac{22}{100})]$

= 4:1

Q. 19

HIV$^-$		HIV$^+$	
0.90		0.10	
0.89	0.11	0.95	0.05
HIV$^-$	HIV$^+$	HIV$^+$	HIV$^-$

Total conditions for HIV$^+$ = 0.90×0.11 + 0.95×0.10 = 0.194

Actual conditions for HIV$^+$= 0.95×0.10 = 0.095

Probability (Actual HIV$^+$) = $\frac{0.095}{0.194}$ = 0.4897

Q. 20

We have $a \square b = \frac{a-b}{a+b}$; $a \odot b = \frac{a+b}{a-b}$; a b = ab.

∴ $66 \square 6 = \frac{66-6}{66+6}; = \frac{60}{72}$ and $66 \odot 6 = \frac{66+6}{66-6} = \frac{72}{60}$

Thus, $(66 \square 6) \rightarrow (66 \odot 6) = \frac{60}{72} \times \frac{72}{60} = 1$

Q. 21

Given that $\log_x(5/7) = -1/3$

$\{\ln(5/7)\}/\{\ln(x)\} = -1/3$ $[\because \log_x y = \ln(y)/\ln(x)]$

⇒ $\ln(x) = 1.0094$

⇒ $x = e^{1.0094} = 2.744$

Option (A) gives 343/125 = 2.744.

Q. 22

x % of y = xy/100

and (y % of x) = xy/100

∴ (x % of y) + (y % of x) = xy/100 + xy/100

= 2xy/100 which is 2% of xy.(i.e. 2xy/100)

Q. 23

Consider, two digit number is xy. The number is represented as 10x + y.

Given, x + y = 12 (i)

Also, 10y + x – (10x + y) = 54

$\Rightarrow$ $9y - 9x = 54$ (ii)

$\because$ Reverse of number $10x + y$ is $10y + x$.

From, equation (i) and (ii)

$x = 3$ and $y = 9$

Hence, number is 39.

Alternate: Directly from options.

For, (A) 39

$3 + 9 = 12$ and $93 - 39 = 54$

Fulfills the conditions given in question.

For, (B) 57

$5 + 7 = 12$ and $75 - 57 = 18 \neq 54$

Doesn't Fulfill the conditions given in question.

For, (C) 66

$6 + 6 = 12$ and $66 - 66 = 0 \neq 54$

Doesn't Fulfill the conditions given in question.

For, (D) 93

$9 + 3 = 12$ and $39 - 93 = -54 \neq 54$

Doesn't Fulfill the conditions given in question.

Q. 24

Interest for company (P) = $8 \times 6 \times T/100 = 0.48\ T$

($\because$ Simple interest = Amount×Rate×Time/100)

Interest for company (Q) = $9 \times 4 \times T/100 = 0.36\ T$

Hence, Required ratio = $0.48\ T/\ 0.36\ T = 4 : 3$

Q. 25

We know

Lateral surface area $= 4\times\left(\left(\frac{1}{2}\times\frac{x}{2}\times\frac{x}{4}\right)\right)$

$= 0.25x^2$

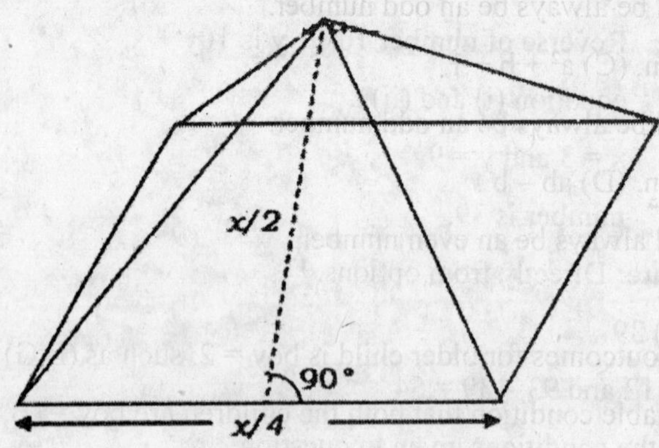

Q. 26

Consider, after x hours, the number of pages to be read by Ananth will twice that to be read by Bharath.

	Ananth	Bharath
Page read in 1 hour	1/6	1/4
Page read in x hour	$x/6$	$x/4$
Page to be read	$1 - x/6$	$1 - x/4$

Hence,

$$1 - x/6 = 2(1 - x/4)$$

$$\Rightarrow \quad x = 3 \text{ hours}$$

Q. 27

According to relation given, a – b is even

It is only possible when, either both the a and b are even or odd.

We know,

Case-I Odd number × Odd number = Odd number

Case-II Even number × Even number = Even number

Case-III Odd number × Even number = Even number

Hence,

Option, (A) ab

May be even number in Case-II and Case-III, but in Case-I, it will be an odd number.

Option, (B) $a^2 + b^2 + 1$

It will be always be an odd number.

Option, (C) $a^2 + b + 1$

It will be always be an odd number.

Option, (D) $ab - b$

It will always be an even number.

Q. 28

Total outcomes for older child is boy = 2, such as (B,G) and (B,B)

Favorable condition that both the children are boy = (B, B)

Hence, The probability that both children are boys = 1/2

Q. 29

Given that

X bullocks + Y tractors → 8 days (i)

Hence, 8X bullocks + 8Y tractors → 1 day (ii)

X/2 bullocks + 2Y tractors → 5 days (iii)

Hence, 5X/2 bullocks + 10Y tractors → 1 day (iv)

Equating, equations (ii) and (iv)

$$8X + 8Y = 5X/2 + 10\,Y$$

$$\Rightarrow \qquad Y = 11X/4$$

Putting it in equation (iii), hence

30X bullocks → 1 day

Hence, X bullocks will plough the field in 30 days.

Q. 30

The difference between the maximum and the minimum pollutant concentrations are such as

In Winter = 8 – 0 = 8 ppm

In Summer = 10.5 – 1.5 = 9 ppm

Therefore, statement (i) is false and (ii) is correct.

Q. 31

We know equilateral triangle = $(\sqrt{3}/4)\ a^2$

$\Rightarrow \quad \sqrt{3} = (\sqrt{3}/4)\ a^2$

$\Rightarrow \quad a = 2$ unit

Hence, Perimeter of the triangle = 3×2 = 6 unit

Q. 32

(i) Volume of cuboids = 10×8×6 = 480 cm^3

(ii) Volume of cube = 8×8×8 = 512 cm^3

(iii) Volume of Cylinder = 3.14×72×7 = 1077.02 cm^3

(iv) Volume of Sphere = (4/3) πr3

= (4/3) × 3.14 ×73 = 1436.03 cm^3

Q. 33

Consider, Distance between city A and B is = x km

Time taken to travel from city A to B = x/60 h

Time taken to travel from city B to A = x/90 h

∴ Average speed = x/(x/60 + x/90) = 72 km/h

Q. 34

To form a parallelogram two parallel lines must intersect two parallel lines.

Hence, the number of parallelograms that are formed when m parallel lines intersect n parallel lines is = ${}^mC_2 \times {}^nC_2$.

Here, 4 parallel lines are intersecting another set of 5 parallel lines.

∴ Number of parallelograms formed = ${}^4C_2 \times {}^5C_2 = 6(10) = 60$

Q. 35

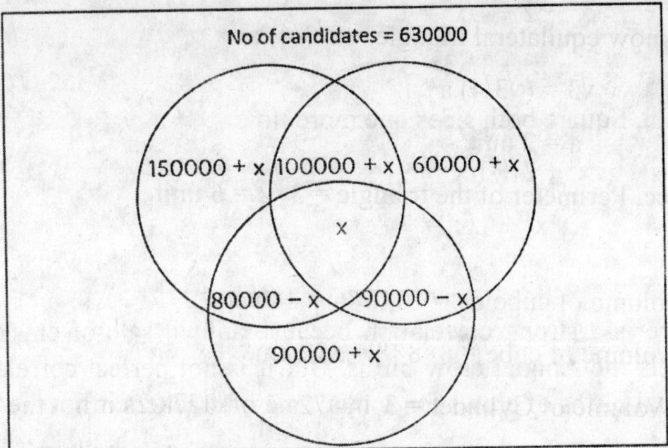

Consider, No of candidates not answered the any of the question = No of candidates

Answered all the question = x

Hence, Total candidates those who answered the questions = 630000 – x

Again, Referring the vain diagram

(150000 + x) + (100000 + x) + (80000 – x) + x + (60000 + x) + (90000 + x) + (90000 – x) = 630000 – x

⇒ x = 30000

Again,

Candidates failed to clear the exam = Candidates answered only one question from A, B and C + Candidates not answered any of the question

= (150000 + x) + (60000 + x) + (90000 + x) + 30000

= 420000

Q. 36

We know that $x \neq 0$, so let's divide by x on both sides:

$$x + 1 - 1/x = 0$$

$$\Rightarrow \quad x - 1/x = -1$$

Again, Square the equation on both sides:

$$x^2 - 2 + 1/x^2 = 1$$

$$\Rightarrow \quad x^2 + 1/x^2 = 3$$

Again, Square both sides one more time:

$$x^4 + 2 + 1/x^4 = 9$$

$$\Rightarrow \quad x^4 + 1/x^4 = 7$$

Q. 37

There is a strong correlation because annual sale of crackers directly affects the Annual crow births. But it is not perfect correlation becase annual births of crows for annual sale of crackers is not the same.

Note:- An example of a situation where you might find a **Perfect positive correlation,**

would be when you compare the total amount of money spent on tickets at the movie theater with the number of people who go. This means that every time that "x" number of people go, "y" amount of money is spent on tickets without variation.

An example of a situation where you might find a **Perfect negative correlation,** would be if you were comparing the amount of time it takes to reach a destination with the distance of a car (traveling at constant speed) from that destination.

An example of a situation where you might find a **Strong correlation** but **not Perfect positive correlation** would be if you examined the number of hours students spent studying for an exam versus the grade received. This won't be a perfect correlation because two students could spend the same amount of time studying and get different grades. But in general the rule will hold true that as the amount of time studying increases so does the grade received.

Q. 38

Volume of right circular cone (V) = $\frac{1}{3}\pi r^2 h$

After increasing radius and height of cone by 10%.

Volume of right circular cone will be (V_1) = $\frac{1}{3}\pi(1.1r)^2(1.1h)$ = 1.331 V

Hence, Percentage change in volume = (1.331 V – V)×100/V = 33.10 %

Q. 39

Relative speed of both trains 80 + 100 = 180 km/hr

Initially they are separated by distance = 540 km

Time taken to meet = 540/180 = 3 hrs

Thus, they meet 3 hrs after 7AM i.e., 10 AM.

Q. 40

Population of country = 1400 million

Population with mobile phones who access internet = 294 million

Population with mobile phones = 1400×70/100 = 980 million

Population with mobile phones who access internet and buy goods from e-commerce portals = 294×50/100 = 147 million

Hence, Percentage of buyers in country = 147×100/1400 = 10.5 %

Q. 42

Just check the options, putting $x_1 = 3$ and $x_2 = 2$

(a) $e^{x_1+x_2} \geq e^{x_1} + e^{x_2} \quad \Rightarrow \quad e^5 > e^3 + e^2$

(b) $\sqrt{x_1 + x_2} \geq \sqrt{x_1} + \sqrt{x_2} \quad \Rightarrow \quad \sqrt{3+2} < \sqrt{3} + \sqrt{2}$

(c) $\frac{1}{x_1 + x_2} \geq \frac{1}{x_1} + \frac{1}{x_2} \quad \Rightarrow \quad \frac{1}{3+2} < \frac{1}{3} + \frac{1}{2}$

(d) $e^{-x_1-x_2} \geq e^{-x_1} + e^{-x_2} \quad \Rightarrow \quad e^{-5} < e^{-3} + e^{-2}$

Q. 44

Quarter past 3:00 is 3:15.

Hours hand moves with a speed of 1/2 degree per min.

Minutes hand moves with speed of 6 degree per min.

Let's start at 3:00,

Minutes hand moves 90 degrees for 15 min and points to 3 and in 15 min hours hand moves 15×(1/2)=7.5 degrees from 3:00.

Q. 45

Shaded area = $\pi a^2 - (\sqrt{2}a)^2$

Q. 46

Given that, $x^2 - (b/a)x + (c/a) = o$

We know,

Sum of roots $(2\beta) = b/a$ (i)

Product of roots $(\beta^2) = c/a$ (ii)

Hence, (i)×(ii) gives $\beta^3 = bc/(2a^2)$

Q. 47

Average number of students in school P = (3+5+5+6+4)/5 = 23/5

Average number of students in school Q = (4+7+8+7+5)/5 = 31/5

Difference of the number of students enrolled in school P and

Q = (31 – 23)/5 = 8/5

Hence, Required ratio = 23:8

Q. 48

We know

Sum of interior angle = (n – 2) × 180 = (10 – 2) × 180 = 1440°

The interior angle between two adjacent sides = 1440/10 = 144°

Q. 49

Let us, multiply $(11^3 + 1)$ with $(11^3 - 1)$ in both numerator and denominator.

$$\Rightarrow \quad \frac{(11^{13}+1)(11^{13}-1)}{(11^{13}-1)}$$

$$\Rightarrow \quad \frac{(11^{26}-1)}{(11^{13}-1)}$$

Again, multiply with $(11^{13} + 1)$ in both numerator and denominator.

$$\Rightarrow \quad \frac{(11^{26}-1)(11^{26}+1)}{(11^{13}-1)(11^{26}+1)}$$

$$\Rightarrow \quad \frac{(11^{52}-1)}{(11^{13}-1)(11^{26}+1)}$$

Hence, $(11^{52} - 1)$ will be divided exactly by $(11^3 + 1)$.

Q. 50

Side of second outer square = $10/\sqrt{2}$

Side of third outer square = $10/\sqrt{2}^2$

Likewise,

Side of fifth outer square or inner smallest square = $10\sqrt{2}^5$

Area of smallest square = $(10/\sqrt{2}^5)^2 = 3.125$

Q. 51

We know,

For a continous random variable function

Mean = $\frac{a+b}{2} = \frac{0+10}{2} = 50$

Q. 52

Total number of student in the year-2 = 60 + 5 = 65

Students who failed in the year and appeared in year-2 = 10

So, the students who appeared first time in year-2 = 65 - 10 = 55

Similarly, Total number of students in year-3 = 53

Students who failed in the year-2 and appeared in year-3 = 5

So, the students who appeared first time in year-3 = 53 - 5 = 48

38

Verbal Reasoning

Graduate Aptitude Test in Engineering - 2010

Q. 1 The question below consists of a pair of related words followed by four pairs of words. Select the pair that best expresses the relation in the original pair.

Unemployed : Worker

(A) Fallow : land | (B) Unaware : sleeper

(C) Wit : jester | (D) Renovated : house

Q. 2 If 137 + 276 = 435 how much is 731 + 672?

(A) 534 | (B) 1403

(C) 1623 | (D) 1513

Q. 3 Hari (H), Gita (G), Irfan (I) and Saira (S) are siblings (i.e. brothers and sisters). All were born on 1st January. The age difference between any two successive siblings (that is born one after another) is less than 3 years. Given the following facts:

i. Hari's age + Gita's age > Irfan's age + Saira's age

ii. The age difference between Gita and Saira is 1 year. However Gita is not the oldest and Saira is not the youngest.

iii. There are no twins.

In what order were they born (oldest first)?

(A) HSIG | (B) SGHI

(C) IGSH | (D) IHSG

Q. 4 Given digits 2, 2, 3, 3, 4, 4, 4, 4 how many distinct 4 digit numbers greater than 3000 can be formed?

(A) 50 | (B) 51

(C) 52 | (D) 54

Graduate Aptitude Test in Engineering - 2011

Q. 5 The question below consists of a pair of related words followed by four pairs of words. Select the pair that best expresses the relation in the original pair:

Gladiator : Arena

(A) Dancer : stage (B) Commuter: train

(C) Teacher : classroom (D) Lawyer : courtroom

Graduate Aptitude Test in Engineering - 2013

Q. 6 All professors are researchers
Some scientists are professors

Which of the given conclusions is logically valid and is inferred from the above arguments:

(A) All scientists are researchers

(B) All professors are scientists

(C) Some researchers are scientists

(D) No conclusion follows

Q. 7 Select the pair that best expresses a relationship similar to that expressed in the pair:

water: pipe::

(A) Cart: road (B) Electricity: wire

(C) Sea: beach (D) Music: instrument

Graduate Aptitude Test in Engineering - 2014

Q. 8 In a group of four children, Som is younger to Riyaz. Shiv is elder to Ansu. Ansu is youngest in the group. Which of the following statements is/are required to find the eldest child in the group?

1. Shiv is younger to Riyaz.
2. Shiv is elder to Som.

(A) Statement (1) is sufficient to recognize.

(B) Statement (2) is sufficient to recognize.

(C) Statement (1) and statement (2) both are required.

(D) Statement (1) and (2) both are insufficient to find.

Q. 9 X is 1 km north-east of Y. Y is 1 km south-east of Z. W is 1 km west of Z. P is 1 km south of W. Q is 1 km east of P. What is the distance between X and Q in km?

(A) 1 (B) $\sqrt{2}$

(C) $\sqrt{3}$ (D) 2

Graduate Aptitude Test in Engineering - 2015

Q. 10 Fill in the missing value

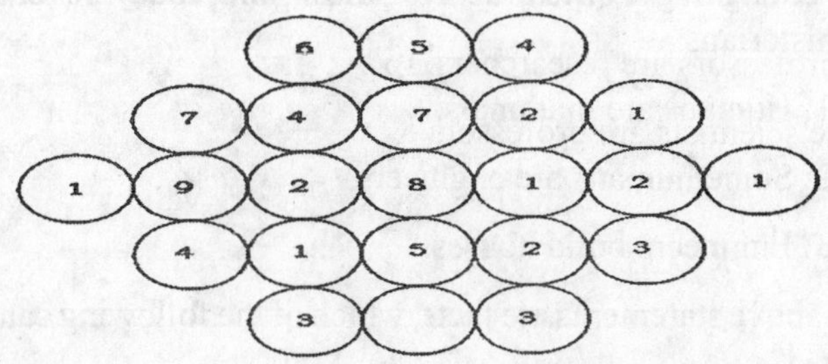

Q. 11 A cube of side 3 units using a set of smaller cubes of side 1 unit. Find the proportion of the number of faces of the smaller cubes visible to those which are NOT visible.

(A) 1 : 4 (B) 1 : 3

(C) 1 : 2 (D) 2 : 3

Q. 12 Humpty Dumpty sits on a wall every day while lunch. The wall sometimes breaks. A person sitting on the wall falls if the wall breaks.

Which one of the statements below is logically valid and can be inferred the above sentences?

(A) Humpty Dumpty always falls while having lunch.

(B) Humpty Dumpty does not fall sometimes while having lunch

(C) Humpty Dumpty never falls during dinner

(D) When Humpty Dumpty does not sit on the wall, the wall does not break

Graduate Aptitude Test in Engineering - 2016

Q. 13 Today, we consider Ashoka as a great ruler because of the copious evidence he left behind in the form of stone carved edicts. Historians tend to correlate greatness of a king at his time with the availability of evidence today. Which of the following can be logically inferred from the above sentences?

(A) Emperors who do not leave significant sculpted evidence are completely forgotten.

(B) Ashoka produced stone carved edicts to ensure that later historians will respect him.

(C) Statues of kings are a reminder of their greatness.

(D) A king's greatness, as we know him today, is interpreted by historians.

Q. 14 Fact 1: Humans are mammals.

Fact 2: Some humans are engineers.

Fact 3: Engineers build houses.

If the above statements are facts, which of the following can be logically inferred?

I. All mammals build houses.

II. Engineers are mammals.

III. Some humans are not engineers.

(A) II only. (B) III only.

(C) I, II and III. (D) I only.

Graduate Aptitude Test in Engineering - 2017

Q. 15 P looks at Q looks at R. P is married, R is not. The number of pairs of people in which a married person is looking at an unnamed person is

(A) 0 (B) 1

(C) 2 (D) Cannot be determined

Q. 16 "If you are looking for a history of india, or for an account of the rise and fall of the British Raj, o for the reason of the cleaving of the subcontinent into two mutually antagonistic parts and the effects this mutilation will have in the respective sections, and ultimately on Asia, you will not find it in these pages; for though I have spent a lifetime in the country, I lived too near the seat of events, and was too intimately associated with the actors, to get the perspective needed for the impartial recording of these matters."

Which of the following is closest in meaning to 'cleaving'?

(A) Deteriorating (B) Arguing

(C) Departing (D) Splitting

Q. 17 There are 4 women P, Q, R and S, and 5 men V, W, X, Y, Z in a group. We are required to form pairs each consisting of one man. P is not to be paired with Z, and Y must necessarily be paired with someone. In how many ways can 4 such pairs be formed?

(A) 74 (B) 76

(C) 78 (D) 80

Q. 18 All people in a certain island are either 'Knights' or 'Knaves' andeach person knows every other person's identity. Knights NEVER lie, and Knaves ALWAYS lie.

P says " Both of us are Knights". Q says "None of us are Knaves".

Which one of the following can be logically inferred from the above?

(A) Both P and Q are Knights.

(B) P is a knight; Q is a Knave.

(C) Both P and Q are Knaves.

(D) The identities of P, Q cannot be determined.

Graduate Aptitude Test in Engineering - 2019

Q.19 Some students were not involved in the strike.

If the above statement is true, which of the following conclusions is/are logically necessary?

1. Some who were involved in the strike were students.
2. No student was involved in the strike.
3. At least one student was involved in the strike.
4. Some who were not involved in the strike were students.

(A) 1 and 2 (B) only 3

(C) only 4 (D) 2 and 3

Q.20 Five numbers 10, 7, 5, 4 and 2 are to be arranged in a sequence from left to right following the directions given below:

1. No two odd or even numbers are next to each other.
2. The second number from the left is exactly half of the left-most number.
3. The middle number is exactly twice the right-most number.

Which is the second number from the right?

(A) 2 (B) 4

(C) 7 (D) 10

Graduate Aptitude Test in Engineering – 2021

Q. 21 A transparent square sheet shown above is folded along the dotted line. The folded sheet will look like.

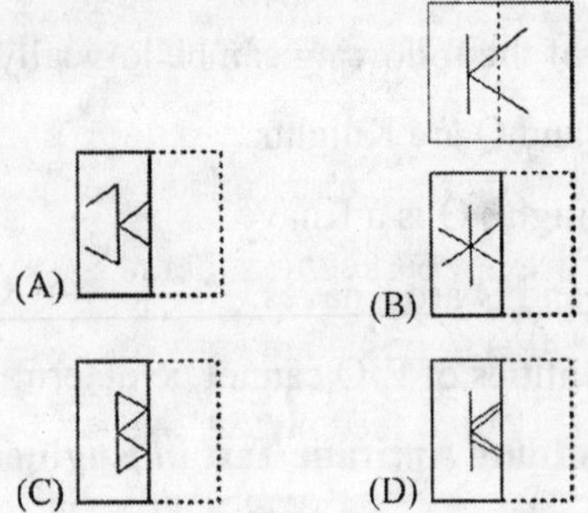

Q. 22 Oasis is to sand as island is to ....

Which one of the following options maintains a similar logical relation in the above sentence?

(A) Stone (B) Land

(C) Water (D) Mountain

Q. 23 The importance of sleep is often overlooked by students when they are reparing for exams. Research has consistently shown that sleep deprivation greatly reduces the ability to recall the material learnt. Hence, cutting down on sleep to study longer hours can be counterproductive.

Which one of the following statements is the CORRECT inference from the above passage?

(A) Sleeping well alone is enough to prepare for an exam. Studying has lesser benefit.

(B) Students are efficient and are not wrong in thinking that sleep is a waste of time.

(C) If a student is extremely well prepared for an exam, he needs little or no sleep.

(D) To do well in an exam, adequate sleep must be part of the preparation.

Q. 24 Seven cars P, Q, R, S, T, U and V are parked in a row not necessarily in that order. The cars T and U should be parked next to each other. The cars S and V also should be parked next to each other, whereas P and Q cannot be parked next to each other. Q and S must be parked next to each other. R is parked to the immediate right of V. T is parked to the left of U.

Based on the above statements, the only INCORRECT option given below is:

(A) There are two cars parked in between Q and V.

(B) Q and R are not parked together.

(C) V is the only car parked in between S and R.

(D) Car P is parked at the extreme end.

Answers Key

1	2	3	4	5	6	7	8	9	10
A	C	B	B	D	C	B	A	C	3
11	12	13	14	15	16	17	18	19	20
C	B	D	B	B	D	C	D	C	C
21	22	23	24						
C	C	D	A						

Explanations

Q. 1

Unemployed – Worker (Opposite Words)

(A) Fallow: Land (Type of land or Undeveloped land)

(B) Unaware: Sleeper (Both are same)

(C) Wit: Jester – Wit means ability to make jokes and jester means joker

(D) Renovated: House – Renovate means to make better and house can be renovated

Q. 2

As given $137_8 + 276_8 = 435_8$

It is an addition of two octal numbers

Thus

$$\begin{array}{r} 731_8 \\ +\ 672_8 \\ \hline 1623_8 \end{array}$$

Rule of Addition:–

(A) If addition of digits is less than 8, then it's ok.

(B) If addition is greater than 8 then subtract 8 from the result and add 1 in forward digits.

Q. 3

As given

1. H+G > I+S
2. G-S = 1 or S-G = 1; G can't be oldest and S can't be youngest.
3. There are no twins thus using statement (2) either GS or SG possible.

(A) HSIG is not possible because of statement 3.

(B) SGHI – SG is possible

S>G>H>I and G+H>I+S (Possible)

(C) IGSH – According to this I>G and S>H. Thus, adding these both we get

H+G < I+S.

(D) IHSG – According to this I>H and S>G. Thus, adding these both we get H+G < I+S.

Q. 4

As the number is greater than 3000. So thousands place can be either 3 or 4.

Case (1) When thousands place is 3

Thus total numbers are $1\times3\times3\times3 = 27$.

In these numbers 3222 and 3333 is invalid because there can only be twice 2 and thrice three.

Thus total valid numbers are $= 27 - 2 = 25$

Case (2) When thousands place is 4

Total numbers are $1\times3\times3\times3 = 27$.

In these numbers 4222 is invalid because there can only be twice 2.

Thus total valid numbers are $= 27 - 1 = 26$

Thus answer is $26 + 25 = 51$

Q. 5

Gladiator: Arena – Gladiator fights in the arena.

(A) Dancer: Stage – Dancer dances (performs) on stage. (Not Appropriate word)

(B) Commuter: Train – A passenger train that is ridden primarily by passengers who travel regularly from one place to another.(Not Appropriate word)

(C) Teacher: Classroom – A teacher teaches in classroom.(Not Appropriate word)

(D) Lawyer: Courtroom – A lawyer fights in courtroom for his client. (Appropriate word)

Q. 6

Referring below figure, only option (C) follows.

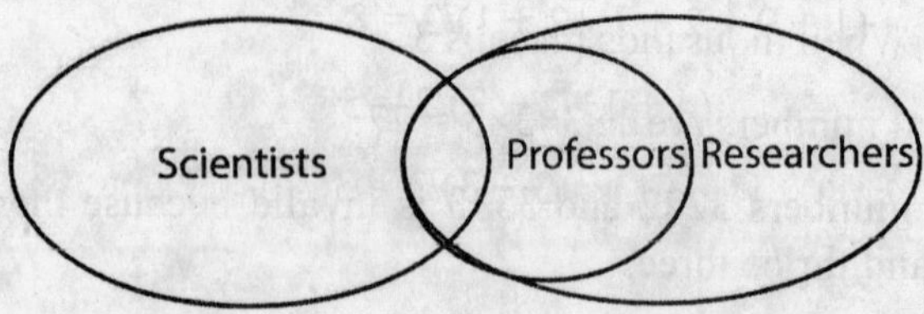

Q. 7

Water flows through pipe, similarly electricity flows through wire.

Q. 8

Given that

Shiv > Ansu

And Riaz > Som

Statement I

If Riaz > Shiv $\Rightarrow$ Riaz is the eldest child in the group.

Statement II

If Shiv > Som

Then, there are two possibilities

1. Riaz > Shiv $\Rightarrow$ Riaz is the eldest child in the group.
2. Shiv > Riaz $\Rightarrow$ Shiv is the eldest child in the group.

Q. 9

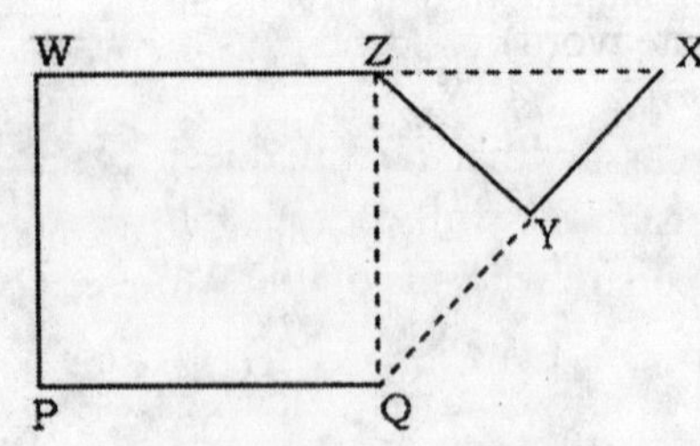

$$XQ = \sqrt{ZX^2 + QZ^2}$$

$$= \sqrt{(\sqrt{2})^2 + 1^2}$$

$$= \sqrt{3}$$

Q. 10

Here, $(6 + 4)/2 = 5$

and $(7 + 4 + 2 + 1)/2 = 7$

and $(1 + 9 + 2 + 1 + 2 + 1)/2 = 8$

and $(4 + 1 + 2 + 3)/2 = 5$

Hence $(3 + 3)/2 = 3$

Q. 11

Such cube of side 3 units can be formed using 27 smaller cubes of side 1 unit.

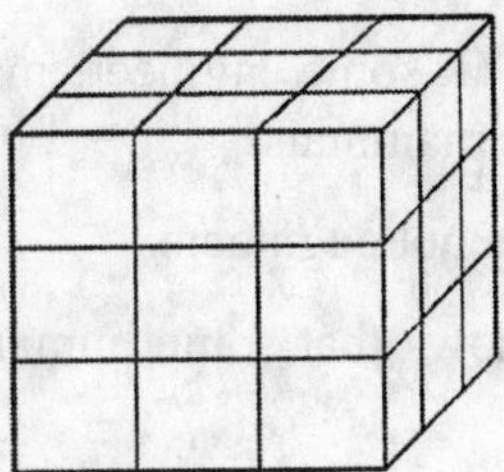

Therefore, total number of faces of the smaller cubes = 27×6 = 162

Faces visible to the cube of side 3 units = 9×6 = 54

Hence, smaller cubes which are not visible = 162 – 54 = 108

∴ Required ratio – 54/108 ═ 1 : 2

Q. 12

(A) It cannot be concluded that Humpty Dumpty **always** falls.

(B) It can be concluded from the above statements.

(C) There is no statement which tells about dinner. Therefore, it cannot be concluded.

(D) It is not necessary that the wall doesn't break when Humpty Dumpty does not sit on the wall. Because, breaking of wall doesn't depend upon the person who sits on the wall.

Q. 14

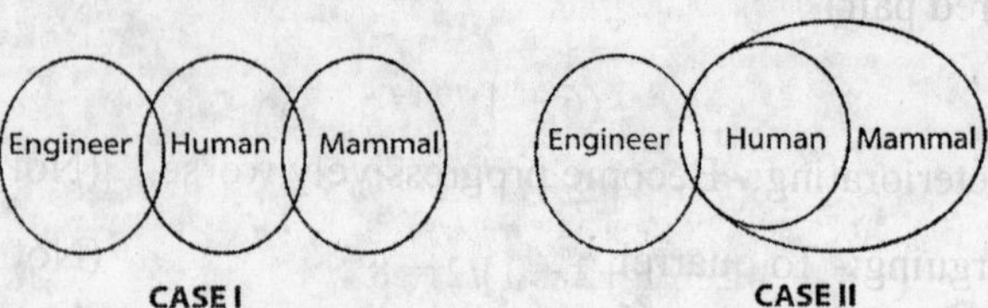

According to the above statements there may be two cases, such as Referring the vain diagrams,

I. All mammals build houses.

Not logically inferred.

II. Engineers are mammals.

According to **CASE II**, some engineers may be mammals. But not sure that all engineers are mammals.

III. Some humans are not engineers.

Both the **CASES** shows that some humans are engineer, surely some won't be engineer.

Q. 15

As per given condition

P* → Q → R

It depends only on marital status of Q.

If Q is unmarried:-

P* → Q is the pair in which a married person (P) is looking at an unmarried person (Q) person.

(Required pair)

Q → R is the pair in which an unmarried person (Q) is looking at an unmarried person (R) person. (Not required pair)

If Q* is married:-

P* → Q* is the only pair in which a married person (P) is looking at a married person (Q) person.

(Not required pair)

Q* → R is the pair in which a married person (Q) is looking at an unmarried person (R) person.

(Required pair)

Q. 16

(A) Deteriorating:- Become progressively worse. (Not appropriate.)

(B) Arguing:- To quarrel. (Not appropriate.)

(C) Departing:- Leave, especially to start a journey. (Not appropriate.)

(D) Splitting:- Divide into two or more groups. (Most appropriate.)

Q. 17

If, P is paired with Y:

So, Q can be paired with any 4 men (V or W or X or Z), R with 3 and S with 2

Hence, Total pairs = 4×3×2 = 24

If, Q is paired with Y:

So, P can be paired with any 3 men (V or W or X), R with 3 and S with 2

Hence, Total pairs = 3×3×2 = 18

If, R is paired with Y:

So, P can be paired with any 3 men (V or W or X), Q with 3 and S with 2

Hence, Total pairs = 3×3×2 = 18

If, S is paired with Y:

So, P can be paired with any 3 men (V or W or X), Q with 3 and R with 2

Hence, Total pairs = 3×3×2 = 18

Therefore, Total number of ways = 24 + 18 + 18 +18 = 78

Q. 18

Case-I, If P is Knight:-

Q will be also knight because P says " Both of us are knights." But is contradicts with Q's statement (None of us are knaves.) Hence both the options (A) and (B) are not appropriate.

Case-II, If P is Knave:-

Q may be knave or knight because P says " Both of us are knights." Hence option (C) is not appropriate.

Hence, The identities of P and Q cannot be determined. Because their qualities contradicts with their characteristics.

Q. 19

4th statement will be true logically according to below venn-diagram.

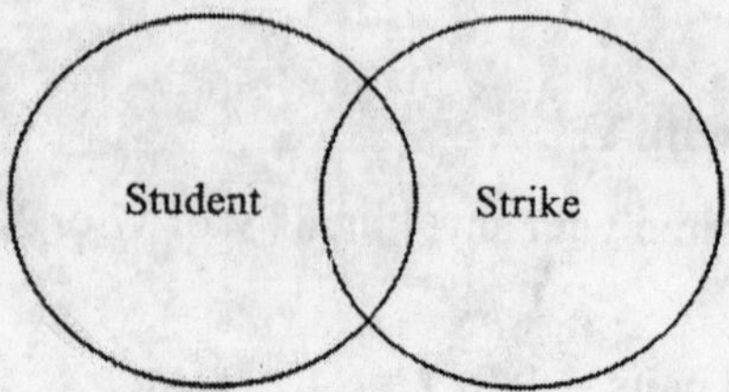

Q. 20

As no two odd or even numbers are next to each other and second number from left is exactly half of the left most number, the only possibility is

Left most 10 5 4 7 2 Right most

Q. 24

Given that, S and V must be parked together, Q and S also must be parked together. From these two conditions, possible parkings may be:

.... Q S V .... V S Q ....

It is observe that in all cases only one car can be parked in between Q and V which is car S. Hence option (a) is incorrect.

VII. Miscellaneous

39

Miscellaneous

UKPSC Combined Assistant Engineering Exam 2013 Paper - I

Q. 1 In building material RCC stands for

(A) Revised cost calculation

(B) Revised construction cost

(C) Reinforced central construction

(D) Reinforced cement concrete

Q. 2 For slabs and beams, the following grade of concrete mix generally used

(A) 1 : 3 : 6 (B) 1 : 4 : 8

(C) 1 : 5 : 10 (D) 1 : 2 : 4

Q. 3 In general, the curing of concrete should be continued for the duration of

(A) 1 to 2 days (B) 3 to 5 days

(C) 15 to 20 days (D) 7 to 10 days

Q. 4 For an average size family of five members including two adults and three children, the capacity of septic tank should be kept about

(A) 1.8 cubic m (B) 2.8 cubic m

(C) 3.8 cubic m (D) 4.8 cubic m

Q. 5 Which of the following is not a method of seasoning of wood ?

(A) Water seasoning (B) Air seasoning

(C) Pneumatic seasoning (D) Kiln seasoning

Q. 6 In a mortar, the common ratio of cement to sand is

(A) 1 : 3 (B) 1 : 6

(C) 1 : 4 (D) 1 : 8

Q. 7 Surkhi as a building material can be used as a substitute for

(A) Sand (B) Steel

(C) Cement (D) Lime

Q. 8 Compressive strength of concrete generally ranges between

(A) 50 – 100 kg/ cm^2 (B) 150 – 200 kg/cm^2

(C) 300 – 700 kg/ cm^2 (D) 800 – 1000 kg/ cm^2

Q. 9 Pure lime contains

(A) 95% or more Calcium oxide

(B) More than 25% Magnesium oxide

(C) Less than 50% Calcium oxide

(D) Less than 25% Magnesium oxide

Q. 10 The Bio chemical oxygen demand (BOD) of domestic sewage ranges from

(A) 100 – 200 mg/litre (B) 150 – 200 mg/litre

(C) 250 – 400 mg/litre (D) 500 – 1000 mg/litre

Q. 11 Flush doors are becoming more popular due to

(A) Greater durability (B) Greater weight

(C) Better look (D) None of the above

Q. 12 The following type of mortar may be used in brick masonary

(A) Lime mortar (B) Mud mortar

(C) Cement – lime mortar (D) All of the above

Q. 13 Rapid hardening Portland cement is manufactured by

(A) Fusing together lime stone and shale

(B) At extremely high temperature cement clinker's

(C) Gypsum is added in small quantities to this clinker

(D) All of the above process

Q. 14 D.P.C is provided throughout the width of plinth for the entire building to prevent

(A) Heating of building

(B) Cooling of the building

(C) To prevent dampness in the building

(D) Stop the theft

Q. 15 A good building stone should have

(A) Strength (B) Good appearance and color

(C) Resistance to fire (D) All above three

Q. 16 The function of the cutting fluid in Lathe machine is to

(A) Cool the tool and work piece (B) Provide the lubrication

(C) Improve surface finish (D) All of the above

Q. 17 Barbed wire is generally made of wire with gauge of about

(A) 8 (B) 10

(C) 12 (D) 14

Q. 18 A thin layer of coal tar is applied on the surface of timber for

(A) Protection from rains (B) Preservation

(C) Insect-resistance (D) All of the above

Q. 19 Farmstead should be located near the source of permanent

(A) Road (B) School

(C) Water supply (D) All above three

Q. 20 Which one of the following is not the sanitary fittings?

(A) Wash basin (B) Sink

(C) Flushing cistern (D) Dustbin

Q. 21 For steel casting following type of sand is better

(A) Medium grain sand (B) Fine grain sand

(C) Coarse sand (D) None of the above

Q. 22 Grouting uses grout which is a mixture of

(A) 1 part cement : 2 part sand : sufficient water

(B) 1 part cement : 3 part sand : 5 part water

(C) 2 part cement : 5 part sand : 10 part water

(D) 2 part cement : 2 part sand : 3 part water

Q. 23 Seasoning of timber is done to

(A) Make it water proof (B) Add moisture in timber

(C) Heat the timber (D) Expel the moisture from timber

Q. 24 Wood for pattern is considered dry when moisture content is

(A) Less than 15 percent (B) More than 50 percent

(C) Zero percent (D) Less than 50 percent

Q. 25 Spread footing is that which support the following

(A) Wall and column (B) Wall only

(C) Column only (D) None of the above

Q. 26 About 90% of full strength of concrete is achieved after curing of

(A) 7 days (B) 14 days

(C) 21 days (D) 28 days

APPSC AEES Agriculture Engineering Exam 2016

Q. 27 Slump test is used for measurement of

(A) Hardness of concrete (B) Plasticity of concrete

(C) Elasticity of concrete (D) All of the above

Q. 28 In stanchion barn, the cows are housed and milked in

(A) Different building (B) Open space

(C) Same buildings (D) None of the above

Q. 29 In deep litter poultry house, the depth of litter in centimeters should be

(A) 15-20 (B) 20-25

(C) 10-15 (D) 30-40

Q. 30 Herring bone is a

(A) Milking parlour (B) Fencing

(C) Barn (D) All of the above

MPPSC Assistant Agricultural Engineer 2013

Q. 31 In arc welding, the arc between work and electrode is produced by ....

(A) Contact resistance (B) Flow of current

(C) Voltage (D) All options are correct

Q. 32 The phenomenon 'Weld decay' is associated with ....

(A) Brass (B) Bronze

(C) Cast iron (D) Stainless steel

Q. 33 The cold chisels are made by ....

(A) Drawing (B) Forging

(C) Piercing (D) Rolling

Q. 34 The term 'Hot tear' is associated with ....

(A) Weathering of non-ferrous materials

(B) Process of fabrication

(C) Process of heat treatment

(D) A fracture in a metal casting during solidification of metal

Q. 35 Inverse quality of toughness is known as ....

(A) Brittleness (B) Friability

(C) Malleability (D) None of these is correct

UKPSC Combined Junior Engineering Exam 2013 Agricultural Engineering Paper-I

Q. 36 Which of the following combination of metals are used to make brass ?

(A) Copper and Zinc (B) Zinc and Tin

(C) Copper and Tin (D) Tin and Antimony

Q. 37 World Food Day is celebrated on which date ?

(A) 15 July (B) 20 July

(C) 16 October (D) 22 October

Q. 38 Cast iron is an alloy of

(A) Iron and carbon (B) Iron and cobalt

(C) Iron and zinc (D) Copper and zinc

Q. 39 Which form of iron is obtained from the blast furnace ?

(A) Pig iron (B) Wrought iron

(C) Cast iron (D) Mild steel

Q. 40 Which of the following is good for making nut-bolts ?

(A) Cast iron (B) Pig iron

(C) Wrought iron (D) High carbon steel

Q. 41 In 24 hour, water absorption by volume is maximum in

(A) Limestone (B) Granite

(C) Gneiss (D) Slate

Q. 42 When water is added to cement

(A) Chemical reaction starts (B) Heat is absorbed

(C) Heat is generated (D) None of the above

Q. 43 Which is the weakest concrete ?

(A) 1 : 2 : 4 (B) 1 : 3 : 6

(C) 1 : 4 : 8 (D) 1 : 5 : 10

Punjab Mechanic Agricultural Machinery Instructor 2014

Q. 44 The metal is cut by

(A) Chisels (B) Hammer

(C) Spanner (D) None of the above

Q. 45 To directly measure the dimension, which tool is used

(A) Foot rule (B) Caliper

(C) Divider (D) None of the above

Q. 46 The principle of working of a dial test indicator is

(A) The linear motion is converted into a rotary motion using rack and pinions

(B) Magnification by electronic means

(C) The linear motion is converted into a reciprocating motion, using slotted link

(D) Magnification of small variation using lenses

Q. 47 The least count of a vernier metric micrometer is

(A) 0.0001 mm (B) 0.001 mm

(C) 0.01 mm (D) 0.1 mm

Q. 48 The point angle for a standard drill is

(A) 108^0 (B) 135^0

(C) 60^0 (D) 118^0

Q. 49 Which caliper can be set quickly

(A) Long point caliper (B) Adjustable caliper

(C) Spring type caliper (D) None of these

Q. 50 Which are permanent fasteners

(A) Welding (B) Rivets

(C) Keys (D) All of the above

Q. 51 Which type of maintenance is most expensive

(A) Routine maintenance (B) Preventive maintenance

(C) Breakdown maintenance (D) Planned maintenance

Q. 52 The automobiles generally utilizes batteries having voltage of

(A) 3 V (B) 6 V

(C) 12 V (D) 24 V

Q. 53 The horn of a tractor is linked with

(A) Its fuel supply system (B) Lubrication system

(C) Electrical system (D) None of the above

RPSC AEn Pre Exam 2013 (Agricultural Engineering)

Q. 54 For which crop the parboiling is more important and beneficial,

(A) Maize (B) Mustard

(C) Rice (D) all of the above

Q. 55 Which of the motors is suitable for farm machinery ?

(A) DC shunt motor (B) DC series motor

(C) Cumulative compound motor (D) Induction motor

Q. 56 Which wiring system is used inside the walls ?

(A) PVC wiring (B) Batten wiring

(C) Concealed wiring (D) Pipe wiring

Q. 57 Three interdependent quantities which characterize direct current are

(A) Potential difference in the circuit, rate of current flow and resistance of circuit

(B) Length of wire, resistance of electric current, and highest voltage

(C) Diameter of electric wire, material of wire and length of wire

(D) A.C. current, voltage, weight of wire

Q. 58 Floor area required per 100 adult hens

(A) 30 to 42 m^2 (B) 7 to 17 m^2

(C) 17 to 27 m^2 (D) 42 to 50 m^2

Q. 59 The base material for distemper is

(A) Lime (B) Chalk

(C) Cement wash (D) Lime putty

Q. 60 The width of flange of a T-beam should be less than

(A) Distance between the centers of T-beam

(B) One third of effective span of the T-beam

(C) Breadth of the rib + 12 times the thickness of the slab

(D) Least of all the above

Q. 61 The most common ratio used in RCC structures for check dams etc. in small stream remains as follows

(A) 1 : 2 : 4 (B) 1 : 4 : 8

(C) 1 : 5 : 10 (D) All are equally applicable

Q. 62 The camber on the pavement is provided by

(A) Circular method (B) Straight line and parabolic at crown

(C) Straight line method (D) Elliptical method

Q. 63 The Drop man holes are provided in sewers for

(A) Cities only (B) Only hilly areas

(C) Large towns (D) Industrial complex

OCS AEn Exam 2011 (Pre) – II (Agricultural Engineering)

Q. 64 A cow need to be kept in an individual .... before calving.

(A) Barn (B) Pen

(C) Farm (D) Hangar

Q. 65 The fraction of grain that is rich in starch is called

(A) Hull (B) Endosperm

(C) Germ (D) Husk

Q. 66 Soyabean is mostly used in India for the production of

(A) Edible oil (B) Pulses

(C) Milk substitutes (D) Processed foods

Q. 67 Parboiling of rice was originated in

(A) Pakistan (B) India

(C) Bhutan (D) Bangladesh

Q. 68 Glazing is done on rice for

(A) Whitening (B) Pearling

(C) Transparent look (D) Polishing

Q. 69 The product derived out of milk after removal of cream is known as

(A) Buttermilk (B) Whey

(C) Skimmed milk (D) Curd

Q. 70 For dairy plant sanitation and hygiene maintenance of related processing machinery, the ……… method of cleaning is mostly followed ?

(A) CIP (B) HACCP

(C) HTST (D) Sterilization

Q. 71 A by-product of rice processing industry is fragile and porous, has bulk density of 90 to 160 kg/m^3 and low nutritional value, identify if amongst the listed below

(A) Bran (B) Polish

(C) Germ (D) Husk

Q. 72 The percentage of husk, bran and bran oil received from rice milling are respectively

(A) 20, 5 and 25 (B) 5, 10 and 30

(C) 20, 5 and 40 (D) 20, 10 and 20

Q. 73 Calorific value of rice husk is approximately

(A) 1.25×10^4 kJ/kg (B) 2.33×10^4 kJ/kg

(C) 4.65×10^4 kJ/kg (D) 4.94×10^4 kJ/kg

Q. 74 Rate of FFA formation in raw bran ……… with increase of the relative humidity of storage.

(A) Increases (B) Decreases

(C) Either increases or decreases (D) Neither increases nor decreases

Q. 75 Baffles are provided in heat exchangers to

(A) Remove dirt (B) Increases heat transfer rate

(B) Reduce vibration (D) Provide better mechanical strength

Q. 76 Salt is a better food preservative than sugar because it

(A) Has low molecular weight

(B) Lowers the vapour pressure of food water by a large extent

(C) Kills microorganisms better

(D) Reduces pH

Q. 77 During processing the maximum amount of food substance lost is

(A) Protein (B) Vitamin

(C) Fats (D) Carbohydrates

Q. 78 The nutrient most sensitive to processing and cooking are

(A) Proteins (B) Carbohydrates

(C) Minerals (D) Vitamins

CGPSC State Engineering Services Exam Paper – I

Q. 79 Cereal is deficient in which of the following amino acid

(A) Lysine (B) Proline

(C) Methonine (D) Serine

(E) Tyrosine

Q. 80 Whole wheat flour is also known as

(A) Perfect flour (B) Straight flour

(C) Graham flour (D) Complete flour

(E) Ideal flour

Q. 81 Specific gravity of skim milk is

(A) More than cow milk (B) Less than cow milk

(C) Less than buffalo milk (D) More than buffalo milk

(E) Both (A) & (D)

Q. 82 Which of the following test is used to detect sesame oil in milk fat

(A) Halphen test (B) Baudouin test

(C) Kreis test (D) Polenske value

(E) Adulteration test

Q. 83 Kraft process of chemical pulping is a

(A) Sulfite process (B) Alkaline process

(C) Acidic process (D) Chemical process

(E) Glass making process

Q. 84 The greatest volume of soft drinks are packed in this type of containers

(A) PET (B) Glass

(C) Metal (D) Corrugated

(E) HDPE

Q. 85 The density of HDPE sheet is more than ……… kg/m³

(A) 840 (B) 940

(C) 1040 (D) 1140

(E) 1240

Q. 86 Major product being produced is static in following layout

(A) Product layout (B) Fixed position layout

(C) Group layout (D) Process layout

(E) Cellular layout

Q. 87 Example for fixed cost

(A) Rent (B) Raw material

(C) Labor (D) Administration

(E) Both (A) & (D)

Q. 88 Sulfite process of pulping is a

(A) Mechanical method (B) Chemical method

(C) Semi chemical method (D) Modification method

(E) Both (A) & (C)

Q. 89 Sliced fruit and sugar juice is kept in the proportion during preparation of vinegar is

(A) 1: 1 (B) 2 : 1

(C) 3 : 1 (D) 4 : 1

(E) 5 : 1

Q. 90 After pretreatment with oil, pulses are kept on floors to diffuse the oil for about

(A) 6 hours (B) 9 hours

(C) 12 hours (D) 15 hours

(E) 17 hours

Q. 91 In deep litter poultry house, the floor area provided per bird is usally

(A) 0.1 m^2 (B) 0.4 m^2

(C) 1.0 m^2 (D) 1.4 m^2

(E) 1.7 m^2

Q. 92 This is suitable for producing single, large, high cost components or products

(A) Group layout (B) Filxed position layout

(C) Process layout (D) Product layout

(E) Functional layout

Q. 93 Lignin softens at following temperature

(A) 100 °C (B) 120 °C

(C) 140 °C (D) 160 °C

(E) 180 °C

Q. 94 The most wide spread consumer package for aseptic products is

(A) Paper board laminate cartoon(B) Bottle

(C) Can (D) Sachet & pouch

(E) PET

Q. 95 Silica is largest constituent of glass and it contains percentage is

(A) 58 – 68 (B) 68 - 73

(C) 78 - 83 (D) 88 - 93

(E) 90 - 95

UKPSC A.En. Exam 2007 Paper – I

Q. 96 The common size of brick which is made in India has the size of

(A) 25.0×35.0×15.0 cm (B) 22.5×11.25×7.5 cm

(C) 20.0×10.0×6.0 cm (D) 22.5×15.0×9.0 cm

Q. 97 Steel/ Iron rods are placed in concrete mixture to put slabs in housing so as to bear

(A) Tensile stress (B) Compressive stress

(C) Shearing stress (D) None of the above

Q. 98 Decomposition of solid sewage in septic tank takes place under conditions which are

(A) Aerobic (B) Anaerobic

(C) Mixed (D) None of the above

Q. 99 The plasticity of concrete is measured by

(A) Slump test (B) Fuller test

(C) Creep test (D) T – test

Q. 100 Mortar for building construction usually consists of a mixture of

(A) Gravel, cement and sand (B) Gravel, sand and water

(C) Cement, sand and lime (D) Cement, sand and water

Q. 101 The detention period in a septic tank is about

(A) 2 – 4 hours (B) 4 – 8 hours

(C) 12 – 36 hours (D) 2 – 4 days

Q. 102 Which of the following is the purest form of iron ?

(A) High carbon steel (B) Mild steel

(C) Wrought iron (D) Cast iron

Q. 103 Curing of concrete is the process done to provide

(A) A moist surface (B) A smooth surface

(C) Degrading surface (D) Strength through hydration

Q. 104 The yield stress of 40 mm mild steel bar of grade I should be greater than

(A) 260 N/m^2 (B) 240 N/mm^2

(C) 200 N/mm^2 (D) 440 N/m^2

Q. 105 Le-Chatelier's device is used to determine

(A) Setting time for concrete

(B) Soundness of concrete

(C) Tensile strength of concrete

(D) Compressive strength of concrete

Q. 106 The main ingredients of Portland cement are

(A) Llime and silica (B) Lime and alumina

(C) Silica and alumina (D) Lime and iron

Q. 107 The most simplest geometrical form of a Truss is

(A) Triangular (B) Circular

(C) Trapezium (D) None of the above

Q. 108 The gradual reduction in the value of property with its age is known as

(A) Extra valuation (B) Revaluation

(C) Depreciation (D) Appreciation

Q. 109 In detailed estimate, the provision for contingency is usually kept

(A) 1 % (B) 3 %

(C) 10 % (D) 12 – 15 %

Q. 110 Screws for wood work are specified by their

(A) Length (B) Diameter

(C) Weight (D) Gauge

Q. 111 The surface to which the first coat of plaster is applied is known as

(A) Base surface (B) Preliminary surface

(C) Ground surface (D) Background surface

Q. 112 The ratio between the total covered area of all floors to the plot area is known as

(A) House area ratio (B) Floor area ratio

(C) Open area ratio (D) Useful area ratio

Q. 113 Which of the following is not the component of Reinforced Bricks (R.B.)?

(A) Reinforced brick cantilever (B) Reinforced brick lintel

(C) Reinforced brick slab (D) Reinforced brick mosaic

Q. 114 In R.C.C., the modulus of elasticity of Steel to that of concrete is called

(A) Concrete ratio (B) Steel ratio

(C) Modular ratio (D) None of the above

Q. 115 If ultimate stress for concrete is 15 N/m^2, its working stress for factor of safety as 3 will be

(A) 18 N/m^2 (B) 180 N/m^2

(C) 5 N/m^2 (D) 45 N/m^2

Q. 116 Steaming paddy during parboiling helps in

(A) Denaturation of protein (B) Gelatinization of starch

(C) Millard reaction (D) None of the above

Q. 117 Pigeon pea is considered as hard to mill due to

(A) High protein content (B) Low husk content

(C) High gum content (D) None of the above

Q. 118 Dry ice is called

(A) Freon-12 (B) Ammonia

(C) Frozen water (D) Solid carbon dioxide

Q. 119 The dimensions of pressure are

(A) ML^2T^{-2} (B) MLT^{-2}

(C) $ML^{-1}T^{-2}$ (D) ML^2T^{-3}

Q. 120 The dimension of power is

(A) $ML^{-2}T^{-2}$ (B) ML^2T^{-2}

(C) ML^2T^{-3} (D) $ML^{-2}T^{-3}$

Where, M is mass, L is length and T is time.

Q. 121 A science devoted to the study of deformation and flow is called

(A) Dynamics (B) Kinematics

(C) Rheology (D) Hydrology

Q. 122 In pressure parboiling method the paddy is soaked for nearly 40 minutes at a temperature of

(A) 45 to 50 °C (B) 60 to 70 °C

(C) 85 to 90 °C (D) 100 to 110 °C

Q. 123 Angle of repose of wheat grain is

(A) 10 – 15 degree (B) 23 – 28 degree

(C) 30 – 35 degree (D) none of the above

Q. 124 The predominant mechanism in dehusking of pigeon-pea is

(A) Shear (B) Compression

(C) Impact (D) Abrasion

Q. 125 Polisher is used in

(A) Wheat milling (B) Corn milling

(C) Rice milling (D) None of the above

Q. 126 Gauge pressure is measured

(A) Above atmospheric pressure (B) Below atmospheric pressure

(C) Above zero pressure (D) None of the above

Q. 127 In sedimentation, particles are separated by

(A) Centrifugal force (B) Gravitational force

(C) Magnetic force (D) None of these

Q. 128 Toluene displacement method is used to determine

(A) Bulk density (B) True density

(C) Moisture content (D) Viscosity

ASCO (RPSC) Exam (Forest) 2011

Q. 129 Which of the following set of crops contains C_4 plants ?

(A) Cotton, rice and wheat

(B) Pearl millet, rice and soyabean

(C) Maize, pearl millet and sorghum

(D) Peanut, rice and wheat

Q. 130 When two or more crops are grown simultaneously but do not have inter-competition effects are called

(A) Turf drops (B) Companion crops

(C) Brake crops (D) Parallel crops

Q. 131 Restorative crops are those crops which

(A) Provide a good harvest along with amelioration of the soil

(B) On growing leave the field exhausted

(C) Are able to protect the soil surface from erosion

(D) Are grown for temporary storing

Q. 132 Growing of four crops per year in such a way that land is passed onto another crop before the previous one is harvested is called

(A) Intercropping (B) Relay cropping

(C) Ratoon cropping (D) Multi-storeyed cropping

Q. 133 If paddy followed by wheat or sorghum followed by gram is cultivated in a succession on a piece of land in an agricultural year is called

(A) Multiple cropping (B) Double cropping

(C) Continuous cropping (D) Rotational cropping

Q. 134 When the seeds of pearl-millet and green gram are mixed and sown by broadcasting method is an example of

(A) Rotational cropping (B) Continuous cropping

(C) Mixed cropping (D) Sequential cropping

Q. 135 Growing only one crop year after year on a piece of land is called

(A) Sole cropping (B) Cropping pattern

(C) Companion cropping (D) Mono-cropping

Q. 136 Central Institute of Arid-Horticulture is situated at

(A) Bangalore (B) Bijapur

(C) Bikaner (D) Bhopal

UKPSC A.En. Exam 2012 Paper – I

Q. 137 In Indian vegetarian diet pulses are major source of

(A) Carbohydrate (B) Protein

(C) Vitamin A (D) Vitamin B

Q. 138 In solvent extraction the quantities of oil left in cake is found as ……. %

(A) 2 to 3 (B) 1.5 to 2

(C) 1.0 (D) 0.75

Q. 139 Percent oil content in parboiled paddy bran is

(A) 10 (B) 20

(C) 30 (D) 40

Q. 140 In brick wall construction when alternative courses have header and stretcher, the bond is called

(A) English bond (B) Flemish bond

(C) Alternate bond (D) None of the above

Q. 141 Grouting is a mixture of

(A) 1 : 2 parts cement and sand + water

(B) 1 : 4 parts cement and sand + water

(C) 2 : 4 parts cement and sand + water

(D) None of the above

Q. 142 Hollow bricks are used for

(A) Earthquake proof building (B) Reducing the cost

(C) Resistant against heat flow (D) Ornamental design

Q. 143 Which one of the following is fluidised solid ?

(A) Flour (B) Fruit juice

(C) Honey (D) CO_2

Q. 144 A thin layer of coal tar is applied on the surface of timber for

(A) Protection (B) Preservation

(C) Insect resistance (D) None of the above

Q. 145 Which constituent of whole milk is not available in skim milk powder ?

(A) Protein (B) Fat

(C) Carbohydrate (D) All the above

Q. 146 Testing of stones is done by

(A) Smith's test (B) Acid test

(C) Absorption test (D) All of the above

Q. 147 Percent dal yield by CFTRI method is about

(A) 60 (B) 70

(C) 80 (D) 90

Q. 148 Which chemical is not used to control insects during grain storage ?

(A) Phosphine (B) Methyle-Bromide

(C) Ethyline-di-bromide (D) Nitrogen

Q. 149 The size of moulds for bricks should be .... the specified size of the bricks

(A) Equal to (B) Smaller than

(C) Longer than (D) None of the above

Q. 150 Lime and sand a good mix is considered to have the ratio of

(A) 1 : 1 (B) 1 : 1.5

(C) 1 : 2 (D) 1 : 2.5

Q. 151 A process in which both, the system and surrounding by any means what so ever return to their initial states after completion of process is known as

(A) Irreversible process (B) Reversible process

(C) Complex process (D) Simple process

Q. 152 Viability of seeds can be increased by

(A) Increasing temperature and decreasing moisture

(B) Increasing both temperature and moisture content

(C) Decreasing both temperature and moisture content

(D) Decreasing temperature and increasing moisture

Q. 153 Quality of grain during drying is dependent on temperature of

(A) Hot air (B) Ambient air

(C) Grain (D) Exhaust air

Q. 154 Seed grain drying temperature is generally recommended as ...... °C

(A) 60 (B) 75

(C) 80 (D) 45

Q. 155 The thinner used for oil paint is

(A) Water (B) Turpentine

(C) Carbon tetra chloride (D) None of the above

Q. 156 The quick hardening cement develops its full strength after setting within

(A) Two days (B) Three days

(C) Four days (D) Five days

Q. 157 The dimension of calorie are

(A) ML^2T^{-2} (B) $ML T^{-2}$

(C) $ML^{-2}T^{-1}$ (D) ML^2T^{-1}

Q. 158 Which of the following glass is used in doors and windows ?

(A) Soda lime glass (B) Lead glass

(C) Boro silicate glass (D) None of the above

Q. 159 Dynamic similarity between model and prototype means

(A) The similarity of forces (B) The similarity of motion

(C) The similarity of shape (D) None of the above

Q. 160 Deposition at the bottom of a septic tank is called

(A) Sewer (B) Scum

(C) Sewerage (D) Sludge

Q. 161 The mass of a substance is measured by

(A) Spring balance (B) Beam balance

(C) Calorimeter (D) Hygrometer

Q. 162 A mechanical device used to increase the pressure energy of a liquid is called as

(A) Machine (B) Engine

(C) Pump (D) All above

Q. 163 The process of strengthening the concrete by hydration of surface is called

(A) Finishing (B) Curing

(C) Mixing (D) Reinforcing

Q. 164 When Mach number is more than 1, the speed of object will be

(A) Sonic (B) Super-sonic

(C) Sub-sonic (D) none of the above

Q. 165 Which of the following will have least value of thermal conductivity :

(A) Copper (B) Water

(C) Glass (D) Air

Kerala AAO Exam 2017

Q. 166 Match the following

Crop	**Origin**
P. Rice	1. Mexico
Q. Sorghum	2. Brazil
R. Maize	3. India and Burma
S. Groundnut	4. Africa

(A) P-3, Q-4, R-1, S-2 (B) P-1, Q-2, R-3, S-4

(C) P-2, Q-3, R-1, S-4 (D) P-4, Q-3, R-2, S-1

Q. 167 Ripening hormone is

(A) Auxin (B) Cytokinin

(C) Ethylene (D) Gibberellin

Q. 168 Match the term with suitable system

P. Silvi – agriculture	1. Pasture is major
Q. Agro – silviculture	2. Tree is major
R. Silvo – pastoral	3. Alley cropping
S. Pastoral – silviculture	4. Shifting cultivation

(A) P-4, Q-3, R-2, S-1 (B) P-1, Q-2, R-3, S-4

(C) P-3, Q-4, R-1, S-2 (D) P-4, Q-1, R-3, S-2

Q. 169 What is the share of forest area in the total geographical area in India ?

(A) 13 % (B) 23 %

(C) 33 % (D) 43%

Graduate Aptitude Test in Engineering - 2016

Q. 170 With n being a positive integer $\sum_{n=1}^{\infty}\frac{1}{n^p}$, the series, for p > 1 is

(A) Convergent (B) Divergent

(C) Asymptotic (D) Oscillatory

Graduate Aptitude Test in Engineering - 2018

Q. 171 The type of the sequence is $a_n = \left(\frac{n}{n-1}\right)^3$

(A) Oscillatory (B) Bounded

(C) Converging (D) Diverging

Q. 172 In refining of the rice bran oil for edible purpose, neutralization process is carried out mainly to remove

(A) Wax and organic impurities (B) Gums and mucilage

(C) Saturated glycerides (D) Free fatty acids

Q. 173 Which one of the following chemicals is NOT used for fumigation in either horizontal or vertical storage structures to kill insects ?

(A) Methyl bromide (B) Phosphine

(C) Phostoxin (D) Silver sulphate

Q. 174 Match the following agricultural/food by-products in Column I with their respective value added products in Column II.

Column I	Column II
P. Banana pseudostem	1. Silicon tetrachloride
Q. Coconut shell	2. Protein food
R. Meat waste	3. Starch
S. Rice husk	4. Furfural

(A) P-2, Q-3, R-1, S-4 (B) P-3, Q-4, R-2, S-1

(C) P-4, Q-3, R-2, S-1 (D) P-3, Q-1, R-2, S-4

Graduate Aptitude Test in Engineering - 2021

Q. 175 In butter, the fishy flavor defect is due to the decomposition of

(A) α-lactalbumin (B) β-lactoglobulin

(C) casein (D) lecithin

Answers Key

1	2	3	4	5	6	7	8	9	10
D	D	D	B	C	A	A	C	A	C
11	12	13	14	15	16	17	18	19	20
A	D	D	C	D	A	D	D	C	D
21	22	23	24	25	26	27	28	29	30
C	A	D	A	A	D	B	C	A	A
31	32	33	34	35	36	37	38	39	40
A	B	B	D	B	A	C	A	A	C
41	42	43	44	45	46	47	48	49	50
A	C	D	A	A	A	B	D	C	A
51	52	53	54	55	56	57	58	59	60
C	C	C	C	D	C	A	A	B	D
61	62	63	64	65	66	67	68	69	70
A	B	C	B	B	A	B	C	C	A
71	72	73	74	75	76	77	78	79	80
D	A	A	A	B	B	B	D	A	C
81	82	83	84	85	86	87	88	89	90
E	B	B	A	B	B	E	B	D	C
91	92	93	94	95	96	97	98	99	100
B	B	D	A	B	B	A	B	A	D
101	102	103	104	105	106	107	108	109	110
C	C	D	B	B	A	A	C	B	A

111	112	113	114	115	116	117	118	119	120
D	B	D	C	C	B	C	D	C	C
121	122	123	124	125	126	127	128	129	130
C	C	B	D	C	A	B	B	C	D
131	132	133	134	135	136	137	138	139	140
A	B	B	C	D	C	B	C	B	A
141	142	143	144	145	146	147	148	149	150
A	C	A	B	B	D	C	D	C	C
151	152	153	154	155	156	157	158	159	160
B	C	C	D	B	B	B	A	A	D
161	162	163	164	165	166	167	168	169	170
B	C	B	B	D	A	C	A	B	A
171	172	173	174	175					
Bonus	D	D	B	D					

Explanations

Q. 70

CIP (Cleaning In Place) system of cleaning the interior surface of pipelines, vessels, filters, process equipment and associated things without dismantling.

Q. 84

Bottles made of polyethylene terephthalate (PET, sometimes PETE) can be "recycled" to reuse the material out of which they are made and to reduce the amount of waste going into landfills.

Q. 115

Working stress = Ultimate stress/ Factor of safety

$= 15/3 = 5 \text{ N/m}^2$

Q. 170

Integral Test

Let $f(x)$ be continious, decreasing and positive for $x \geq 1$, Then, $\sum_{k=1}^{\infty} f(k)$ converges if and only if $\int_1^{\infty} f(x)\ dx$ converges.

Since,

$$\sum^{\infty} \frac{}{} = \begin{cases} \frac{}{} x \Big|^{\infty}, p \quad 1 \\ ln|x|\Big|\frac{\infty}{} \quad , p = 1 \\ \frac{}{} x \Big|^{\infty}, 0 < P > 1 \end{cases} = \begin{cases} 1/(1 \quad) \\ \infty \\ \infty \end{cases}$$

Hence the series converges for P>1 and diverges for 0<P>1.

Q. 171

We have $a_n = \left(\frac{n}{n-1}\right)^3$ and $a_{n+1} = \left(\frac{n+1}{n}\right)^3$

Apply D' Alembert's ratio test

$$\therefore \lim_{n\to\infty} \frac{a_{n+1}}{a_n} = \lim_{n\to\infty} \left(\frac{(n+1)(n-1)}{n^2}\right)^3 = 1$$

Thus, D' Alembert's ratio test fails.

Apply Comparison test

Let, A convergent series $(V_n) = 1/n^3$

n	$(V_n) = 1/n^3$	$a_n = \left(\frac{n}{n-1}\right)^3$
0	∞	0
0.5	8	-1
0.8	1.95	-64
0.9	1.37	-729
1	1	∞
2	0.125	8
3	.037	3.375
100	0.000001	1.010

Here, an $\geq V_n$ for all values of $n \geq 1$.

Hence, the given sequence is diveregent.

Q. 173

Fumigation:- Fumigation is a process of gaseous sterilisation which is used for killing of micro-organisms and prevention of microbial growth in air, surface of wall or floor.

Widely used fumigants includes:- Phosphine, Formaldehyde, 1,3-dichloropropene, Chloropicrin, Methyl isocyanate, Hydrogen cyanide, Sulfuryl fluoride, Lodoform, Methyl bromide, Phostoxin etc.

Q. 174

(A) The rhizome, pseudostem of the banana have a high starch content. Its concentration is higher in the middle fleshy leaf sheaths and increases gradually towards the rhizome downward along the length of the pseudostem.

(B) Coconut shell is rich in pentosans and is a good source of furfural via acid hydrolysis.

(C) Meat is food source rich in protein traditionally made from animal flesh.

(D) The rice husk is a by-product of the rice processing that contains substantially silica and carbon compounds. For this reason, it has been used as raw material to obtain silicon tetrachloride employed in several kinds of industry applications.